AF752183

GEOTHERMAL ENERGY ENGINEERING

GEOTHERMAL ENERGY ENGINEERING

Technology Transfer with the Oil and Gas Industry

Edited by

SILVIU LIVESCU, PH.D.
Co-founder and Chief Technology Officer, Bedrock Energy, Austin, TX, United States
Formerly Professor, Department of Petroleum Engineering, University of Texas at Austin, Austin, TX, United States

BIROL DINDORUK, PH.D.
AADE Endowed Professor, Department of Petroleum Engineering, University of Houston, Houston, TX, United States

ELSEVIER

Elsevier
Radarweg 29, PO Box 211, 1000 AE Amsterdam, Netherlands
125 London Wall, London EC2Y 5AS, United Kingdom
50 Hampshire Street, 5th Floor, Cambridge, MA 02139, United States

Copyright © 2025 Elsevier Inc. All rights are reserved, including those for text and data mining, AI training, and similar technologies.

Publisher's note: Elsevier takes a neutral position with respect to territorial disputes or jurisdictional claims in its published content, including in maps and institutional affiliations.

No part of this publication may be reproduced or transmitted in any form or by any means, electronic or mechanical, including photocopying, recording, or any information storage and retrieval system, without permission in writing from the publisher. Details on how to seek permission, further information about the Publisher's permissions policies and our arrangements with organizations such as the Copyright Clearance Center and the Copyright Licensing Agency, can be found at our website: www.elsevier.com/permissions.

This book and the individual contributions contained in it are protected under copyright by the Publisher (other than as may be noted herein).

Notices
Knowledge and best practice in this field are constantly changing. As new research and experience broaden our understanding, changes in research methods, professional practices, or medical treatment may become necessary.

Practitioners and researchers must always rely on their own experience and knowledge in evaluating and using any information, methods, compounds, or experiments described herein. In using such information or methods they should be mindful of their own safety and the safety of others, including parties for whom they have a professional responsibility.

To the fullest extent of the law, neither the Publisher nor the authors, contributors, or editors, assume any liability for any injury and/or damage to persons or property as a matter of products liability, negligence or otherwise, or from any use or operation of any methods, products, instructions, or ideas contained in the material herein.

ISBN: 978-0-443-21662-6

For information on all Elsevier publications
visit our website at https://www.elsevier.com/books-and-journals

Publisher: Megan Ball
Acquisitions Editor: Fran Kennedy-Ellis
Editorial Project Manager: Ellie Barnett
Production Project Manager: Sharmila Kirouchenadassou
Cover Designer: Greg Harris

Typeset by STRAIVE, India

ELSEVIER Book Aid International
Working together
to grow libraries in
developing countries
www.elsevier.com • www.bookaid.org

Contents

6. High-temperature geothermal well cementing

Axel-Pierre Bois, Simon James, and Anthony Badalamenti

7. Geothermal production, injection, and storage engineering

Gioia Falcone, Christopher S. Brown, Isa Kolo, David Banks, and Orestis Angelidis

8. Stimulation in enhanced geothermal systems

Carlos A. Fernandez, Zihao Li, Chao Zeng, Guoqing Jian, and Cheng Chen

9. Production and operation of geothermal systems

Pejman Shoeibi Omrani and Govert de With

10. Simulation of geothermal resources management

Denis Voskov

11. Artificial intelligence in the geothermal energy systems

Pejman Shoeibi Omrani, Jonah Poort, and Sadegh Shahmohammadi

12. Global screening for superhot rock geothermal energy: Geodynamic settings, prospective heat endowment and extraction techniques

Philip J. Ball, Graham Banks, Mikenna Montgomery, and Juan Carlos Afonso

13. Geothermal energy and fluid properties with a special focus on geothermal noncondensable gases

Niyazi Aksoy, Silviu Livescu, Ram Ratnakar, and Birol Dindoruk

Contributors

Khizar Abid MPGE, University of Oklahoma, Norman, OK, United States

Juan Carlos Afonso University of Twente, Enschede, the Netherlands

Niyazi Aksoy NTU Geothermal Consulting Ltd., Ankara Cad, Izmir, Türkiye

Orestis Angelidis University of Glasgow, Glasgow, United Kingdom

Anthony Badalamenti CURISTEC, Neuville-sur-Saône, France

Philip J. Ball School of Geography, Geology and the Environment, Keele University, Newcastle, United Kingdom; Geothermal Energy Advisors, Houston, TX, United States

David Banks University of Glasgow, Glasgow, United Kingdom

Graham Banks Route To Reserves Consulting Inc, Vancouver, BC, Canada

Axel-Pierre Bois XGS Energy, Houston, TX, United States

Christopher S. Brown University of Glasgow, Glasgow, United Kingdom

Cheng Chen Stevens Institute of Technology, Hoboken, NJ, United States

Sayyam Chopra MPGE, University of Oklahoma, Norman, OK, United States

Govert de With NRG, Arnhem, The Netherlands

Birol Dindoruk Department of Petroleum Engineering, University of Houston, Houston, TX, United States

Gioia Falcone University of Glasgow, Glasgow, United Kingdom

Carlos A. Fernandez Pacific Northwest National Laboratory, Richland, WA, United States

Hans-Christian Freitag Freitag Energy Advisory, Houston, TX, United States

Ghazal Izadi XGS Energy, Palo Alto, CA, United States

Simon James CS8 Consulting, Le Plessis-Robinson, France

Guoqing Jian Aramco Americas, Houston, TX, United States

Isa Kolo University of Glasgow, Glasgow, United Kingdom

Zihao Li Pacific Northwest National Laboratory, Richland, WA, United States

Silviu Livescu Bedrock Energy; Department of Petroleum Engineering, University of Texas at Austin, Austin, TX, United States

Mikenna Montgomery M3 Groundwork, San Diego, CA, United States

Jonah Poort Heat Transfer and Fluid Dynamics Department, TNO, The Hague, The Netherlands

Ram Ratnakar Shell, Houston, TX, United States

Sadegh Shahmohammadi Healthy Living & Work, Netherlands Organisation for Applied Scientific Research (TNO), The Hague, The Netherlands

Aditya Sharma MPGE, University of Oklahoma, Norman, OK, United States

Pejman Shoeibi Omrani Heat Transfer and Fluid Dynamics Department, TNO, The Hague; Environmental Technology, Wageningen University and Research, Wageningen, The Netherlands

Catalin Teodoriu MPGE, University of Oklahoma, Norman, OK, United States

Denis Voskov Department of Geoscience & Engineering, Delft University of Technology, Delft, The Netherlands; Department of Energy Science and Engineering, Stanford University, Stanford, CA, United States

Katya Waters MPGE, University of Oklahoma, Norman, OK, United States

Ken Wisian Bureau of Economic Geology, University of Texas at Austin, Austin, TX, United States

Meisong Yan WE Resources, LLC, Houston, TX, United States

Chao Zeng Pacific Northwest National Laboratory, Richland, WA, United States

CHAPTER

1

Overview of geothermal systems

Meisong Yan[a], Silviu Livescu[b,c], Birol Dindoruk[d], and Ken Wisian[e]

[a]WE Resources, LLC, Houston, TX, United States [b]Bedrock Energy, Austin, TX, United States [c]Department of Petroleum Engineering, University of Texas at Austin, Austin, TX, United States [d]Department of Petroleum Engineering, University of Houston, Houston, TX, United States [e]Bureau of Economic Geology, University of Texas at Austin, Austin, TX, United States

1 Introduction to geothermal energy

Scientists classify the structure of the Earth into four fundamental parts: the Earth's crust, the mantle enveloping it, the outer core, and the inner core at its heart (Fig. 1). Each part has unique chemical compositions and physical properties:

- The inner core, which is comprised of solid iron, spanning a diameter of ~1500 miles.
- The outer core, which is a layer of scorched molten rocks, known as magma in its liquid state, extending to a thickness of roughly 1500 miles.
- The mantle, which is composed of both magma and rock, surrounding the outer core with a thickness measuring around 1800 miles (or 2900 km; Stober and Bucher, 2021).
- The solid rock crust, forming both continents and ocean floors, which exhibits varying thicknesses. Beneath the continents, it ranges from 15 to 35 miles in depth, while below the oceans, it measures between 3 and 5 miles (Stober and Bucher, 2021).

As illustrated in Fig. 2, the temperature within the Earth's interior increases nonlinearly as we delve deeper. Geothermal gradient is defined as the amount that the Earth's temperature increases with depth. In normal continental crust, a typical geothermal gradient within the first 3–5 km (2 or 3 miles) of the Earth's surface is about 25°C/km, or 1 °F/70 ft. This gradient, however, decreases to no more than about 16°C/km at a depth of 40 km (25 miles; DiPietro, 2013).

At the base of the crust, the temperature is roughly 727°C, signifying a considerable heat presence. Diving further, the base of the mantle reveals an even more elevated temperature, reaching ~3227°C. Finally, at the very heart of our planet, the core radiates a formidable heat, with temperatures hovering around 4727°C.

Geothermal Energy Engineering
https://doi.org/10.1016/B978-0-443-21662-6.00001-3

Copyright © 2025 Elsevier Inc. All rights are reserved, including those for text and data mining, AI training, and similar technologies.

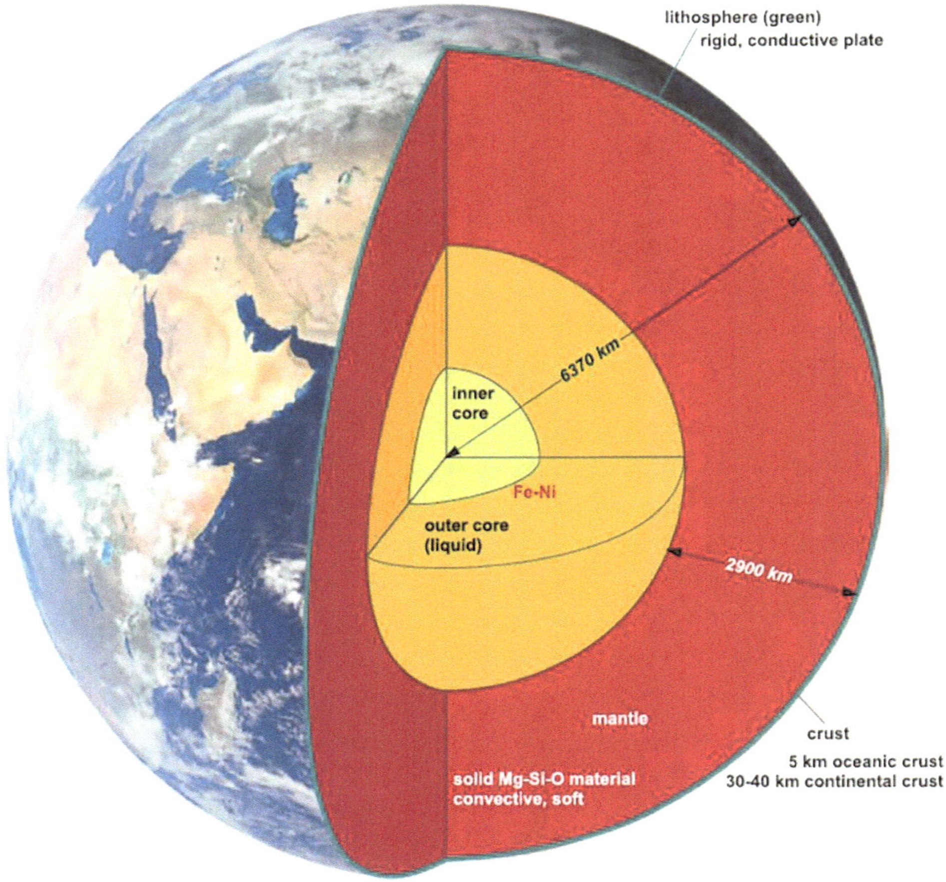

FIG. 1 The Earth's interior (Stober and Bucher, 2021).

For further insight, consider these fun facts: a staggering 99% of the Earth is hotter than 1000°C, with a mere 0.1% cooler than 100°C—a testament to the planet's enormous heat resource. As we look at the sky, the sun's scorching surface bears a temperature of about 5800°C, corresponding to the temperature at the inner core of the Earth.

This abundant source of naturally occurring, ubiquitous, and clean energy beneath us exists residually as a result of the formation of our planet and finds its way to the surface, most typically around plate boundaries and in Earth's volcanic regions. The use of geothermal energy for power production is widely utilized in many regions of the world today, where core heat finds such a conduit to the surface, including in locales like Iceland, Hawaii, and areas within the Ring of Fire, which is mostly located along the Pacific Ocean. When magma flows toward the surface, it heats the groundwater trapped in porous rock or water running along natural fractures and faults (Livescu et al., 2023). Radioactive decay of elements in the Earth's crust is another abundant source of subsurface heat. This frequently overlooked but significant heat source, estimated to be present near the surface in the range of 40 TW, is the target of many geothermal concepts that seek to harvest geothermal "anywhere," as opposed to in volcanic and boundary regions as discussed earlier.

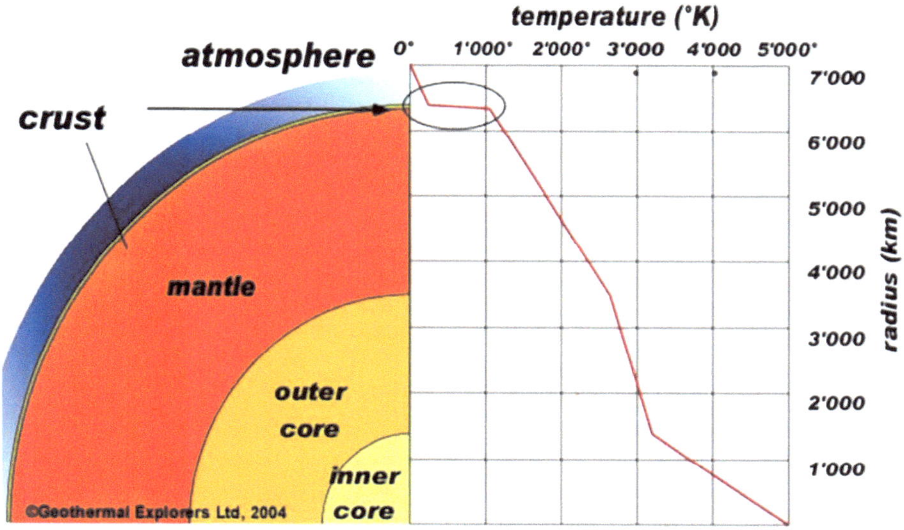

FIG. 2 The Earth's temperature profile (Krauter, 2021).

2 Types of geothermal systems for power production

Geothermal systems are classified differently based on the purpose of the definition. Typically, geologists categorize geothermal systems based on their geological settings and how these systems were formed, each with unique geological characteristics that influence the extraction and utilization of geothermal energy. When we look into the final applications of geothermal energy, geothermal systems could then be categorized into geothermal power plants for electricity generation and direct heat use systems, such as geothermal heat pumps using the constant temperatures near the surface of the Earth to provide both heating and cooling to a building.

Geothermal systems for electricity production can be divided into four categories (Livescu et al., 2023):

a. Conventional hydrothermal systems
b. Enhanced (or engineered) geothermal systems
c. Advanced geothermal systems
d. Multisystem hybrids

In this book, we adopt the commonly accepted reservoir engineering approach. The methodology is based on the mechanism of heat transfer, the evaluation technique during well testing, and the system's long-term production. The reservoir engineering basis could deem the technical and financial success of a geothermal project.

a. Conventional hydrothermal systems

Conventional hydrothermal systems comprise nearly all geothermal electrical power generations existing today (IRENA, 2017). They exist where geothermal reservoir temperature and production flows are naturally sufficient to produce electricity, meaning that the sufficient porosity in the subsurface, the abundant transfer of heat into the system, and the natural presence of water combine to produce a near-surface, developable resource. The transfer of heat from the mantle to shallow porous rocks, and to fluids present within the pore space of the rocks, relies primarily on convection and secondarily on conduction. Temperatures above 225°C (437°F) are optimal for higher power plant efficiency.

Historically, geothermal energy systems are selected at the locations with naturally stored heat and water inside the permeable rocks, so that the extraction of energy requires less human intervention. The Old Faithful Geyser is a well-known geothermal feature in the Yellowstone National Park. Each of its eruption lasts 1.5–5 min, expelling 3700–8400 gal (14,000–32,000 L) of boiling water while reaching as high as 106–184 ft. (30–55 m) (Fig. 3).

FIG. 3 Library of Congress (1917).

Therefore, the development of a conventional geothermal system at the early stage would be drilling a well to produce a natural reservoir of heated water for the geothermal power facility. A desirable geological location, but poses a huge limitation on the scale and the expansion of the geothermal power plant, is typically near the tectonic plates (Fig. 4).

Exploring conventional hydrothermal reservoirs for electricity generation relies on identifying the surface indicators such as geysers or steam vents. Traditional technology used in petroleum exploration, i.e., seismic and gravity magnetics, struggles to differentiate between good and poor reservoirs due to the absence of impedance and fluid density contrasts.

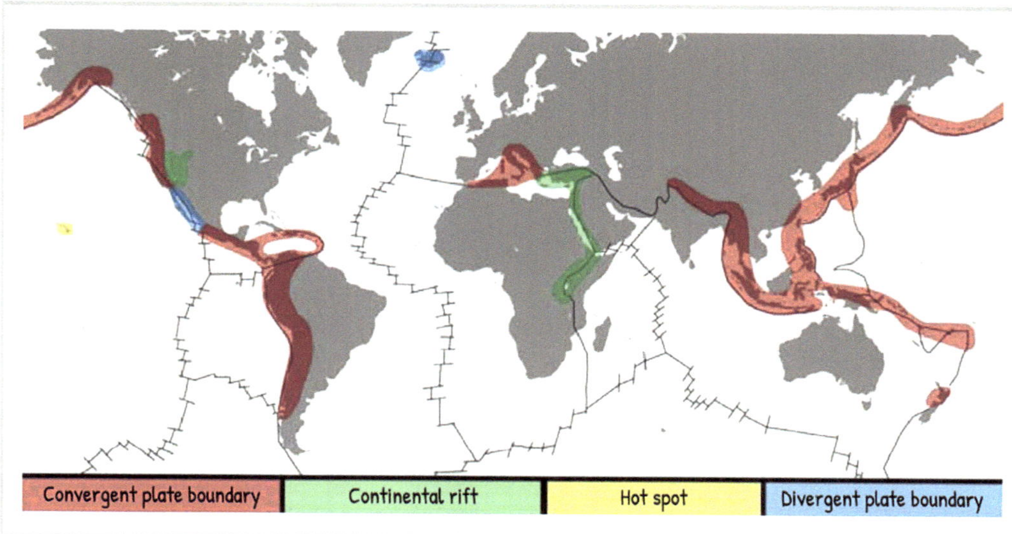

FIG. 4 High flow of heat to the surface and conventional hydrothermal regions (Ball, 2020). Map lines delineate study areas and do not necessarily depict accepted national boundaries.

Additionally, these reservoirs often experience rapid pressure changes from tectonic shifts, and deeper, hotter ones undergo diagenetic processes, reducing their porosity and permeability, which makes the drilling challenging.

b. Enhanced geothermal systems

Engineered (or enhanced) geothermal systems, also known as "EGS," are scalable geothermal technologies where one or more wells are drilled, either via natural or hydraulically stimulated fractures; the wells are connected to one another in the subsurface, creating an engineered reservoir. Water is then injected into the reservoir, where it absorbs heat from hot rocks it is circulating through. It is then produced to the surface, where the fluid or steam is passed through a turbine and used to generate electricity. Thanks to recent oil and gas technological advancements, such as deep well drilling, logging, and construction, as well as improvements in materials, including cement and well casing, untapped geothermal resources in hot, dry rocks with little or no permeability or naturally occurring fluids are now accessible (Livescu et al., 2023). That is, unlike conventional hydrothermal systems, EGS leverage the cutting-edge drilling and completion techniques advanced by the oil and gas industry to drill through significantly deeper caprocks and fracture them to artificially create a circulation network for the energy extraction.

In a "traditional" EGS design, multiple wells are drilled vertically, and then, hydraulic fractures are employed between them to enable the circulation of fluids through the subsurface formation. Hydraulic fracturing, a critical technique for EGS, is used to access resources in low-permeability underground formations. This process involves injecting a substantial volume of high-pressure fluid into the formation to fracture the rock, either creating new cracks or extending existing fractures. When preexisting fractures are present in the formation, the newly created crack can intersect with these fractures, forming interconnected conductive fracture networks.

EGS signify a breakthrough in utilizing geothermal resources stored inside less-permeable formations that are largely nonproducing. Discovering or generating adequate porosity and permeability in deep basement rocks poses significant challenges, demanding more time, energy, and financial resources compared with hydrocarbon wells at similar depths. Stimulations through hydraulic forces, thermal energy, or chemicals could make the field development eventually economically feasible.

However, the stimulated fracture may close after the pressure of the reservoir reduces, due to the rock mechanics or the interactions with the circulating fluid. Additionally, the hydraulic fracturing process may ignite seismic activities. Therefore, the Frontier Observatory for Research in Geothermal Energy (FORGE) launched a comprehensive pilot program aiming to improve geothermal energy technology. It has drilled not only the production, injection, and stimulated wells but also three observation wells to continuously track reservoir pressure, fluid flow, temperature distribution, and seismic activities (FORGE, 2020, 2021).

According to Robins et al. (2021) and Livescu et al. (2023a), there have been several EGS projects over the past decades before FORGE, including Fenton Hill (United States); Rosemanowes (United Kingdom); Le Mayet, Soultz, and Strasbourg (France); Hijiori (Japan); and Cooper Basin (Australia; Calpine, 2024; Cyrq, 2024; Li et al., 2021; Kneafsey et al., 2018; Richter, 2017; Allis et al., 2013; Garcia et al., 2012). In addition, Calpine's EGS project is in Middletown, California (Calpine, 2024); Ormat's Desert Peak and Brady field projects are located in Churchill County, Nevada (Richter, 2013); and, formerly owned by the US Geothermal, Ormat's Raft River EGS project is located in Raft River, Idaho (Richter, 2016). Among them, Ormat's Desert Peak and Raft River (Richter, 2013) and Calpine's Geysers EGS operations are commercially active (Calpine, 2024).

The longest-operating commercial EGS project currently generating power is the Soultz EGS project in Alsace, France (Koelbel and Genter, 2017). A pilot power plant began operation at Soultz in 2009, with an installed gross capacity of 1.7 MWe. It has two stimulated reservoirs within fractured granite: one at 2.2 miles (3.5 km) depth and the other at 3.1 miles (5 km) depth, with commercial electricity production beginning in 2016 (Ravier et al., 2019).

At the FORGE site, numerous preexisting natural fractures were identified, and then, a total of four hydraulic fracturing experiments were conducted in different segments of the wellbore in 2017 and 2019 (FORGE, 2020, 2021). The initial hydraulic fracturing trial took place in the open-hole section of the wellbore in 2017, and this section was subsequently refractured in 2019. Two additional hydraulic fracturing tests occurred in 2019 within the cased sections of the wellbore, each targeting preexisting fractures with distinct orientations. In one region, the preexisting fractures ran parallel to the maximum horizontal stress, an ideal configuration for shearing and dilation. In the other region, fewer fractures were oriented at a higher angle to the maximum horizontal stress, requiring higher injection pressure for stimulation. Both pressure responses from the tests and observations from the Formation MicroScanner image log confirm the initiation and extension of hydraulic fractures as well as the shearing of preexisting fractures. Preliminary results from the FORGE project show promise for EGS as a viable source of energy supply (Fig. 5).

EGS could benefit from the best practices and the lessons learned from the oil and gas industry for a sustainable extraction of the geothermal resources. Hydraulic fracturing techniques can create a network of fractures in the hot rock formation, improving heat transfer and increasing geothermal fluid production. Directional drilling and rotary steerable systems

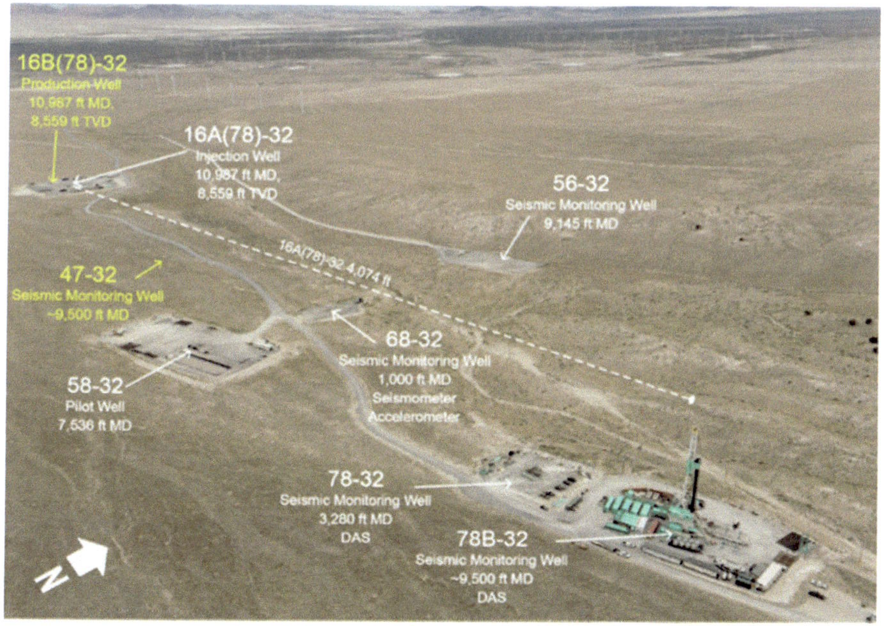

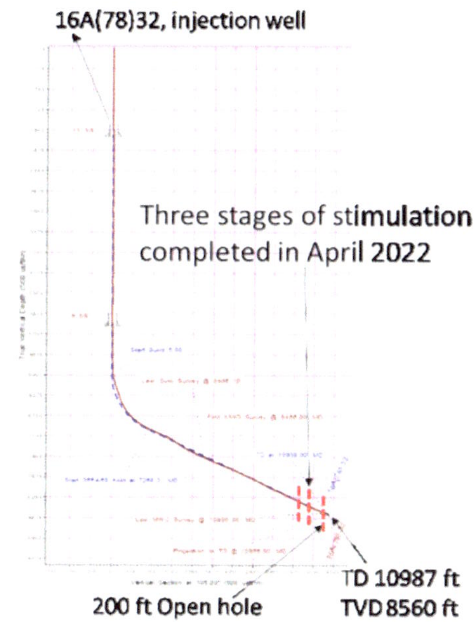

FIG. 5 FORGE pilot program (FORGE, 2021).

have the potential to optimize wellbore placement in geothermal reservoirs, maximizing contact with high-temperature zones and improving overall resource recovery. Innovations in casing and cementing practices can improve the integrity of the geothermal well. Real-time reservoir monitoring, downhole sensors, and data analytics can be transferred to EGS projects to better understand the behavior of the reservoir, optimize production, and identify potential issues, therefore reducing the risk of seismic hazard.

However, this transfer of technologies needs to be carefully evaluated and adapted to suit the unique characteristics of geothermal resources, considering the distinctive differences between two extraction processes in terms of the geology of the reservoirs, the fluid properties, and the environmental considerations.

c. Advanced geothermal systems

Advanced geothermal systems, in many circles, have become a catch-all term that includes a variety of next-generation and emerging geothermal concepts, including closed-loop geothermal systems (CLGS). Even some concepts in the traditional hydrothermal space are now referred to as advanced geothermal systems. Closed-loop geothermal systems can have any configuration that allows the circulation of fluid without direct contact between the working fluid and reservoir. In the so-called closed-to-reservoir concepts, as opposed to "open-to-reservoir" concepts like the enhanced geothermal systems, fluid is pumped from the surface to the subsurface, picks up heat from the surrounding formation through conduction, and is then returned to the surface, bringing with it heat from the formation. Because these systems function in a closed loop, and thus theoretically do not exchange fluids with the subsurface, the use of engineered, nonwater working fluids, such as supercritical carbon dioxide (sCO_2), is an area of fast-moving innovation. Because rock is a low-conductive medium, lacking significant contributions from convection, CLGS experience inefficient transfer of heat from the subsurface to the circulating fluid, and innovation in this area is needed.

Some examples of advanced geothermal systems are pipe-in-pipe systems (concentric pipes), U-shaped wells, or geothermal probes for heat pumps aiming for power generation and direct heat use. Recent studies on concentric pipe-in-pipe CLGS have revealed how various well parameters, including fluid flow rate, well dimensions, temperature, working fluid type, and heat transfer coefficients, affect fluid temperature at the surface. These parameters have intricate relationships with thermal and electric energy production, with overall heat transfer coefficients being particularly influential. Adjusting these coefficients while keeping other factors constant can significantly impact the outlet temperature, affecting the economic viability of geothermal projects.

Due to the limiting heat conductivity of rocks and low convection effects, advanced geothermal systems face difficulties in economically transferring heat from the subsurface to the circulating fluid. Achieving commercial viability for power generation with advanced geothermal systems requires technological breakthroughs (Figs. 6–7).

In recent years, advanced geothermal systems have gained renewed interest because of their potential for simplicity and versatility, particularly when utilizing engineered nonwater working fluids. They also offer the possibility of avoiding hydraulic fracturing while producing both power and direct heat use. These technologies are undergoing testing across various temperature and rock conditions, spanning from low-temperature sedimentary zones to high-temperature dry rock formations.

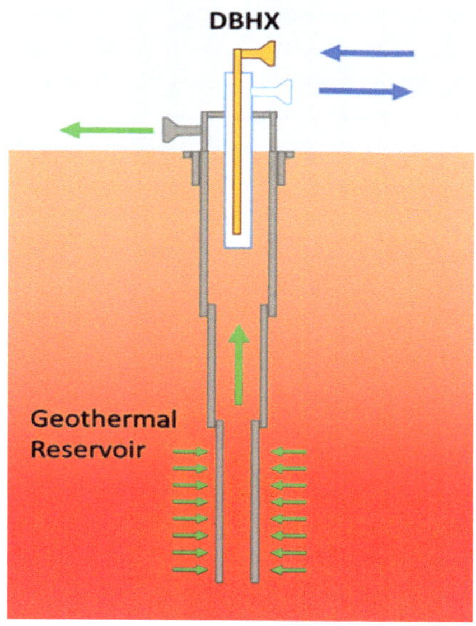

FIG. 6 Example of an advanced geothermal system (Amaya et al., 2020).

Nonetheless, the key to success lies in conducting field trials to demonstrate heat exchange processes within the wellbore and the surrounding rock. These trials aim to establish a stable output suitable for commercial power generation. Successful initial field trials have the potential to inspire innovative subsurface well designs, which could provide long-term heat exchange capacity for sustainable operations (Livescu et al., 2021; Livescu and Dindoruk, 2022a,b).

d. Multisystem hybrids

Multisystem hybrids, also known as hybrid geothermal systems, are systems that combine two geothermal systems, such as enhanced and advanced geothermal systems, or two different systems, e.g., solar photovoltaic (PV) and geothermal; concentrated solar power (CSP) and geothermal; direct air capture (DAC) and geothermal (Li et al., 2020; Kuru et al., 2023a, b, 2024); and carbon capture, usage, and storage (CCUS) and geothermal. Fig. 7 could best describe the difference among these four geothermal systems.

One example of a hybrid system that combines advanced geothermal systems and EGS concepts within sedimentary formations has been developed by Sage Geosystems (Sage, 2024). The focus is to achieve wellbore temperatures ranging from 100°C (212 °F) to 250°C (482 °F) at depths of 1.9–3.7 miles (3–6 km). These depths are accessible using conventional drilling technology, equipment, and service providers.

Based on "no-one-size-fits-all" principle, Sage has developed three innovative geothermal concepts: HeatRoot, HeatFlood, and HeatLoop for different geological conditions.

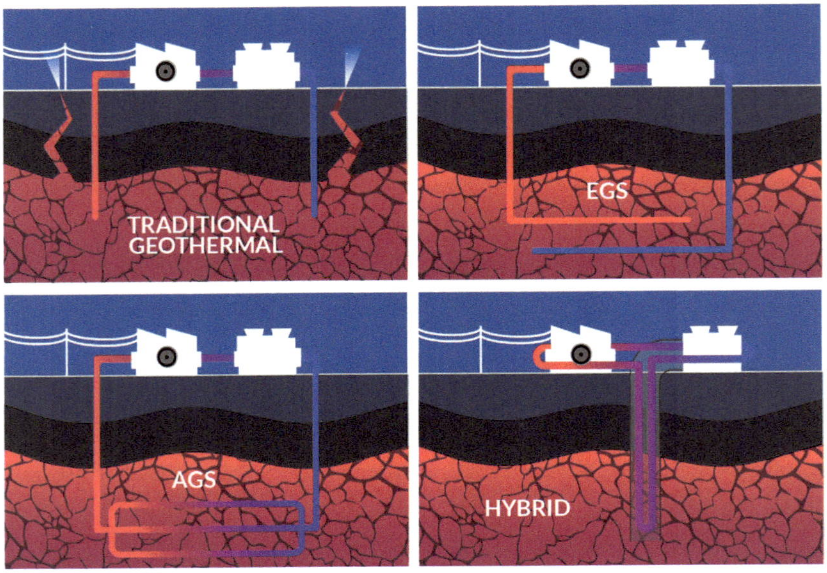

FIG. 7 Different geothermal systems (Westwood Energy, 2024).

- HeatRoot is designed for deeper hot dry rock formations with temperatures at or above 150°C (302 °F). It involves a downward-oriented fracture acting as a heat chimney from deeper, hotter formations. Brine circulates inside to transport heat to the downhole heat exchanger.
- HeatLoop is a variation of HeatRoot where multiple lateral well sections are drilled and fractures interconnect them.
- HeatFlood utilizes wells to extract heat from porous sand formations capable of flowing produced hot fluids at high rates.

Sage also employs sCO_2 as the working fluid in combination with a custom sCO_2 turbine developed in collaboration with their partner, Southwest Research Institute (SWRI). This approach offers several significant advantages, including an anticipated doubling of efficiency compared with traditional geothermal plants. These benefits stem from a notably smaller and more cost-effective turbine, reduced energy losses due to mechanical friction, and decreased pumping expenses facilitated by natural thermosiphon processes.

All the innovative thinking and technology advancements are building a solid foundation to efficiently extract this energy resource with commercial success.

3 Geothermal heating and cooling systems

Direct heat use systems, also referred to as geothermal heating and cooling systems (GHCS), utilize geothermal energy directly, as opposed to the use of a power turbine to generate electricity. GHCS can be shallow, i.e., 500–2000 ft. deep, or deep, such as, deeper than 2000 ft. The shallow GHCS, also referred to as ground-source or geothermal heat pumps, use shallow geothermal energy for a variety of applications, the most common being the heating and cooling of buildings. However, as shown in Fig. 8, there is no consensus about the depth definition for different GHCS; they can be better categorized by temperature and not depth (Kaleghi and Livescu, 2023).

As the energy required by buildings and industrial processes for heating and cooling is more than any other sectors (EIA, 2022), direct heat use (cooling and heating) could generate significant operational cost savings for users, relieve grid strain, and can make a substantial and material contribution to the decarbonization of the electrical grid.

GHCS are a mature, well-established technology that will benefit from further spillover of oil and gas technologies and know-how. With the technologies available today, GHCS help building owners save between 30 and 70% in heating costs and between 20% and 25% in cooling costs compared with conventional systems, such as air-source heat pumps, natural gas furnaces, and traditional air conditioning systems (EPA, 2022).

Traditional air conditioning methods, using air-source heat pumps, do not remove or eliminate heat, but rather move it from the building to the hotter air outside. When a building interior is cooled, that thermal energy is transferred to the exterior environment. In dense urban areas, this effect increases local temperatures, exacerbating summer heat waves in places that are already heat islands as a result of urbanization. The solution to mitigate this public health issue is to transfer that heat to the subsurface, using GHCS instead.

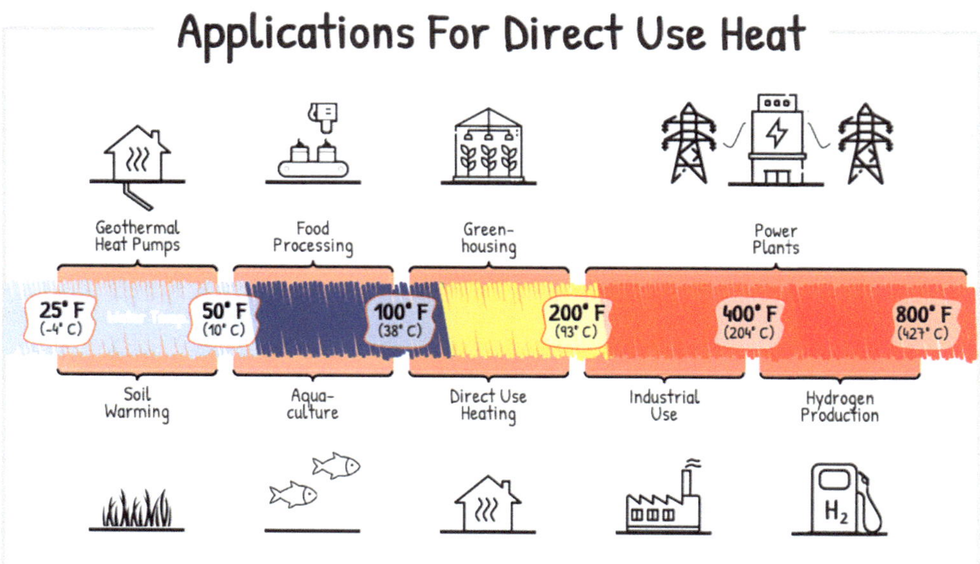

FIG. 8 Direct heat use applications of geothermal energy (Kapusta et al., 2023).

4 Development history of geothermal systems

Historically, geothermal energy has been utilized to serve our daily needs before modern technologies emerged. The ancient Romans used hypocaust systems that circulated the geothermal water to warm buildings and baths. The Maori people in New Zealand also made ingenious use of geothermal steam for cooking food in earth ovens, known as "hangi." In North America, the usage of the geothermal energy dated back to at least the early nineteenth century.

In 1830, Asa Thompson introduced a business model by charging a dollar for access to three spring-fed baths within a wooden tub in Hot Springs, Arkansas, marking the first documented case of the use of commercial geothermal energy.

In 1892, the world's first geothermal district heating system was built in Boise, Idaho, to warm more than 200 residences and 40 commercial buildings, demonstrating the early potential of geothermal energy for large-scale heating applications.

It was not until 1960 that the first large-scale geothermal electricity generation plant began operating in the United States, at the Geysers in California by Pacific Gas and Electric. The first turbine achieved a remarkable output of 11 MW of net power and maintained its efficient operation for more than three decades. It currently produces over 1500 MW of power, the largest geothermal power plant in the world.

Worldwide, the first geothermal power generator was built in 1904 in Tuscany, thanks to the Italian scientist Piero Ginori Conti of Trevignano. Flash plants take high-pressure hot water from deep inside the Larderello field and convert it into steam that drives generator turbines (Fig. 9).

FIG. 9 First geothermal power plant in Larderello, Italy (Lafayette College, 2014).

Back in the 1950s, a significant breakthrough emerged with the inception of the first heat pumps utilizing water sources, specifically designed for residential use. Subsequently, in the early 1960s, these heating and cooling systems were introduced to commercial and institutional places. In the 1980s, engineers and scientists designed the early version of the geothermal heat pumps that are more popular in today's world.

Several oil and gas companies have publicly announced their conventional hydrothermal systems effort across various locations worldwide since 2021. An example would be

Repsol. The Government of the Canary Islands have granted them two geothermal exploration permits in the south of Tenerife. One of the initiatives, known as Tajao, is planned to cover 576 mining squares located within the municipalities of Granadilla de Abona and Arico. The second one, referred to as Isora, will cover an area of 630 mining squares situated in the territories of Adeje, Santiago del Teide, and Guía de Isora.

According to the Energy Information Administration (EIA), the year 2021 witnessed the substantial role that geothermal energy plays in electricity generation across the world. Approximately 27 nations collectively generated 92 billion kWh of electricity from geothermal sources. The United States led the effort with about 16 billion kWh, while Indonesia was in the second place. About 5% of Indonesia's total electricity output was extracted from geothermal resources.

The top 10 geothermal countries as of year-end 2023 were (Cariaga, 2024) (1) United States, 3900 MW; (2) Indonesia, 2418 MW; (3) Philippines, 1952 MW; (4) Turkey, 1691 MW; (5) New Zealand, 1042 MW; (6) Kenya, 985 MW; (7) Mexico, 976 MW; (8) Italy, 916 MW; (9) Iceland, 754 MW; and (10) Japan, 576 MW.

5 Opportunities and challenges of geothermal energy

Geothermal energy is recognized as a renewable and environmental-friendly energy source that could be continuously produced and has a low carbon footprint.

The energy provided by geothermal is considered reliable because the resource is always available to be tapped into. Unlike wind or solar power, which depends on weather conditions, geothermal energy is available 24/7, providing a constant and consistent source for power. This uninterrupted availability makes it an attractive option for both baseload power generation and direct heating applications. The output of a geothermal power plant is also highly consistent, as the Earth's heat remains relatively stable over time. This reliable predictability simplifies the energy production planning.

Geothermal energy is categorized as "green power" because of its minimal environmental impact. It has significantly lower CO_2 emissions per kilowatt hour of electricity produced, even when compared with the cleanest fossil fuel option, natural gas. Bloomfield and Moore (1999) study showed that CO_2 emissions from various power sources are as follows: geothermal at 0.18, coal at 2.13, petroleum at 1.56, and natural gas at 1.03 (unit: pound of CO_2 per kWh). This positions the geothermal power as an eco-friendly natural resource to mitigate the CO_2 emissions.

The National Renewable Energy Laboratory (NREL) published a map of the geothermal resources of the United States highlighting both the existing hydrothermal locations and the prospects of deep enhanced geothermal systems. However, the United States only tapped less than 0.7% of geothermal resources. The vast majority of this resource can be unlocked through the development and the application of EGS technology to meet our future demands for heating, cooling, and electricity (Fig. 10).

The GeoVision launched by the Geothermal Technologies Office (GTO) in 2019 provided a comprehensive analysis on the possibilities of expanding geothermal utilization to shape America's energy future (DOE, 2019, Fig. 11). The key findings of the GeoVision report clearly established the role of geothermal energy as critical to the nation's aggressive, zero-carbon

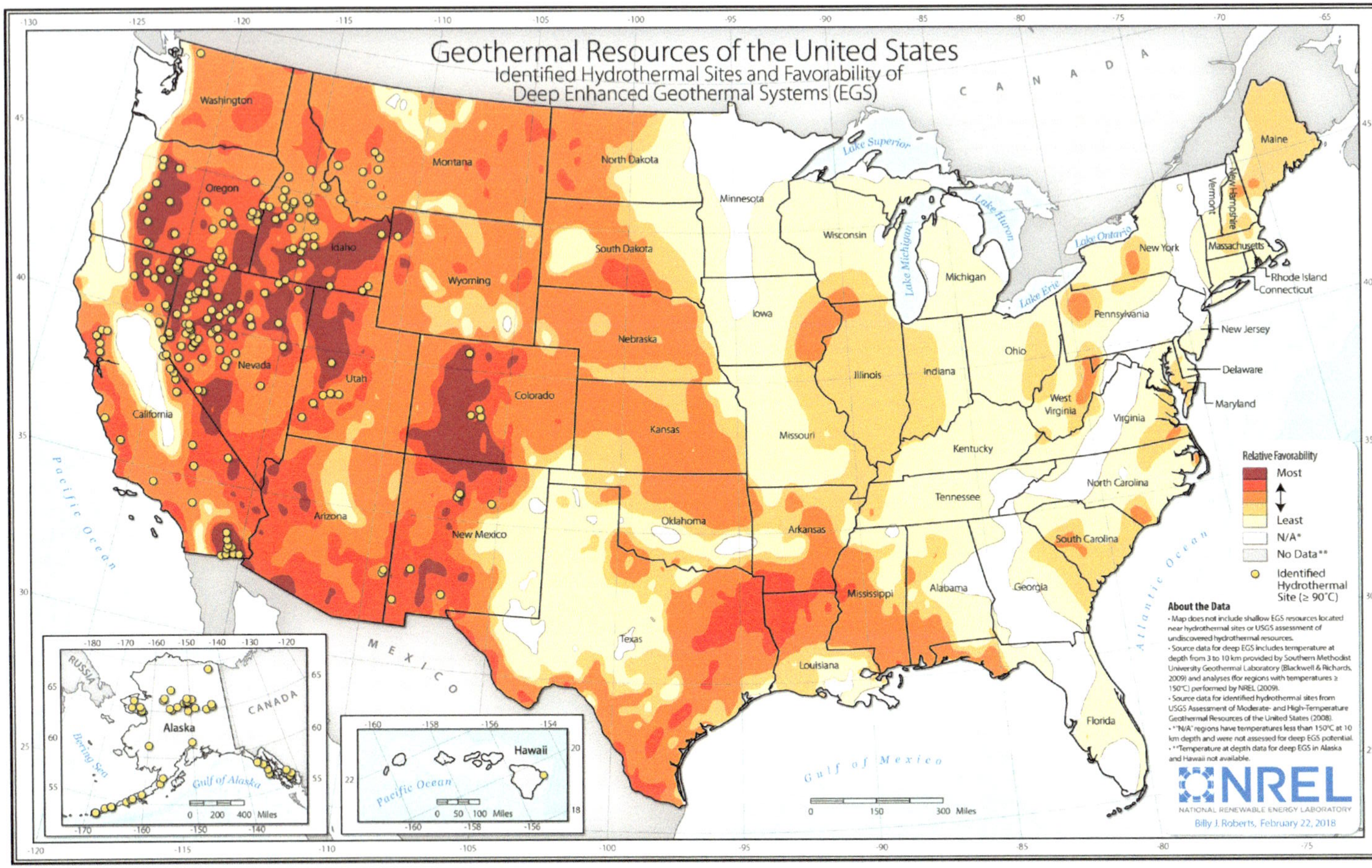

FIG. 10 US geothermal resource map by NREL (Roberts, 2018). Map lines delineate study areas and do not necessarily depict accepted national boundaries.

energy transformation. Three strategic goals were proposed in order to fully utilize the potential of geothermal energy.

- Strategic Goal 1: Drive toward a carbon-free electricity grid by supplying 60 GW of EGS and hydrothermal resource deployment by 2050.
- Strategic Goal 2: Decarbonize building heating and cooling loads by capturing the economic potential of 17,500 geothermal district heating installations and by installing geothermal heat pumps in 28 million households nationwide by 2050.
- Strategic Goal 3: Deliver economic, environmental, and social justice advancements through increased geothermal technology deployment.

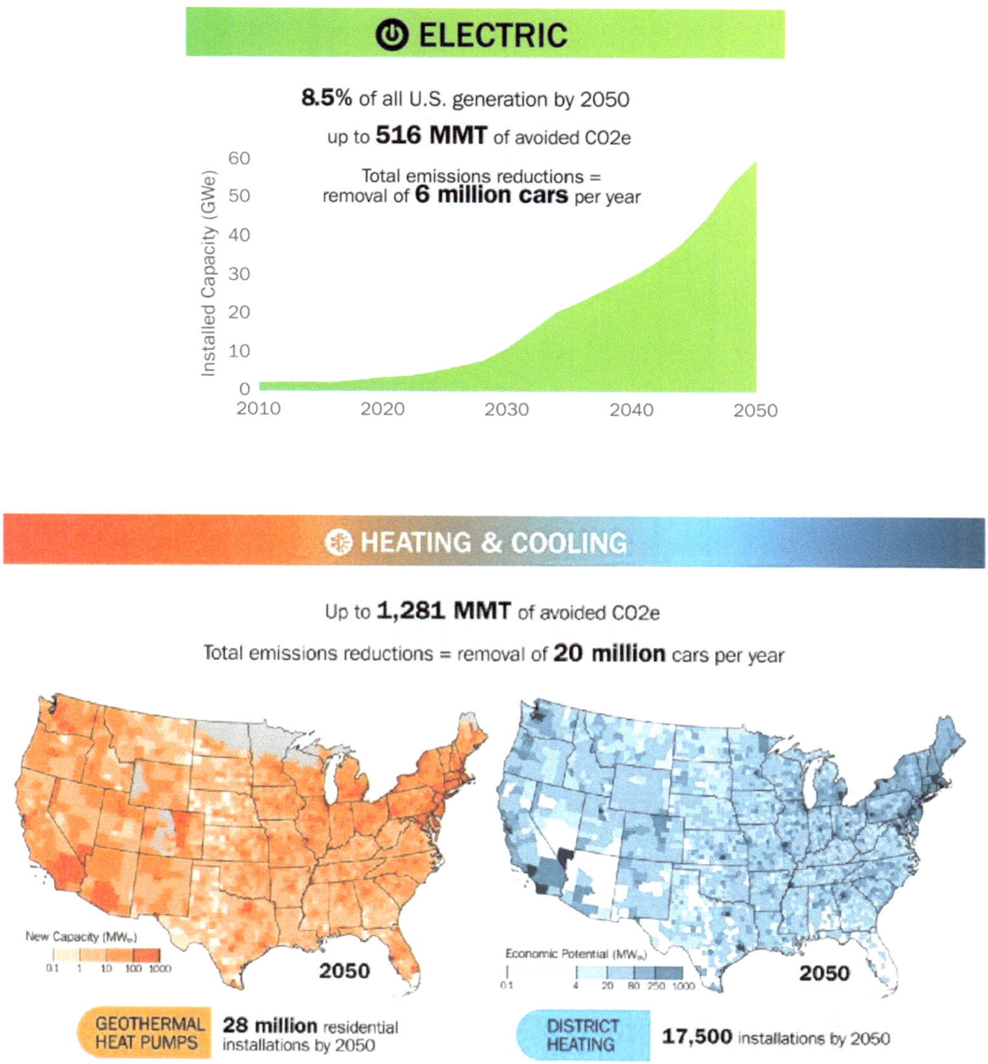

FIG. 11 US DOE's GeoVision strategic goals by 2050 (DOE, 2019).

According to an analysis released by Third Way and Boston Consulting Group (BCG), through 2050, the US companies are expected to capture a substantial global geothermal market, conservatively estimated to be $1.5 trillion and around 100,000 job opportunities. The actual market size could be further expanded when the pivotal advancements happen in EGS, deep drilling techniques, and well materials (Fig. 12).

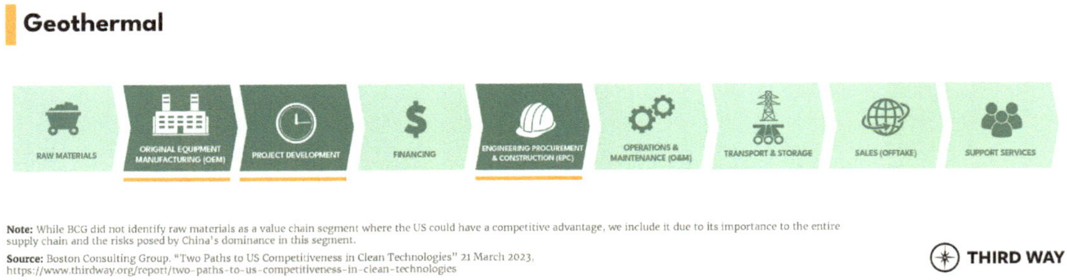

FIG. 12 Geothermal value chain between 2020 and 2050 (Third Way, 2023).

Nevertheless, several significant challenges need to be addressed to fully leverage the potential of the geothermal energy. These challenges include not only the considerable initial investment costs, but also the intricate technical complexities associated with tapping into subsurface heat resources. Furthermore, geothermal projects also struggle with potential environmental risks, requiring meticulous planning and execution to minimize any adverse ecological impacts. Beyond the technical and environmental aspects, social acceptance plays a pivotal role. Engaging with local communities and securing their support are crucial to ensure a successful development of the geothermal projects.

One of the primary limitations of geothermal projects is its inherent geographical restriction. Geothermal power generation is contingent on the presence of adequate heat reservoirs beneath the Earth's surface. Consequently, geothermal power plants must be strategically situated in regions where these heat sources are accessible. Furthermore, even in geothermal-rich regions, the feasibility of the development can be influenced by factors such as land ownership, land use regulations, and environmental considerations. These considerations could further narrow down the available sites for geothermal projects.

Induced seismicity is another concern in geothermal energy development, especially with enhanced geothermal power plants when water is injected into the Earth's crust to increase the permeability of the hot rock formations to facilitate the circulation of fluids and heat exchange. Several factors would influence the occurrence of this induced seismicity in geothermal projects, including the rate and volume of the fluid injection, the geological properties of the site, and the preexisting stress conditions in the Earth's crust. Therefore, industry practices and real-time monitoring technologies are required to mitigate this risk. Responsible reservoir management and adherence to best practices help ensure the safety of the nearby communities and infrastructure.

However, predicting fluid flow dynamics in EGS is challenging due to limited data on downhole conditions. In reservoirs that rely on natural or induced fractures for fluid movement, it is challenging to determine the flow direction beforehand, leading to an exploratory approach similar to conventional hydrothermal systems. Additionally, fractures may open or close based on operational factors like changes in injection fluid composition or injection pressure differences.

To address these challenges, the development of a reliable numerical simulator capable of modeling complex and interconnected physical processes, including non-isothermal multiphase flow, geomechanics, and reactive transport with chemical reactions, is essential. Although a few simulators have been created for modeling such coupled processes, further advancements are necessary.

Furthermore, geothermal systems also face several operational challenges and constraints. High-temperature and high-pressure conditions can cause materials used in geothermal systems to degrade much faster over time. This includes well casings, pumps, joints, and cementing. Specialized materials that can withstand extreme conditions are often required, which can be costly.

As geothermal fluids contain minerals and gases, they have the potential to cause scaling and corrosion in the system. Scaling occurs when minerals precipitate out of the fluid and accumulate on surfaces, reducing flow efficiency. Corrosion can damage equipment and pipelines, leading to costly maintenance and replacement.

The biggest barrier for a geothermal project would be its productivity to justify the economic return to drill to deep depths. The subsurface complexity and the geological exploration risks add substantial upfront capital requirement to the life cycle of a project. The price tags start from around \$2–7 million for a plant with a 1 MW capacity. Understanding and characterizing these subsurface properties are critical to make accurate predictions about the resource's energy output. Advanced exploration techniques and robust data analysis could reduce the uncertainties in the geothermal exploration.

Hence, the expenses associated with geothermal electricity generation are impacted by factors such as the reservoir's thermal characteristics, the expenditures related to site investigation and well drilling, the quantity and depth of wells, and the type of power plant utilized. In order to make this natural resource extraction financially viable, the US DOE announced Energy Earthshots Initiative in 2022 and set a department-wide goal of reducing the average national cost of EGS projects by 90%, to \$45/MWh by 2035.

In 2023, the US DOE Geothermal Energy from Oil and Gas Demonstrated Engineering (GEODE) initiative awarded up to \$165 million to a consortium comprised of Project InnerSpace, Society of Petroleum Engineers, and Geothermal Rising, to expand US geothermal energy deployment by leveraging technologies and workforce from the oil and gas industry (SPE, 2023). Leveraging the extensive knowledge, technology, skill, and experience of the oil and gas sector will allow the geothermal industry to tackle barriers to geothermal deployment while also including fossil fuel–based communities and workers in the transition to clean energy. Although the US geothermal resource is vast, only a small portion of it has been developed due to unique challenges associated with subsurface environments, along with process issues of geothermal projects, such as long permitting timelines. The oil and gas and geothermal industries have numerous similarities that provide new opportunities for geothermal expansion—from advances in drilling and well construction to coproduction possibilities in existing oil and gas basins.

In summary, geothermal energy presents valuable opportunities as a reliable, low-emission, and sustainable energy source. However, it also faces challenges related to its geographical limitations, upfront costs, technical complexity, environmental considerations, and social acceptance. Addressing these challenges while capitalizing on its strengths is essential to fully realize the potential of geothermal energy for our energy transition.

6 Future prospects

The energy landscape has been evolving rapidly, especially after the 2020–2022 COVID-19 pandemic. The shift toward renewable energy sources, like solar, wind, hydro, and geothermal power, was expected to continue. Falling costs of renewable technologies and increased awareness of environmental concerns have been driving this energy transition.

The US Energy Information Administration projected in its Annual Energy Outlook 2023 that the US energy–related CO_2 emissions will drop 25%–38% below the 2005 level by 2030, while the renewable generating capacity grows in all regions of the United States in 2030, supported by growth in installed battery capacity EIA, 2023. Apparently, as one of the cleanest and most dependable forms of renewable energy, geothermal could be one essential component in the renewable world for a sustainable energy supply in the future.

Throughout history, energy crises have acted as catalysts for the advancement of geothermal exploration and development. In 1973, the members of the Organization of Arab Petroleum Exporting Countries (OPEC) announced an oil embargo, pushing many countries to look for alternative and sustainable energy sources. Looking ahead, the geothermal energy development is still closely linked to the traditional oil and gas industry.

By leveraging shared knowledge, technical expertise, and infrastructural resources, the oil and gas industry can potentially forge synergistic relationships that drive the growth of the geothermal energy solution. There have been a lot of exciting talks regarding the potential transformation of the aging and depleted oil and gas wells into operational geothermal power plants. Repurposing these low-productivity wells would reuse the preexisting infrastructure, reduce the upfront capital requirement, and expedite the timeline of the geothermal projects. This concept demonstrates a tangible step toward harnessing our existing resources in novel and environmentally conscientious ways for a cleaner energy future (Fig. 13).

The by-products generated during the geothermal extraction process also carry values to enhance the economic returns of a geothermal field development. An example of this is the condensate from geothermal resource posttreatment, which could be used for various industrial applications. Additionally, certain geothermal brines might contain valuable minerals, such as lithium, zinc, high-grade silica, and even gold. Extracting and utilizing these valuable minerals could add revenue streams to geothermal projects.

Recovering lithium from the heated saline water found deep underground and channeling it to geothermal power plants bring a dual benefit—renewable energy production coupled with the potential extraction of rare mineral that is essential for rechargeable batteries in electric vehicles, pacemakers, cell phones, and more. This also aligns with environmentally responsible practices because lithium is acquired from the geothermal water already brought to the surface rather than through traditional mining operations.

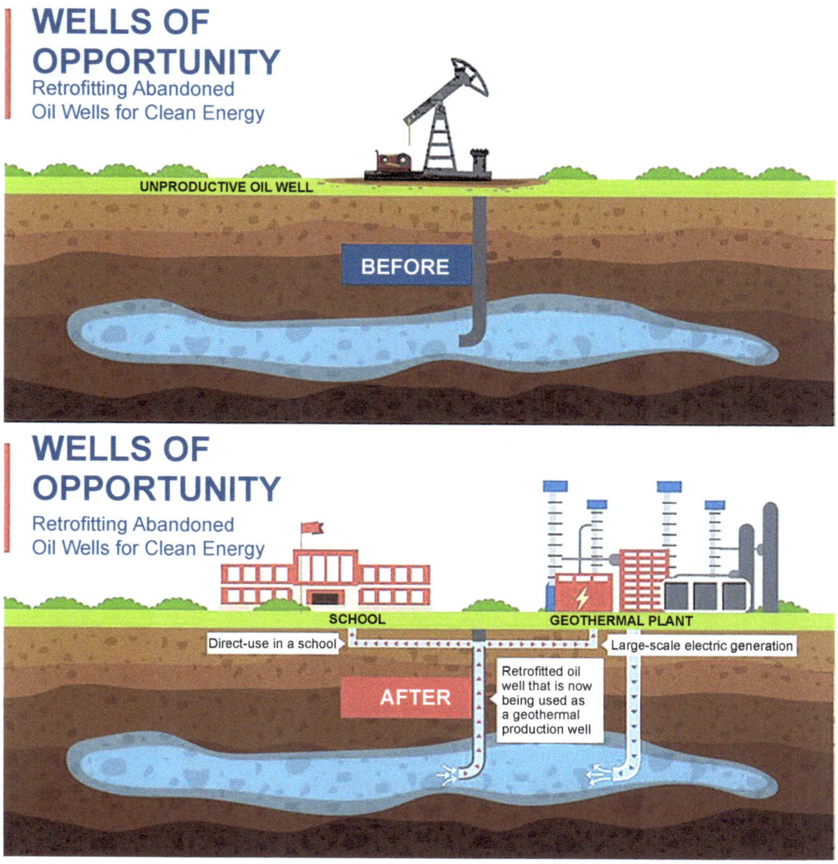

FIG. 13 Repurposing vintage wells into a functional geothermal power plant (MacGregor, 2023).

Technology improvement, innovation, and integration have played a key role in driving down the production costs associated with geothermal energy and led to better energy recovery, extended field life, and enhanced reliability. Progress in exploration, drilling, and assessment technologies has revolutionized the understanding of subsurface geology, facilitating the identification of optimal locations to drill geothermal wells, and the efficient production of geothermal power. All these enhance the utilization efficiency and overall economic viability of geothermal projects.

The latest machine learning (ML) and artificial intelligent (AI) technology advancements create opportunities for innovation and optimization in the geothermal industry. The Open Energy Data Initiative is dedicated to enhancing data accessibility, thereby facilitating comprehensive analysis that could not be accomplished before (OEDI, 2024). This initiative introduces a robust analysis platform that improves the accessibility through the establishment of a data lake, data catalog, and a collection of high-value open datasets, all hosted across various cloud computing platforms.

All of these innovative tools better characterize the geologic properties, significantly boost the accuracy of geothermal exploration, predict the future temperature outputs in geothermal systems, optimize the geothermal operations, and eventually lower the overall costs associated with geothermal energy development.

In summary, geothermal energy is not only a source of energy for power and direct heat use but also a reservoir for energy storage, as well as a resource for rare minerals. However, it also faces several challenges including significantly high upfront capex requirements, technical difficulties, environmental concerns, and social acceptance. Advancements in exploration and drilling techniques revolutionize the resource identification and utilization. The development of EGS and advanced geothermal systems pivots a pioneering approach to harnessing geothermal energy that could not be tapped before. Emerging technology in AI with open data and open code will advance the efficiency and performance of geothermal systems.

Ultimately, through knowledge sharing; collaboration among industry stakeholders, policymakers, and researchers; and technology advancement, geothermal energy could be integrated with other renewable energy sources as part of the smart grid for a collective energy future.

References

Allis, R., Moore, J.N., Anderson, T., Deo, M., Kirby, S., Roehner, R., Spencer, T., 2013. Characterizing the power potential of hot stratigraphic reservoirs in the Western US. In: Proceedings, 38th Workshop on Geothermal Reservoir Engineering. Stanford University, Stanford, California, February 11–13, 2013, SGP-TR-198. https://pangea.stanford.edu/ERE/pdf/IGAstandard/SGW/2013/Allis.pdf.

Amaya, A., Scherer, J., Muir, J., Patel, M., Higgins, B., 2020. GreenFire energy closed-loop geothermal demonstration using supercritical carbon dioxide as working fluid. In: Proceedings, 45th Workshop on Geothermal Reservoir Engineering Stanford University, Stanford, California, February 10–12, 2020. SGP-TR-216., https://doi.org/10.3390/ani11010010. https://pangea.stanford.edu/ERE/db/GeoConf/papers/SGW/2020/Higgins.pdf.

Ball, P.J., 2020. A review of geothermal technologies and their role in reducing greenhouse gas emissions in the USA. J. Energy Resour. Technol. 143 (1). https://doi.org/10.1115/1.4048187.

Bloomfield, K.K., Moore, J.N., 1999. Geothermal electrical production CO2 emission study. In: Geothermal Resource Council 1999 Annual Meeting., https://doi.org/10.1039/a906243b. https://www.osti.gov/servlets/purl/10996#:~:text=This%20study%20estimates%20that%20the,dioxide%20per%20kilowatt%2Dhour%20respectively.

Calpine Corporation, 2024. https://doi.org/10.1080/14737140.2024.2391364. https://geysers.com/egsGeysers. Retrieved January 27, 2024.

Cariaga, C., 2024. ThinkGeoEnergy's Top 10 Geothermal Countries 2023 – Power Generation Capacity. https://www.thinkgeoenergy.com/thinkgeoenergys-top-10-geothermal-countries-2023-power-generation-capacity/. Retrieved January 27, 2024.

Cyrq Energy, 2024. https://www.cyrqenergy.com/. Retrieved January 27, 2024.

DiPietro, J., 2013. Landscape Evolution in the United States. An Introduction to the Geography, Geology, and Natural History. Elsevier, https://doi.org/10.1016/C2011-0-05551-5.

Garcia, J., Walters, M., Beall, J., Hartline, C., Pingol, A., Pistone, S., Wright, M., 2012. Overview of the northwest Geysers EGS demonstration project. In: Proceedings, 37th Workshop on Geothermal Reservoir Engineering Stanford University, Stanford, California, January 30–February 1, 2012. SGP-TR-194., https://doi.org/10.4244/EIJV8SQA9. https://pangea.stanford.edu/ERE/pdf/IGAstandard/SGW/2012/Garcia.pdf.

International Renewable Energy Agency—IRENA, 2017. Geothermal Power: Technology Brief. Abu Dhabi, UAE.

Kaleghi, K., Livescu, S., 2023. A review of vertical closed-loop geothermal heating and cooling systems with an Emphasis on the importance of the subsurface. J. Pet. Sci. Eng. 220 (Part B), 111137. https://doi.org/10.1016/j.petrol.2022.111137.

Kapusta, S., Livescu, S., Dindoruk, B., Schulz, R., Webber, M., 2023. Chapter 2: Direct use applications: decarbonization of industrial processes, and heating and cooling scenarios. In: Beard, J., Jones, B. (Eds.), The Future of Geothermal in Texas—The Coming Century of Growth & Prosperity in the Lone Star State., https://doi.org/10.26153/tsw/44082.

Kneafsey, T.J., Dobson, P., Blankenship, D., Morris, J., Knox, H., Schwering, P., Valladao, C., 2018. An overview of the EGS Collab project: field validation of coupled process modeling of fracturing and fluid flow at the Sanford Underground Research Facility. In: Proceedings, 43rd Workshop on Geothermal Reservoir Engineering Stanford University, Stanford, California, February 12–14, 2018. SGP-TR-213. https://pangea.stanford.edu/ERE/pdf/IGAstandard/SGW/2018/Kneafsey.pdf.

Koelbel, T., Genter, A., 2017. Enhanced geothermal systems: the Soultz-sous-Forêts project. In: Towards 100 Percent Renewable Energy. Springer, Cham, pp. 243–248, https://doi.org/10.1007/978-3-319-45659-1_25.

Krauter, S., 2021. Lecture 09 Energy Transition—Geothermal Energy. https://www.youtube.com/watch?v=KdpMWaf94qA&list=PLpgi7D_IhqlrD37mBky0fSoKb9hvfutE9&index=12. Retrieved January 26, 2024.

Kuru, T., Khaleghi, K., Livescu, S., 2023a. Solid sorbent direct air capture using geothermal energy resources (S-DAC-GT)–region specific analysis. Geoenergy Sci. Eng. 224, 211645. https://doi.org/10.1016/j.geoen.2023.211645.

Kuru, T., Khaleghi, K., Livescu, S., 2023b. Solid Sorbent Direct Air Capture Using Geothermal Energy Resources (S-DAC-GT)-Model for Region Specific Economic Analysis. Presented at the SPE Energy Transition Symposium, Houston, Texas, USA, August. SPE-215735-MS., https://doi.org/10.2118/215735-MS.

Kuru, T., Khaleghi, K., Livescu, S., 2024. Solid sorbent direct air capture using geothermal energy resources (S-DAC-GT)-model for region-specific economic analysis. Geoenergy Sci. Eng. 2024, 213346. https://doi.org/10.1016/j.geoen.2024.213346.

Lafayette College, 2014.https://doi.org/10.3768/rtipress.2024.rb.0035.2402. https://sites.lafayette.edu/egrs352-sp14-geothermal/general-information/history/ Retrieved January 28, 2024.

Li, K., Liu, C., Jiang, S., Chen, Y., 2020. Review on hybrid geothermal and solar power systems. J. Clean. Prod. 250, 119481. https://doi.org/10.1016/j.jclepro.2019.119481.

Li, K., Garrison, G., Moore, M., Zhu, Y., Liu, C., Hepper, J., Petty, S., 2021. Field test of thermoelectric generators at bottle rock geothermal power plant. J. Power Sources 485, 1–9. https://doi.org/10.1016/j.jpowsour.2020.229266.

Library of Congress, 1917. Old Faithful Geyser. Yellowstone National Park, Wyoming. https://www.loc.gov/item/92521945/.

Livescu, S., Dindoruk, B., 2022a. Subsurface technology pivoting from oil and gas to geothermal resources: a sensitivity study of well parameters. SPE Prod. Oper. 37 (02), 280–294. SPE-209227-PA https://doi.org/10.2118/209227-PA.

Livescu, S., Dindoruk, B., 2022b. Coupled Well-Reservoir Heat Modelling for Closed-Loop Geothermal Wells—A Feasibility Study. Presented at the SPE Improved Oil Recovery Conference, Virtual, April 2022. SPE-209437-MS., https://doi.org/10.2118/209437-MS.

Livescu, S., Ng, J., Schakleton, D., Vragov, V., Keay, J., Oreilly, C., Chan, T., Valenciano, A., 2021. Technology transfer to Geothermal from oil & gas and other subsurface disciplines. In: Geothermal Rising Conference. San Diego, CA, USA.

Livescu, S., Dindoruk, B., Schulz, R., Boul, P., Kim, J., Wu, K., 2023. Chapter 1: Geothermal and electricity production: scalable geothermal concepts. In: Beard, J., Jones, B. (Eds.), The Future of Geothermal in Texas—The Coming Century of Growth & Prosperity in the Lone Star State., https://doi.org/10.26153/tsw/44083.

MacGregor, K., 2023. Full Steam Ahead: Unearthing the Power of Geothermal. https://www.nrel.gov/news/features/2023/full-steam-ahead-unearthing-the-power-of-geothermal.html. Retrieved on January 27, 2024.

Open Energy Data Initiative (OEDI), 2024. https://doi.org/10.1080/14737140.2024.2391364 https://data.openei.org/ Retrieved January 27, 2024.

Ravier, G., Seibel, O., Pratiwi, A.S., Mouchot, J., Genter, A., Ragnarsdóttir, K.R., Sengelen, X., 2019. Towards an optimized operation of the EGS Soultz-sous-Forêts power plant (Upper Rhine Graben, France). In: European Geothermal Congress 2019. Den Haag, The Netherlands, 11–14 June 2019. https://europeangeothermalcongress.eu/wp-content/uploads/2019/07/198.pdf.

Richter, A., 2013. Successful EGS Power Production at Ormat Desert Peak 2 Plant. Think GeoEnergy—Geothermal Energy News. https://www.thinkgeoenergy.com/successful-egs-power-production-at-ormat-desert-peak-2-plant/. (Retrieved January 27, 2024).

Richter, A., 2016. U.S. Geothermal Starts Drilling to Improve Output at Raft River Plant. Think GeoEnergy—Geothermal Energy News. https://www.thinkgeoenergy.com/u-s-geothermal-starts-drilling-to-improve-output-at-raft-river-plant/. (Retrieved January 27, 2024).

Richter, A., 2017. Cyrq Energy Opens Solar Plant at its Patua Geothermal Facilities. Think GeoEnergy—Geothermal Energy News. https://www.thinkgeoenergy.com/cyrq-energy-opens-solar-plant-at-its-patua-geothermal-facilities/. (Retrieved January 27, 2024).

Roberts, B., 2018. U.S. Geothermal Resource Maps., https://doi.org/10.1186/s12887-024-04986-7. https://www.nrel.gov/gis/geothermal.html. Retrieved January 27, 2024.

Robins, J.C., Kolker, A., Flores-Espino, F., Pettitt, W., Schmidt, W., Beckers, K., Pauling, H., Anderson, B., 2021. US Geothermal Power Production and District Heating Market Report. National Renewable Energy Laboratory (NREL). https://www.nrel.gov/docs/fy21osti/78291.pdf.

Sage Geosystems, 2024. https://doi.org/10.1080/14737140.2024.2391364. https://www.sagegeosystems.com/. Retrieved January 27, 2024.

Society of Petroleum Engineers (SPE), 2023. DOE awards SPE and consortium partners $165 million grant to advance geothermal. J. Pet. Technol. https://jpt.spe.org/doe-awards-spe-and-consortium-partners-165-million-grant-to-advance-geothermal. Retrieved January 27, 2024.

Stober, I., Bucher, K., 2021. Geothermal Energy: From Theoretical Models to Exploration and Development. Springer (ISBN 978-3-030-71684-4.).

Third Way, 2023. https://www.thirdway.org/memo/geothermal-policies-to-help-america-lead. Retrieved January 28, 2024.

U.S. Department of Energy—DOE, 2019. GeoVision: Harnessing the Heat Beneath our Feet. https://www.energy.gov/sites/default/files/2019/06/f63/GeoVision-full-report-opt.pdf. Retrieved January 27, 2024.

U.S. Energy Information Association (EIA), 2022. U.S. Energy Consumption by Source and Sector, 2022., https://doi.org/10.3390/ijerph192416424. https://www.eia.gov/energyexplained/us-energy-facts/images/consumption-by-source-and-sector.pdf. Retrieved January 27, 2024.

U.S. Energy Information Association (EIA), 2023. Annual Energy Outlook 2023. https://www.eia.gov/outlooks/aeo/. Retrieved January 27, 2024.

U.S. Environmental Protection Agency (EPA), 2022. Geothermal Heating and Cooling Technologies., https://doi.org/10.12688/openreseurope.14665.2. https://www.epa.gov/rhc/geothermal-heating-and-cooling-technologies. Retrieved January 27, 2024.

Utah FORGE, 2020. Drilling Progress of Well 16A(78)-32. https://utahforge.com/2020/11/09/drilling-progress-of-well-16a78-32/. Retrieved January 27, 2024.

Utah FORGE, 2021. Current and Planned Activities., https://doi.org/10.12688/f1000research.73347.3. https://utahforge.com/data-dashboard/current-and-plannedactivities/. Retrieved January 27, 2024.

Westwood Energy, 2024. https://www.westwoodenergy.com/sectors/energy-transition/explaining-geothermal-energy. Retrieved on January 27, 2024.

CHAPTER

2

Resource assessment and management for different geothermal systems (hydrothermal, enhanced geothermal, and advanced geothermal systems)

Ghazal Izadi[a] *and Hans-Christian Freitag*[b]

[a]XGS Energy, Palo Alto, CA, United States [b]Freitag Energy Advisory, Houston, TX, United States

1 Introduction

1.1 Energy demand and the environment

The demand for energy will increase over the next decades. Even considering very modest average annual growth rates of 1.1%, by 2050, the global consumption of energy, is predicted by the International Energy Authority (IEA) to increase from currently 638 to 855 quadrillion BTU, as published in their report International Energy Outlook 2023 (https://www.eia.gov/outlooks/ieo/data.php) in October 2023 (Fig. 1).

These estimates are based on forecasts related to economic (manufacturing and trade) and population growth, per capita demand for improvements in standard of living and healthcare, or, as another example, access to modern energy services, such as dependable electricity or clean cooking.

Geothermal Energy Engineering
https://doi.org/10.1016/B978-0-443-21662-6.00003-7

Copyright © 2025 Elsevier Inc. All rights are reserved, including those for text and data mining, AI training, and similar technologies.

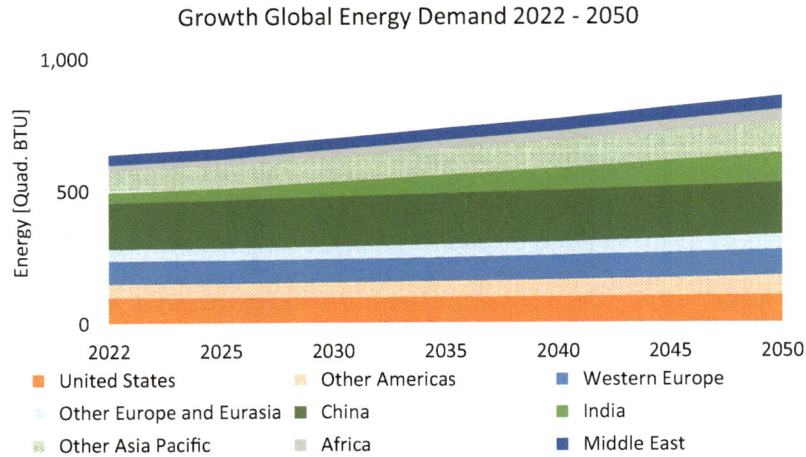

FIG. 1 Growth in energy demand from IEO 2023, Oct. 2023. *Taken from https://www.eia.gov/outlooks/ieo/data.php.*

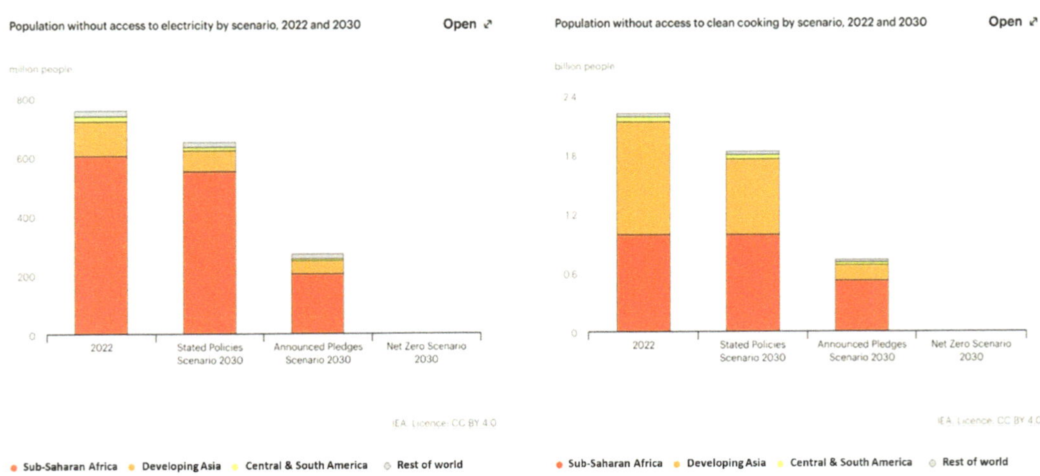

FIG. 2 World population without access to electricity (left graph) and clean cooking (right graph) in 2022, as well as how those numbers develop by 2030 with three different scenarios considered—STEPS, APS, and NZE. *Taken from IEA Global Energy and Climate Model (GEC)—https://www.iea.org/reports/global-energy-and-climate-model/understanding-gec-model-scenarios.*

As mentioned earlier, the demand for energy will increase significantly over the next few decades, driven by a variety of factors: demographic factors, i.e., global population growth, and socioeconomic factors, such as required improvements in health and living standards for large percentages of the global population, as highlighted in Fig. 2 There is a third, political factor that will drive a strong increase in demand for dispatchable, affordable, and clean energy. This is the stated ambition by the global community to achieve a significant reduction in CO_2 emissions. To achieve progress in this domain, i.e., to achieve a net-zero emission (NZE) future by 2050, the development and implementation of energy transition measures are needed, including the introduction of completely new technologies. As highlighted in the closing statements issued at the most recent COP28 in Dubai, in December 2023, one of the agreed-upon goals is the phase-out or phase-down of fossil fuels, to meet the Paris climate agreement.

Energy supply in the future must be dependable, secure, and sustainable. Fig. 3 reviews the changes forecast by the IEA for various energy sources from 2021 to 2030 and 2050. In this context, it is important to note that many countries, companies, and other institutions have pledged to achieve NZEs by 2050. The other two scenarios modeled here are stated policy scenarios (STEPS) and announced pledges scenarios (APS).

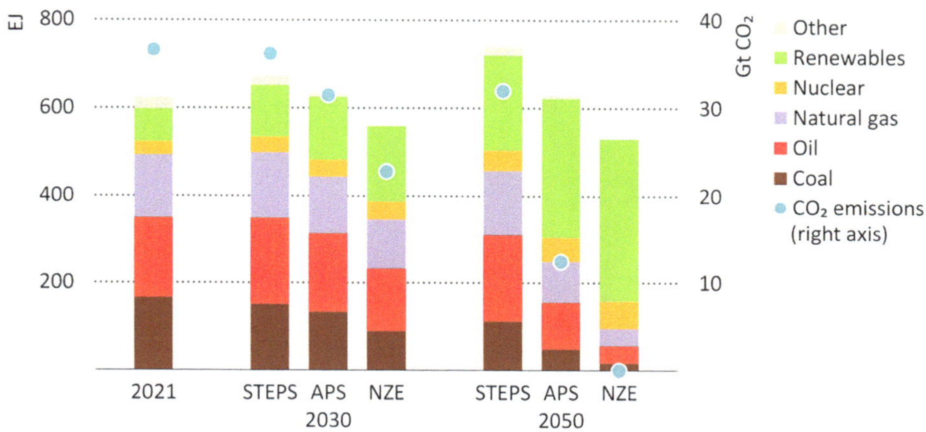

IEA. CC BY 4.0.

FIG. 3 IEA modeling of energy mix by scenario (STEPS, APS, and NZE) from 2021 to 2030 and 2050. EJ, exajoule; $GtCO_2$, gigatonnes of CO_2 emissions.

1.2 An introduction to geothermal

The exploitation (provision) of any source of energy comes at a cost; therefore, energy mix in the future will have to be sourced from a variety of technologies and implementations, to satisfy the demands of the world. One of these sources of energy is geothermal.

Geothermal energy is a form of clean, renewable, baseload energy that is harnessed from the heat energy generated in the earth's core and mantle and is accessible from the earth's crust (Fig. 4).

Heat energy that can be accessed in the earth's crust is composed of two sources—the heat generated in the solid inner core of the earth and transmitted through the liquid outer core to the mantle and the radiogenic heat generated in the mantle itself.

Depending on the local geothermal gradient encountered in different geographical areas, the depths at which elevated temperatures may be encountered vary significantly. This heat energy is brought to the surface through different geothermal systems, which will be investigated thoroughly in this chapter (Fig. 5).

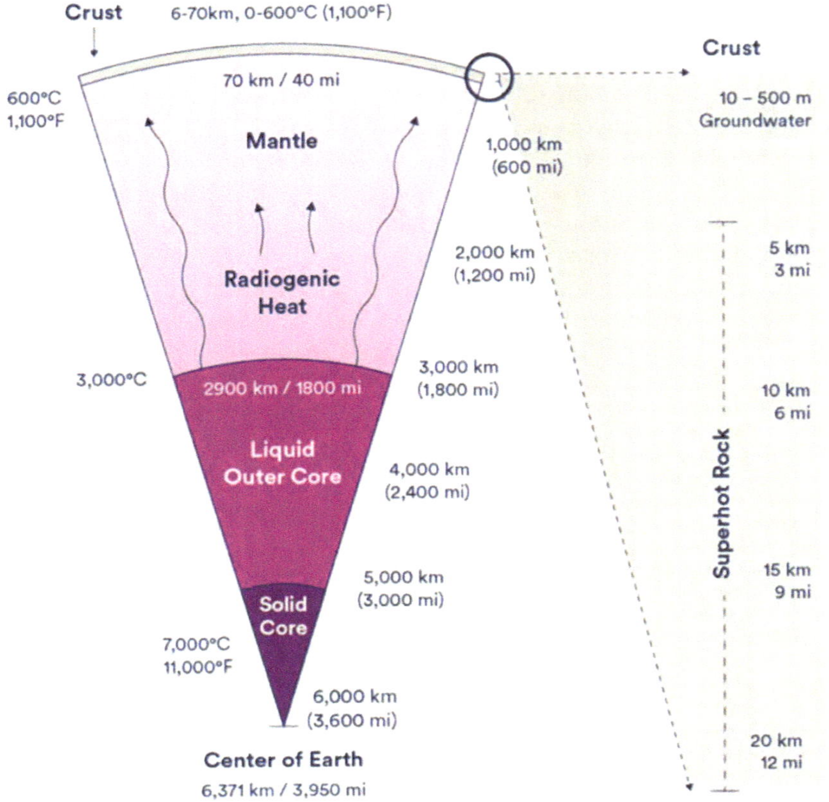

FIG. 4 Heat distribution in the sub-surface, showing the spatial relationships between the dimensions of the earth from core to crust, which is the area of geothermal energy extraction activity. Note from the magnified view of the earth's crust on the right-hand side of the diagram that very high temperatures may be encountered at depths of 5 km or less. *From the CATF report.*

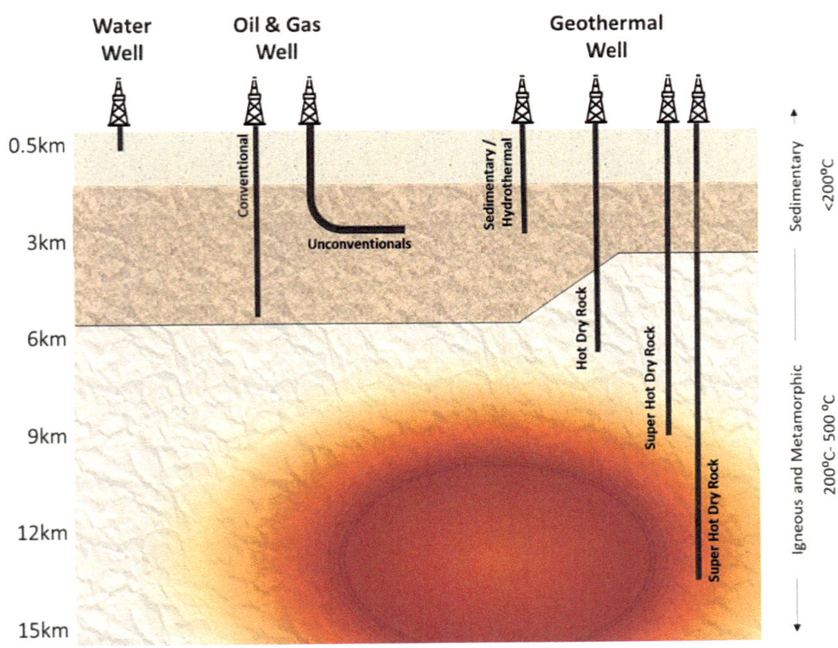

FIG. 5 Idealized view of depths and temperatures accessed by different categories and types of wells. *Figure courtesy XGS Energy.*

In geothermal systems, geothermal energy from the subsurface is brought to the surface through a combination of five essential components: a heat source, i.e., hot rock; a heat exchanger in the subsurface, such as a porous, liquid-filled sediment; a heat carrier fluid, which is circulated through the heat exchanger to transfer the heat from the subsurface to the surface; conduits for controlled fluid flow from subsurface to surface and vice versa; and lastly, a surface facility that extracts the heat from the carrier fluid.

In hydrothermal systems, a porous and permeable rock formation, which contains brine (water with dissolved solids), is accessed through drilling the wells from the surface, allowing the brine or geothermal fluids to be brought to the surface for heating and power generation. The cooled liquid is reintroduced into the rock at a distal injector well to create an injection/production cycle. The volumes of liquid that need to be handled are substantial.

In enhanced geothermal systems (EGS), which are suitable for hot rock formations with insufficient permeability for natural liquid flow, hydraulic fracturing is utilized to create fluid flow pathways between an injector and a producer well. A carrier fluid can then be circulated between these two wells to extract heat and energy from the hot rock. Large volumes of water are required to create the hydraulic fracturing network between the two wells during the construction phase. It is not yet clear how much water will be needed during the operation phase, as the injected water will not only move between the injector and producer well, but also dissipate into other areas in the rock. Operating fluid loss is currently expected to be on the order of 10% of total.

Closed-loop geothermal systems (CLGS) are a novel approach in extracting heat from hot rocks. They utilize a single wellbore to access heat resources and extract heat by circulating a carrier fluid from the surface into the wellbore where it is heated as it circulates past the hot rock and brought back to the surface through a coaxial tubing in the well. Fluid volumes for CLGS are significantly less than for the other methods described, which make them deployable wherever there is suitably hot rock in the subsurface. The different geothermal systems are compared graphically in Fig. 6.

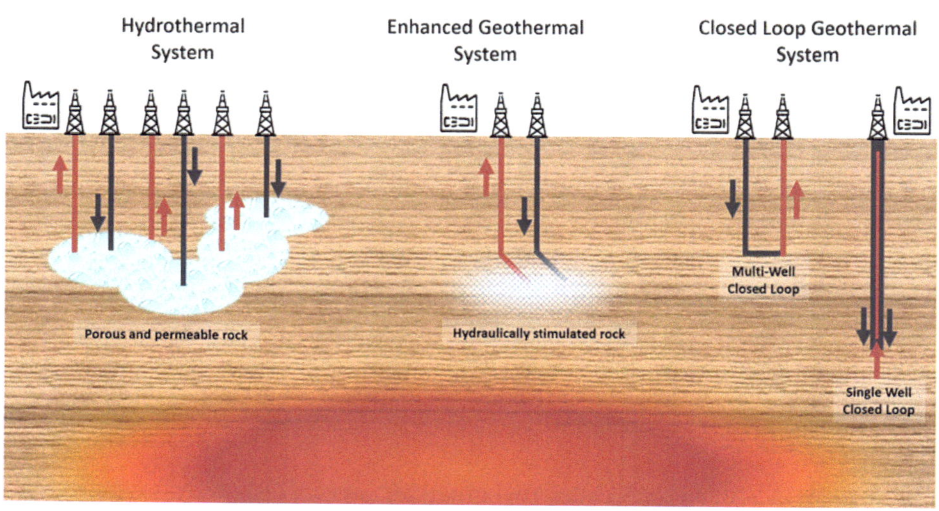

FIG. 6 Different geothermal systems. Note that depth is not to scale. *Figure courtesy XGS Energy.*

Geothermal plants utilize heat either for district heating purposes, for combined heating and power generation, or for power generation. Resource temperature, proximity of resource to off-takers, and the presence of suitable infrastructure, such as pipe networks and/or electricity grids, all have to be taken into account in terms of permitting, development cost, and timeline to implementation.

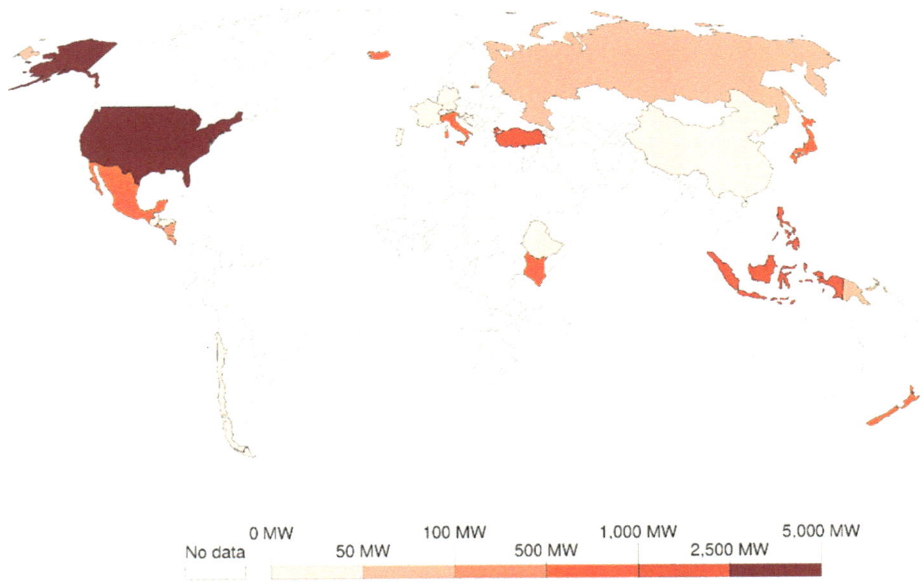

FIG. 7 BP's statistical review of world energy (2021). Map lines delineate study areas and do not necessarily depict accepted national boundaries.

The feasibility of geothermal projects depends on several factors, including the presence, accessibility, and characteristics of the geothermal resources and surface infrastructures. During operation, geothermal projects have minimal environmental impact compared with conventional fossil fuel–based power plants. However, potential environmental issues, such as water usage, the need for infill drilling, and chemical inhibition to mitigate against scale and corrosion, and induced seismicity, need to be carefully evaluated. The economic feasibility of geothermal projects depends on various factors, including capital costs, operating expenses, heat or electricity prices, and government incentives. A comprehensive economic analysis is essential to determine the financial viability of a project.

Taking a global view, as published in BP's 2021 "Statistical Review of World Energy" (Fig. 7), there are a number of key takeaways:

- A number of countries exhibit excellent potential in the development of geothermal energy projects.
- Based on the countries highlighted in Fig. 7, it may be assumed that the option to develop traditional hydrothermal was implemented.
- There is yet a lot to learn about geothermal potential, as over 60% of the world map falls into the "no data" category.

- More activity is required in this research domain. Moreover, the geothermal potential map will change to show a significantly larger potential, as some of the novel engineering approaches introduced are implemented to supplement the traditional hydrothermal approach.

A good example is Larderello in Tuscany, Italy, the oldest geothermal plant in the world which has been operating since 1913. It should be noted that all these plants are hydrothermal developments.

The questions to be addressed are why geothermal development has not grown at a more rapid pace over the last few decades and what type of framework needs to be established to foster geothermal development at pace.

2 Geology

Geology is one of the key disciplines that assesses the geothermal potential of a particular geographical area and the subsurface. It involves understanding the processes that shaped the rock formations that comprise the subsurface today in terms of lithology, i.e., rock type, mineralogy, porosity, permeability, fluids, in situ stress magnitude and orientation, and structural elements, such as anticlines, faults, and intrusions.

One geological concept to explore in more detail is "the rock cycle," which describes how the three main rock types—igneous, sedimentary, and metamorphic—are intricately linked and transform into one another through a series of processes, such as erosion and subsequent deposition, as well as tectonic events and burial depth, with associated heat and pressure changes (Fig. 8).

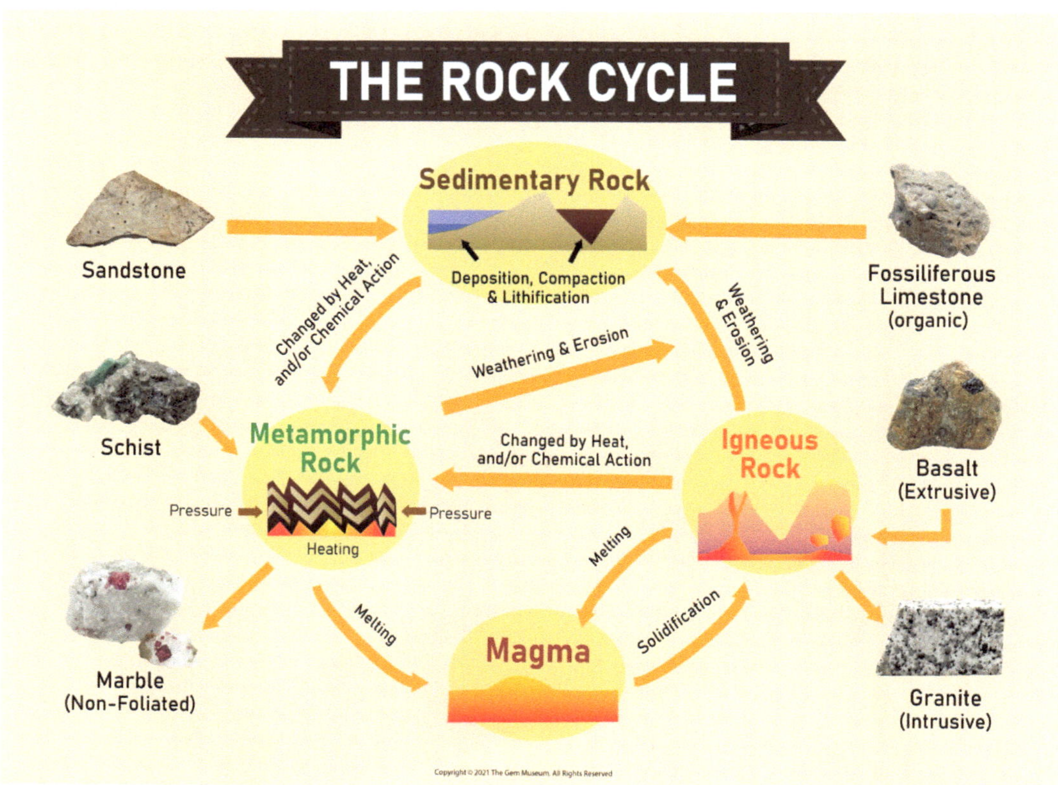

FIG. 8 The rock cycle, as published by The Gem Museum Singapore. *Source: https://thegemmuseum.gallery/wp-content/uploads/2021/04/The-Rock-Cycle.jpg.*

From the top of the diagram, we see basalt, an igneous rock, created through melting of rock materials at very high temperature, producing magma. Once this magma is deposited on the surface of the earth, erosion sets in, breaking down the basalt into smaller and smaller pieces called clastics, or siliciclastics, which may be transported off by a stream or river or wind, then accumulate on the sea floor, in a beach front, in a lake, in a river, or in a dune. When these sediments are buried and subjected to sufficient pressure, they will form a sedimentary rock, such as sandstone or shale or similar. In contrast, limestone, another type of sedimentary rock, is composed of broken shell and coral fragments, rather than siliciclastic materials. Depending on the burial depth, temperature, and chemical composition of fluids within this "sedimentary slurry," a number of processes will take place that will affect the primary (original) porosity or pore space volume and the transmissivity or permeability of this rock formation. During further diagenesis of sedimentary rocks, the primary porosity may increase or decrease. Should this sedimentary rock experience increased temperature and pressure through further deeper burial, it will undergo further change, possibly resulting in the formation of a metamorphic rock, such as marble, as per the diagram. Should this marble be exposed to the elements, erosion will take its toll, and a new sediment supply will be generated. Alternatively, should the marble be exposed to high heat, generated, for example, by a magmatic intrusion, it will melt and become magma itself. Once this magma cools, either at shallower depths or on the surface of the earth, igneous rocks will form, such as basalt.

It should be noted that, contrary to sedimentary rock, neither igneous nor metamorphic rocks have primary porosity, i.e., an initial porosity. However, these rocks may be exposed to tectonic forces, which may cause the rocks to break, allowing fracture porosity and permeability to develop. The introduction of fluids may then further increase the pore space as well as the rock's permeability through dissolution processes. However, depending on the mineralization of the fluids in these rocks, minerals may also crystallize, reducing porosity and permeability (Fig. 9).

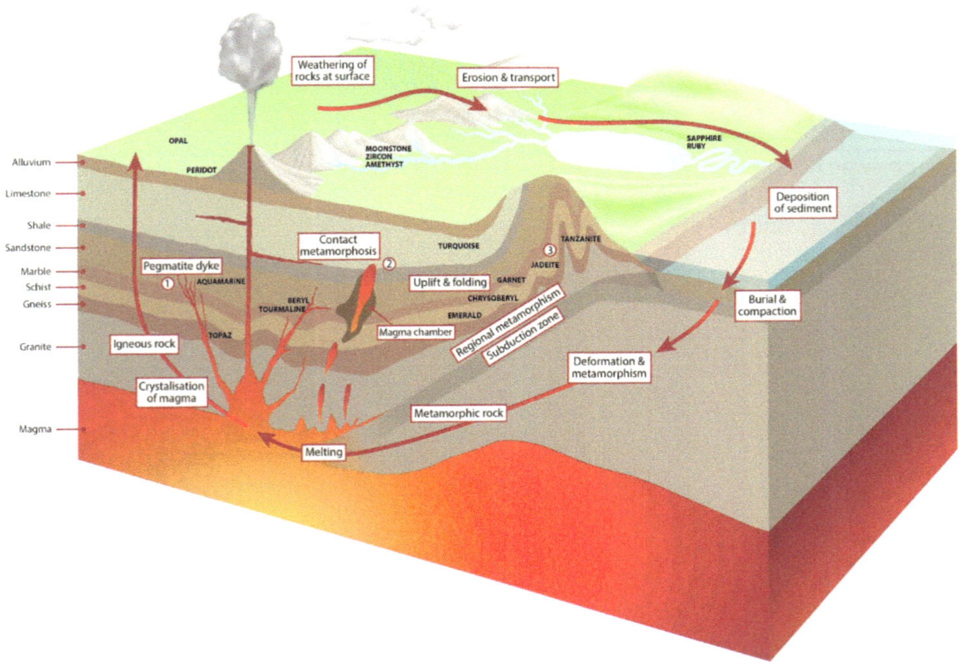

FIG. 9 The rock cycle (http://www.markfranklinarts.co.uk/wp-content/uploads/2015/09/Rock-Cycle.jpg).

It should be noted that metamorphic and igneous rocks, once exposed to in situ weathering as well as erosion on the earth's surface, may develop notable porosity.

The figure illustrates these key concepts as well as highlights that multiple types of geothermal systems may coexist at different depths, pressures, and temperatures in a certain geographic area.

Depending on the local geological setting, there may be several or only one geothermal development option to choose from. We can differentiate between a sedimentary setting, where clastic deposits, such as sandstones or limestones, have formed over time, with porosity, permeability, and pore fluids, and other settings where volcanic activity is prevalent and close to the surface, heating sedimentary or fractured igneous formations and associated fluids. Lastly, there are hot dry rock (HDR) scenarios, where granite or basalt may display high temperatures but do not display any fracture permeability. Each type has its advantages and considerations, and site-specific assessments are crucial for successful project planning.

Examples of various rocks:

- Igneous rocks: basalt, granite, gabbro, obsidian (lava glass), pumice (volcanic ash).
- Sedimentary rocks: sandstone, shale, chalk, and limestone (marine origin, composed mainly of shells and coral fragments)
- Metamorphic rocks: marble, slate, schist, gneiss

2.1 Sedimentary geothermal resource

- Sedimentary geothermal reservoirs are subsurface formations containing hot water or steam. They are formed when fluids are trapped and heated in porous and permeable layers of sedimentary rock. Sedimentary rocks are either clastic sedimentary, such as sandstones, or non-clastic sedimentary rock, such as limestones. Geologists have developed criteria and naming conventions for these sedimentary rocks, with respect to the grain size distribution, the physical shape of the clasts, their mineral composition, and so on. Inferences related to the depositional environment of sedimentary rock can be made based on detailed observations but will not be explored in this chapter.
- Diagenetic processes such as compaction, cementation, and recrystallization affect the pore volume and the transmissivity to fluid flow of sedimentary rocks. These processes are illustrated in the Fig. 10. Note that cementation and recrystallization may also take place during the operational phase of a hydrothermal resource, as production and injection water circulate through the rock.
- Sedimentary geothermal reservoirs can be found in a variety of locations, including basins, valleys, and coastal areas. As a rule of thumb, the deeper the depth of the rock formation, the more compaction the rock will probably experience, reducing porosity and either expelling or over-pressuring liquids contained in the

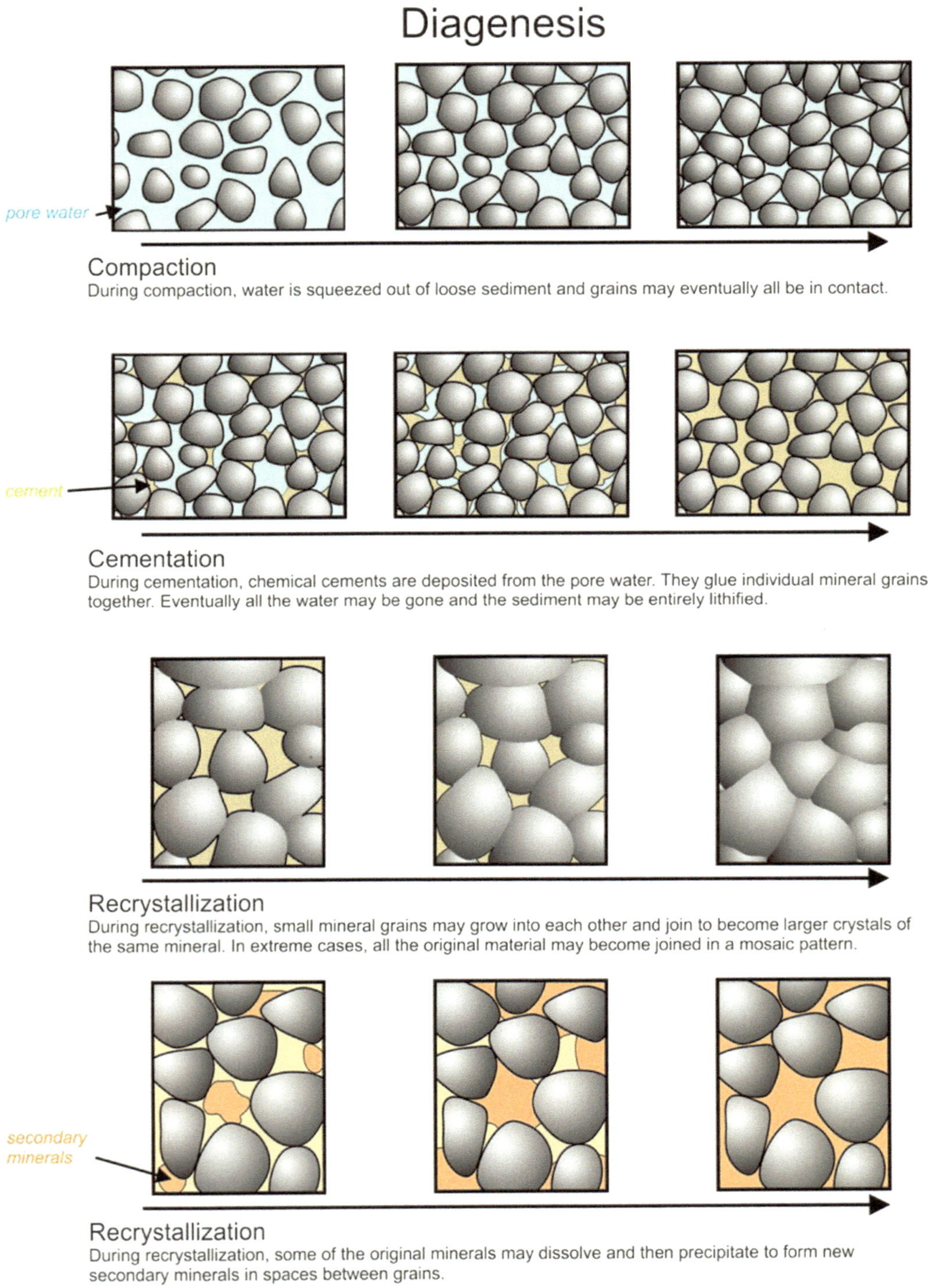

FIG. 10 Schematic illustration of diagenetic processes in sedimentary rocks. *Taken from https://opengeology.org/Mineralogy/7-sedimentary-minerals-and-sedimentary-rocks/#siliciclastic.*

pore space. However, with a deeper depth of the resource, the sedimentary rock formation may be at a higher temperature, potentially improving heat recovery and energy efficiency.

- **Formation**: Sedimentary geothermal resources are created through the deposition of clastic material or through the deposition of shell and coral fragments.
- **Permeability**: Depending on the type of sedimentological composition, the depositional environment, secondary solution, and/or cementation processes, these reservoirs may display a wide range of permeability. For the so-called tight rock, i.e., sedimentary rock with permeabilities <1 mD, hydraulic fracturing methodology successfully developed and implemented for shale oil and gas reservoirs, particularly in the United States, can be deployed to create fluid flow paths.
- **Fluid characteristics**: Typically, fluids in sedimentary reservoirs will have lower temperatures than igneous reservoirs due to their shallower burial depth.
- **Geothermal potential**: Can be suitable for lower-temperature hydrothermal applications, such as direct and district heating and combined power and heating (CPH). These types of applications have been in use for decades in many countries. As an example, Paris and Munich municipalities employ this approach, laying out hundreds of miles of special insulated piping within the cities for district heating.

2.2 Igneous and metamorphic geothermal resource

Igneous rocks are important for the production geothermal energy, because they are often directly associated with heat sources, such as magma chambers and other high-temperature intrusions. Hot springs and fumaroles on the surface are often reliable indicators of high-temperature resources in the subsurface. In addition, fractured igneous rocks will allow fluids to circulate from the surface to the heat source and then migrate back to the surface as hot fluids or even steam. Examples of such hydrothermal systems can be found in areas where active volcanics can be observed—at continental margins, for example. Hydrothermal, EGS, and CLGS development options can be used based on the resource temperature, pressure, fracture systems, permeability, and reservoir fluid.

- **Formation**: Originates from melting processes and transformation of rocks under high heat. This may stem from direct exposure to the heat from volcanic activity or heat generated by the presence and expansion of underground magma chambers.
- **Permeability**: Highly variable. As mentioned, the natural, primary permeability and porosity of igneous and metamorphic rocks are extremely low. However, secondary porosity and permeability may have been created through a variety of diagenetic processes. For these types of rocks, hydraulic fracturing methodology successfully developed and implemented for shale oil and gas reservoirs, particularly in the United States, can be deployed to create fluid flow paths between injector and producer wells.
- **Fluid characteristics**: Higher temperatures due to proximity to volcanic heat sources. However, it involves injecting water through closed loop or, for EGS, requires hydraulic fracturing to extract heat in the absence of reservoir fluid or if resource pressure is not sufficient. CLGS may be a suitable technology, specifically for areas where water availability is an issue.
- **Geothermal potential**: Suitable for electricity generation and higher-temperature direct-use applications.

2.3 Hot dry rock (HDR) as a geothermal resource

The interest in hot dry rock or HDR, defined as rock with in situ temperatures in excess of 300°C, has rapidly grown over the last several decades. This is mainly due to the extremely efficient heat-to-power conversion coefficient, making it up to five times more effective than rock formations at temperatures of 105°C or less. HDR, which has been accessed successfully in several parts of the world at depths of 5000 m or less, has the potential to be a significant source of geothermal energy.

However, in the past, several HDR projects, specifically targeting granite, have relied on a hydrothermal approach, i.e., an approach requiring the circulation of fluids between injector and producer wells. Efforts were made to create fluid pathways between vertical injection and production wells through hydraulic stimulation. While this approach worked in the short term, it often led to finding that the high in situ stress would lead to these hydraulic fractures closing over short time periods, crushing the proppant, and significantly reducing water flow between the wells, essentially rendering the project un-economical. However, some success has been achieved in a joint project between France and Germany in Soultz-sous-Forets, which was initiated in 1991, with a demonstration power plant of 1.7 MW established in 2016.

Now, with the novel engineering and heat harvesting approach that CLGS offer, HDR promises to be a key target resource.

- **Formation**: Impermeable, nonporous rocks at 300°C or higher. Natural fractures, either cemented or open, may be present.
- **Permeability**: Primary porosity and permeability of these rocks are extremely low. While it may be enhanced through hydraulic stimulation or hydraulic fracturing for EGS-type designs, efforts in the past have not been graced with success. It is also worth noting that EGS designs call for a pair of parallel, horizontal wells to be constructed to enable hydraulic fracturing between them. This requires precise wellbore placement utilizing advanced measurement while drilling (MWD) and logging while drilling (LWD) technology. The current operating limit for such tools is at best $\leq$200°C. However, novel engineering ideas around single well or U-shaped CLGS-type designs should be tested and implemented, which do not require hydraulic stimulation required.
- **Fluid characteristics**: Typically, HDR systems involve circulating relatively small volumes of a heat harvesting fluid within closed loop systems. Losses are expected to be minimal, allowing such systems to be deployed in areas with little to no access to water.
- **Geothermal potential**: Suitable for high-efficiency electricity generation, due to the significantly better heat-to-power conversion efficiency. Very attractive for areas where water scarcity makes hydrothermal and/or EGS approaches untenable.

2.4 Super-hot rock geothermal resources

With the interest in high-efficiency geothermal resources growing rapidly, new research is underway to determine the viability of accessing super-hot rock, defined as rock with in situ temperatures in excess of 450°C, with the primary aim of baseload power generation.

Two arguments that are frequently heard in the context of super-hot rock are that (a) one has to drill to extreme depths to access such resources, which has a detrimental impact on project cost and that (b) drilling technology still requires significant advances to be able to access such high-temperature rocks.

However, careful literature review shows that both of these arguments can be dispelled with facts. The Clean Air Task Force (CATF) has published a number of reports as well as created an interactive research tool on their website (www.CATF.com) that allows visualization of relevant data.

- **Formation**: Impermeable rocks at 450°C or higher.
- **Permeability**: Extremely low, may potentially be enhanced through hydraulic stimulation or hydraulic fracturing, but will suffer from the same in situ stress considerations and proppant crush as HDR resources. However, CLGS-type design may be a solution approach to be implemented.
- **Fluid characteristics**: See section about HDR systems, which require injecting a suitable heat transfer fluid within a closed-loop system to extract heat.
- **Geothermal potential:** Suitable for high-efficiency electricity generation, due to the significantly better heat-to-power conversion efficiency. Very attractive for areas where water scarcity makes hydrothermal and/or EGS approaches untenable. However, new drilling technology, such as plasma drilling, will need to be developed to access super-hot rock resources at temperatures exceeding 500°C.

3 Geothermal resource classification

Geothermal resources are categorized based on their temperature and thermodynamic characteristics. Low-temperature geothermal resources, with temperatures up to ~90°C, find applications in residential and commercial heating, greenhouse heating, and industrial processes. Moderate-temperature geothermal resources, with temperatures ranging between 90°C and 150°C, offer opportunities for electricity generation through binary cycle power plants, district heating, and various industrial applications. High-temperature geothermal resources, between 150°C and 300°C, and super-high-temperature resources, exceeding 300°C, are predominantly utilized for electricity generation through flash steam or dry steam power plants, making them well-suited for large-scale power production. Beyond temperature, factors such as depth, fluid type, and resource permeability contribute to the comprehensive classification of geothermal resources.

Evaluating the temperature of the heat resource is crucial to determine its optimal utilization, whether for power generation, domestic heating, district heating, agricultural applications, or electricity generation. Proximity of the heat source to the usage center becomes a pivotal factor in the assessment. For district heating or combined heat and power plants, establishing a pipeline network from the heat source to the targeted towns or cities is essential, ensuring the efficient delivery of hot water to individual buildings. Conversely, electricity generated from geothermal energy can be transmitted over extensive distances leveraging existing state or national grids (Fig. 11).

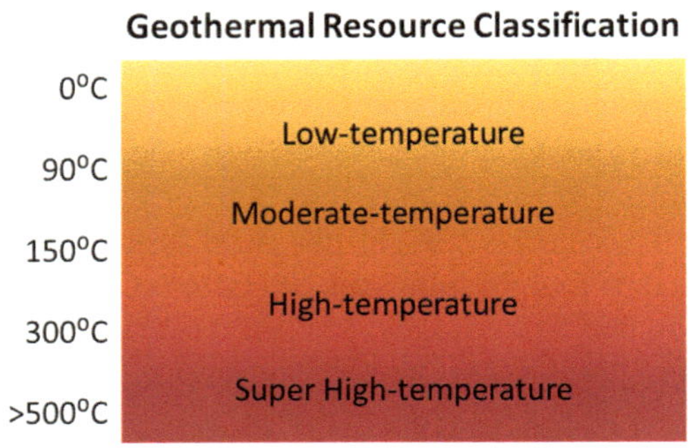

FIG. 11 Geothermal resource classification based on resource temperature. *Figure courtesy XGS Energy.*

4 Geothermal resource exploration and assessment

In an early exploration stage, parties interested in geothermal resources can utilize existing data sources, such as geological, hydrological, and heat maps published by a variety of institutions on a global and national basis. As an example, the International Heat Flow Commission (IHFC) shares heat flow data via an interactive visualization tool on their website https://ihfc.org (Fig. 12). The clean air task force (CATF), a US-based NGO, focuses on hot and super-hot rock and published this information on their https://catf.us website (Fig. 13).

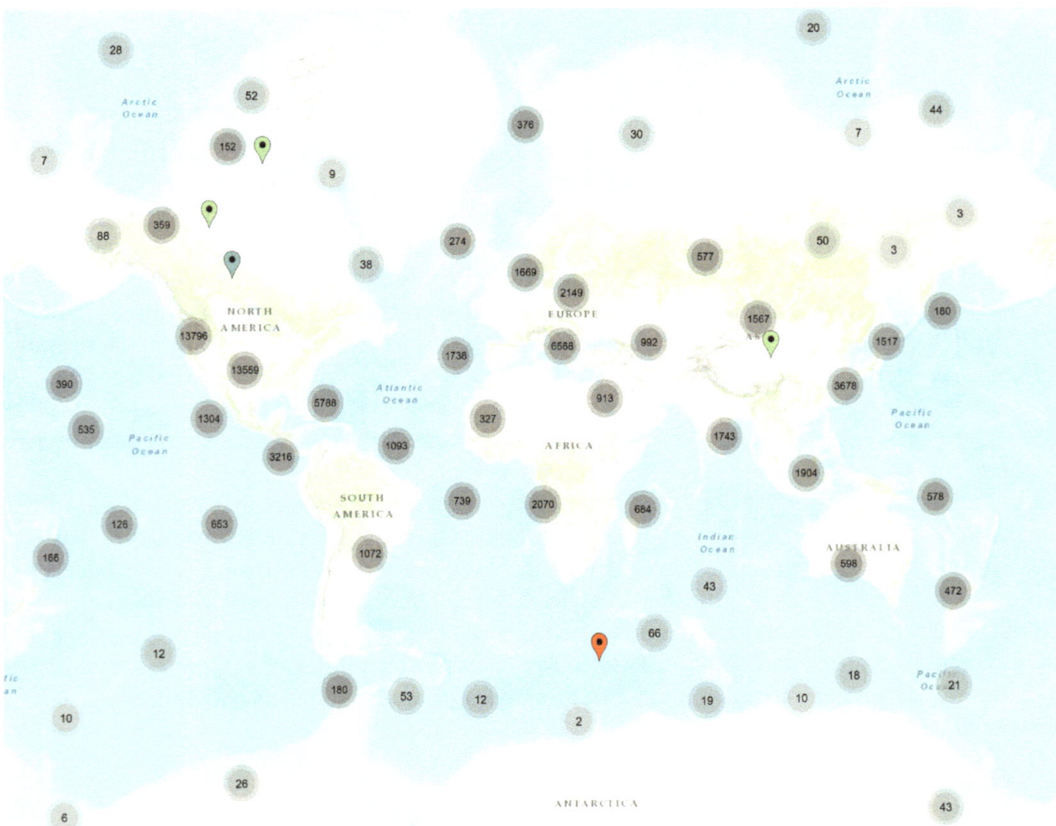

FIG. 12 Interactive global heat flow map—numbers in the circles denote location and number of measurements available. *Taken from the International Heat Flow Commission (IHFC) https://ihfc-iugg.org/viewer/.* Map lines delineate study areas and do not necessarily depict accepted national boundaries.

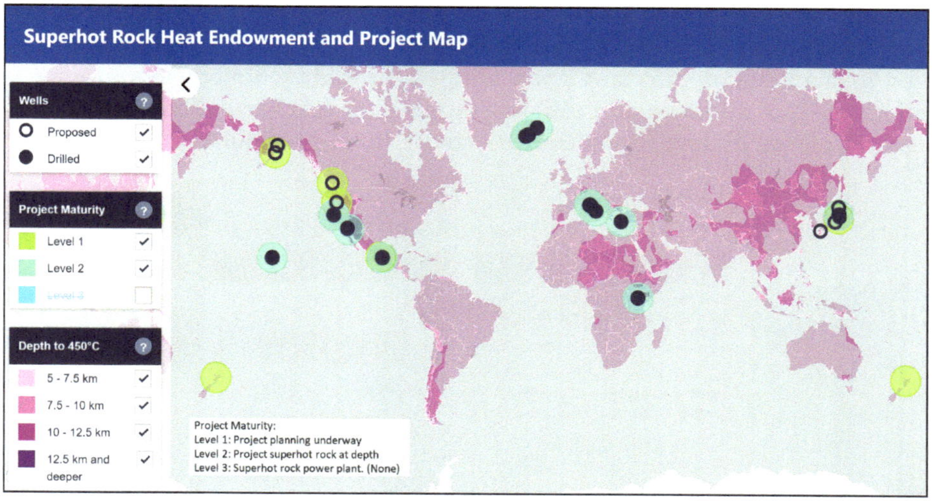

FIG. 13 Global hot rock projects—executed and planned—from the CATF interactive viewer https://www.catf.us/shrmap/. Map lines delineate study areas and do not necessarily depict accepted national boundaries.

Other state or national institutions, such as the National Renewables Energy Laboratory (NREL) in the USA, as well as other organizations collect, manage, and publish information that can serve as a starting point to identify areas of interest and then further evaluate, analyze, and potentially gather additional information (Fig. 14).

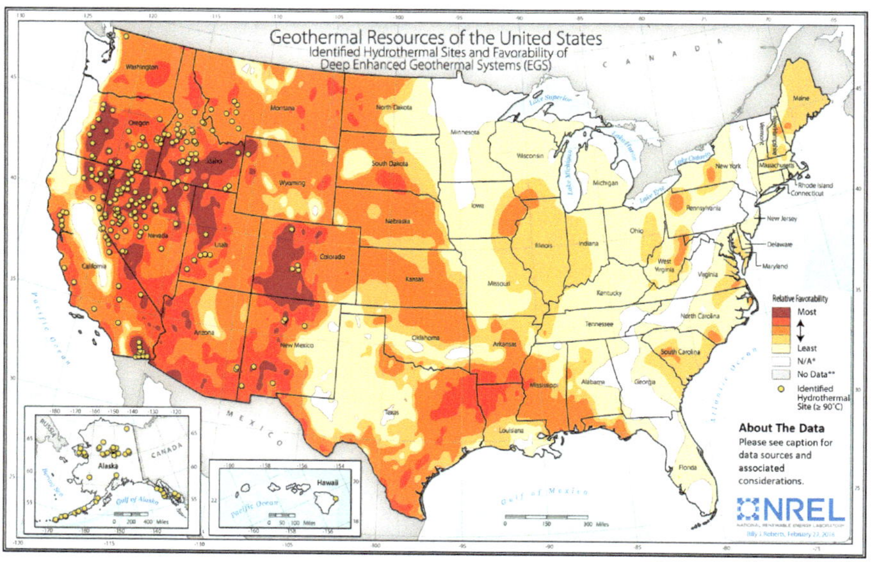

FIG. 14 From the National Renewables Energy Laboratory (NREL) https://www.nrel.gov/gis/assets/images/geothermal-identified-hydrothermal-and-egs.jpg. For this map, the following considerations apply: Map does not include shallow EGS resources located near hydrothermal sites or USGS assessment of undiscovered hydrothermal resources. Source data for deep EGS includes temperature at depths from 3 to 10 km provided by Southern Methodist University Geothermal Laboratory (Blackwell et al., 2011) and analyses (for regions with temperatures >150°C) performed in Figure 1 on p. 3 by Augustine et al. (2023). Source data for identified hydrothermal sites from USGS Assessment of Moderate- and High-Temperature Geothermal Resources of the United States (2008). *"N/A" regions have temperatures less than 150°C at 10 km depth and were not assessed for deep EGS potential. **Temperature at depth data for deep EGS in Alaska and Hawaii not available. Map lines delineate study areas and do not necessarily depict accepted national boundaries.

In addition, temperature measurements from mineral mining activities as well as oil and gas drilling and production activities will prove useful. While these datasets will, in general, be limited to shallower depths and lower temperatures, they nonetheless provide very detailed information to augment and calibrate larger-scale geothermal maps.

It is good practice to develop a detailed activity timeline, similar to the one that follows (Fig. 15).

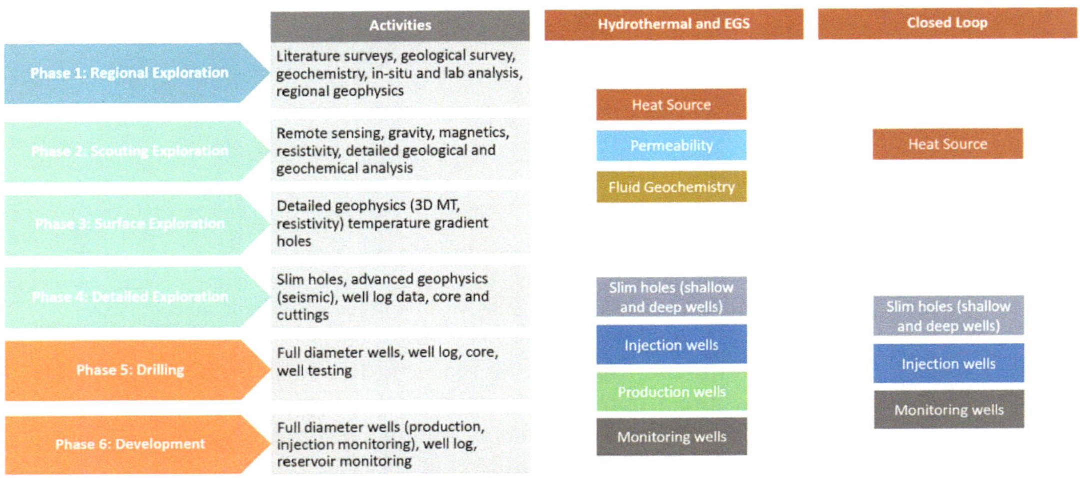

FIG. 15 Timeline of exploration activities and required/recommended types of measurements. *Figure courtesy XGS Energy.*

Several well-known and proven techniques, measurements, and tools applied in other extraction industries are increasingly being used in geothermal exploration to identify resource rock types, resource extent, and properties to deliver accurate subsurface models.

In the following overview, we review different subsurface investigation methods. These range from remote sensing, such as satellite-based imaging and infrared temperature mapping, side-looking radar (SLAR) as well as aircraft- or drone-conveyed sensing of gravimetric and magnetic properties and anomalies all the way to seismic data acquisition, processing and interpretation. Furthermore, the drilling of investigation or pliot wells to enhance, calibrate and validate the evolving subsurface model have proven extremely useful as a project progresses. These different investigation methods and technologies provide information and insights at different spatial and vertical resolution and detail. Depending on the type of geothermal resource that is being investigated, a customized data acquisition, data integration, interpretation, and analysis program must be designed and executed to reduce project risk to an acceptable level, ensuring project success (Table 1).

Fig. 16 illustrates different investigation and exploration methods, which are utilized to obtain a first subsurface model. Subsequently, exploration wells are drilled to calibrate the existing subsurface model.

TABLE 1 Applicability of different exploration methods with respect to conveyance method.

Measurement methodology	Remote sensing		Surface	Subsurface
	Satellite	Airborne		
Infrared surface temp	X	X		
InSAR—earth's surface deformation	X	X		
Gravimetry	X	x	X	
Magnetics	x	x	X	
Magneto-tellurics	x	x	X	
Resistivity profiling			X	X
Geological mapping	X	X	X	
Geochemical sampling—minerals/water				X
Seismic data acquisition			x	X
Temperature profiling—in situ				X
Drill cutting analysis				x
Geophysical well logging				X
Well testing—flow testing				X

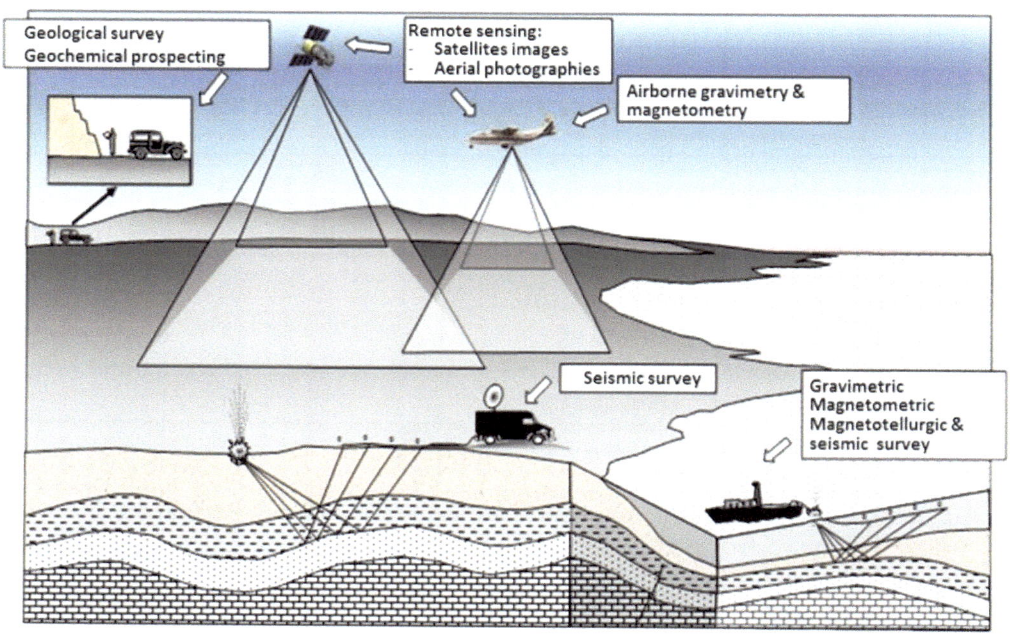

FIG. 16 Illustrative example of various exploration methods. *Taken from https://www.oil-gasportal.com/upstream/petroleum-exploration.*

4.1 Remote sensing

Purpose: Uses satellites, aircraft, or drones to obtain a wide variety of data covering large areas of interest with relatively low-cost and easy to deploy "scouting measurements" to identify surface and subsurface features and potential geothermal indicators.

Tools: Satellite, aircraft, drones, etc. for deploying suitable remote sensing measurements, such as radar, infrared sensing, gravimetry, magnetics, and electromagnetics, to cover a large area which may then be augmented with methodologies and techniques that are deployed on the surface of the earth (Fig. 17).

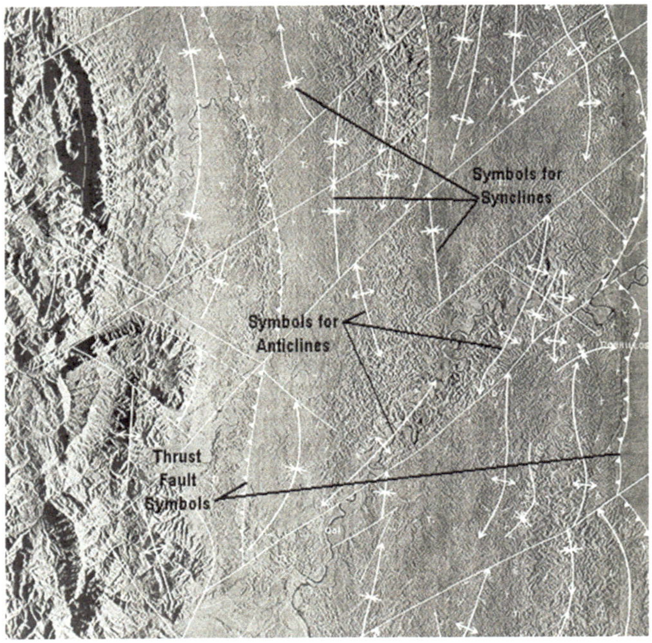

FIG. 17 Example of a sideway-looking radar (SLAR) image of an area in South America. Prior to SLAR technology, it was virtually impossible to explore this area due to almost perpetual dense cloud cover. *Taken from https://www.dnr.louisiana.gov/assets/TAD/education/BGBB/5/techniques.html.*

Recently, with the Landsat 8 and Landsat 9 missions, new remote sensing measurements have been added to the explorationist's toolbox:

- Thermal imaging using infrared sensing methodologies to determine temperature-related anomalies of the earth's surface.
- InSAR (Interferometric Synthetic Aperture Radar) which yields information on deformation of the earth's crust based on making time lapse measurements of two-way of radar waves from the satellite to the area of interest.

It should be noted that remote sensing measurements may also be used during the operational phase of a geothermal plant, to address and better understand such phenomena as subsidence and the heat flow/cooling effect, as can be seen in the Fig. 18, showing time lapse InSAR data.

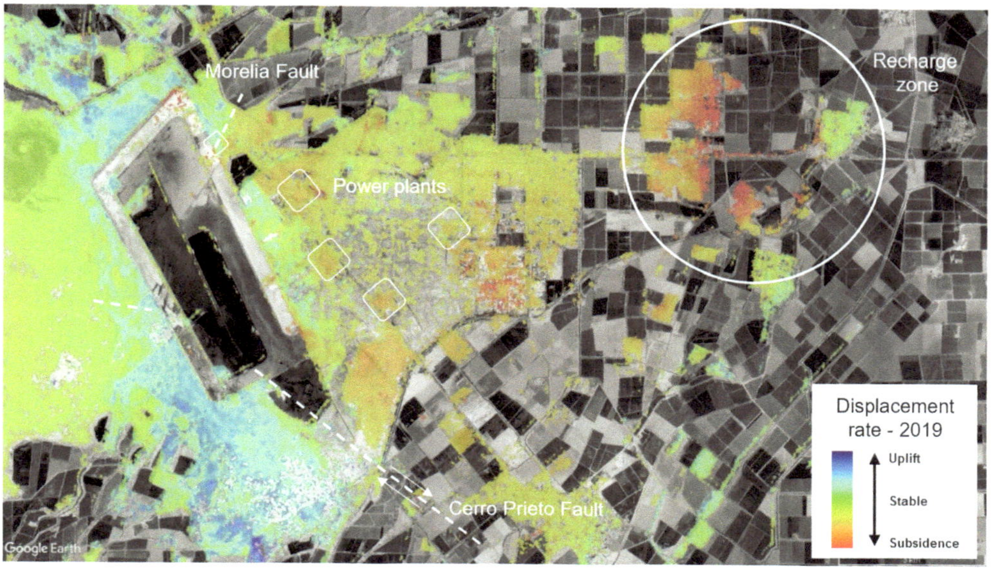

FIG. 18 InSAR displacement rate map across the Cerro Prieto geothermal complex, Mexico. This highlights areas of subsidence (*red colors*) associated with changes in fluid balance in areas of extraction and recharge, and lineaments correlating with known faults. *Image © CGG Satellite Mapping, contains modified Copernicus Sentinel data (2019), background image © Google (2021).*

While moderate subsidence can be observed in the direct vicinity of the plants, stronger subsidence can be observed in the water injection or recharge zone, indicating that the fluid volumes being injected are insufficient to replace the produced hot water.

4.2 Geological surveys and interpretation

Purpose: Infer subsurface geology from remote sensing methods mentioned. This will lead to a first identification of potential "sweet spots," which can then be investigated through direct observation of surface rock strata exposure and structural elements, such as faults and erosion surfaces, which can be observed on the surface or in quarries. Specifically for geothermal exploration, surface temperature measurements in hot springs and streams are as valuable as geochemical analysis of minerals and water.

Tools: Field mapping tools, geological compass, and satellite and aerial imaging; see examples of remote sensing in the following.

4.3 Gravity surveys

Purpose: Determine subsurface rock body size and extent by measuring variations in naturally occurring gravitational forces caused by different rock densities. The density of a rock formation in the subsurface is determined by three main factors: first, the mineral components that compose the matrix or the solid part of the rock; second, the volume of pore space within that rock; and, lastly, the type of fluid in the poor space. Therefore, a porous rock, such as sandstone, for example, with a porosity of 20% and oil or gas within that pore space will exhibit a different response to a gravimetry measurement than a rock that has no porosity at all. Following this train of thought, a rock with porosity and gas in the pore space will have an even lower density than a porous rock with fluids in it. Thus, gravity variations in the subsurface can be used to identify porous from nonporous rock, with the added benefit of shedding some light on the fluids in the pore space. Obviously, in the case of hydrothermal systems, the interest would primarily be in whether or not water is in the pore space (Fig. 19).

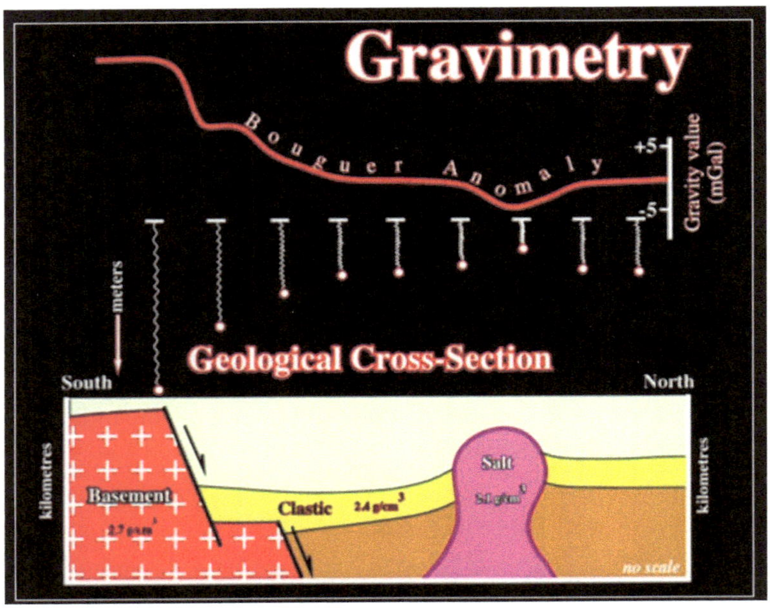

FIG. 19 Schematic response of a gravity survey to a left-to-right change in subsurface materials, ranging from high-density basement (high density rock) on the left to a sedimentary sequence (medium density) with a salt intrusion (low density). *Taken from https://www.oil-gasportal.com/upstream/petroleum-exploration.*

The example shows how explorationists utilize gravimetry data to model changes in density brought about by rock lithology, rock thickness, and rock density (and porosity). Note how the gravimetry response can also be used to identify faults, dislocations, and intrusions, such as plutons or magma chambers.

Tools: Gravimeters—both airborne and surface.

4.4 Magnetic surveys

Purpose: Infer subsurface rock types and structures by measuring the naturally present magnetic susceptibility and changes in these properties. Different types of rocks exhibit varying degrees of magnetism, which is primarily a function of their content of magnetic minerals. Minerals with the highest degree of magnetic susceptibility are made of iron, titanium, sulfur, and oxygen. Magnetite, which essentially is iron oxide, exhibits the highest magnetic susceptibility, closely followed by ilmenite, an iron titanium oxide, and parasite, an iron sulfide mineral. These minerals tend to be more abundant in igneous and metamorphic rocks than in sedimentary rocks. However, there are areas of overlap in in terms of magnetic susceptibility of different rock types, depending on their respective content of magnetic minerals, as shown in the Table 2.

TABLE 2 Average magnetic susceptibility of different rocks and minerals, compiled and adapted from various sources, including T.M. Boyd, Colorado School of Mines.

Rock group	Material	Susceptibility $\times 10^{-3}$ (SI)
Magnetic minerals	Magnetite	1200–19,200
	Ilmenite	300–3500
	Pyrrhotite	1500
Igneous rocks	Basalt	0.2–175
	Granite	0–50
	Gabbro	1–90
Metamorphic rocks	Slate	0–35
	Gneiss	0.1–25
	Schist	0.3–3
Sedimentary rocks and other	Sandstones	0–20
	Limestones	0–3
	Shales	0.01–15
	Salt	<0.01

Fig. 20 shows how explorationists utilize magnetic susceptibility data to model the thickness of sedimentary rock layers and the transition from sedimentary to igneous rock along a fault line in this case. Likewise, changes in magnetic susceptibility can also be extremely helpful in identifying igneous or metamorphic intrusions, such as magma chambers or plutons.

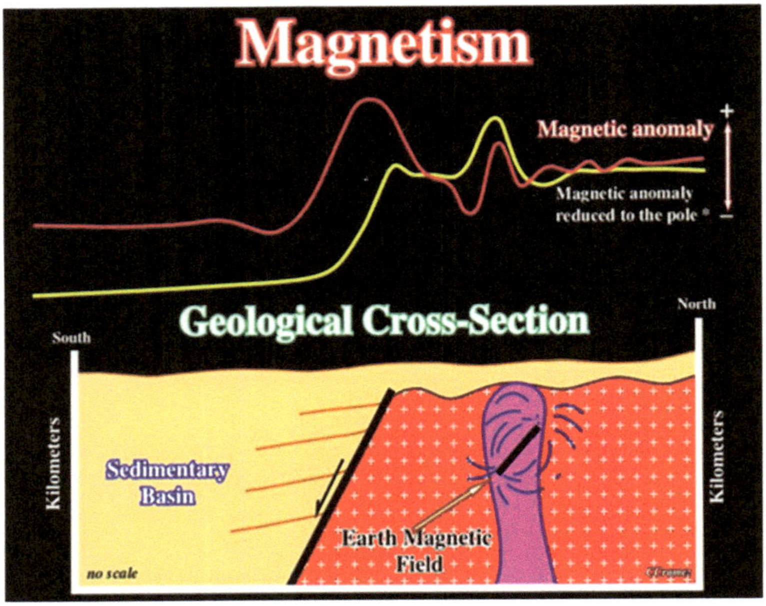

FIG. 20 Schematic response of a magnetic survey to delineate a left-to-right change in subsurface materials, ranging from low magnetic susceptibility in sediments on the left to higher magnetic susceptibility metamorphic or igneous rock on the right. *Taken from https://www.oil-gasportal.com/upstream/petroleum-exploration.*

Tools: Magnetometers, both airborne and surface.

4.5 Surface electrical/resistivity surveys

Purpose: Identify electrically conductive subsurface rock strata. The natural electrical conductivity of subsurface formations is indicative of the presence of conductive material, such as clays, brine (salt water), or geothermal weathering products that can form on the upper boundary of hydrothermal reservoirs. The methodology employed consists of utilizing a four-electrode setup, using two pairs of electrodes each. Current is injected in the outer electrodes and measured. Between these two current electrodes, two potential electrodes are installed to measure the resulting potential. With the knowledge of current (I) and potential (U), Ohm's law can be applied to derive the resistance of the subsurface. This methodology is one of the most widely used in groundwater and geothermal exploration. The depth of investigation of this method depends on the spacing between injector and return electrode as well as on the conductivity of water bearing rock formations in the subsurface (Fig. 21).

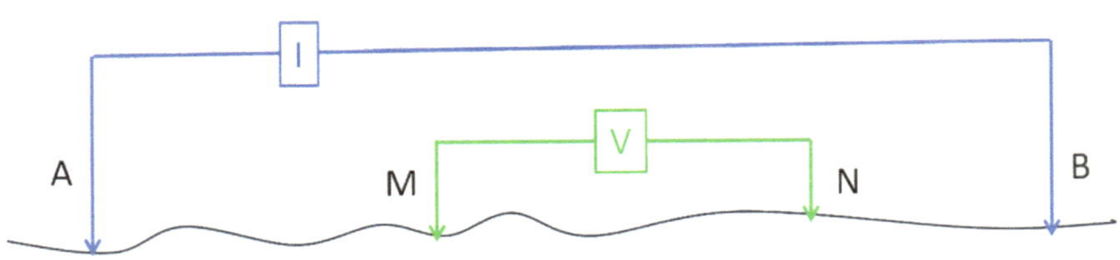

FIG. 21 Schematic of an inline surface resistivity measurement layout. An AC current is injected at electrodes A and B, and electrodes M and N measure the voltage. *After https://www.researchgate.net/profile/Rhett-Herman/publication/228876236_An_introduction_to_electrical_resistivity_in_geophysics/links/53d6bc580cf220632f3dce72/An-introduction-to-electrical-resistivity-in-geophysics.pdf.*

Tools: Surface resistivity measurement instruments as per the diagram presented.

4.6 Electromagnetic (EM) surveys

Purpose: Identify electrically conductive subsurface features using electromagnetics to induce electrical currents in the subsurface, which in turn induce a magnetic field in the rock, which subsequently induces a current that can be measured at a return electrode. This methodology enables the identification and depth determination of rock layers with high conductivity, such as porous, water-bearing rock formations, shales, or geothermally altered zones. Contrary to passive electrical or resistivity surveys.

Tools: Surface EM instruments can be grouped into four essential different methods: Frequency domain electromagnetic (FDEM); transient/time domain electromagnetic (TDEM); ground penetrating radar (GPR); and, lastly, very–low-frequency electromagnetic (VLF).

4.7 Magneto-telluric (MT) surveys

Purpose: Identify subsurface features based on their electrical resistivity. Magneto-telluric surveys are a passive geophysical methodology simultaneously acquiring measurements of the spatial and time-based variations of the earth's electrical and magnetic fields to determine the resistivity of subsurface rock formations. As stated earlier, the resistivity of such rocks can vary significantly from very-low-resistivity water-filled porous rock to highly resistive crystalline rocks (Fig. 22).

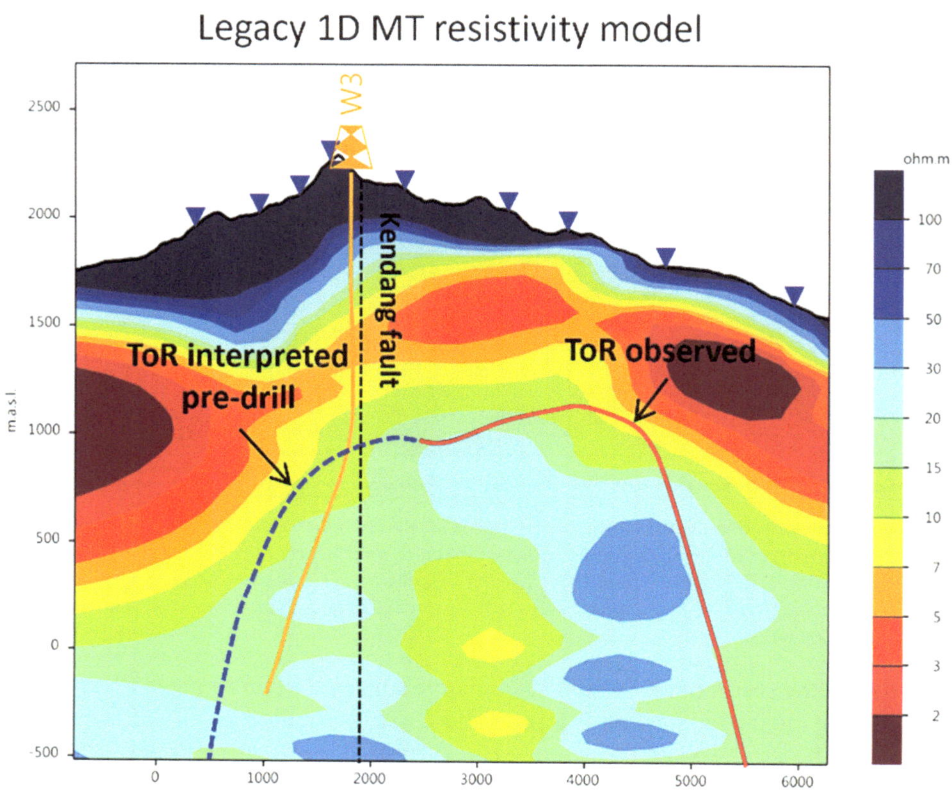

FIG. 22 1D magneto-telluric interpretation results, showing areas of high and low resistivity in the subsurface. The top of reservoir (TOR) is illustrated as the *dashed line*.

By measuring time variations of the earth's magnetic and electrical fields over different frequencies, it is possible to obtain information from a few hundred meters (higher frequencies, shorter wavelength) to many kilometers depth (lower frequencies, longer wavelengths). This methodology was introduced in the 1950s to aid in mineral and oil and gas exploration. Especially in its airborne form, it provides an easy to deploy methodology to efficiently cover basins and areas hard to access with other geophysical methods, such as surface measurements, including seismic. Due to the diffusive propagation of electromagnetic fields in the subsurface, the vertical resolution of magneto-tellurics is in the order of 100 m.

Tools: Airborne and surface MT instruments.

4.8 Seismic surveys

Purpose: Determines subsurface structure, such as rock layers and large-scale faults, as well as rock formation thickness and depth, rock types, and acoustic processing power of super-computers. The figure that follows shows some complex subsurface structures with a variety of oil and gas impedance, a measure of rock density by analyzing the propagation of seismic waves.

Tools: Active and passive seismic reflection and refraction surveys, which are widely used in oil and gas exploration. "Passive" seismic refers to utilizing the mechanical energy delivered and exerted into rock formations primarily through earthquakes and similar events. "Active" seismic, on the other hand, refers to mechanical energy created by a variety of source being used to generate acoustic waves in the subsurface at controlled frequencies and amplitudes to accurately investigate subsurface structures based on their physical properties, namely, acoustic velocity and impedance.

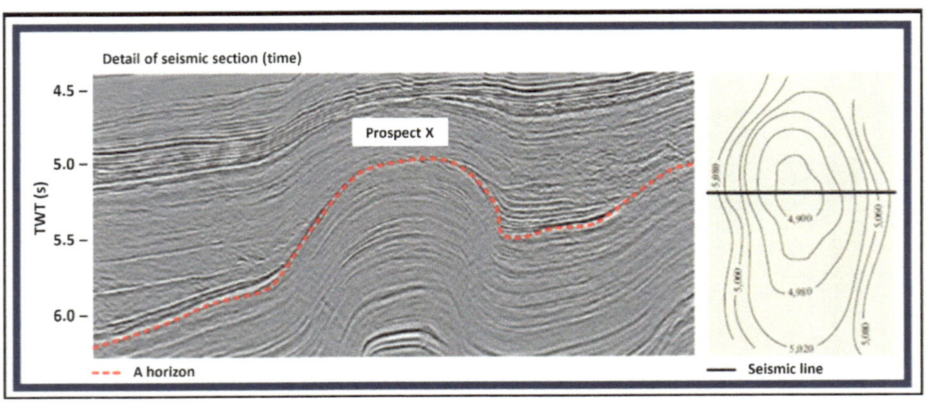

FIG. 23 Seismic cross-section. Note that the depth scale is labeled two-way travel time (TWT), i.e., the time for the acoustic signal to travel from the acoustic energy source at the earth's surface to the reflector and back to the recording geophones on the surface. *Giniaux et al., Monitoring Ground Deformation at Geothermal Fields Using InSAR, an Insight to Subsurface Fluid Dynamics presented at European Space Agency Fringe 2021 conference. From https://www.oil-gasportal.com/upstream/petroleum-exploration.*

Fig. 23 shows the result of an "active" 2D seismic survey, i.e., an image of the subsurface in two dimensions: surface extent and depth. The z-axis is labeled as "TWT/two-way travel time", i.e., the time for the acoustic signal to travel from its surface origin to the bed boundary in the sub-surface and back to the surface. The continuous-line patterns on this seismic image are the bed boundaries of subsurface rock formations with different acoustic impedance, which is determined by the density of a particular layer, and its acoustic velocity. The uplift in the center labeled "Prospect X" can be interpreted as an anticline.

Acoustic energy can be propagated from the earth's surface through a variety of methods, such as air guns set off in specially constructed water pits, seismic vibrator trucks ("thumpers"), and similar devices. This acoustic energy will travel through the subsurface until it encounters an acoustic impedance contrast. At this point, part of the acoustic energy is reflected back to the surface where it can be measured with geophones, while the remaining acoustic energy travels to greater depths where it interacts with subsequent, deeper boundaries/acoustic impedance contrast. These types of surveys have been used in the oil and gas and other extraction industries to identify different rock layers, their burial depth, and their thickness. In this example, we can see how deeper rock formations are pushing the rocks upward and creating a "dome" structure. In oil and gas, this might be interpreted as an anticline, which may be a hydrocarbon trap or what in geothermal might be interpreted as an intrusion from a magma chamber. From the early days of seismic data acquisition and processing, technology has moved on significantly, enabling more detailed 3D seismic surveys to map subsurface structures aided primarily by the processing power of super-computers. Fig. 24 shows some complex subsurface structures with a variety of oil and gas traps.

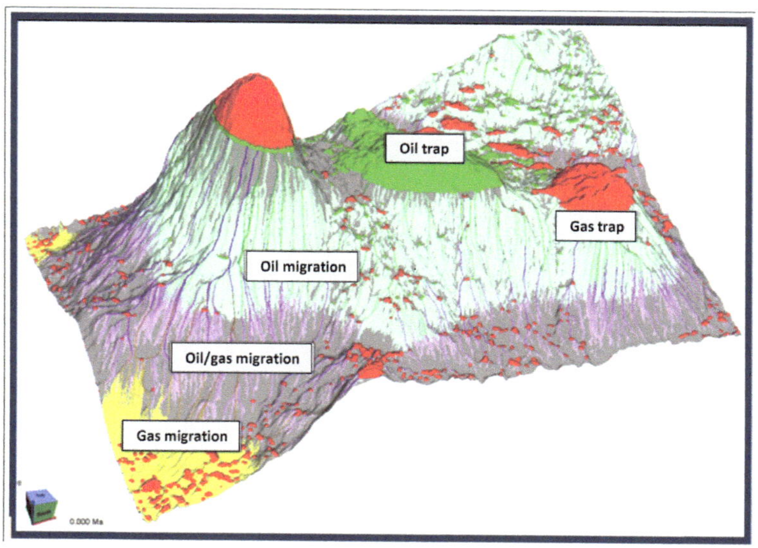

FIG. 24 Sophisticated 3D seismic interpretation and subsurface modeling result for an oil and gas resource. *Taken from https://www.oil-gasportal.com/upstream/petroleum-exploration.*

Acquisition of high-resolution 3D seismic data is a key element of subsurface characterization for oil and gas resources. Advanced seismic signal processing and 3D modeling has significantly assisted in de-risking and ranking oil and gas prospects in terms of geological complexity and petrophysical detail. Following the first exploration well(s), subsequent delineation wells are drilled, and data on high-resolution geophysical, core, and fluid obtained from these wells are integrated with the seismic data to extrapolate subsurface characteristics, such as porosity, clay content, and even permeability between wells to enable operators to more accurately determine resources and reserves.

However, despite the proven success of an integrated subsurface analysis that utilizes interpreted seismic data as a key component, the industry does not yet see a significant uptake of seismic data acquisition for geothermal prospects. This is surprising, specifically for hydrothermal prospects, which share significant commonality with oil and gas prospects in terms of the importance of accurately characterizing lithology, porosity, and fluid flow. Further to this, as geothermal exploration and exploitation start to focus on deeper and hotter resources, seismic data acquisition will become more important to provide more detail to supplement the more common geothermal exploration methodologies described before. For geothermal energy to rise to significantly higher levels than the paltry 0.1% of total global energy production it currently has, utilizing data to de-risk projects and provide a sound basis for a well-rounded techno-commercial analysis will be essential. Today, there are numerous cases where a lack of subsurface understanding has led to geothermal power plants producing <50% of nameplate capacity due to insufficient fluid support, or where up to 30% of all wells drilled for a hydrothermal project have been "dry," i.e., incapable of supplying the required fluid to replenish the porous reservoir, with a significant negative impact on project CAPEX as well OPEX.

Some examples of a combined, multi-physics data analysis and interpretation utilizing high-quality seismic data will be discussed in Section 4.13.

A seismic technique that utilizes acoustic energy to characterize the subsurface by localizing mechanical failure events caused by rock fracturing is micro-seismic. With arrays of geophones at the surface or at shallow depth, the sounds generated by even minimal rock material displacement can be recorded and the source of the event can be tri-angulated. This is particularly useful for operations monitoring of hydrothermal geothermal resources. The data can be used to ensure that withdrawal and/or injection of water is executed at the appropriate rate and pressure to avoid inadvertently cracking or fracturing the reservoir rock. For EGS, which rely on creating a network of fractures between an injection and a producing well, micro-seismic monitoring of the hydraulic stimulation operation can provide a real-time understanding of the fracture network being created, which offers the opportunity to modify and optimize the hydraulic stimulation operation in real time.

4.9 Drilling of slim hole wells to validate the local geothermal gradient

Purpose: Characterize the local geothermal gradient in detail by measuring temperature vs depth at discrete points along the well trajectory.

Tools: Specialized slim hole drilling equipment and temperature measurement equipment to determine temperature vs depth, such as thermocouples and/or optic fiber for distributed temperature sensing (DTS).

4.10 Drill cuttings and core analysis

Purpose: Develop a detailed understanding of rock lithology and diagenesis based on mineralogy and chemistry examination of drill cuttings and rock cores obtained from drilling. Can also provide a better approximation of formation depth than seismic data on its own. Drill cuttings are an invaluable source of subsurface information, with specialized service companies providing wellsite expertise and equipment. From the size and shape of the cuttings, additional information on geomechanics as well as borehole and formation stability may be gleaned.

Tools: Drilling rigs, mud logging equipment and expertise, core sampling tools, and laboratory analysis equipment and expertise.

4.11 Geochemical analysis

Purpose: Examines the composition of rocks, soil, or water to identify geothermal indicators, such as water diagenesis, dissolved minerals with a potential for scaling or corrosion, etc.

Tools: Rock and water sampling, laboratory analysis.

4.12 Borehole geophysics and well logging

Purpose: Fine-tuning of subsurface characterization to identify and characterize rock types, sequence, formation depth and thickness, porosity, in situ fluids, formation pressure, structural and sedimentological details, presence and orientation of fractures, in situ stress field characterization, geomechanics, and temperature gradients by collecting data from boreholes using well-established borehole logging technology and interpretation methodology.

Tools: Various downhole wireline tools such as gamma ray, resistivity, acoustic, density and neutron, borehole imaging, formation pressure tester and sampling, pulsed neutron, and other tools. Table 3 shows the wireline logging tools and their recommended applications for rock formation and fluids characterization.

TABLE 3 Geophysical logging tools and tool combinations recommended for geothermal well logging.

Information required	**Geophysical logging tools**
Basic information	Basic logging suite
• Formation depth and thickness • Lithology • Porosity • Temperature gradient • Fluid types and water saturation	• Gamma ray log • Resistivity log • Density/neutron logs • Temperature log
Additional petrophysical information	Additional logging tools
• Calibration of seismic data for formation depth and thickness • Permeability • Mineralogy	• Acoustic log • Nuclear magnetic resonance log • Density log with PE measurement or Pulsed neutron log
Additional geological information	Additional logging tools
• Faults, fractures, dykes, and other structural elements • Sedimentology • Geomechanics	• Borehole imaging tool—acoustic or resistivity imaging or both • Dipole acoustic • Deep shear imaging
Additional reservoir engineering information	Additional logging tools
• Formation pressure • Formation fluid chemistry	• Formation pressure tester • Formation fluid sampling tool

Geothermal resource assessment requires a combination of geological, geophysical, and geochemical data to determine the potential for harnessing geothermal energy. Each geothermal system requires a different methodology and novel technology from early prospecting and exploration all the way to the development phase, which come with certain risks and costs. Table 4 summarizes the data required for a thorough resource assessment for different types of geothermal systems.

4.13 The benefits of integrated, multidisciplinary subsurface analysis

In the preceding sections, a variety of geophysical measurement techniques and methodologies were introduced, which can be used to characterize the subsurface based on resistivity, magnetic, gravity, density, and acoustic responses. These various methods can also be deployed differently, ranging from satellite to airborne to surface deployed. It should, however, be noted that combining data from various geophysical measurements enhances the subsurface model considerably and reduces uncertainty and ambiguity.

The following example shows the evolution of the subsurface model of the Darajat geothermal reservoir in Indonesia, from a simple 1D magneto-telluric resistivity model to a better defined and detailed subsurface model using 3D joint inversion and constraining it with the geological model (Fig. 25).

The initial model shown on the left, based on 1D magneto-telluric data, shows the interpreted pre-drill top of reservoir (ToR) at a reasonable distance to the left of the planned well trajectory for well W3. By comparison, the new 3D joint inversion model on the right shows the ToR in very close proximity to the planned well trajectory, offering the development team the opportunity to reassess and revise their drilling plans.

In the following example, we will showcase the benefits of an integrated analysis of seismic and well log data—one from a hydrothermal resource, a porous limestone in the Paris Basin (Fig. 26), and the other from a HDR resource in the Upper Rhine Graben (Fig. 27).

The following examples are taken from the Upper Rhine Graben area which straddles France and SW Germany. This area exhibits a temperature suitable for hot rock projects at reasonably shallow depths and has been explored on the French and the German side (Fig. 27).

As can be seen from the examples, the integrated, multidisciplinary analysis, and interpretation of data acquired at different resolutions and with different methodologies provides additional, detailed insight, reducing subsurface uncertainty, enabling improved scenario definition and modeling. An example can be seen in Fig. 28, providing a combined global view of geothermal resources and geographic areas where this this potential is already being accessed with geothermal plants.

TABLE 4 Data requirements for geothermal resource evaluation, depending on the type of geothermal system to be deployed.

Information required		Hydrothermal Systems	Enhanced Geothermal Systems	Closed Loop Geothermal Systems
Geological Information	Rock formations, formation depth and thickness	Essential	Essential	Essential
	Faults, fractures, dykes and other structural elements	Essential	Essential	Essential
	Geomechanical information: in-situ stress magnitude and orientation	Recommended: For well construction	Essential: For hydraulic stimulation Recommended: For well construction	Recommended: For well construction
Heat and Energy Information	Temperature gradient and anomalies	Essential	Essential	Essential
	Target formation temperature	Essential	Essential	Essential
	Resource rock thermal conductivity	Essential	Essential	Essential
	Heat recharge mechanism	Essential	Essential	Essential
Petro-physical Information	Lithology	Essential	Essential	Essential
	Porosity	Essential	Very Useful	Not Required
	Permeability	Essential	Very Useful	Not Required
	Mineralogy	Essential	Essential	Not Required
	Water Saturation	Essential	Nice to have	Not Required
Reservoir Engineering Information	Presence of fluids, water saturation	Essential	Nice to have	Not Required
	Fluids chemistry	Essential	Nice to have	Not Required
	Flow pathways (porosity, fractures, etc)	Essential	Recommended	Not Required
	Fluid recharge mechanism	Essential	Essential	Not Required
	Presence and integrity of cap rock	Essential	Essential	Not Required

Legend

Essential Recommended Very Useful Nice to have Not Required

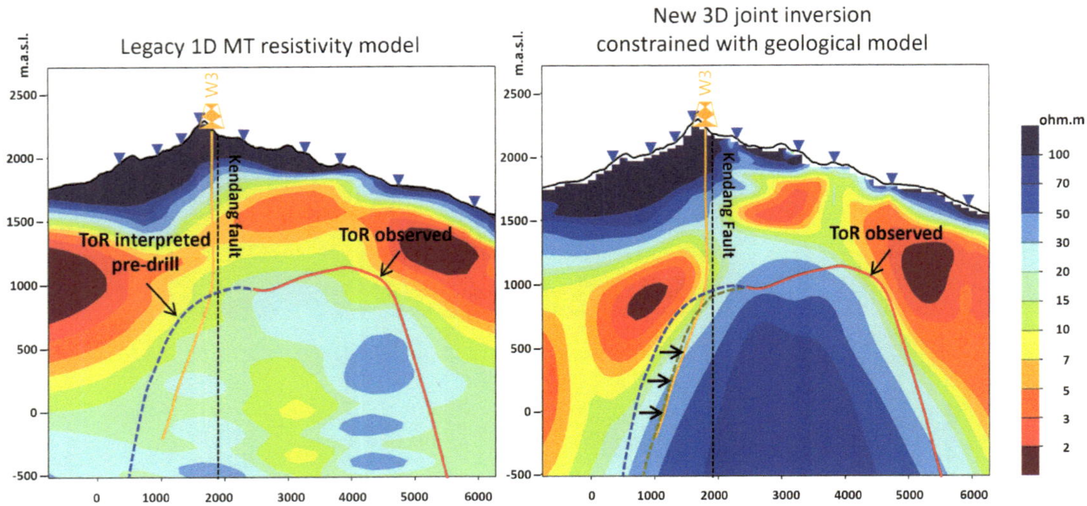

FIG. 25 Geologically consistent joint inversion of multiple geophysics datasets, as applied to the Darajat, Indonesia, geothermal reservoir boundary definition (Soyer et al. 2018).

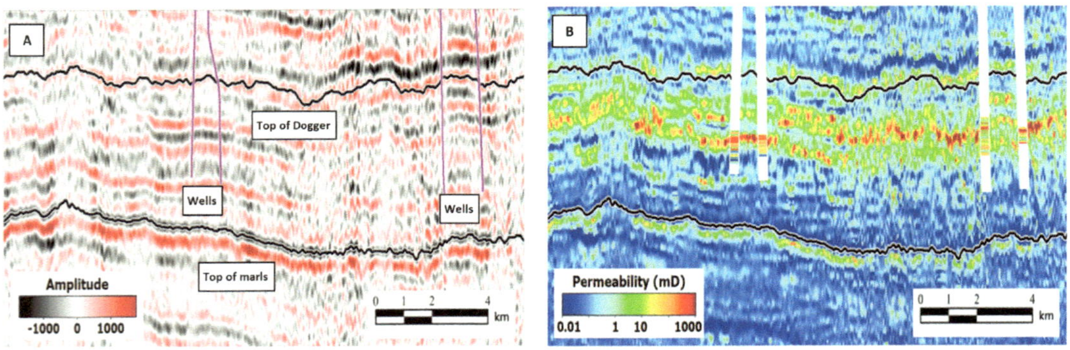

FIG. 26 (A) Input seismic section crossing the study area from west to east. The top horizon represents the top of the Dogger Formation. The bottom horizon corresponds to the top of the marls with Ostrea acuminata (Middle Bajocian). (B) Effective porosity obtained by combining the estimated total porosity and clay volume highlighting layers with high connected porosity.

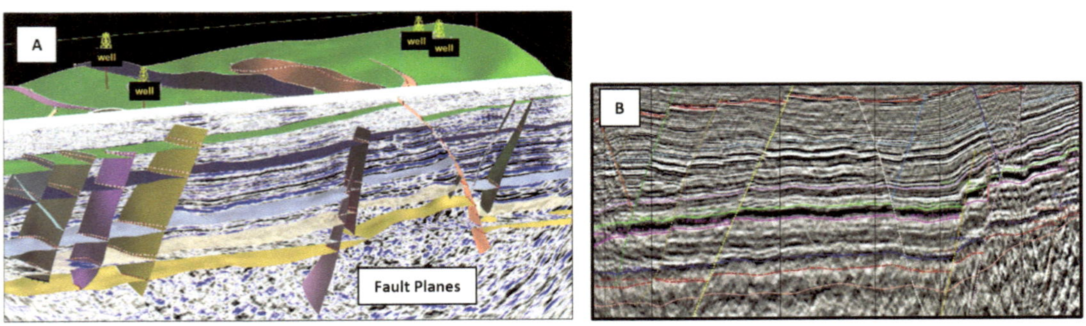

FIG. 27 (A) Three-dimensional structural model containing wells, horizons, and fault planes iteratively built from (B) various interpreted seismic volumes delivered after different steps of preprocessing and imaging. The model was also used to refine the velocity model in order to locate future production wells. The *dark blue* line shows the Permian horizon. The *red line* shows the fractured granite. *Soyer, W., Mackie, R. L., Hallinan, S., Pavesi, A., Nordquist, G., Suminar, A., Intani, R., Nelson, C., 2018. Geologically consistent multiphysics imaging of the Darajat geothermal steam field, First Break 36(6), 77–83. https://doi.org/10.3997/1365-2397.n0102*

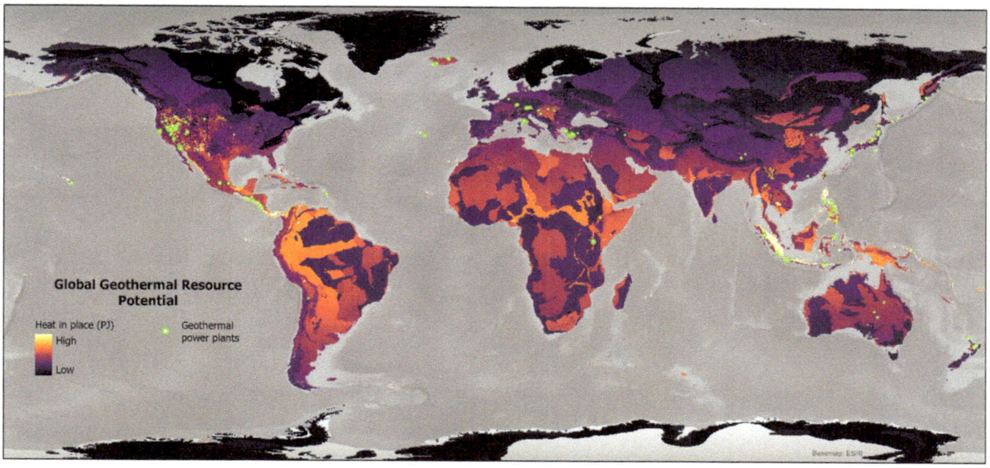

FIG. 28 Global geothermal resource map overlain with geothermal power plant locations. The geothermal resource map example shows total heat in place (PJ) within production temperatures 40–350°C to a maximum 5km depth, created by using temperature and tectonic regime data. *Credit: CGG Earth Data Global Geothermal Resource Assessment product 2021.* Map lines delineate study areas and do not necessarily depict accepted national boundaries.

5 A comparison of different geothermal systems

Geothermal resource assessment involves evaluating the subsurface conditions to determine the potential for harnessing geothermal energy. In this chapter, we will break down the assessments for three different systems: hydrothermal, Enhanced Geothermal Systems, and Closed Loop Geothermal Systems (Fig. 29).

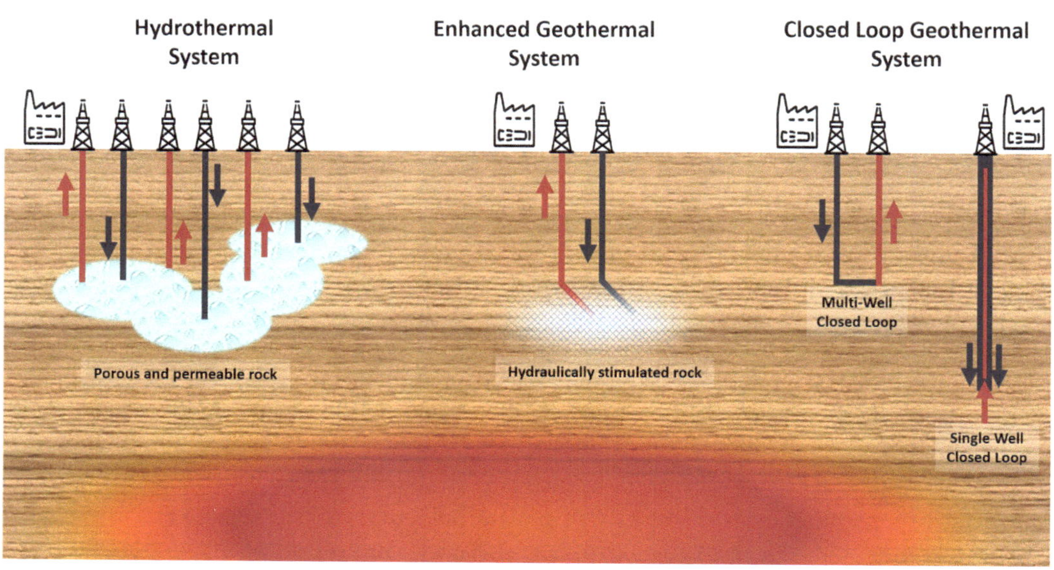

FIG. 29 Schematic overview of different geothermal systems—initially introduced in Fig. 6. *Figure courtesy XGS Energy.*

5.1 Hydrothermal system

Hydrothermal systems are currently the most widely utilized method of harvesting geothermal energy from the subsurface. They are comprised of a geological system where water or steam circulates through heated porous and permeable rock formations, producing the heated liquids to the surface via boreholes drilled into these rock formations, thereby transferring energy to the surface. There are four essential conditions that need to be met for a hydrothermal system to work: a heat source, a water source, permeable rocks that allow the hydrothermal fluids to circulate through the system, and a caprock that efficiently seals the hydrothermal reservoir and prevents the unplanned escape (migration) of hydrothermal fluids, to maintain the pressure and temperature regime of the system (Fig. 30).

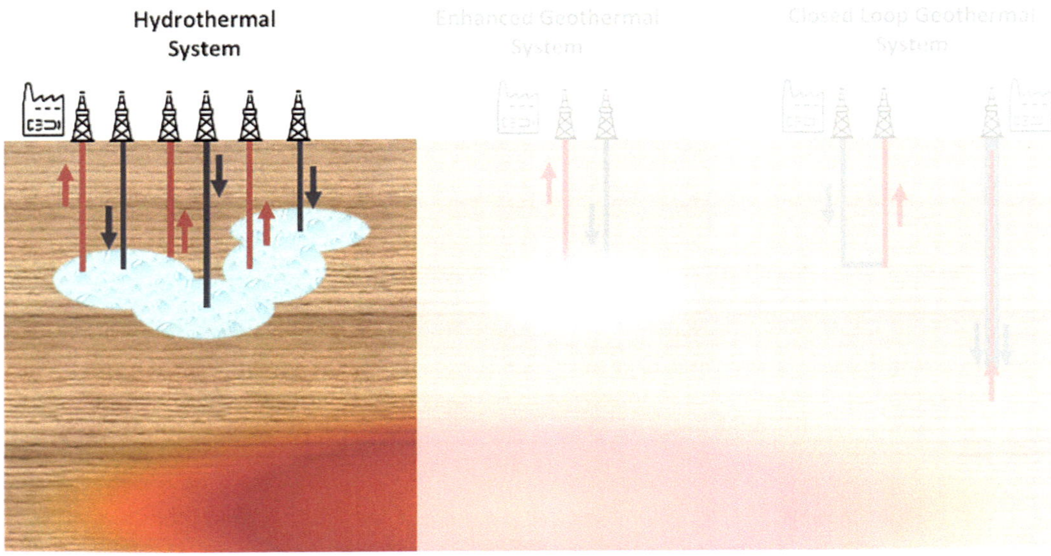

FIG. 30 Focus on hydrothermal systems. *Soyer, W., Mackie, R.L., Hallinan, S., Pavesi, A., Nordquist, G., Suminar, A., Intani, R., Nelson, C., 2018. Geologically consistent multiphysics imaging of the Darajat geothermal steam field, First Break 36(6), 77–83. https://doi.org/10.3997/1365-2397.n0102. Figure courtesy XGS Energy.*

- **Accessibility**: Access to hydrothermal resources is limited to certain geographical areas where all the following four conditions are met: areas where water availability, low-reservoir pressure, and poor permeability are encountered are unsuitable for exploitation of geothermal energy using a hydrothermal system design.
- **Sustainability**: A significant hurdle in harnessing hydrothermal resources lies in the decline of reservoir pressure. As geothermal fluids are withdrawn from the reservoir, the internal pressure diminishes, impeding the flow of fluids to the power plant. Consequently, this decline in pressure can result in diminished power production and elevated costs associated with fluid extraction. This issue can be addressed either by drilling new production wells or by using water injection wells (see paragraph on "Water Consumption" below). Secondly, corrosion and scale deposition pose notable operational challenges for hydrothermal systems. The corrosive nature of the hot acidic fluids in these

systems can cause deterioration of pipes, pumps, turbines, and other equipment, resulting in issues such as leaks, breakdowns, and the need for expensive and time-consuming repairs. Scaling presents a challenge for hydrothermal systems as the minerals dissolved in the fluids have the potential to precipitate and create deposits on pipes and other equipment. This accumulation can significantly diminish fluid flow to the point where production tubulars must be replaced, both a costly and time-consuming undertaking. Scaling and corrosion can be mitigated by using appropriate chemical inhibitors, which do increase operating costs. Both water withdrawal and water injection need to carefully be controlled and monitored to mitigate induced seismicity, i.e., mini-earthquakes, which indicate the breaking of the reservoir rock formation—which quickly gets widespread attention and can easily jeopardize the entire project's future.

- **Water consumption**: As mentioned before, decline in reservoir pressure in a hydrothermal system may necessitate the injection of water to extract heat, and the volume of required injection water can be substantial. As an example, for the geothermal power plants in The Geysers in California, 20 million gallons of treated wastewater from Lake County and the City of Santa Rosa are injected daily to maintain geothermal energy production (https://geysers.com/water).

5.2 Enhanced geothermal systems (EGS)

EGS are based on an engineering approach to create flow pathways in hot rock where natural fluid flow is impossible, due to rock compaction, cementation, or other reasons. Examples of such rocks would be igneous and metamorphic rocks or any highly compacted sedimentary rock with significant cementation or non-connected pore space. The concept is to artificially create fluid flow pathways, which will be described in more detail, and then circulate water between wells. This places EGS close to hydrothermal systems, with the potential benefit of lesser dependence on water. However, as these systems are still at an early technology readiness level (TRL), even proof-of-concept results are not yet available (Fig. 31).

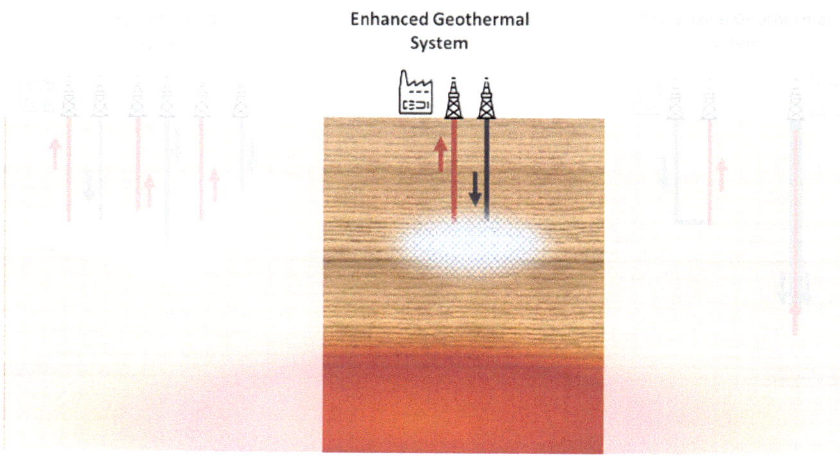

FIG. 31 Focus on enhanced geothermal systems. *Allo, F., Coulon, J.P., Formento, J.L., Rebou,l R., Capar, L., Darnet, M., Issautier, B., Marc, S., Stopin, A., 2021. Characterization of a carbonate geothermal reservoir using rock-physics guided deep neural networks. Leading Edge 40, 751–758. Figure courtesy XGS Energy.*

- **Accessibility**: The EGS approach utilizes two concepts specifically developed for the successful development of unconventional or shale oil and gas resources, which essentially are extremely low-permeability rocks, i.e., rocks where no fluids can flow naturally. However, these unconventional rocks hold large volumes of hydrocarbons. Over the course of several decades, small- to mid-size North American operators, in close collaboration with the service companies, devised ways and methods to create permeability in these unconventional resources and access the hydrocarbons stored there, namely, horizontal drilling and hydraulic fracturing. Because of the success of the unconventionals, the United States today is the largest oil and gas producer globally. The concept has been adapted for extremely low-permeability geothermal resources, which consists of two precisely placed parallel wells, one injector and one producer. Flow paths between these two wells are created through hydraulic stimulation, to enable the circulation of water from the injector to the producer. Currently, this approach is restricted to resource temperatures of 200°C or less due to technical limitations of wellbore positioning tools. Additionally, the lack of hydraulic stimulation tools and equipment for high-temperature rocks, along with stringent federal, state, and country regulations and public opinion and resistance, poses significant barriers for hydraulic fracturing operation in certain geographies.
- **Sustainability**: Like conventional hydrothermal systems, corrosion and scale deposition issues will pose a challenge for EGS. As water circulates through the fracture system, minerals present in the rock may be dissolved and subsequently re-deposited, altering the flow characteristics of the fracture system. Specifically, uranium salts dissolve particularly well in water, such that radio-active water may be produced. Apart from the closing of flow paths, due to deposition of minerals, the geomechanical stresses encountered at depth may result in proppant crush and hydraulic fractures closing partially or completely, significantly reducing fluid flow between wells. As has been stated earlier for hydrothermal systems, water injection rates and pressures need to carefully be monitored to avoid induced seismicity.
- **Water consumption**: In EGS, significant volumes of water and proppant are required during the processes of well construction and hydraulic fracturing. During operation of the system, a substantial volume of water will be circulated between injection and production wells. Some of this water will be lost, as it migrates and escapes through fractures and natural fluid flow paths, necessitating the replenishment of the system with makeup water. The oil and gas industry has numerous examples of unexpected and undesired waterflood behavior in fractured reservoirs, so learnings from such examples should be incorporated in the design of enhanced geothermal systems.

5.3 Closed-loop geothermal systems (CLGS)

The concepts of CLGS have evolved to virtually eliminate the need for large volumes of water inherent to hydrothermal and EGS concepts. As the name implies, in this engineering concept, the working fluid, i.e., the fluid used to extract heat from the rock, is never in direct contact with the rock, thereby enhancing system sustainability and availability and reducing system cost (Fig. 32).

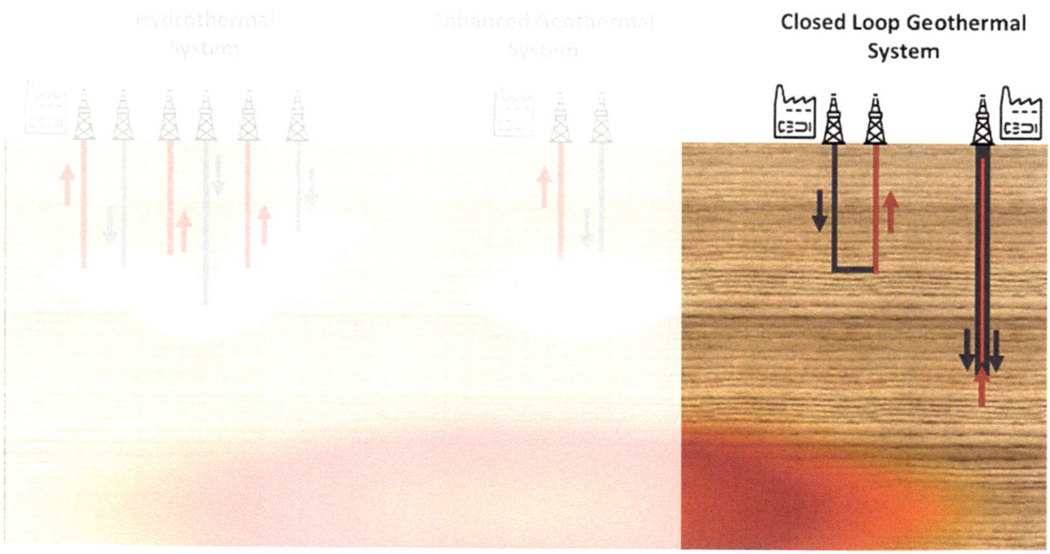

FIG. 32 Focus on closed-loop geothermal systems. *Salaun, N., Toubiana, H., Mitschler, J.-B., et al., 2020. High-resolution 3D seismic imaging and refined velocity model building improve the image of a deep geothermal reservoir in the Upper Rhine Graben, Leading Edge 39 (12), 857–863. https://doi.org/10.1190/tle39120857.1. Figure courtesy XGS Energy.*

- **Accessibility**: CLGS can be implemented in any hot rock formation and any geological setting, irrespective of the presence of porosity, permeability, or water. Thus, from a subsurface assessment point of view, the only parameters to determine are resource temperature, resource depth, and resource thermal conductivity. Compared with the previously discussed engineering approaches, where large volumes of water are circulated through hot rock formations to extract the heat, for CLGS, the key engineering challenge is to create an efficient heat exchanger in the subsurface.
- **Sustainability**: In CLGS, the heat carrier fluid that is continuously circulated in the system does not interact with the rock formations itself, thereby eliminating the risks of scaling or corrosion, improving system uptime, and reducing maintenance costs.
- **Water consumption**: CLGS require a comparatively small amount of carrier fluid within the closed-loop architecture, eliminating subsurface water withdrawal or injection and significantly reducing the risk of induced seismic activity.

6 Geothermal feasibility evaluation

Geothermal project feasibility is a comprehensive analysis conducted to assess the viability of developing a geothermal project. It involves evaluating technical, economic, and environmental aspects to determine whether the project is economically feasible and environmentally sustainable.

6.1 Resource assessment

Within the domain of resource assessment, the primary objective is to evaluate both the quality and quantity of the geothermal resource. This undertaking involves a series of tasks aimed at comprehensively understanding the geological characteristics and potential of the resource. The process includes conducting thorough geological and geophysical surveys to understand the subsurface structure, variations in rock density, and subsurface magnetic properties and identify potential resources. Geochemical assessments analyze the composition of fluids emanating from the subsurface. This includes studying the chemical and isotopic properties of fluids from wells. It also requires analyzing data from existing wells and studying parameters such as temperature, pressure, and fluid composition. Finally, evaluating the capacity of geothermal resources helps us understand their energy-producing potential. This multifaceted approach ensures a holistic assessment for informed decision-making in geothermal energy utilization.

6.2 Exploration drilling

Exploration drilling is a strategic process that entails drilling exploratory wells to pinpoint geothermal resources, ascertain their spatial extent, and evaluate their viability for energy production. The primary objective is to collect comprehensive information about the subsurface system. This encompasses an in-depth analysis of subsurface geology, assessment of resource temperature and pressure, and examination of geothermal fluid characteristics and flow dynamics. Additionally, the drilling aims to evaluate the geothermal resource's capacity to facilitate the efficient flow of fluid or heat, thereby influencing the efficiency of energy extraction. Through this process, exploration drilling serves as a crucial step in understanding and harnessing geothermal potential.

6.3 Well testing

Geothermal well testing involves injection or production from geothermal resources and measuring various parameters to assess the well's productivity and determine the optimal parameters for long-term production. Well testing provides valuable insights into the characteristics of the geothermal resources, including measuring the rate at which fluids can be extracted from, monitoring changes in reservoir pressure during pumping and recovery, analyzing the temperature profiles of the produced or working fluids, and evaluating the properties of the produced fluids, such as viscosity, density, and chemical composition to assess any potential scaling or corrosion issues.

There are different types of test for geothermal wells. Injection tests involve injecting fluids into the reservoir to assess its ability to accept fluids and determine its injectivity index. In case of creating artificial fractures in the subsurface to enhance fluid flow and heat exchange, primary objectives of well testing include the determination of the ability to hydraulically stimulate or fracture the rock, measuring the permeability of the fractured reservoir, and identifying the orientation and extent of the fractures. This allows to optimize injection/production strategies to maximize energy extraction. Production tests involve pumping fluids from the reservoir, or pumping fluids from one well and reinjecting them into another well to assess its productivity and determine its productivity index.

Well testing results provide the basis for improved geothermal system performance through a deeper understanding of resource characteristics and well productivity. Well testing can optimize well design and operation, leading to increased energy production and reduced operating costs.

6.4 Power plant technology, design, and engineering

Temperature, pressure, flow rate, and the type of working fluid are critical factors influencing the choice of power plant technology in geothermal applications. High-temperature resources with high-pressure fluids are typically conducive to the operation of flash steam power plants, while moderate-temperature resources may find compatibility with binary cycle systems. Low-temperature resources, characterized by their lower heat levels, are often more suitable for direct heating applications rather than electricity generation. Aligning the power plant's capacity with the inherent capacity of geothermal resources is imperative, along with considering the energy demand and financial aspects of the project. Additionally, if the project site holds the potential for future resource expansion or increased power generation capacity, the chosen technology should exhibit adaptability to accommodate such growth. Factors such as modularity, scalability, and integration with additional power generation units become crucial in ensuring the flexibility and sustainability of the chosen geothermal power plant technology (Fig. 33).

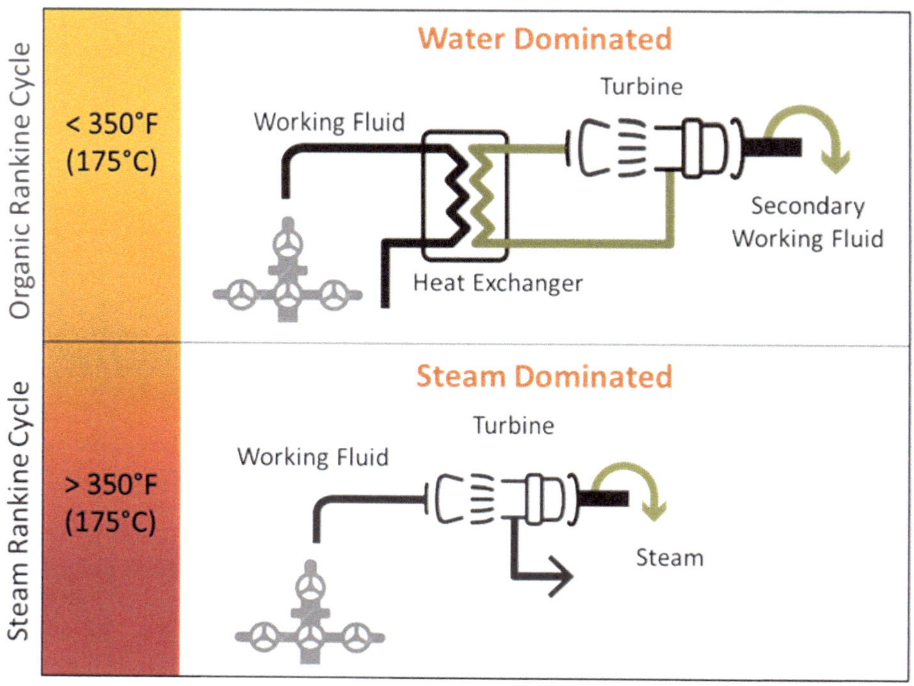

FIG. 33 A steam Rankine cycle, which is suitable for high-temperature geothermal systems >350 °F, directly utilizes the steam from the subsurface to drive a surface turbine and generate power. For a temperature environment <350 °F, an organic Rankine cycle (ORC) is utilized, which incorporates a heat exchanger, where the hot water from the subsurface creates steam from organic fluids with a lower boiling point than water. This organic fluid steam is then used to drive a turbine to generate power. *Credit: CGG Earth Data Global Geothermal Resource Assessment product 2021. Figure courtesy XGS Energy.*

6.5 Economic analysis and financial modeling

Geothermal economic models are crucial for assessing the financial viability of geothermal projects and guiding informed decisions about resource development. Geothermal projects typically have a long lifespan, often exceeding 30 years. These models account for the time value of money and project depreciation over the extended lifespan, to evaluate the project's long-term profitability. Several factors are incorporated into these models. Well construction costs are influenced by the depth of the resource, the type of drilling technology required, and the geological conditions. Additionally, the choice of power plant technology depends on the resource characteristics and their capacity. These models assess the capital costs, operating costs, and efficiency of different power plant technologies to determine the most cost-effective option. Financing options, government incentives, and renewable energy credits can significantly impact the economic feasibility of geothermal projects. These models incorporate these factors to evaluate the project's financial attractiveness to investors. Geothermal economic models also include metrics such as levelized cost of electricity (LCOE), net present value (NPV), internal rate of return (IRR), and payback period. Sensitivity analysis evaluates the impact of changes in key project parameters, such as resource characteristics, drilling costs, and electricity prices, on project outcomes. By carefully considering these factors and applying geothermal economic models, developers can identify viable geothermal projects.

6.6 Risk analysis

Conducting risk analysis is essential for identifying, assessing, and ranking the potential risks associated with geothermal projects and developing strategies to mitigate those risks. The uncertainty of geothermal resource characteristics, such as resource size, temperature, depth, and other geological features, can significantly impact project feasibility and energy output. Well construction operations can involve potential technical challenges, including encountering unexpected geological formations, facing lost circulation of drilling fluids, addressing wellbore stability issues, and dealing with unsuccessful completion and stimulation operation. Environmental regulations, permitting, and policies can also influence the feasibility and cost of projects. Utilizing advanced numerical models to simulate the subsurface environment, conducting sensitivity analysis of operational parameters, and implementing real-time monitoring to quantify the likelihood and potential impact of various risk events can significantly contribute to project success.

6.7 Environmental and social/socioeconomic impact assessments

As with any industrial, construction, energy extraction, or similar project, an environmental impact statement (EIS) has to be drawn up prior to or as part of the permitting process. An EIS consists of two components, an environmental impact assessment (EIA) and a social or socioeconomic impact assessment (SIA). The general public is well aware of the term EIA, usually understanding this to be an assessment of how wildlife and vegetation may be impacted by a new project. In the United States, EIAs have become key parts of project planning documentation after the passing of the National Environmental Policy Act (NEPA) of 1969.

However, from around the 1990s onward, it became apparent that attention also needed to be focused on social or SIAs of public or private projects to provide interdisciplinary assessments. Such assessments take into account the potential changes that might affect how people in the project area live, find work, commute, get schooling and training, and are impacted by additional truck traffic during the construction and operational phase of a project, the influx of a temporary workforce, the associated potential increase in cost of living due to higher demand, and more.

While it may appear easy to win "hearts and minds" within a community with a geothermal energy project that promises to deliver clean, affordable, dependable baseload power, it is still essential to engage with the community early on to ensure faster permitting and acceptance within the community. While there can be no doubt about the benefits of geothermal energy generation for households and industry alike, there is an opportunity to win hearts and minds through thoughtful dialogue with the community. This can come in the form of townhall meetings, with active engagement of all associated parties.

The shale gas and shale oil development in North America over the last 20 years is an excellent learning opportunity for geothermal development: in the early days, operators did not engage with communities, leaving an information void, which in turn gave rise to speculation, rumors, and even movies with a negative message about the "inherent dangers" of hydraulic stimulation, with notable celebrities from the entertainment industry entering the discussion and shaping public opinion. In some cases, this created significant project delays; in some countries, "fracking" was even banned completely in response to public sentiment. Moreover, the logistics associated with high-intensity well construction and hydraulic fracturing, which requires moving millions of pounds of sand and millions of gallons of water, created extraordinarily high wear on tear on roads. The influx of a well-compensated workforce, some of it temporary, created demand for lodging, food, and entertainment, oftentimes driving local inflation, to the detriment of the original inhabitants of the area.

The geothermal industry can and must learn from those mentioned to, on the one hand, avoid the initial missteps of the unconventional oil and gas industry and to accelerate the rate of adoption of geothermal energy on the other.

7 Geothermal resource management

Geothermal energy systems harness a variety of heat sources to produce sustainable and renewable power. The specific heat sources in a given location are contingent upon the geological and tectonic characteristics of that region. Radiogenic heat results from the natural radioactive decay of isotopes like uranium, thorium, and potassium in the Earth's crust, releasing heat into the surrounding rocks. Primordial heat, a residual from the planet's formation, stems from the Earth's initial accretion and subsequent layer differentiation, generating substantial heat. Another significant heat source is convective heat transfer, facilitated by the movement of fluids (water or steam) circulating within the Earth's crust. This process is influenced by temperature gradients in the subsurface and fluid movements due to geological processes. Magma, molten rock beneath the Earth's surface, serves as a robust source of heat, especially in volcanic regions where magma chambers or conduits can bring high-temperature heat close to the surface.

Frictional heat is generated by tectonic plate movement, faulting, and other geological processes, as stress and strain on rocks during these movements create heat.

Despite the existence of these heat sources, temperatures from geothermal resources will diminish over time, due to several factors. Continuous extraction of geothermal fluids at a rate surpassing natural replenishment may result in a decline in temperature. If fluid extraction exceeds the rate at which heat naturally replenishes in the reservoir, the resource may cool progressively. Over time, as heat is extracted and the reservoir is depleted, the remaining resource may exhibit a lower temperature, particularly in reservoirs with limited recharge or slow heat regeneration. In certain cases, geothermal wells situated at a distance from the primary heat source may experience lower fluid temperatures as extraction continues. This can be influenced by the geological structure and fluid pathways within the reservoir. The thermal conductivity of the surrounding rock matrix plays a role in the rate at which heat is conducted away from the reservoir. Additionally, the presence of nearby geothermal wells or other extraction activities in the same reservoir can lead to thermal interference. When multiple extraction points are in close proximity, they may compete for available heat, causing a decline in temperature for each well.

Fig. 34 illustrates the temperature degradation profile of the thermal resource as a function of time and working fluid flow rates. In the initial phase, there is an accelerated thermal interaction among the fluid, rock, and the surrounding environment, leading to rapid cooling of the rock. The cooling rate is directly proportional to the flow rate, with higher rates leading to faster cooling. Subsequently, the second stage is characterized by a gradual cooling process. Within this stage, the temperature change curve adopts a nearly linear trend indicative of a more moderate cooling rate. Finally, the third stage signifies a period of steady and consistent cooling.

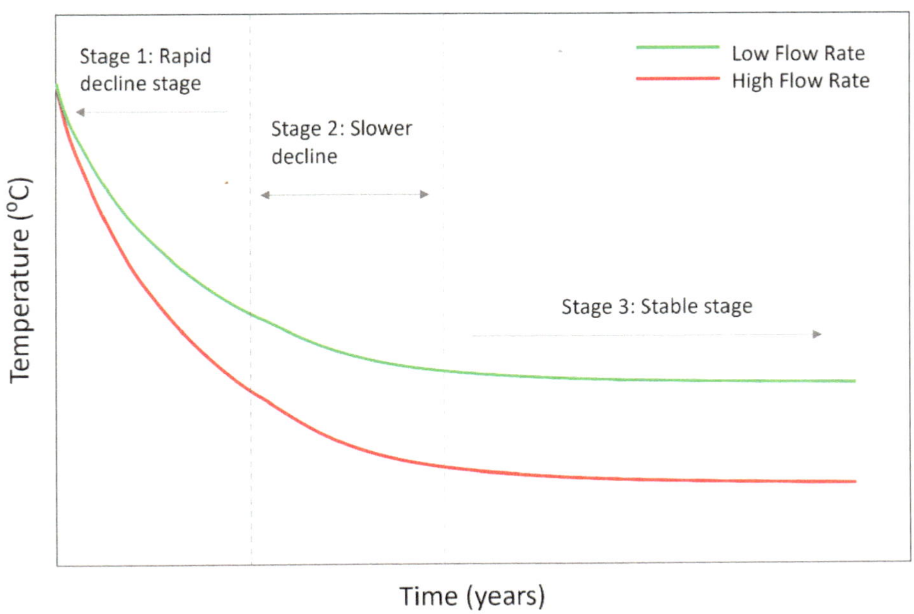

FIG. 34 Temperature decline of the resource as a function of time and flow rate. *Figure courtesy XGS Energy.*

Effective geothermal resource management integrates a comprehensive process and workflow to unlock the complete potential of sustainable geothermal energy. The method involves strategic measures to optimize reservoir pressure, enhance fluid circulation, control flow rates, and extract heat efficiently. This entails a continuous cycle of monitoring, strategic in-fill drilling to enhance reservoir performance, and specific measures to mitigate corrosion and manage scale.

The resource management process is uniquely tailored to each system—be it hydrothermal, EGS, or closed-loop—addressing system-specific challenges and considerations. This strategic customization ensures the sustainable utilization of geothermal resources while minimizing environmental impacts. Continuous refinement of this process is imperative to adapt to evolving conditions and enhance overall geothermal energy efficiency.

7.1 Hydrothermal system

- **Monitoring**: Monitoring a hydrothermal system is pivotal for comprehending its dynamics, predicting potential changes, and ensuring sustainable utilization. This entails tracking diverse parameters that offer insights into the system's health, productivity, and environmental impacts. Temperature monitoring is fundamental, reflecting the energy content and fluid movement within the hydrothermal system. Temperature changes may signal resource depletion, disruptions in fluid circulation, or the emergence of new heat sources. Pressure monitoring provides crucial information about the reservoir's energy potential and the overall state of the hydrothermal system. A decline in pressure may indicate resource depletion, while excessive pressure can pose safety concerns. Continuous monitoring of fluid flow rates and quality is essential for assessing productivity, identifying potential scaling or corrosion issues, and understanding the composition of geothermal fluids. Monitoring the chemical composition of groundwater serves as an indicator of changes in the hydrothermal system, such as the presence of contaminants or alterations in fluid flow patterns. Ground subsidence, an environmental impact of geothermal operations, can be identified and mitigated through continuous ground deformation monitoring. Induced seismicity, another potential impact, can be monitored to identify and address any seismic events linked to geothermal operations.
- **Operational control:** Adjusting the flow rate of geothermal fluids is crucial for balancing energy extraction with reservoir sustainability. Reducing the extraction rate can help extend the life of the reservoir, while increasing the extraction rate can lead to depletion and environmental impacts. Maintaining appropriate reservoir pressure is essential for ensuring proper fluid circulation, preventing leaks, and minimizing induced seismicity. Controlling the temperature of geothermal fluids can help prevent scaling and corrosion in equipment. This can be done by adjusting the flow rate, using different working fluids, or implementing heat exchange strategies. Maintaining water quality is essential for preventing environmental contamination and ensuring the long-term viability of geothermal resources. This involves treating fluids to remove contaminants, monitoring water quality parameters, and implementing responsible waste disposal practices. Minimizing induced seismicity is crucial for protecting the surrounding environment and ensuring the safety of geothermal operations. This can be achieved by monitoring seismic

activity, optimizing well placement and injection parameters, and implementing mitigation strategies. Regular maintenance and monitoring of geothermal wells are essential to ensure their integrity, prevent leaks, and optimize performance. This includes inspections, pressure testing, and cleaning and repair work.

- **In-fill drilling strategies**: Strategically placing additional wells within an existing geothermal field, known as in-fill drilling, aims to access untapped areas of the reservoir initially beyond the reach of existing wells. This approach offers a substantial increase in the energy potential of hydrothermal resources. In-fill drilling enhances fluid circulation within the reservoir by establishing additional pathways for fluid movement. Consequently, this can result in heightened fluid flow rates, translating to a higher energy output from the geothermal power plant. This strategy plays a pivotal role in maintaining reservoir pressure by replenishing fluids and averting pressure decline, particularly crucial in mature geothermal fields where pressure depletion may be a concern. Decisions regarding in-fill drilling should consider economic factors, encompassing the cost of drilling new wells, the anticipated increase in energy output, and the potential revenue generated from additional electricity production.

7.2 Enhanced gothermal systems (EGS)

- **Monitoring**: Ensuring the successful operation, minimal environmental impact, and optimal contribution to renewable energy production in EGS necessitates comprehensive monitoring. Given the complexity of EGS systems and their various components, tracking key parameters is essential for performance assessment and issue identification. Monitoring fluid flow rates and temperatures in production and injection wells is crucial for evaluating the efficiency of fluid circulation and heat transfer within the reservoir. Sustaining reservoir pressure is critical for EGS sustainability, and monitoring pressure changes can indicate potential fluid depletion, alterations in the fracture network, or stress changes induced by injection. Temperature changes along production and injection wells offer insights into heat distribution, fluid circulation patterns, and potential reservoir anomalies. Microseismic monitoring is instrumental in detecting and locating microearthquakes, providing indicators of fracture formation, fluid movement, or potential induced seismicity. Analyzing the chemical composition of geothermal fluids reveals changes in fluid chemistry, potential scaling or corrosion issues, and the presence of contaminants. Ground subsidence or uplift, resulting from fluid extraction or injection in EGS systems, can be monitored for timely identification and mitigation. Environmental parameters, including water quality, air quality, and noise levels, must be continuously monitored to assess and minimize environmental impacts associated with EGS operations.
- **Operational control:** Maintaining appropriate injection rates and pressures is critical for maintaining fracture networks, enhancing heat transfer, and optimizing energy extraction from the EGS reservoir. Excessive injection can lead to over-pressurization and induced seismicity, while insufficient injection may not adequately create or maintain fracture networks. Managing the properties of the working fluid, such as viscosity, density, and temperature, is essential for ensuring efficient heat transfer and preventing operational issues like scaling or corrosion. Adjusting fluid composition or implementing filtration and

treatment processes can optimize fluid performance. Tracking temperature changes within the EGS reservoir and around injection and production wells provides insights into heat transfer patterns and potential areas for improvement. This data can guide decisions on injection strategies, well placement, and heat extraction optimization. Continuously monitoring microseismic activity, which indicates the formation and movement of fractures, is crucial for understanding the reservoir's response to injection and identifying potential induced seismicity. This data can guide adjustments to injection parameters and inform seismic mitigation strategies.

- **In-fill drilling strategies**: In EGS, in-fill drilling is strategically employed to target areas where new fracture networks can be established or existing fractures can be enhanced. The primary goal is to increase the reservoir's heat exchange surface area, thereby improving fluid circulation. By focusing on specific regions, in-fill drilling expands the heat exchange zones within the reservoir. This expansion facilitates more efficient heat extraction, contributing to an overall increase in the energy potential of the EGS system. Furthermore, it aids in maintaining connectivity between existing fracture networks, optimizing injection strategies to ensure efficient fluid flow and heat transfer throughout the reservoir.

7.3 Closed-loop geothermal systems (CLGS)

- **Monitoring**: Ensuring the efficient operation of CLGS relies on comprehensive monitoring of various components. Key aspects include temperature monitoring at critical points in the closed loop, such as the heat exchanger, working fluid, and resource. This is vital in assessing heat transfer efficiency and detecting potential blockages or anomalies. Maintaining appropriate pressure levels throughout the closed-loop is crucial to ensure proper fluid circulation and prevent leaks or ruptures. Continuous monitoring of the working fluid's flow rate is essential for evaluating the system's performance and identifying potential issues like flow restrictions or pump malfunctions. Detailed analysis of the working fluid's properties, including viscosity, density, and specific heat capacity, aids in identifying potential concerns related to fluid degradation or contamination. These monitoring measures collectively contribute to the effective and reliable operation of CLGS.
- **In-fill drilling**: In-fill drilling serves as a strategic approach to access deeper regions within geothermal resources, unlocking the potential of higher-temperature reservoirs and elevating the overall heat potential of the system. This process not only delves into untapped depths but also extends the heat exchange zones within the resource. By creating additional pathways for heat movement, in-fill drilling enhances the efficiency of heat extraction, facilitating increased heat transfer and contributing to a higher overall energy output for the closed-loop system. This expansion of heat exchange areas plays a pivotal role in optimizing system performance and elevating its capability for sustainable energy production.
- **Operational control:** Optimizing the performance of the downhole heat exchanger is essential for maximizing heat transfer between the geothermal resource and the working fluid. This may involve adjusting flow rates and controlling fluid temperatures and surface area. Balancing the flow rate ensures efficient heat extraction while preventing excessive

temperature drops. Optimizing the operation of pumps is crucial for maintaining efficient fluid circulation and ensuring adequate heat transfer. This may involve adjusting pump speeds, scheduling pump operation, and implementing energy-efficient pump technologies. Maintaining appropriate injection temperature throughout the closed-loop system is essential for maximizing heat extraction. By effectively managing these key operational control parameters, we can optimize closed-loop system performance and ensure the long-term sustainability of this valuable renewable and dispatchable energy source.

8 Analogies between oil and gas and geothermal resources

As the demand for energy production from geothermal resources is increasing, it is important to highlight the commonalities between geothermal resources on the one hand and oil and gas resources on the other hand. It is envisaged that the learnings and technologies from oil and gas exploration and (exploitation) production can be applied to geothermal resources. Thus, a workforce trained in assessing, modeling, and accessing the subsurface safely and in a cost-efficient manner to produce hydrocarbons will be instrumental to rapidly expand the utilization of geothermal resources. In the following paragraphs, we will review established procedures and methodologies from the oil and gas industry and how they may effectively be employed for geothermal resources.

Oil and gas and geothermal resource assessments share many similarities due to the geological and geophysical processes involved during the genesis of these resources, on the one hand, and the subsurface exploration, characterization, and modeling techniques involved, on the other hand. Both industries face challenges in accurately estimating resources due to subsurface uncertainty and economic considerations. However, technological advancements and the increasing demand for renewable energy are driving innovation and improving the feasibility of geothermal projects.

8.1 Exploration techniques

Exploration techniques, ranging from remote sensing, including satellite and airborne methods, to surface methods, have been described in detail in the earlier sections of this chapter. In this overview, we will focus on identifying the methods that are more commonly used in oil and gas exploration. Seismic surveys and well logging are common exploration techniques in both industries to understand subsurface features and identify potential resources.

8.2 Well testing

Well testing in oil and gas assesses the flow rate, pressure, and composition of hydrocarbons to determine reservoir properties. In geothermal, well testing evaluates parameters, such as fluid flow rates and temperatures, to assess the characteristics of the geothermal resource.

8.3 Reservoir characterization, static, and dynamic modeling

In oil and gas exploration and exploitation, reservoir characterization involves developing a detailed understanding of subsurface rock properties, the compositions of hydrocarbons and other fluids, reservoir compartmentalization, and pressure. Subsequently, a static subsurface model is constructed by a multidisciplinary team. This static subsurface model is essential in the simulation of reservoirs, a step in which reservoir engineers utilize sophisticated dynamic modeling software to predict reservoir performance over time, as oil and gas are being produced. During the reservoir simulation exercise, a multitude of scenarios are evaluated to deliver a field development plan (FDP), which will take into account the number of wells to be drilled, well completions, the surface facilities, and off-take infrastructure (such as pipelines), the production support required (such as water flooding or gas lift), and the production rates and volumes to be expected over the life of the field. As such, the FDP provides the basis to assess the commercial viability of a particular prospect, tying together the investment required to develop the field as well as the revenue to be generated during the production life of the field. In geothermal resource assessments, understanding subsurface geological features, rock permeability, and the temperature gradient aids in estimating the geothermal potential. During the static modeling, data from different sources (such as geological and geophysical surface surveys), well data (such as drill cuttings and cores), borehole geophysics, and in-well temperature/pressure surveys versus depth and time.

In geothermal plays, similar methodologies as in oil and gas are utilized to derive a static subsurface model. Geothermal reservoir simulation employs advanced dynamic multiphysics computer modeling techniques to predict the dynamic behavior of geothermal reservoirs over time. This approach encompasses a comprehensive understanding of various physical processes within the reservoir, including fluid flow, heat transfer, geomechanics, and chemical reactions between fluids and rock formations. By gaining insights into the reservoir's behavior, operators can make informed decisions about fluid production and injection strategies. This not only aids in optimizing geothermal energy production efficiency but also helps identify and mitigate potential risks associated with geothermal development, such as induced seismicity. Furthermore, the technique assists in pinpointing optimal locations for new wells. The development of accurate and efficient numerical methods, coupled with the incorporation of relevant physical processes, is paramount for achieving realistic simulations of geothermal reservoirs. Additionally, the integration of real-time data enables operational decisions based on the latest information and simulations. Enhancing the synergy between data and models is crucial for achieving more accurate and reliable simulations in geothermal reservoir management.

Through data integration, we can derive 1D (along wellbore), 2D (fence diagrams from well to well), and full 3D models, to better characterize the potential of the geothermal resource.

8.4 Resource estimation

Oil and gas resource estimation involves calculating the recoverable reserves based on geological and reservoir engineering assessments. Geothermal resource estimation involves calculating the available heat in the subsurface based on temperature, rock properties, and fluid content.

8.5 Environmental impact assessment (EIA)

Both industries conduct EIAs to evaluate and mitigate the environmental impact of resource extraction. Geothermal projects also undergo EIAs to assess potential environmental effects and implement mitigation measures. See section xyz on SIAs.

8.6 Drilling challenges

Oil and gas wells face challenges such as drilling through hard rock formations or encountering unexpected subsurface conditions. Geothermal wells encounter similar drilling challenges, and the choice of drilling techniques and equipment becomes crucial. This includes the type of rig, the types of drill bits for the different hole sections, the drilling fluid system, and, of course, the design of the bottom hole assembly (BHA). Through the rate of penetration (ROP) analysis of offset wells by lithology and hole section, a drilling plan can be derived for subsequent wells. Such a drilling plan typically consists of a day versus depth curve, the anticipated risks on a hole size by hole size breakdown, and risk mitigation measures, such as managing the mud weight to ensure more stability as well as optimum ROP in wells.

8.7 Operational risk management

Both industries deal with operational risks related to reservoir performance, well integrity, and equipment reliability. Geothermal projects require effective risk management to address uncertainties in reservoir behavior and operational conditions.

8.8 Infrastructure planning

Infrastructures that process and transport oil and gas from the field to off-takers is essential for economic success—without a connection between supply and demand, value creation is impossible. However, for long-term project economics to be positive, oil and gas surface facilities have to be right-sized. If they are under-dimensioned, production will have to be deferred, negatively impacting cash flow. If they are over-dimensioned, CAPEX will have been wasted, and they will never operate at capacity.

Geothermal projects require infrastructure planning for power plants, transmission lines, and fluid handling systems. Currently, there are quite a number of hydrothermal power plants running at 50% of nameplate capacity or less, due to significant shortfalls of injection water.

While analogies exist between oil and gas on the one hand and geothermal on the other hand, it is important to note that each energy resource has unique characteristics and considerations. The renewable nature of geothermal energy and distinct subsurface conditions differentiate it from conventional oil and gas exploration and production.

8.9 Automation and optimization

Automation has sparked significant transformative changes across diverse industries, and the areas of well construction, downhole resource monitoring and optimization, and surface plant operations and maintenance (O&M) are experiencing such changes at an ever-increasing pace. The field of geothermal energy development and utilization will be of no exception. Instrumented drilling rigs, 24/7 monitoring systems for subsurface and surface equipment machine health, coupled with enhanced computing power to deliver machine learning (ML) and artificial intelligence (AI) at scale and pace will augment human expertise to deliver differentiating outcomes.

Since well construction activities are one of the main cost drivers for geothermal projects, additional attention needs to be focused on optimizing these activities. In the oil and gas industry, significant time and cost savings have been achieved on larger-scale projects, where 20 or 30 wells had to be delivered for a specific asset, i.e., an area with similar subsurface characteristics. Detailed analysis of the time spent on individual operational steps offered insights into improvement opportunities. Implementing individual measures or "best practices" to reduce the well construction time over the next few wells has invariably resulted in considerable time savings of 50% or more compared with the initial wells in the project. Similarly, this detailed operation breakdown has also identified repetitive rig activities where process automation would deliver more consistent results faster and more safely. Machine learning (ML) and artificial intelligence (AI) have also been used successfully to analyze data from thousands of wells, such as weight on bit (WOB), ROP, and drill bit type to deliver optimum performance. The adoption of automated technologies has not only expedited the drilling process but has also elevated the precision in well placement. This has allowed us to navigate intricate geological formations more efficiently, resulting in significantly reduced drilling times and minimal operational disruptions. In summary, analysis of historical drilling data, implementation of best practices, and larger-scale drilling projects, rather than one-off wells, have been keys to success in oil and gas. They will also be the keys to success for geothermal development.

The role of data analysis and automation is equally pivotal in reservoir monitoring to drive continuous improvement. A recommended starting point is to "treat data as an asset" through the analysis of existing operational data to derive a more detailed understanding of the subsurface, culminating in predictive models. Combined with real-time data collection and analysis, this methodology will provide valuable insights into reservoir performance, facilitating proactive decision-making to dynamically optimize fluid and heat extraction and injection rates, ensuring sustained productivity.

The cost-effective operation of surface equipment, such as power plants, power distribution grids, and heating and cooling infrastructure, is also essential to the commercial success of geothermal development projects. Power plant operations have experienced substantial benefits through the integration of advanced machine health monitoring and control systems, which are now seeing the introduction of ML and AI to harness the vast amounts of data and information collected and the intricate interdependencies that links the operational parameters. Automation in turbine control, heat conversion processes, and overall plant management has led to enhanced operational reliability and efficiency through better informed scheduling of O&M tasks. The optimization of these processes has not only increased energy output but has also contributed to a reduction in operational costs (Fig. 35).

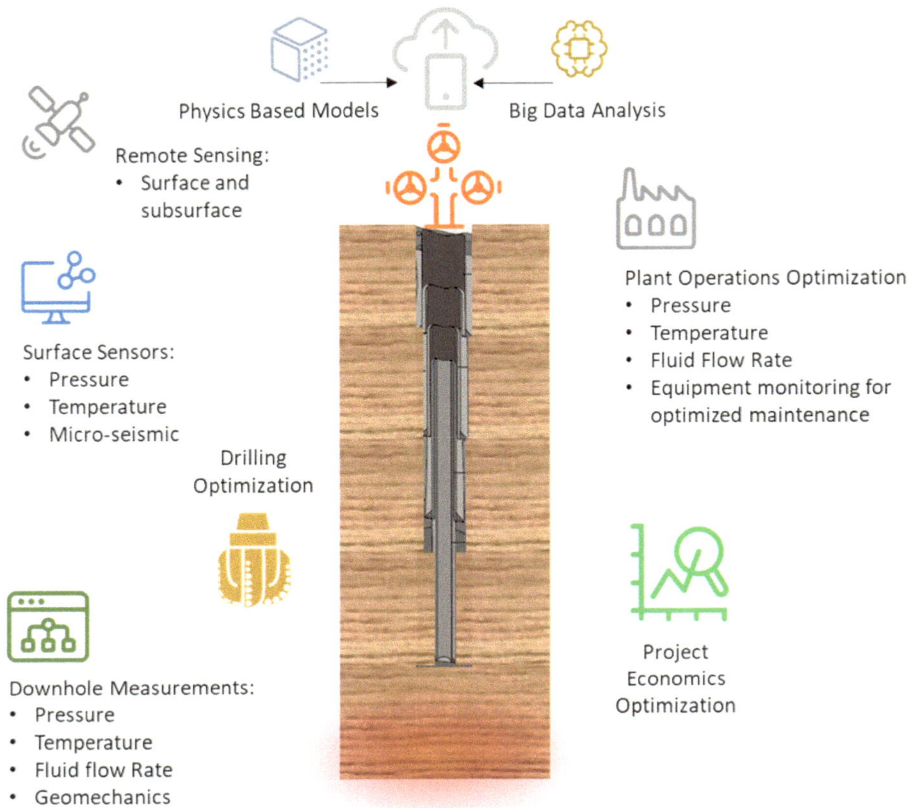

FIG. 35 Automation opportunities for geothermal development and operation.

In summary, through the strategic implementation of automated systems, we will achieve streamlined processes, heightened operational efficiency, and a reduction in human intervention in routine tasks. Geothermal development will benefit from rapidly harnessing the power of automation in various facets, including drilling operations, reservoir monitoring, and power plant control and operations systems.

9 Summary and outlook

In this chapter, we have outlined the opportunity and necessity for adding geothermal energy to the existing global energy mix and to do so at scale—from a climate point of view as well as from the point of view of providing access to affordable baseload power to a larger percentage of the global population, where hundreds of millions of people have neither access to electricity nor to clean cooking fuels.

The benefits of geothermal energy are numerous: clean, baseload, dispatchable, and distributable energy that can be made available virtually anywhere in the world, provided some of the new concepts, which we have reviewed in detail in this chapter, are implemented. Our technical review presented a wide variety of geothermal systems: hydrothermal systems, EGS, and CLGS.

To deploy geothermal energy successfully and at scale, both the subsurface, i.e., the hot rock, fluid flow pathways, and heat recovery mechanisms, and the framework settings, i.e., grid connections, off-take contracts, power generation, and district heating, as well as the regulatory framework must be well understood.

Technical and commercial de-risking is essential to project success—both from an economic, environmental, and socioeconomic point of view. Positive public opinion, "winning hearts and minds," will be important to the growth of the sector—and transparency, education, and open dialogue will prove to be key elements to success.

Geothermal energy still is very much a niche energy source—restricted geographically because of the legacy focus on hydrothermal systems—and very much seen to be in the academic arena, rather than in the commercial arena. Therefore, the transfer of technology, expertise and best practices from other industries will be essential to growth.

For this, we have reviewed a number of geophysical and geological exploration and subsequent multidisciplinary data integration, which have proven to be successful in the oil and gas industry. The similarities between oil and gas exploration, de-risking, and production, on the one hand, and the same process steps for geothermal energy, on the other hand, are manifold. Utilizing proven technology and expertise from oil and gas in the geothermal environment will lead to a faster and more widespread adoption and development of geothermal energy. Likewise, transitioning oil and gas expertise into the geothermal realm will provide new and meaningful employment opportunities for oil and gas experts.

Social impact assessment

Guidelines and Principles For Social Impact Assessment, Prepared by The Interorganizational Committee on Guidelines and Principles for Social Impact Assessment, May 1994; U.S. Department of Commerce; National Oceanic and Atmospheric Administration; National Marine Fisheries Service.

Listed alphabetically, the paper was prepared by Burdge, Fricke, Finsterbush, Freudenburg, Gramling, Holden, Llewwellyn, Petterson, Thompson and Williams. Comments were received from Hobson Bryan, Tom Greider, Lambert Wenner, and Richard Stoffle. A previous draft of the paper was given with the title, "Social Impact Assessment: Principles and Standards for U.S. Federal Agencies and U.S. Sponsored Donor Agencies," as a parallel plenary session at the 13th Annual Meeting of the International Association for Impact Assessment, Qian He Hotel, Shanghai, China, June 12–15, 1993 and included in the Abstracts; (p. 15–16).

Reference

Augustine, C., Fisher, S., Ho, J., Warren, I., Witter, E., 2023. Enhanced Geothermal Shot Analysis for the Geothermal Technologies Office. National Renewable Energy Laboratory. Technical Report: NREL/TP-5700-84822, January 2023.

Blackwell, D., Richards, M., Frone, Z., Batir, J., Ruzo, A., Dingwall, R., Williams, M., 2011. Temperature-at-depth maps for the conterminous U.S. and geothermal resource estimates. In: Geothermal Resources Council Transactions. Geothermal Resources Council, San Diego, CA, pp. 1545–1550. https://publications.mygeoenergynow.org/grc/1029452.pdf.

Soyer, W., Mackie, R.L., Hallinan, S., Pavesi, A., Nordquist, G., Suminar, A., Intani, R., Nelson, C., 2018. Geologically consistent multiphysics imaging of the Darajat geothermal steam field. First Break 36 (6), 77–83. https://doi.org/10.3997/1365-2397.n0102.

CHAPTER

3

Geothermal well drilling: The drilling rig

Aditya Sharma, Khizar Abid, and Catalin Teodoriu

MPGE, University of Oklahoma, Norman, OK, United States

1 Overview of drilling costs

The "upfront" cost of geothermal plants is much higher than that in conventional power plants. Moreover, geothermal projects may only break-even many years after the wells have been completed, whereas drilling for hydrocarbons yields a fast cash return. The cost of drilling a well is a key concern when determining the economic viability of a geothermal plant, because it is a substantial part of the total value of developing any geothermal project (Abid et al., 2024). Thorhallsson and Sveinbjornsso (2012) estimated that about 40% of the budget of the project goes to the drilling of the geothermal well and completion operations. Numerous factors, like the depth of the geothermal reservoir and the characteristics, composition, and hardness of the rock formation, impact the cost associated with these wells. Additionally, these variables affect the diameter of the well, number of casing installed, and penetration rate, thereby influencing the total duration and cost in completing the well. Moreover, geographical location is another factor that affects the costs involved. Figs. 1 and 2 show the breakdown of the costs involved in a geothermal project in Iceland and Indonesia, respectively.

Over the years, oil and gas drilling has been optimized, resulting in affordable drilling costs in terms of dollars per foot. However, the tubulars used for complex well designs may double the cost of the well. The diameter of the production casing in a geothermal well, i.e., the last casing string, is usually 9 5/8 in. (Teodoriu and Falcone, 2008). Hence, a larger surface casing is required to accommodate these large-diameter production casings. In the United States and Japan, it is common to use a 13 3/8 in. diameter surface casing (Hidashi, 2000). While in Europe, surface casings with a diameter of 18 5/8 in. or greater are used for most wells drilled to depths

Geothermal Energy Engineering
https://doi.org/10.1016/B978-0-443-21662-6.00011-6

Copyright © 2025 Elsevier Inc. All rights are reserved, including those for text and data mining, AI training, and similar technologies.

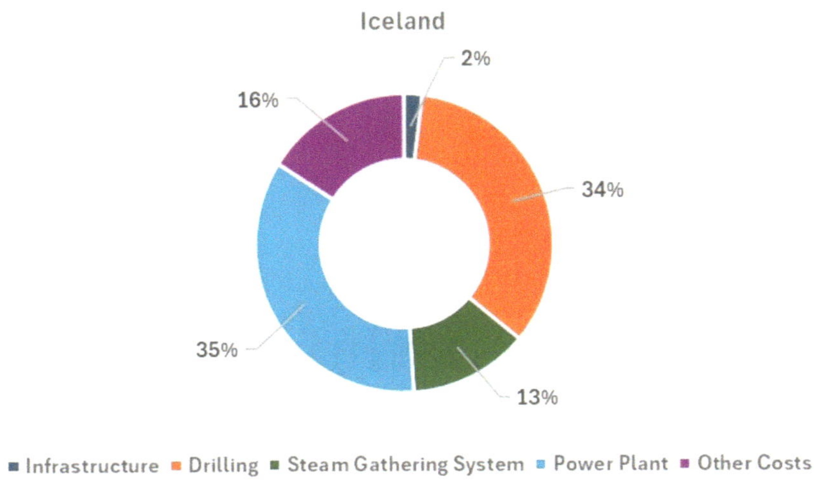

FIG. 1 Breakdown of the development of a geothermal project in Iceland (Gehringer and Loksha, 2012).

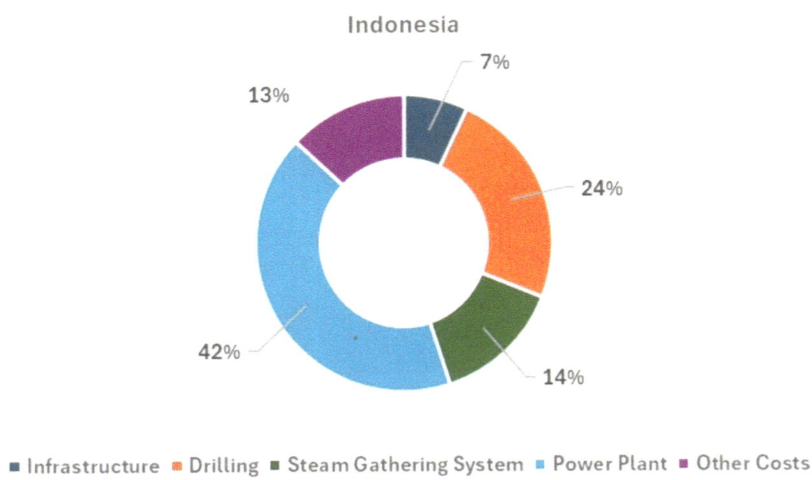

FIG. 2 Breakdown of the development of a geothermal project in Indonesia (IRENA: Renewable Power Generation Cost in 2014/2015).

greater than 4000m (Tenzer, 2001). The need for production casings with a larger diameter arises from the volume of fluids (and the accompanying heat) that must be extracted from geothermal wells to make the project economically feasible. In the case of bigger power systems, a minimum production casing diameter of 13 3/8in. is necessary, but as the depth of the well increases, these diameter specifications significantly impact the overall expenses associated with the well.

Therefore, drilling engineers face a fundamental quandary when dealing with larger casing strings and deeper drilling objectives: how can expenses be minimized while ensuring a safe and precise drilling to reach the target? One solution involves utilizing nontraditional drilling equipment, a topic that will be explored in the subsequent section.

2 The drilling rig

A drilling rig is an integrated system of machinery used for creating holes or wells in the earth's surface, typically for the purpose of extracting natural resources. Rotary drilling is a widely favored and effective method employed for boring or drilling a hole in the earth's surface, allowing access to subsurface hydrocarbon and geothermal resources. Typically, a drilling rig comprises of five main subsystems: pipe rotation, drilling fluid circulation, hoisting, power supply, and blow-out preventer. These subsystems are engineered to function in conjunction with each other.

Drilling rigs can be categorized as either mobile (able to be transported on roads, trailers, or trucks) or constructed on-site. The mobility of a traditional drilling rig is determined by the maximum hook load it can endure. Modern generations of drilling rigs aim to reduce the time needed for setup and dismantling, but this often results in the need for transporting heavy individual rig components.

All drilling rigs are characterized by five main systems: rotary, hoisting, circulation, power, and well control system. Modern drilling rigs have mostly kept the five systems in a conventional way: rotary is, today, generated by top drive, hoisting uses the block and tackle system, while circulation relies on triplex piston pumps. To achieve maximum efficiency, conventional rigs have been improved in terms of higher torque and hook load capacities as well as incorporating high level of sensors and measurements.

While advanced technologies enable the targeting of hydrocarbon reservoirs from considerable distances, thereby avoiding environmentally sensitive areas, drilling geothermal wells in urban regions necessitates the use of specialized rigs that are compatible with the surrounding environment. To achieve this, an adaptation of existing drilling equipment is required, which can be referred to as an "urbanized drilling rig." These urbanized rigs should focus on reducing noise, occupying minimal space, and having an aesthetically pleasing appearance, in order for the local community to accept their presence in the vicinity. Furthermore, geothermal drilling implies deeper and larger wells, which requires higher hook loads and faster tripping capabilities.

Table 1 shows a comparison of the conventional and unconventional rigs for oil and gas versus geothermal (Bello et al., 2013).

2.1 From conventional drilling rigs to urban geothermal applications

For geothermal projects, such as HDR drilling, conventional rigs commonly used in the oil and gas industry have been proven to be the most efficient choice, due to their ability to reach the desired drilling depth rapidly. However, in periods of high oil and gas prices, there is an increased demand for these rigs, leading to higher day rates, which renders the use of oil and gas drilling rigs for geothermal purposes economically unfeasible. In response to this challenge, several service companies have designed their rig fleets with specialized geothermal drilling capabilities.

TABLE 1 Drilling rig parameters as a comparison between conventional, improved, and unconventional rigs.

Technical parameters	Conventional rig	Improved conventional rig	Unconventional rig
Drilling depth (m)	<4000	>4000	>5000
Hook load capacity	150–400 t	200–450 t	225–370 t
Power packs	Diesel–electric driven	Electrically driven (including AC–AC technology)	Electrically driven (AC generators and drivers) or fully hydraulic transmission
Structural height	Double or triple stand	Double or triple stand	Single or double stand
Drilling rig automation	Absent/low	Present/high	Present/high
Automated pipe handling equipment	Absent (Keller spinner)	Present	Present
Monitoring system	Absent	Present	Present
Noise pollution	High	Low	Low
Carbon emission	High	Low	Low
Automatic braking system	Absent	Present	Present
-. Rig up-/downtime	Longer time		Shorter time
-. Mobilization cost	Higher cost		Lower cost
-. Rig move for Transportation	35–40 loads		22–30 loads
Drilling rig pipe used	3 drill pipes		1 or 2 drill pipes
Braking system	Mechanical main brake and hydraulic auxiliary brake	Mechanical disk main brake and hydraulic auxiliary brake	Regenerative AC or hydraulic brake

3 ITAG Rig 40

The ITAG Rig 40 (Gutsche, 2005), depicted in Fig. 3, offers faster transportation and setup compared with equivalent oil rigs. Originally designed for easy transport through narrow streets in southern Germany and northern France, as a dedicated rig for underground storage activities, its modular design allows adaptability for geothermal drilling applications. Despite its relatively compact footprint (3500 m^2), ITAG Rig 40 compares favorably with traditional mobile rigs in terms of its maximum hook load (2241 kN) and mud tank capacity.

FIG. 3 ITAG Rig 40. *Credit: ITAG.*

4 Drilltec rig VDD370

This rig has a hydraulically driven top drive and pipe handling systems. Its 3700 kN (831,793 lbf) maximum hook load capacity makes it suitable for deep drilling applications. The power packs, which are electrically driven, provide hydraulic energy, offering the flexibility to draw power from a city network or, in cases of insufficient electrical supply, employ a diesel motor–driven generator. The compact footprint and relatively low height of just 31 m make it advantageous for urban environments. The rack and pinion–type rig is designed with noise protection in mind, with the gears fully insulated to absorb the noise generated by the top drive, equipped with six hydraulic motors for movement and four for rotation. Additionally, most noise-emitting equipment, such as pumps, generators, and power packs, are enclosed in insulated containers (Fig. 4).

FIG. 4 Drilltec Rig VDD370. *Source: Drilltec.*

5 Herrenknecht Vertical Terra Invader 350 (TI-350) rig

Specifically designed for urban drilling tasks, the Terra Invader 350 (TI-350) rig built by Herrenknecht Vertical, as depicted in Fig. 5, excels in terms of noise reduction. The primary innovation in this rig's design is its hoisting system, which is hydraulically driven through linear hydraulic cylinders and includes noise-insulated containers for the main drivers and mud pumps. It has been reported that the expected noise levels at a distance of ~150m from the rig are comparable to that of a typical household TV or radio playing at a normal volume. The rig has a hook load capacity of 3487kN (784,000lbs) or 4085kN (903,895lbf) and a derrick height of 51.8m. To achieve optimal drilling performance and safety, the new rig design incorporates extensive automation of processes, minimizing manual intervention. This innovative deep drilling rig has been successfully put into use in multiple locations in southern Germany (Mauerstetten, Dürrnhaar, Kirchstockach, Kirchweidach, and Traunreut) for geothermal applications. Compared with the traditional drilling systems, the Herrenknecht Vertical rig incorporates various innovative features that bring several technical advantages and exceptional drilling capabilities. One notable innovation is the "X-Y-Stepping" system, where hydraulic cylinder supports can lift and shift the entire drilling rig by several meters. This enables the rig operator to drill multiple boreholes, creating multiwell pads, which eliminates time-consuming and expensive rig disassembly and reassembly procedures while minimizing the ecological impact. The hydraulic pipe handling system offers additional benefits as it automates the guidance of drill strings and casing from the pad to the drilling location. This significantly enhances workplace safety throughout the drilling process, as there is no need for personnel to remain in hazardous areas. The rig operator can access real-time monitoring of all the processes through cockpit displays, and data are recorded for future analysis, forming the basis for continuous project monitoring and optimization.

FIG. 5 Herrenknecht Vertical Terra Invader 350 (Teodoriu et al., 2009).

6 Bauer TBA 440 M2

The TBA 440 M2 (Fig. 6) is a deep drilling unit that follows a modular design and is constructed on safety, environmental protection, efficiency, reliability, and quality principles. The rig boasts a hook load capacity of up to 4384 kN (970,033) and a height of 46 m. In addition to its automated vertical pipe handling system, the TBA 440 M2 offers a 2000 HP drawworks, a 500 t electric top drive, advanced hands-free technology, and modern automated systems. The modular design enhances efficiency and simplifies transportation. It also features an integrated walking system, allowing on-site mobility, even when the pipe storage system is fully loaded. Automating repetitive operational sequences, ensuring redundant key systems, and improving rig availability significantly enhance the efficiency and overall drilling performance of the rig. Additionally, eliminating tasks performed at unsafe heights (e.g., no fingerboard, self-erecting systems), introducing hands-free or remotely controlled operations, implementing anti-collision systems, and reducing required manpower collectively contribute to achieving the highest level of possible safety. Furthermore, the smaller footprint of the unit (24.9 m × 18.8 m), compared with conventional rigs with similar capacity, leads to cost savings in well site preparation and reduces environmental impact.

FIG. 6 Bauer TBA 440 M2. *Source: bauer.de.*

7 Huisman LOC 400

The need for swift operations, adaptability, safety, and efficient rig relocations to various global sites prompted the creation of the Huisman LOC 400. It has a block capacity of 3108 kN (698,865 lbf) and a mast height of 38.1 m. Among the key attributes of the LOC 400, its compact dimensions and the capability to disassemble the entire rig into 17 primary modules stand out. In a span of 48 h, which includes minimal transportation time, a crew can dismantle the entire rig and reassemble it at a different location. Another advantage of the rig is its small footprint, roughly 65 × 30 m, making it well suited for drilling operations in constrained urban areas. When connected to the power grid, the rig has a low noise average of under 50 db at 300 m. The LOC 400 offers an efficient, secure, and eco-friendly alternative to traditional drilling rigs for drilling geothermal wells. Its minimal footprint and containerized design ensure outstanding mobility, making it a perfect choice for operations in densely populated urban regions. The LOC 400 has already been employed in several notable geothermal projects in The Netherlands, including a geothermal well in the heart of The Hague, Pijnacker, and two geothermal wells in Bergschenhoek (Fig. 7).

FIG. 7 Huisman LOC 400 in The Hague, The Netherlands. *Credit: Huisman.*

8 Precision Drilling Super Triple 1200 (ST 1200)

The ST 1200 ensures secure and high-performing operations by incorporating automated elements like mechanized pipe handling and an iron roughneck. Its hydraulic-driven BOP handling system, single-piece substructure, integrated top drive, and modular design contribute to a safe and efficient rig setup, teardown, and relocations. It can transition between wells without the need to disconnect from the backyard complex, complete with a full setback and transfer tank. This system permits the backyard to be situated up to 300 ft. away from the rig without any decrease in efficiency. This rig has a hook load capacity of 2224 kN (500,000 lbf) and a height of 41 m. Precision Drilling collaborated with Eavor Technologies to drill the world's first closed-loop geothermal system in Alberta, Canada.

9 NOV Rapid Rig

The NOV Rapid Rig is a high-efficiency land rig designed for "single" drilling, offering optimal speed, safety, and performance in a compact, road-legal drilling package. It has a hook load capacity of 2491 kN (551,155 lbf) and is designed for shallow-to-moderate-depth wells. The mast height is 24 m, when fully scoped for operation. The Rapid Rig's smaller dimensions and self-deploying structure make transportation and on-site setup quicker. This rig design decreases crew requirements and the risk of accidents and creates a comfortable and efficient working environment. The Amphion drilling control system automates drill floor operations, providing the driller with greater control and awareness of the entire drilling process. Additionally, the Rapid Rig's compact components and primary AC-powered systems contribute to reduced environmental impact at the drilling site (Fig. 8).

FIG. 8 NOV Rapid Rig. *Credit: NOV.*

10 Essential criteria for an urban drilling rig

To mitigate the adverse effects of a drilling rig on the surrounding urban architecture, the most effective approach is to lower the height of the drilling rig to facilitate better blending with the surroundings. Fig. 9 illustrates a comparison of heights between a traditional deep drilling rig and a contemporary or unconventional drilling rig, which significantly reduces the need for excessive overhead working space.

In order to meet the evolving demands of urban drilling, the next generation of drilling rigs must incorporate the following design features:

- Modular construction: This design allows for easier mobilization and transportation through narrow city streets. Many rig components are housed in small containers, reducing noise emissions through sound insulation.
- Low exterior noise: Reducing noise levels is imperative, which can be achieved through lowering the height of the rig by using single rather than triple drilling rigs and employing non-conventional hoist systems, like rack and pinion or linear hydraulic motors.
- High hook load capacity: Despite their smaller size, these drilling rigs must have the ability to handle substantial loads that can be made possible by non-conventional technologies.
- Camouflage capabilities: The rigs should be easily concealed for extended drilling projects.
- Low-emission smart power supply: Utilizing on-site diesel generators or connecting to the city's electrical power grid while maintaining low noise, vibrations, and exhaust gas emissions.

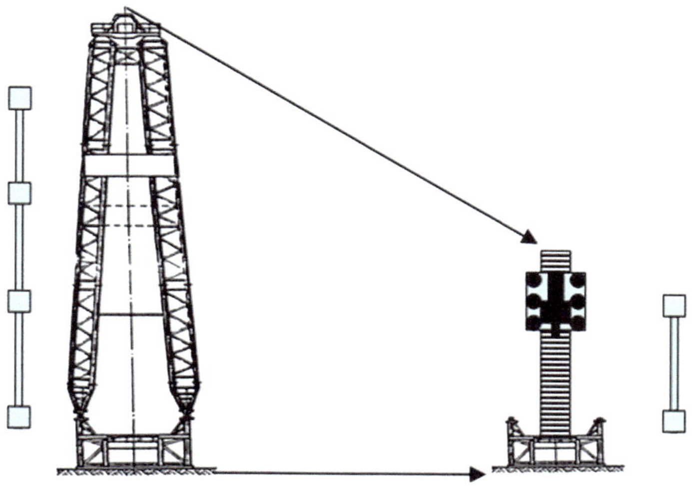

FIG. 9 Traditional drilling rig on the left (measuring 45–55m in height and designed to accommodate three drill pipes) compared with an unconventional drilling rig on the right (standing at 15–18m in height and suitable for a single drill pipe; Teodoriu et al., 2009).

- Closed-loop spill and mud processing: This system prevents contamination of the rig site.
- Compact rig size: Achieved through the integration of automated processes and computer-controlled rig floor equipment.
- Enhanced safety systems: Including non-conventional BOP systems to accommodate the low substructure.

Although most non-conventional rigs meet these criteria, it is worth noting that some of them have slower tripping times than conventional rigs. This may result in lower daily rates, potentially affecting the profits of rig owners. Unfortunately, no statistical information is available to compare the daily rates of non-conventional rigs.

References

Abid, K., Velasquez, A.F.B., Sharma, A., McSheridan, A.N., Sirivastava, S., Teodoriu, C., 2024. Risk assessment through feature, event, and process for repurposing suspended oil and gas wells for geothermal purposes. Renew. Energy.

Bello, O., Teodoriu, C., Falcone, G., 2013. Performance Assessment of Drilling Rigs Used in Geothermal Applications. Pisa, Italy, European Geothermal Congress.

Gehringer, M., Loksha, V., 2012. Geothermal Handbook: Planning and Financing Power Generation. The World Bank Group, Washington, USA, pp. 16–100.

Gutsche, W., 2005. A new generation AC-driven light land rigs. In: Paper Presented at Institute of Petroleum Engineering. TU Clausthal, Germany.

Hidashi, J., 2000–2015. High angle directional drilling at Takigami geothermal field. In: IRENA: Renewable Power Generation Cost in 2014. Proceedings World Geothermal Congress. International Renewable Energy Agency.

Tenzer, H., 2001. Development of hot dry rock technology. Geo-Heat Center Q. Bull. 22 (4), 14–22.

Teodoriu, C., Falcone, G., 2008. Comparing completion design in hydrocarbon and geothermal wells: The need to evaluate the integrity of casing connections subject to thermal stresses. Geothermics 38 (2), 238–246.

Teodoriu, C., Falcone, G., Fichter, C., 2009. Drilling beyond oil and gas: a discussion about how technology will change to enhance drilling in urban areas. In: DGMK/ÖGEW-Frühjahrstagung 2009, Fachbereich Aufsuchung und Gewinnung. DGMK, Celle.

Thorhallsson, S., Sveinbjornsso, B.M., 2012. Geothermal Drilling Cost and Drilling Effectiveness.

CHAPTER

4

Geothermal well construction and completion: Overview of casing materials, sizes, depth, and solutions

Katya Waters, Sayyam Chopra, Khizar Abid, and Catalin Teodoriu

MPGE, University of Oklahoma, Norman, OK, United States

1 Introduction

The first step in the process of classifying a geothermal well is to understand what geothermal energy is. It is a type of renewable energy that comes from the earth's crust and is a thermal energy that is stored within the earth's subsurface rocks and fluids. Geothermal energy is produced by, first, drilling a well to a certain depth from which the heat can be gathered with the help of a working fluid. The following are different geothermal systems that exist in the subsurface (Abid et al., 2022):

1. Hydrothermal systems, where a high volume of water is present.
2. Geopressurized systems, in which the pressure of water is higher than the depth of their location. The temperature in these systems lies in the middle range.
3. Magnetic systems, which have a temperature that ranges from 600°C to 1400°C (112–2552 °F).
4. Hot dry rocks, a system in which the volume of fluid and the permeability of the formation are low. The temperature in these systems ranges from 200°C to 300°C (92–572 °F).
5. Enhanced geothermal systems (EGS), which are gaining importance and have characteristics that lie between HDR and hydrothermal systems. In the EGS, water is pumped into the HDR system through injection wells, and then the fluid is converted into steam/hot water, where it is returned to the surface through a production well. The heat is then extracted at the surface, and the rest of the fluid is reinjected back into the system.

Geothermal Energy Engineering
https://doi.org/10.1016/B978-0-443-21662-6.00010-4

Copyright © 2025 Elsevier Inc. All rights are reserved, including those for text and data mining, AI training, and similar technologies.

The energy harnessed from these resources may be used for domestic (heating home) and commercial purposes (power generation). Many countries utilize geothermal energy for power generation, at the same time promoting the use of renewable energy (Fig. 1). It is also worth noting that the quality of the energy depends upon the heat collected at the wellhead.

The major challenges faced in geothermal wells are the high-temperature gradient, depths, casing fatigue, high flow rates, selection of casing and tubing materials, issues in well integrity, and coupling preference for tubing/casing connection, which can affect the efficiency of the geothermal project. An attempt has been made to define and classify geothermal wells based on their completion, since these components, if properly analyzed, can save a lot of costs in the construction of a geothermal system. A set of papers concerning the completion of geothermal wells was analyzed, and different parameters were recorded. The information on the completion with respect to different geothermal wells presented in this chapter is as follows:

1. Total vertical depth versus open hole length
2. Well type with casing type/size and material
3. Well type with liner diameter and material
4. Well type and depths
5. Well type and temperature
6. Well type with flow rates

Knowledge on these aspects could help mitigate the issues on completion and understand the common characteristics of a well. Moreover, this characterization could assist in designing the completion for different geothermal wells.

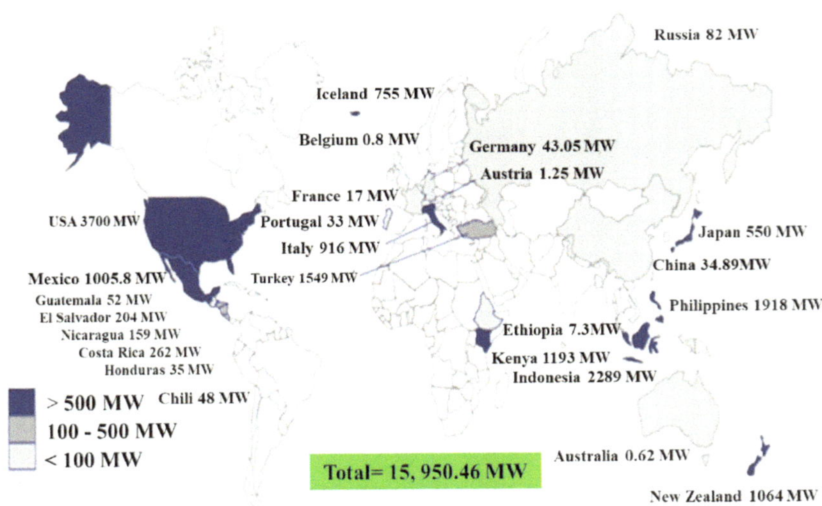

FIG. 1 Geothermal power generation in the world (installed capacity; Benti et al., 2023). Map lines delineate study areas and do not necessarily depict accepted national boundaries.

2 State of the art

In geothermal wells, the production is mainly from the casing instead of the tubing, due to the **high flow rates**. Therefore, the **diameter** of the casing has to be considered during the completion design phase. As for the geothermal wells, the diameters of the surface casing and bottom hole are comparatively larger than those of conventional wells, since the flow rates and temperatures are high. The geometry of the bottom hole in geothermal wells is also studied because issues with temperature and flow begin at the bottom. In that respect, the selection of liners on different types of well will be presented. A liner is used at the bottom of the well in order to prevent collapse of the well due to the presence of unconsolidated formation.

The **depth of the well** is another criterion upon which the completions of the geothermal well are analyzed. **Total vertical depth (TVD) versus open hole length** was examined to distinguish the depths between wells. Another important aspect to consider is the source of the geothermal energy. The completions are designed according to the type of **well,** which is dependent on the depth. Thus, the depth of the well is studied based on the type of well. Generally, higher-grade materials are considered in the completion of a well due to high flow rates, frictional losses, temperature gradient, well integrity, and casing fatigue. Therefore, the **grade of material** chosen for casing, liners, and anchors is another criterion used for the evaluation of the completions.

The existing classification of wells based on the geothermal system does not provide enough information on the parameters of the completion which are critical to the design of the well. Because of this, the completions must be characterized separately to understand the design of the well (Fig. 2).

After an initial review of the selected papers, the data presented in each paper were used to create a series of spreadsheets that compared the results. The main spreadsheet included the title of the paper; type of well, whether the well was vertical or deviated; TVD; BHP; temperature; type of completion; pH of the completion fluid; and information on conductor, surface, intermediate/tieback, production, and hang down/anchor casings, liner, tubing, packer, and pump. From this initial spreadsheet, the data were separated and grouped with respect to flow rate, TVD and open hole depth, diameter of the surface and bottom, size and length of casing, depth, material, tubing, and liner. These categories were compared with each other based on the type of well from a completion standpoint. After the spreadsheets were created and analyzed, it was decided to classify the geothermal wells based on their type, which included hot dry rock (HDR), hot water, supercritical steam, and dry steam. The data were also used to prepare various charts and graphs to further compare, differentiate, and classify the wells.

Table 1 shows the TVD and open hole length with respect to the type of well, including the temperature and depth of the liner for each study. Table 2 analyzes the diameter of the surface and bottom hole of different types of well.

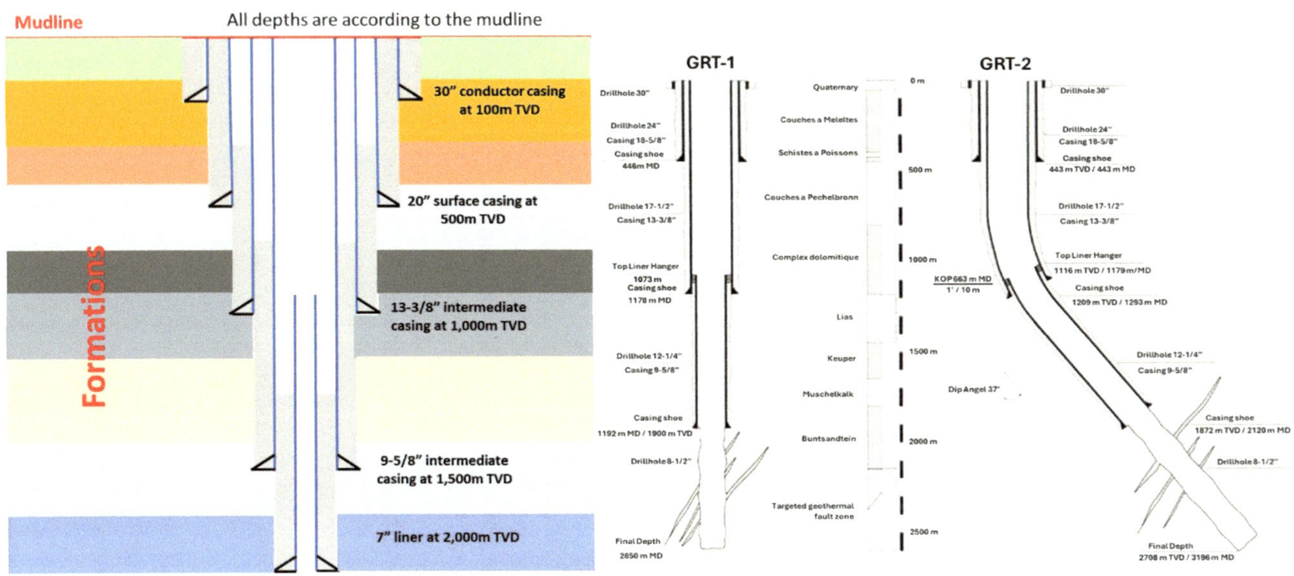

FIG. 2 Different completions of a geothermal well (Baujard et al., 2017).

TABLE 1 TVD and open hole length with respect to the type of well, including temperature and liner depth.

Reference	TVD (ft)	Open hole length (ft)	Well type	Temperature (°F)	Liner depth (ft)
Nicholson and Snyder (1982)	6000	3000	Dry steam	500	3000
Snyder (1979)	7000	3000	Dry steam	200–400	5800–7000
Snyder (1979)	7500	1500	Dry steam	200–400	1300–5300; 5100–7500
Snyder (1978)	7000	2000	Dry steam	<500	3000
Snyder (1978)	9619	2431; 1604	Hot water	400	8973–9581
Reinsch et al. (2015)	14,437	2432	Hot water	293	7814–12,723; 12,523–14,400
Baujard et al. (2017)	8464.567	1463.25	Hot water	338	n/a
Sveinbjornsson and Thorhallsson (2013)	7332.677	572	Hot water	n/a	n/a
Marbun et al. (2019)	8113.517	1443.57	Hot water	536–626	7995.407
Petrică (2017)	10,171	3625.328	Hot water	178	9 5/8, 3625.32–6840.55; 7, 6840.55–7644.35
Karlsdottir et al. (2014)	6889	285	Supercritical steam	248	n/a
Karlsdottir et al. (2019)	15,255.91	360	Supercritical steam	800	n/a
Carpenter (2021)	8500	2800–3000	HWR low pressure	620	2850–6454; 6200–8500
Akhmadullin (2016)	14,720	3510	Downhole heat exchanger	258	n/a

TABLE 2 Diameter of the surface and bottom hole of different types of well.

Reference	Well type	Surface casing dia. (in.)	Bottom hole/liner dia. (in.)
Reinsch et al. (2015)	Hot water	18.625	5–7
Carpenter (2021)	Hot water	13.375	7–9.625
Petrică (2017)	Hot water	13.375	7–9.625
Nicholson and Snyder (1982)	Dry steam	13.375	6.625–9.625
Snyder (1979)	Dry steam	13.375	7
Holligan et al. (1989)	Dry steam	24	9.625 (hangdown casing)
Okech et al. (2015)	Dry steam	7	5.5
Snyder (1978)	Dry steam	(13.375)–20	7.625–9.625

Continued

TABLE 2 Diameter of the surface and bottom hole of different types of well—cont'd

Reference	Well type	Surface casing dia. (in.)	Bottom hole/liner dia. (in.)
Karlsdottir et al. (2014)	Supercritical steam	32.5	9.625
Karlsdottir et al. (2019)	Supercritical steam	22.5	7
Teodoriu and Falcone (2008)	HDR	9.625	7
Carden et al. (1985)	HDR	9.625	4.5
Falcone and Teodoriu (2008)	HDR	18.625	7
Omrani et al. (2021)	HDR	18.625	6.625–9.625

The next parameter analyzed was the depth of the well, which is shown in Table 3, that has been divided into three groups, i.e., ≤4000 ft, 4001–9000 ft, and >9001 ft. Table 4 presents the length of the surface, production, and intermediate casings in different types of well. Table 5 is organized based on liner OD, liner material, and coupling connection used in different types of well.

The type of well and the different casing materials used in the completion of a geothermal well are grouped in Table 6. Analysis of flow rate must be critically examined due to high frictional losses in the system. Therefore, the type of well and the analyzed flow rates are given in Table 7.

TABLE 3 Type of well arranged with respect to the depth.

Reference	Depth (ft)	Well type
Nicholson and Snyder (1982)	4000	Hot water
Petrică (2017)	10,171	Hot water
Baujard et al. (2017)	8464.567	Hot water
Sveinbjornsson and Thorhallsson (2013)	7332.677	Hot water
Marbun et al. (2019)	8113.517	Hot water
Snyder (1978)	9619	Hot water
Snyder (1978)	16,500	Hot water
Reinsch et al. (2015)	14,437	Hot water
Nicholson and Snyder (1982)	9000	Dry steam
Snyder (1979)	7000	Dry steam
Snyder (1979)	7500	Dry steam
Holligan et al. (1989)	8000	Dry steam
Okech et al. (2015)	7545	Dry steam

TABLE 3 Type of well arranged with respect to the depth—cont'd

Reference	Depth (ft)	Well type
Snyder (1978)	7000	Dry steam
Karlsdottir et al. (2014)	6889	Supercritical steam
Karlsdottir et al. (2019)	15,255.91	Supercritical steam
Carden et al. (1985)	15,000	HDR
Pettitt (1979)	10,053	HDR
Falcone and Teodoriu (2008)	9187	HDR

TABLE 4 Type of well and length of the surface, production, and intermediate casings.

Reference	Well type	Surface casing length (ft)	Production casing length (ft)	Intermediate casing length (ft)
Snyder (1979)	Dry steam	3000	6000	
Snyder (1979)	Dry steam	1500		
Holligan et al. (1989)	Dry steam			
Nicholson and Snyder (1982)	Dry steam	3000		
Snyder (1978)	Dry steam	2000		
Nicholson and Snyder (1982)	Hot water	1200		
Nicholson and Snyder (1982)	Hot water	1500		
Petrică (2017)	Hot water	3625		
Reinsch et al. (2015)	Hot water	2432	7814	
Baujard et al. (2017)	Hot water	1463.25	6305	3864.83
Sveinbjornsson and Thorhallsson (2013)	Hot water	572	2624	
Marbun et al. (2019)	Hot water	1443.57		2739.20
Snyder (1978)	Low pressure	2000		
Snyder (1978)	Low pressure	2431		6420
Snyder (1978)	Low pressure	1604		2535
Carpenter (2021)	Low pressure	2800–3000	2800; 3200–4100	3000
Karlsdottir et al. (2014)	Supercritical steam	285		833; 2575
Karlsdottir et al. (2019)	Supercritical steam	360	5905	1476

TABLE 5 Type of well and liner data.

Reference	Well type	Liner (OD; in.)	Liner material	Liner connection
Nicholson and Snyder (1982)	Hot water	8.625		
Nicholson and Snyder (1982)	Hot water	6.625		API Buttress Joint Flush
Snyder (1978)	Hot water	5		
Petrică (2017)	Hot water	9.625		
Marbun et al. (2019)	Hot water		L-80	
Carpenter (2021)	Hot water	9.625		
Carpenter (2021)	Hot water	7		
Reinsch et al. (2015)	Dry steam	9.625	N89	API Buttress or premium connection
Reinsch et al. (2015)	Dry steam	7.625		
Reinsch et al. (2015)	Dry steam	7.625		
Nicholson and Snyder (1982)	Dry steam	9.625		
Snyder (1979)	Dry steam	7	N80	Hydril Seu
Karlsdottir et al. (2014)	Superheated steam	9.625	K55, BTC	
Karlsdottir et al. (2019)	Supercritical steam	7		
Falcone and Teodoriu (2008)	HDR	7		API Buttress or premium
Omrani et al. (2021)	HDR	8.625		API Buttress or premium
		6.625		
		9.625		
Snyder (1979)	HDR	8.625	K55 K55	API Buttress Joint Flush
Carden et al. (1985)	HDR	4.5	N-80	
Teodoriu and Falcone (2008)	HDR	7		

TABLE 6 Type of well and casing data.

Reference	Well type	Surface casing material	Production casing	Anchor casing	Intermediate casing
Petrică (2017)	Hot water	K-55			
Marbun et al. (2019)	Hot water	K-55			K-22
Snyder (1978)	Dry steam	K-55			
Holligan et al. (1989)	Dry steam	K-55	K-55		K-55
Snyder (1979)	Dry steam	K-55	N-80		
Karlsdottir et al. (2014)	Superheated steam	X56, welded			K55, welded + K55, BTC
Karlsdottir et al. (2019)	Superheated steam	X 56	K55	T 95	X-56
Carden et al. (1985)	HDR		P110		

TABLE 7 Type of well and flow rate data.

Reference	Well type	Casing or liner at the bottom	ID (in.)	Area (in.^{2})	Max flow rate (in.^{3}/s)
Nicholson and Snyder (1982)	Hot water	Liner	8.625	58.43	6900.73
Nicholson and Snyder (1982)	Hot water	Liner	6.625	34.47	4071.44
Snyder (1978)	Hot water	Liner	9.625	72.76	8593.66
Snyder (1978)	Hot water	Liner	7.625	45.66	5393.32
Reinsch et al. (2015)	Hot water	Liner	7	38.48	4545.41
Reinsch et al. (2015)	Hot water	Liner	5	19.63	2319.08
Petrică (2017)	Hot water	Liner	9.625	72.76	8593.66
Teodoriu and Falcone (2008)	HDR	Liner	7	38.48	4545.41
Carden et al. (1985)	HDR	Liner	4.5	15.90	1878.46
Pettitt (1979)	HDR	Production casing	7.625	45.66	5393.32
Okech et al. (2015)	HDR	Liner	5.5	23.76	2806.09
Falcone and Teodoriu (2008)	HDR	Liner	7	38.48	4545.41
Omrani et al. (2021)	HDR	Liner	9.625	72.76	8593.66
Nicholson and Snyder (1982)	Dry steam	Liner	9.625	72.76	8593.66
Snyder (1979)	Dry steam	Liner	7	38.48	4545.41
Snyder (1979)	Dry steam	Liner	6.625	34.47	4071.44
Holligan et al. (1989)	Dry steam	Production casing	13.375	140.50	16,594.50
Karlsdottir et al. (2014)	Supercritical steam	Liner	9.625	72.76	8593.66
Karlsdottir et al. (2019)	Supercritical steam	Liner	7	38.48	4545.41

From the tables that have been presented, different observations were made and are presented in the following section.

2.1 Well type and (open hole length vs total vertical depth)

The first plot generated from the data compared the TVD and open hole length at the bottom of the well. One plot is created for each type of well. For the dry steam wells (Fig. 3), a significant difference between TVD and open hole length can be noted. For the hot water wells (Fig. 4), the difference between the TVD and open hole lengths was further increased. The supercritical steam wells had the greatest difference (Fig. 5), which is understandable, since supercritical steam wells are drilled to very deep depths and have high-temperature gradients and flow rates. From this observation, it can be concluded that the ratio of the open hole length to total vertical depth is higher in geothermal wells.

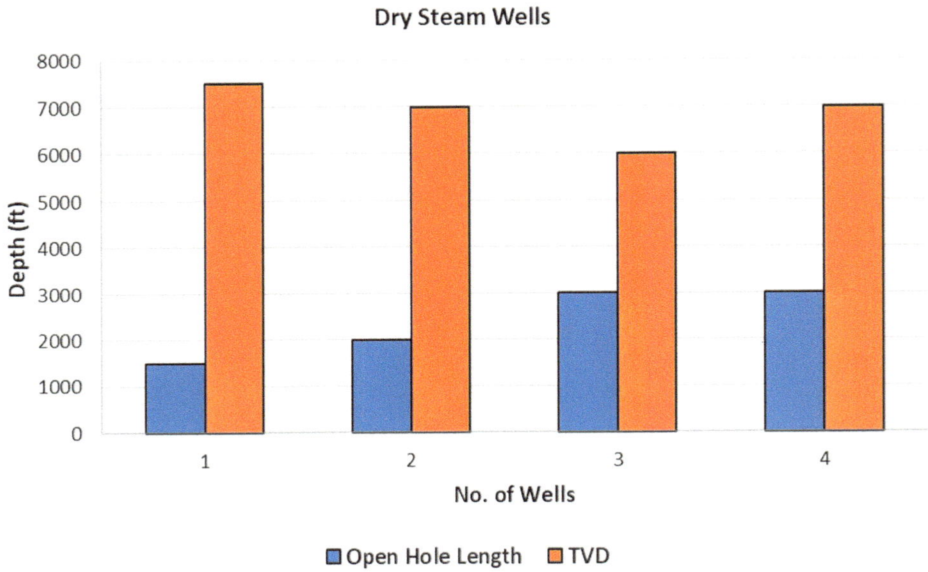

FIG. 3 Open hole length and TVD for dry steam wells.

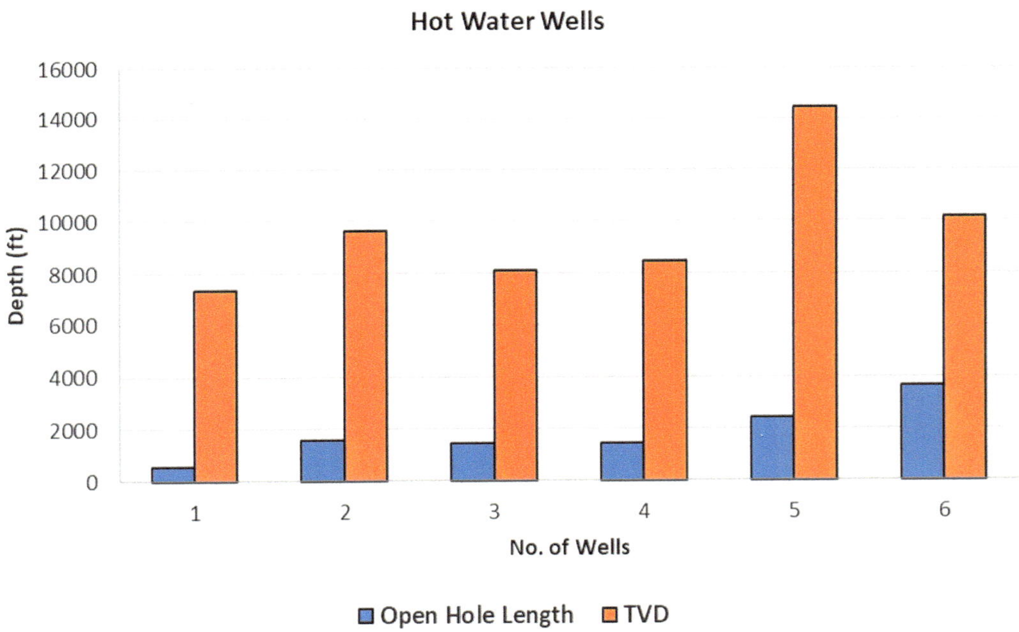

FIG. 4 Open hole length and TVD for hot water wells.

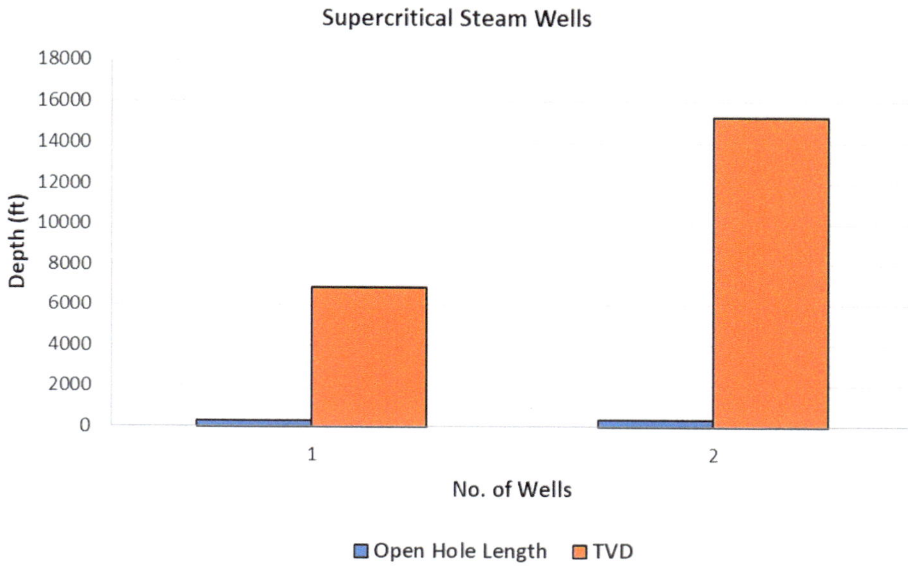

FIG. 5 Open hole length and TVD for supercritical steams.

2.2 Geothermal well type and its temperature range

The next parameter analyzed is the temperature found in the different types of well (Fig. 6). Temperature was divided into two groups, <499 °F and >499 °F. The analysis of the dry steam wells revealed that two of the wells fall under the category of <499 °F, while the other two wells belong to >499 °F. The supercritical steam wells were evenly split. Most of the hot water wells were <499 °F, with a few having a temperature >499 °F.

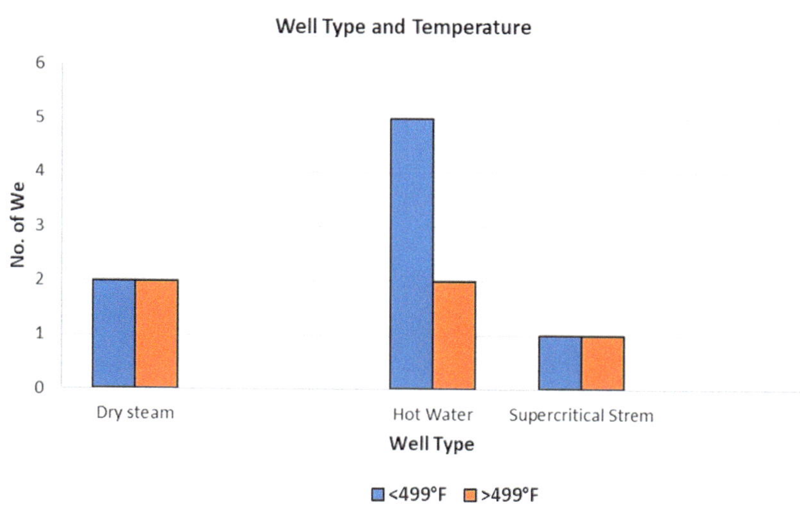

FIG. 6 Temperature found in different types of well.

2.3 Well type and (surface diameter vs bottom hole diameter)

Another parameter analyzed was the diameter of the surface and bottom hole found in geothermal wells. These are important parameters, as the diameter of the borehole in the geothermal wells is bigger than that in the conventional oil and gas wells.

Hot water and supercritical steam wells that were studied had a larger surface diameter compared with the bottom hole (Figs. 7 and 8). It is worth noting that the size of the surface casing is quite large, which is common in geothermal wells. Fig. 9 illustrates the diameter of the surface and bottom hole for dry steam wells, indicating that the difference between their diameters is not big, with surface holes remaining the larger one. Two of the four HDR wells analyzed (Fig. 10) were close in diameter, while the other two significantly had larger surface diameters. From these plots, it can be said that supercritical wells had the largest and most significant difference in diameters because of their greater depths.

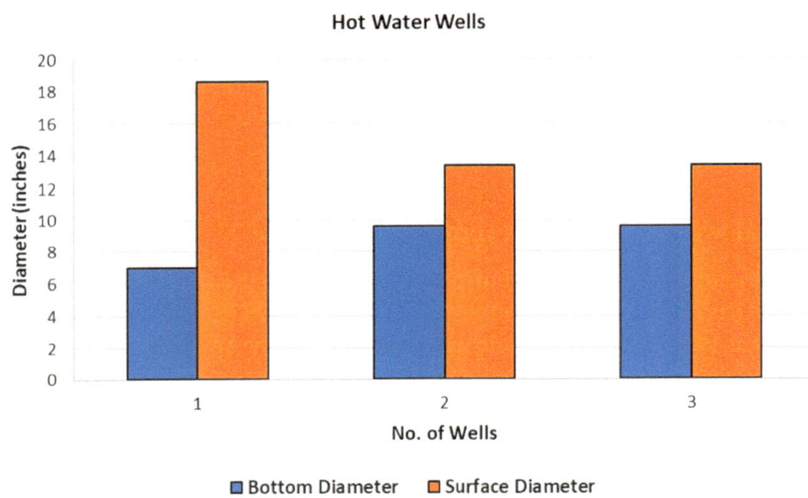

FIG. 7 Bottom and surface diameter for hot water wells.

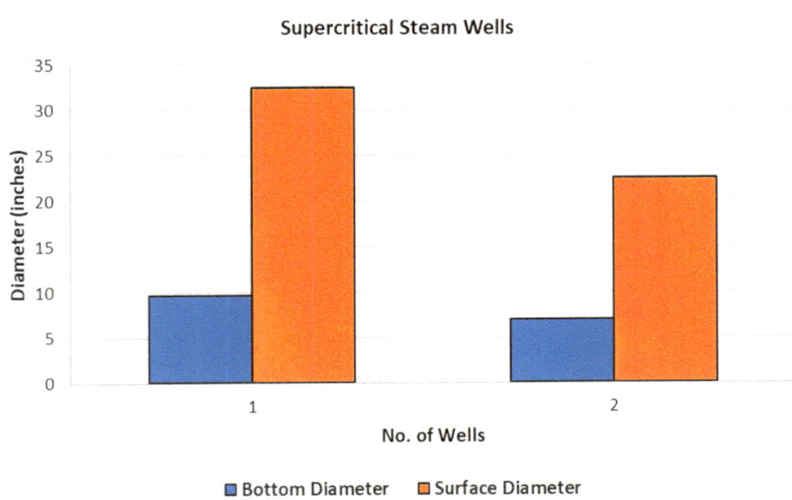

FIG. 8 Bottom and surface diameter for supercritical steam wells.

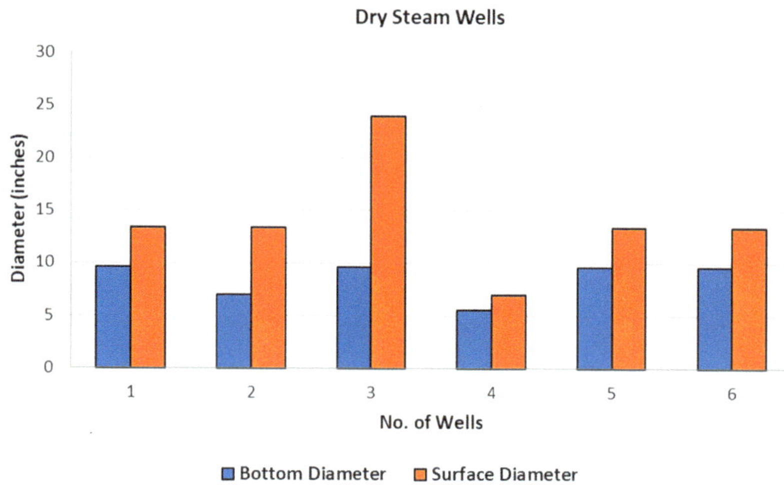

FIG. 9 Bottom and surface diameter for dry steam wells.

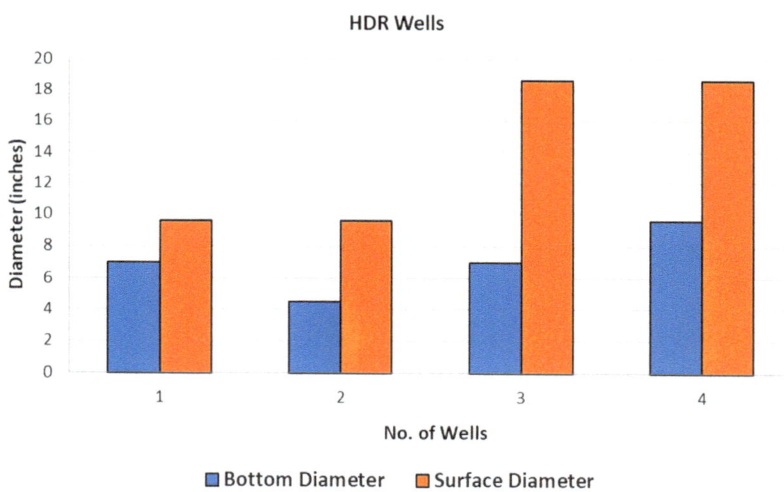

FIG. 10 Bottom and surface diameter for hot dry rock wells.

2.4 Well type and total vertical length of casing

Lengths of the surface, production, and intermediate casings were plotted, with respect to the type of well. Fig. 11 shows the length of the surface casing for dry steam, hot water, and supercritical steam wells. Most of the dry steam wells had surface casing lengths between 2000 and 2500 ft. Dry steam wells had surface casing lengths of 500–1500 ft. The hot water wells showed variation when it comes to surface casing length. Most of the surface casing lengths for the hot water wells were between 1000 and 1500 ft, with the rest falling between 500 and 1000 ft, 1500 and 2000 ft, and 2500 and 4000 ft. The supercritical steam wells are the most interesting because the surface casing lengths for all wells were <500 ft. Fig. 12 shows the types of well and production casing length. All of the production casing lengths for the dry steam and supercritical steam wells were between 3000 and 6000 ft. The production casing lengths for the hot water wells were evenly divided into <3000 ft, 6000–7000 ft, and 7000–8000 ft. The intermediate casings for the hot water wells were > 2000 ft (Fig. 13). Most intermediate supercritical steam casings were < 1000 ft, and some were > 2000 ft. It can be inferred that different types of casing and lengths are used in geothermal wells depending on the condition of the wells.

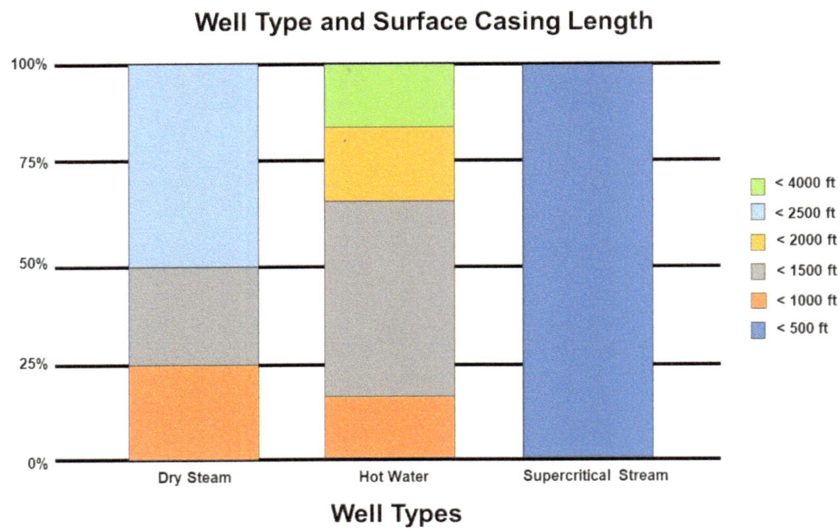

FIG. 11 Types of well and surface casing length (increment length, 500 ft).

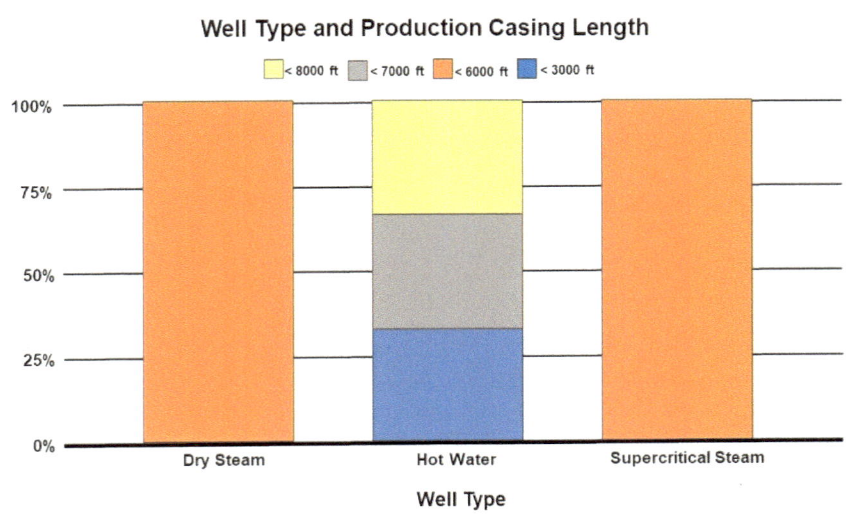

FIG. 12 Type of well and production casing length.

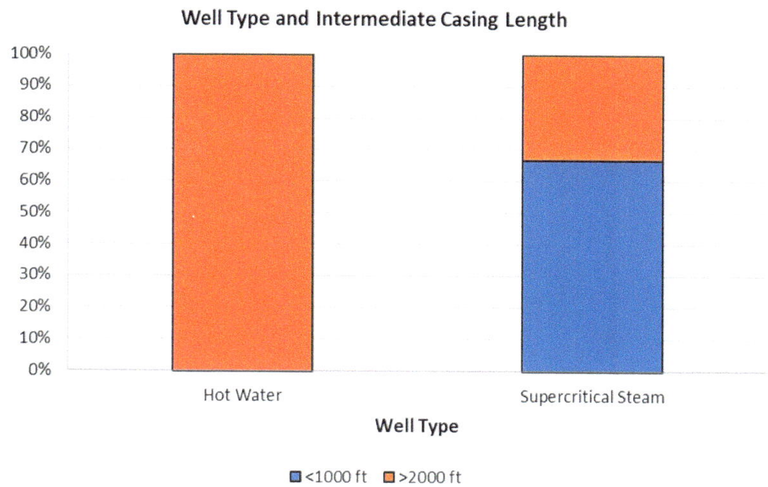

FIG. 13 Type of well and intermediate casing length.

2.5 Geothermal well type and total vertical depth

Total vertical depth plays a critical role in the completion of a geothermal well. To compare the analyses, the well is divided into three categories (Fig. 14): <4000 ft, 4001–9000 ft, and >9000 ft.

All the hot water wells were found to have a depth of >9000 ft, while that of the dry steam wells was between 4001 and 9000 ft. The depth of two supercritical steam wells was analyzed, and it was noted that one was between 4001 and 9000 ft and the other was >9000 ft. All the HDR wells analyzed in this study had a depth >9000 ft. Hence, it can be said that the HDR wells studied in this paper belong in the deepest depth category.

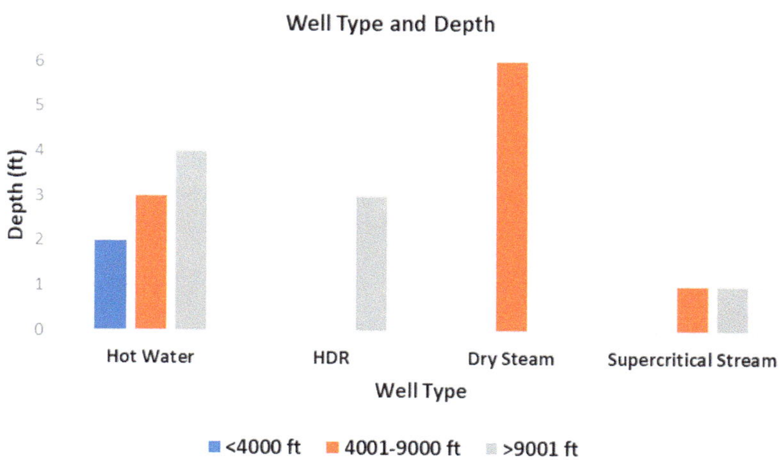

FIG. 14 Type of well and total vertical depth.

2.6 Well type and liner outer diameter

The liner ODs were gathered from various research papers and arranged according to the types of well. Fig. 15 shows hot water and HDR wells utilizing liners with smaller diameters (4.5–6.625″). However, it is worth noting that the HDR wells use liners of every diameter except for 5″ and 7.625″. Dry steam wells were the only type with a liner diameter of 7.625″. However, most dry steam wells used a 9.625″ liner. The two examined supercritical steam wells used a 7″ and 9.625″ liners. It can be inferred that supercritical steam wells use higher OD liners due to higher flow rates and higher-temperature gradients. Fig. 15 also shows that most geothermal wells use a 9.625″ liner.

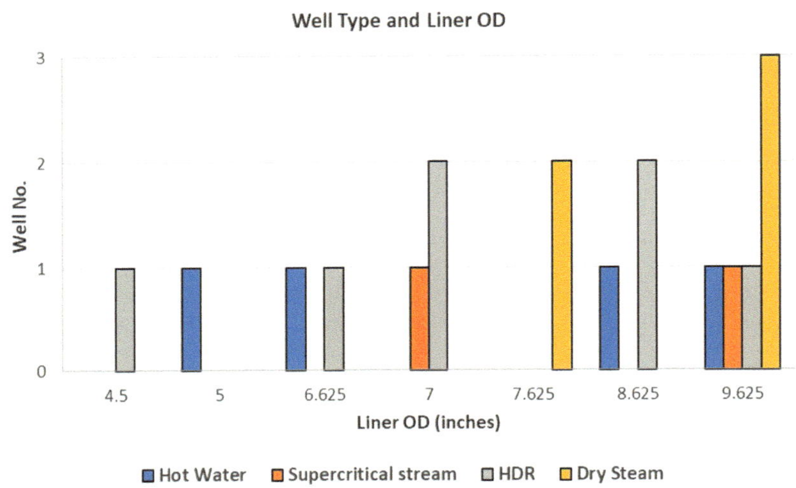

FIG. 15 Type of well and total vertical depth.

2.7 Well type and liner material

The next parameter examined was the liner material (Fig. 16). Hot water wells only use L-80–grade liners. Dry steam wells use N-89 or N-80–grade liners. Supercritical steam wells only use K-55–grade liners. Interestingly, this reveals a discrepancy, since these wells generally make use of higher-grade materials due to higher-temperature gradients and flow rates. HDR wells use either K-55– or N-80–grade liners. Based on these findings, it can be said that N-80– and K-55–grade liners are the most common liners in the geothermal well.

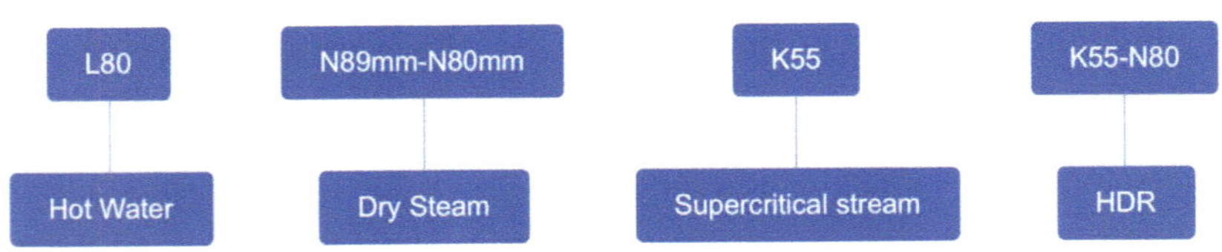

FIG. 16 Type of well and liner material.

2.8 Well type and flow rates

The flow rate for each well is calculated by finding the liner's or production casing's area at the bottom of the well and then multiplying it by the assumed velocity of 118.11 in./s (3 m/s). The flow rates were divided into different ranges, which can be found in Fig. 17. Based on the findings, hot water and HDR wells vary the most in terms of flow rate. Most of the hot water wells had a flow rate between 4000 and 6000 in.^{3}/s. The hot water wells also have flow rates <4000 in.^{3}/s and up to 10,000 in.^{3}/s. Most of the HDR wells also have flow rates between 4000 and 6000 in.^{3}/s. However, the HDR wells also had flow rates of <2000 in.^{3}/s, and some had a flow rate as high as 10,000 in.^{3}/s. The dry steam wells showed the highest flow rate, which was >10,001 in.^{3}/s, and other wells had a flow rate between 8000 and 10,000 in.^{3}/s and 4000 and 6000 in.^{3}/s. Of the two supercritical steam wells, one had a flow rate between 8000 and 10,000 in.^{3}/s and the other had a flow rate between 4000 and 6000 in.^{3}/s. Based on these findings, it can be inferred that the flow rates of the geothermal wells are generally quite high.

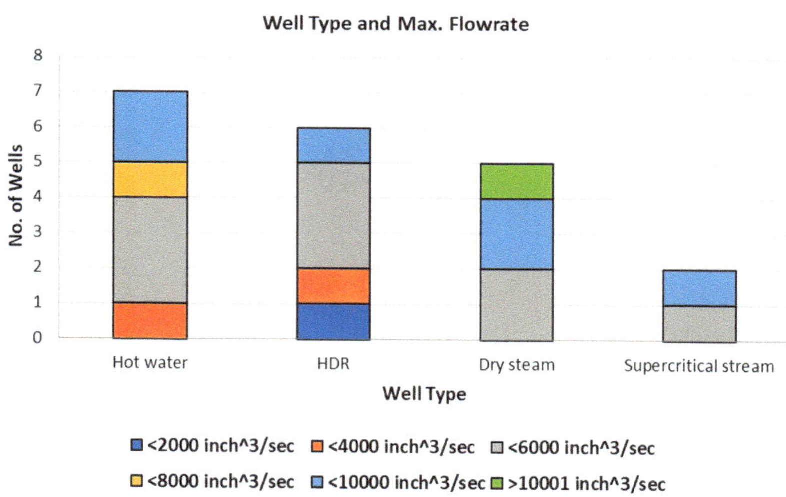

FIG. 17 Type of well and calculated maximum flow rate.

2.9 Well type and (surface casing/production casing/anchor casing/intermediate casing material)

The data gathered from the papers show that the surface casing used in both hot water and dry steam wells was made up of K-55–grade casing (Fig. 18). Supercritical steam wells only used X-56–grade surface casing, a higher-grade material, which is indicative of the high flow rates and temperature gradients. For the production casing, K-55 or N-80 is used for the dry steam wells (Fig. 19). Supercritical steam wells use only K-55 grade, and HDR wells use only P-110 grade for production casings. The only available information regarding anchor casings stated that supercritical steam wells use T-95–grade casings. Fig. 20 shows that hot water wells use K-22 grade, dry steam wells use K-55 grade, and supercritical steams use K-55 or X-56 grade for intermediate casings. As shown in Figs. 18, 19, and 20, K-55–grade casing is the most utilized casing material and is subject to change based on depth, temperature gradient, and well conditions.

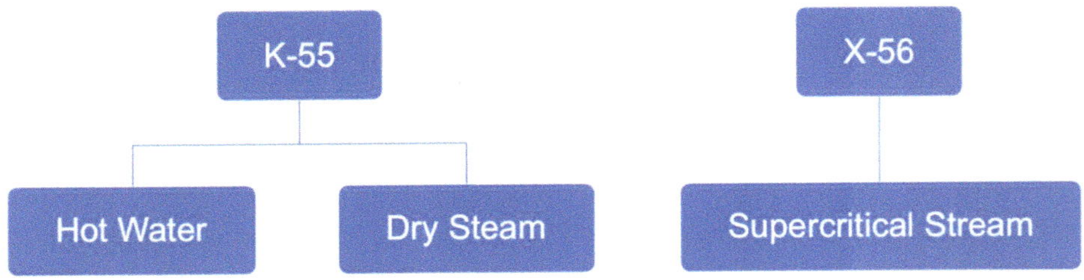

FIG. 18 Surface casing grades used for different types of wells.

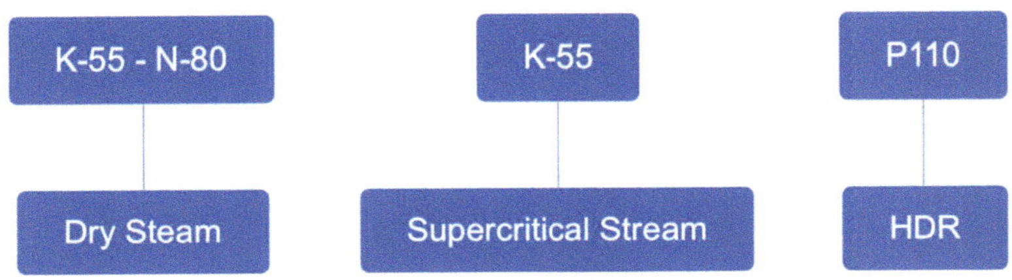

FIG. 19 Production casing grades used for different types of wells.

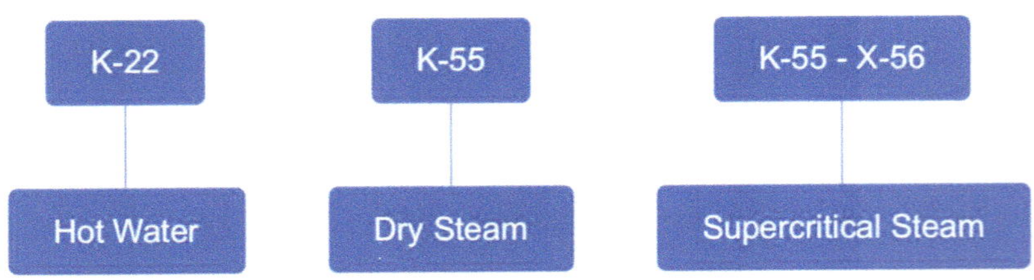

FIG. 20 Intermediate casing grades used for different types of wells.

2.10 Unconventional materials used for geothermal well construction

Titanium and corrosion-resistant alloys are used for geothermal wells in extreme situations. Titanium is considered a good alternative to stainless steel when the chloride levels and temperature exceed 5000 ppm and 100°C, respectively. Titanium is also compatible with oxygen, which makes it the first choice in geothermal systems where the fluid is re-circulated. The common titanium grade used for tubulars is Ti grade 29, which was used in the Salton Sea, California, geothermal project. Pye, Holligan, Cron, and Love (1989) first mentioned titanium usage and described beta-C titanium as a cost-effective material for hot wells with hypersaline brines rich in H_2S and CO_2.

3 Conclusions

This chapter presented information regarding the completion characteristics of geothermal wells from which their classification was made. This characterization can help overcome different challenges faced in geotherm wells, such as high-temperature gradient, depths, casing fatigue, high flow rates, borehole geometry, selection of casing and tubing materials, issues in well integrity, and coupling preference for tubing/casing connection, which are incurred in the design of geothermal wells.

References

Abid, K., Sharma, A., Ahmed, S., Srivastava, S., Velazco, A.T., Teodoriu, C., 2022. A review on geothermal energy and hpht packers for geothermal applications. In: Energies.

Akhmadullin, I., 2016. Design and Analysis of Geothermal Wellbore Energy Conversion Design and Analysis of Geothermal Wellbore Energy Conversion System Working on Zero Mass Withdrawal Principle System Working on Zero Mass Withdrawal Princip. Louisiana State University Louisiana State University.

Baujard, C., Genter, A., Dalmais, E., Maurer, V., Hehn, R., Rosillette, R., et al., 2017. Hydrothermal characterization of wells GRT-1 and GRT-2 in Rittershoffen, France: Implications on the understanding of natural flow systems in the rhine graben. In: Geothermics, pp. 255–268.

Benti, N., Woldegiyorgis, T., Geffe, C., Gurmesa, G., Chaka, M., Mekonnen, Y., 2023. Overview of geothermal resources utilization in Ethiopia: Potentials, opportunities, and challenges. In: Scientific African.

Carden, R., Nicholson, R., Pettitt, R., Rowley, J., 1985. Unique aspects of drilling and completing hot, dry rock geothermal Wells. J. Pet. Technol., 821–834.

Carpenter, C., 2021. Design and safety considerations for coiled tubing operations in geothermal Wells. J. Pet. Technol., 51–52.

Falcone, G., Teodoriu, C., 2008. Oil and gas expertise for geothermal exploitation: the need for technology transfer. In: Europec/EAGE Conference and Exhibition. Rome.

Holligan, D., Cron, C., Love, W., Buster, J., 1989. Performance of Beta titanium in a Salton Sea field geothermal production well. In: SPE/IADC Drilling Conference. New Orleans.

Karlsdottir, S., Thorbjornsson, I., Ragnarsdottir, K., Einarsson, A., 2014. Corrosion testing of heat exchanger tubes in steam from the IDDP-1 exploratory geothermal well in Krafla, Iceland. In: CORROSION 2014. SPE, San Antonio.

Karlsdottir, S., Krogh, B., Sæther, S., Þórólfsson, G., 2019. Corrosion damage of injection string from the deep geothermal well IDDP-2 in Reykjanes Iceland. In: CORROSION 2019. SPE, Nashville.

Marbun, B., Ridwan, R., Sinaga, S., Pande, B., Purbantanu, B., 2019. Casing failure identification of long-abandoned geothermal wells in Field Dieng, Indonesia. In: Geothermal Energy.

Nicholson, J., Snyder, R., 1982. Geothermal-Well Completions: A Survey and Technical Evaluation of Existing Equipment and Needs. DOE.

Okech, R.R., Liu, X., Falcone, G., Teodoriu, C., 2015. Unconventional completion design for deep geothermal wells. In: SPE Latin American and Caribbean Petroleum Engineering Conference. SPE, Quito, Ecuador.

Omrani, P., der Valk, K., Bos, W., Nizamutdinov, E., der Sluijs, L., Eilers, J., et al., 2021. Overview of opportunities and challenges of electrical submersible pumps esp in the geothermal energy production systems. In: SPE Gulf Coast Section Electric Submersible Pumps Symposium. SPE, The Woodlands.

Petrică, V., 2017. Common Geothermal Well Design and a Case Study of the Low-Temperature Geothermal Reservoir in Otopeni, Romania. United Nations University; Orkustofnun.

Pettitt, R., 1979. Completion of hot dry rock geothermal well systems. In: SPE Annual Technical Conference and Exhibition. SPE, Las Vegas.

Pye, D., Holligan, D., Cron, C., Love, W., 1989. The use of beta-c titanium for downhole production casing in geothermal wells. Geothermics, 259–267.

Reinsch, T., Regenspurg, S., Feldbusch, E., Saadat, A., Huenges, E., Erbas, K., et al., 2015. Reverse cleanout in a geothermal well: analysis of a failed coiled-tubing operation. SPE Prod. Oper, 312–320.

Snyder, R., 1978. Geothermal Well Completions: An Overview of Existing Methods in Four Types of Developments. Sandia Laboratories.

Snyder, R., 1979. Geothermal well completions: A critical review of downhole problems and specialized technology needs. In: SPE Annual Technical Conference and Exhibition. SPE, Las Vegas.

Sveinbjornsson, B., Thorhallsson, S., 2013. Drilling performance and productivity of geothermal wells - case history from Hengill geothermal area in Iceland. In: ARMA 2013 47th U.S. Rock Mechanics/Geomechanics Symposium.

Teodoriu, C., Falcone, G., 2008. Comparison of well completions used in oil/gas production and geothermal operations: a new approach to technology transfer. In: Thirty-Third Workshop on Geothermal Reservoir Engineering Stanford University. Stanford.

CHAPTER

5

Well integrity and future construction of geothermal wells

Catalin Teodoriu and Khizar Abid

MPGE, University of Oklahoma, Norman, OK, United States

1 Introduction

With each passing day, as the world population increases, the demand for energy production also increases. As reported by Dudley (2017), by 2035, the demand for energy will increase by 35%, and the world population will be 8.8 billion by that time. In addition, the report published by the Energy Information Administration (EIA, 2016) projected that, by 2040, the world population will increase by 48%. Therefore, to meet the high demand for energy production, both conventional and nonconventional methods have to be utilized. Furthermore, in the coming future, the contribution of renewable energies in the production of energy will be four times more than that of fossil fuel energy sources (Dudley, 2017). In that respect, geothermal energy plays an important role, as it is the only source of renewable energy that is not affected by any metrological conditions. Geothermal is considered a continuous source of energy, as it remains in operational mode, 98% of the time (Budisulistyo and Krumdieck, 2015). Moreover, geothermal energy is available in many countries at certain depths and remains unexplored. Hence, geothermal has a huge potential to produce energy that has a low carbon footprint and is environmentally friendly (Moya et al., 2018; Shortall et al., 2015).

Geothermal comes from the Greek words "geo," which means earth, and "therme," meaning energy. Therefore, geothermal energy is the energy that is collected from the earth's subsurface. It is not a new source of energy, as it dates back to thousands of years when Paleo-Indians used the water from hot springs for their daily use and utilized the minerals from these hot water for medication purposes (DOE, 2022). In 1904, the first commercial power plant operated by geothermal energy was set in Tuscany, Italy (EIA, 2016).

Geothermal energy has a wide range of applications, from small to large scale. On a small scale, geothermal energy is mostly utilized for the heating of domestic buildings, while in the

Geothermal Energy Engineering
https://doi.org/10.1016/B978-0-443-21662-6.00007-4

Copyright © 2025 Elsevier Inc. All rights are reserved, including those for text and data mining, AI training, and similar technologies.

large scale, it is used in power plants for the production of electricity. However, the applicability of geothermal energy depends on the heat gathered from the subsurface. Table 1 shows the different applications of geothermal wells with respect to their temperature ranges. It can be seen from the table that geothermal systems with higher temperatures are more widely used for commercial purposes. According to the geothermal gradient, the greater the depth, the higher the temperature that will be gathered. Moreover, it is important to understand that the geothermal gradient changes from place to place, depending on geological activities such as volcanic activities and the movement of the tectonic plates. Nonetheless, in order to achieve high temperatures, deep geothermal wells have to be drilled. It was reported by Goetzl (2020) that a deep geothermal well has a thermal capacity of 1 to >50 MW_t, while on the other hand, the maximum thermal capacity that can be produced from a shallow geothermal well is 5 MW_t.

TABLE 1 Different applications of geothermal energy with respect to temperature ranges (Abid et al., 2022).

Temperature	Range	Fluid in domination	Usage
More than 150°C	High	Liquid and vapor	Power and fuel generation
Close to 95°C	Medium	Mostly liquid	Production of electricity through the binary power plant
Between 90°C and 150°C			Temperature control in the building and drying of the timber
Between 30°C and 90°C	Low	Liquid	Food processing, aquaculture, and warming of the greenhouses
Lower than 30°C			Soil warming

2 Well construction

The drilling phases for geothermal or conventional oil and gas wells are the same. First, the conductor casing is rammed, and a diverter is installed over it. Next, the drilling for the next section starts and goes to the depth of the freshwater aquifer. Then, drilling will be stopped, and the surface casing will be installed and cemented to protect the freshwater zone. Subsequently, the blowout preventer (BOP) will be installed over the surface casing. Then, to reach the desired target in the subsurface, different formations have to be drilled, which might have abnormal or subnormal pressure, and to manage these lithologies, intermediate casing will be deployed and cemented at different depths. When the pay zone is drilled, the last casing, known as the production casing, will be installed and cemented. It is also possible to install a production liner instead of a production casing. It is worth mentioning that after every cement job, the cement bond log test is conducted to make sure that the cement is properly placed in the annulus and is properly bonded with the formation and the casing. At the end of drilling, the schematic of the well will look like a telescope, in which different casings have been installed, from the largest casing at the top to the smallest at the bottom hole (Fig. 1).

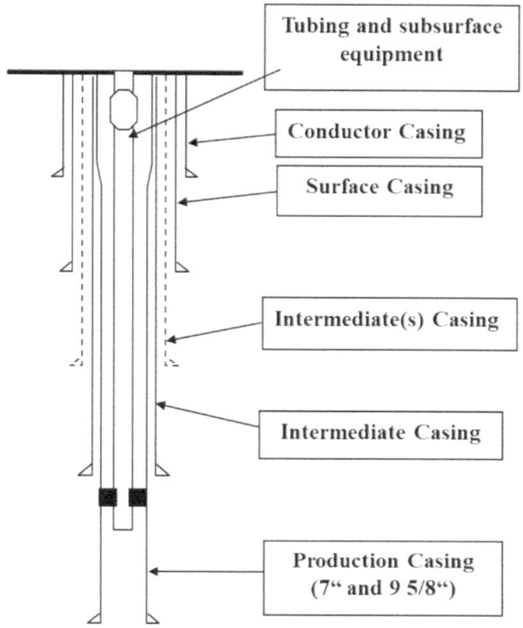

FIG. 1 Well design of conventional oil and gas wells (Teodoriu and Falcone, 2009).

Though the drilling phases of geothermal and oil and gas wells are the same, there are still differences, as both of these wells are drilled in different formations. For instance, oil and gas wells are drilled in sedimentary rocks, whereas geothermal wells are drilled in hard volcanic igneous rock having fractures and fissures. Moreover, the temperature, pressure, casing, and design of the geothermal well are totally different from the conventional oil and gas wells (Moya et al., 2018). On the economic side, it was reported by DiPippo (2016) that the cost of developing a geothermal well is about 4.5–5.5 million USD per installed MW, and about 50% of the budget is spent on its completion and drilling. On the other hand, according to Bronicki (2017), about 43% of the cost of the geothermal project is intended for drilling and exploration operations. Therefore, for an effective utilization of the economical budget for deep geothermal wells, it is important to reduce the cost of drilling. In that respect, many studies are being carried out and new designs for the construction of geothermal wells have been presented.

In order to minimize the cost of drilling and completion, one of the concepts presented is monobore drilling. It is described as the wellbore that has casing strings the same as those of the ID, from top to bottom, regardless of the fracture or formation gradient of the lithologies. Monobore is considered to be effective in saving cost and drilling time, and the integrity of the well is ascertained. DeMong and Rivenbark (2003) presented the concept of optimizing monobore designs using expandables (Fig. 2), which help reduce the size of the well, while Kumar et al. (2010) introduced monobore well designs using shell proprietary expendable tubulars, which are shown in Fig. 3. A monobore well design was proposed by Marinovation (2011) using a foldable steel technology for the casing. Another monobore design using expandables was proposed by Siddiqui and York (2012) (Fig. 4), which is shown in Fig. 5. Varahram et al. (2011) presented a monobore design using a combination of foldable steel and expandable tubular. The conductor and surface casing are drilled, cemented, and tested like conventional wells. Afterward, the rest of the well is completed with the deformable pipe consisting of either foldable steel or expandables. Fig. 6 shows the mobore well design proposed by Teodoriu et al. (2012) for the Gebo project. They proposed the use of an 8 5/8 in. casing, but unlike the conventional oil and gas wells, the ID of the well will be kept constant, from top to bottom, regardless of the required intermediate casing. The smallest ID used should be 8 in., so that the next section of the well can be drilled.

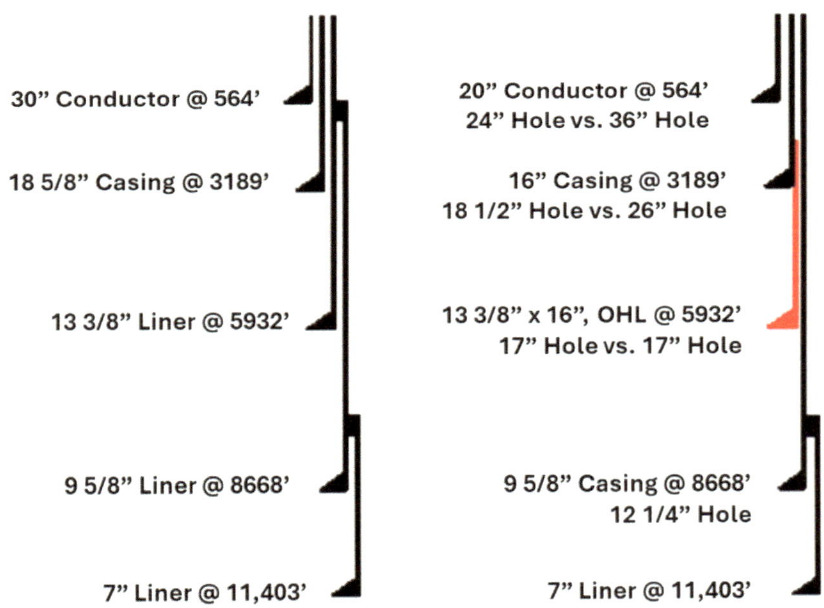

FIG. 2 Monobore design using expandable to reduce the size of the well (DeMong and Rivenbark, 2003).

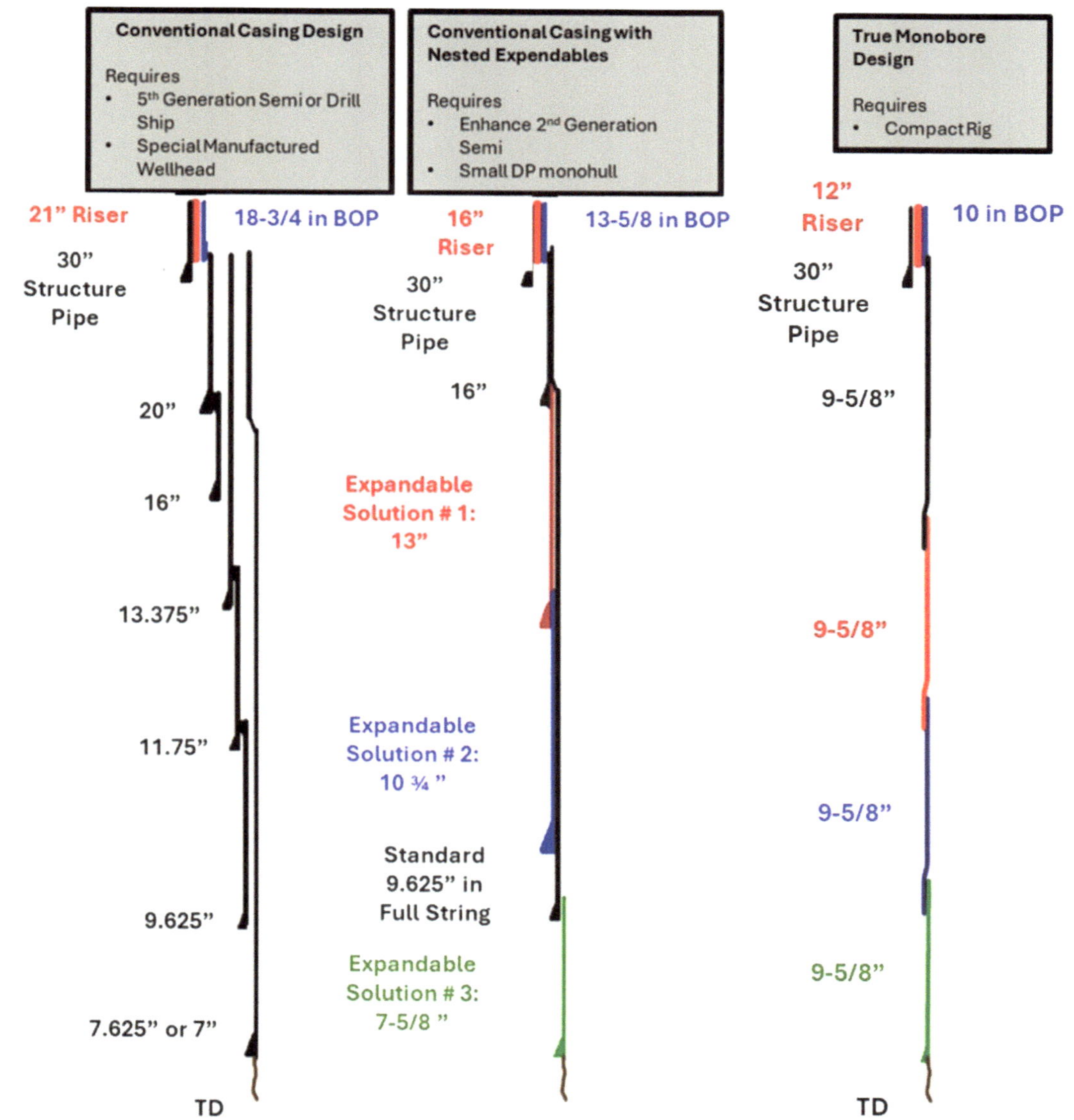

FIG. 3 Shell proprietary expandable technology tubular used for the monobore well design (Dupal et al., 2000).

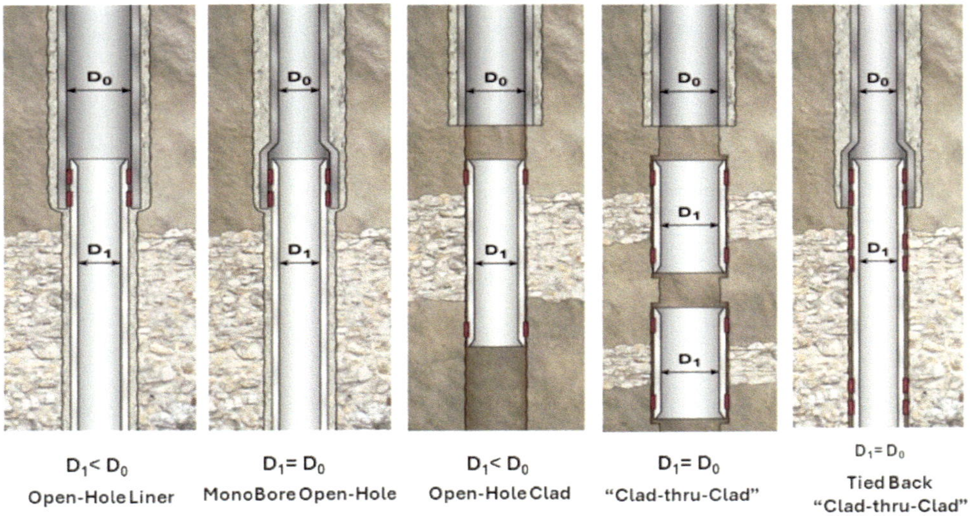

FIG. 4 Expandable tubular used for monobore well design (Siddiqui and York, 2012).

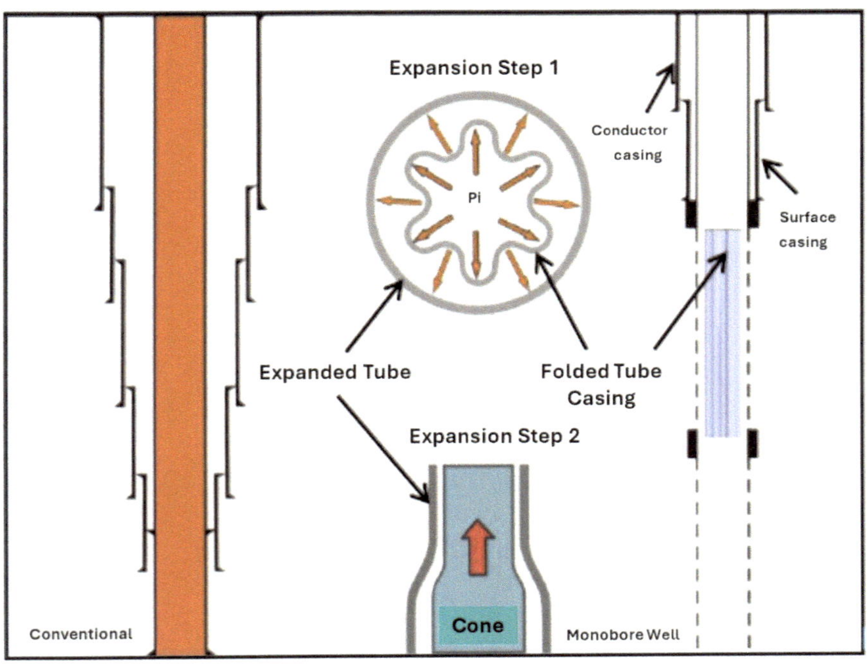

FIG. 5 Conventional and monobore well (Varahram et al., 2011).

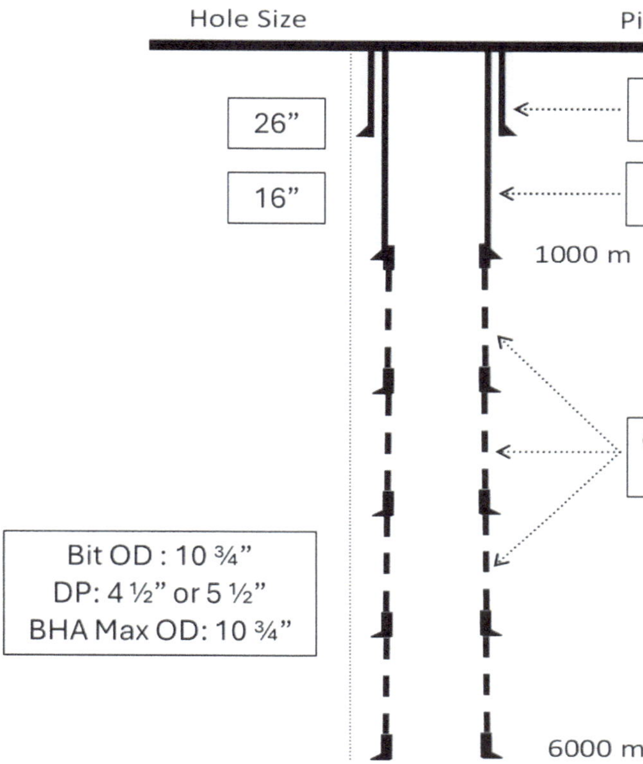

FIG. 6 Proposed monobore for the GEBO project (Teodoriu, 2015b).

According to Teodoriu (2015a,b), the following steps should be taken to have a quality monobore well: (a) the well should be drilled at 300–1000 m (open hole) depths, depending upon the efficiency of the expansion process; (b) after drilling to the desired depth, the drill string is retrieved, and the expandable/foldable tubular is lowered in the well with the help of a working string; and (c) subsequently, the tubular is expanded and cemented.

For a deep geothermal well to be economically feasible, the production rates should be high. It was reported by Fichter et al. (2009) that the rate of production in a power plant should be in the range of 60–120 L/s or 10,000 m^3/day for a geothermal system that supplies temperatures of 120°C at the surface. On the opposite spectrum, the production rate of oil is only a few hundred cubic meters per day, and the wells are drilled in the undisturbed geological segments. However, if conventional oil and gas wells have to be repurposed for a geothermal well, a high flow rate will result in high velocities, which will increase frictional losses and can also cause solid particle erosion. Therefore, Srivastava and Teodoriu (2018) put forward the calculation in which different flow rates were related to the size of the casing, velocity, and pressure drop due to the frictional losses (Table 2). For the purpose of calculating frictional losses, a vertical well with a depth of 10,000 ft. was considered.

TABLE 2 Flow rate, pressure drop, and velocities with respect to different casing sizes.

Flow rate	Diameter (ID) of production tubing													
	4-1/2 in.		5 in.		5-1/2 in.		6-5/8 in.		7 in.		8-5/8 in.		9-5/8 in.	
	Velocity	Pressure drop	Velocity	Pressure drop	Velocity	Pressure drop	Velocity	Pressure Drop	Velocity	Pressure drop	Velocity	Pressure drop	Velocity	Pressure drop
L/s	*u* (m/s)	Bar	*u* (m/s)	Bar	*u* (m/s)	Bar	*u* (m/s)	Bar	*u* (m/s)	Bar	*u* (m/s)	Bar	*u* (m/s)	Bar
30	3.54	29.64	2.85	17.16	2.33	10.35	1.57	3.91	1.38	2.85	0.90	0.99	0.72	0.57
40	4.72	51.49	3.80	29.73	3.10	17.89	2.10	6.72	1.85	4.89	1.20	1.70	0.96	0.97
50	5.90	79.22	4.75	45.65	3.88	27.42	2.62	10.25	2.31	7.46	1.51	2.58	1.20	1.48
60	7.08	112.83	5.70	64.91	4.65	38.92	3.15	14.51	2.77	10.54	1.81	3.63	1.44	2.08
70	8.26	152.29	6.64	87.51	5.43	52.41	3.67	19.49	3.23	14.15	2.11	4.86	1.68	2.77

Modified from Srivastava, S., Teodoriu, C., 2018. Gas well conversion to geothermal: a case study for Oklahoma. 43rd Workshop on Geothermal Reservoir Engineering Stanford University, Stanford University, Stanford.

From Table 2, it can be concluded that the smaller the ID of the casing, the higher the velocity and the pressure drop of the respective flow rates. For example, the highest velocity and pressure drop among all the flow rates and casing are from the 4 1/2 in. production casing. Moreover, in the table, pressure drops are highlighted both with red (above 40 bar) and green (below 40 bars). According to the author, a pressure drop below 40 bar is economical because it is good for producing power through the use of a binary system from geothermal wells. However, casings with a size of 8 5/8 in. and 9 5/8 in. are marked in red, even though the pressure drop is less than 40 bars, because, if the preexisting oil and gas wells have to be repurposed for geothermal wells, they are not used in the predrilled wells. Moreover, it is not economically feasible to do a re-completion job on these old wells. Therefore, according to Fichter et al. (2009), for the wells to be used for geothermal purposes, the following parameters should be taken into consideration:

- The velocity of the production fluid should be smaller than the critical velocity that causes erosion.
- Completion should be designed in such a way that it should prevent formation instability and sand/solid production.
- The dimensions of the well should be selected in such a way that the maximum payoff of the project is assured while keeping in mind the tradeoff relation between pumping requirement and the size of the well.

In that respect, Fichter et al. (2009) presented a well design that was optimized for the deep geothermal well in Rhine Graben Area in Germany (Fig. 7). The geological setting of the Rhine Graben Area is very challenging and consists of troublesome and unconsolidated formations. It can be seen in Fig. 7 that four sections of the well were designed to overcome troublesome formations. In order to reach the top of the reservoir, three sections are used, while the last section penetrates through the pay zone that consists of faults and fractures. However, one more section was included for contingency, following good drilling practice. The purpose of the contingency section is to tackle any geological problem that may arise due to the instability of the formation and to install one or two slotted liners to overcome such problems. Therefore, the whole well is comprised of five sections. According to the author, this deep geothermal design can accommodate a high flow rate of 60–120L/s which can make the whole geothermal project economically feasible.

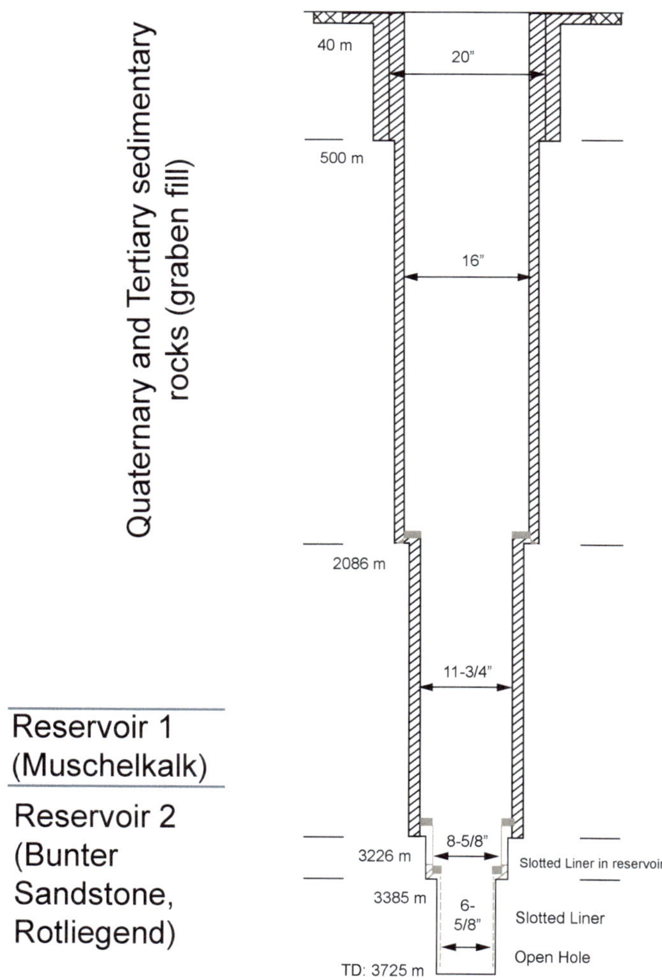

FIG. 7 Proposed well design for geothermal well in Rhine Graben Area (Fichter et al., 2009).

3 Well integrity

The integrity of the well is one of the most important factors that will determine the success or failure of ageothermal project. Even with the growing popularity of geothermal wells, no particular set of well integrity standards has been devised. Instead, it relies on the integrity parameters set for oil and gas wells. However, it is important to note that the subsurface conditions and completion design of geothermal wells are completely different from those in conventional oil and gas wells. Well integrity barriers have to be defined, in order to prevent the influx of undesired fluid, stop cross-flow to the shallow formation, and avoid the leak of the fluid to the surface. All these barriers should be properly certified, and their performance with time should be recorded. Most importantly, barriers should be independent of each other. For example, the primary barrier should be autonomous from the secondary barrier. All the barriers should have the capability to endure different loads and should maintain their integrity throughout the life of the well. As for the geothermal well, the flow takes place through the casing instead of the production tubing; therefore, the barriers of the geothermal well will not be the same as those of the oil and gas wells and require proper attention.

The hot water from the geothermal well flows continuously from the bottom to the surface where it is utilized for different applications. Therefore, primary and secondary barriers should ascertain that there is no cross-flow and that the flow can be controlled or even stopped in cases of any emergency, as stated in the ISO/TS 16530-2 (2014)—there should be at least two independent well barriers. For conventional oil and gas wells, the primary barrier consists of production casing, packers, cement, formation, and production tubing along with the safety valve. On the other hand, the secondary barrier consists of an intermediate casing, christmas tree, and formation.

The bottom hole temperature of the geothermal well is very high; therefore, it is categorized under the high-pressure high-temperature (HPHT) well, and the completion design for the HPHT well is shown in Fig. 8 (NORSOK D-010, 2013). For a better control of the well, production tubing and packer have to be used; however, for the geothermal well, the production is taken directly from the casing instead of the tubing to increase the flow rates. Hence, in the geothermal wells, only primary barriers will be available, which consist of production casing, cement, formation, and wellhead. Fig. 9 shows the well barriers of a geothermal well. Therefore, the integrity of the casing and cement must be maintained as they are the most important parameters in the primary barrier of geothermal wells. It is also presented in different studies that casing and cement suffer degradations in geothermal environments because of the presence of high temperature, corrosive fluid, and thermal cyclic loading. In some instances, casing and cement are compromised separately, while in other scenarios, they might be compromised in tandem, which exacerbates the situation and adds to its complexity.

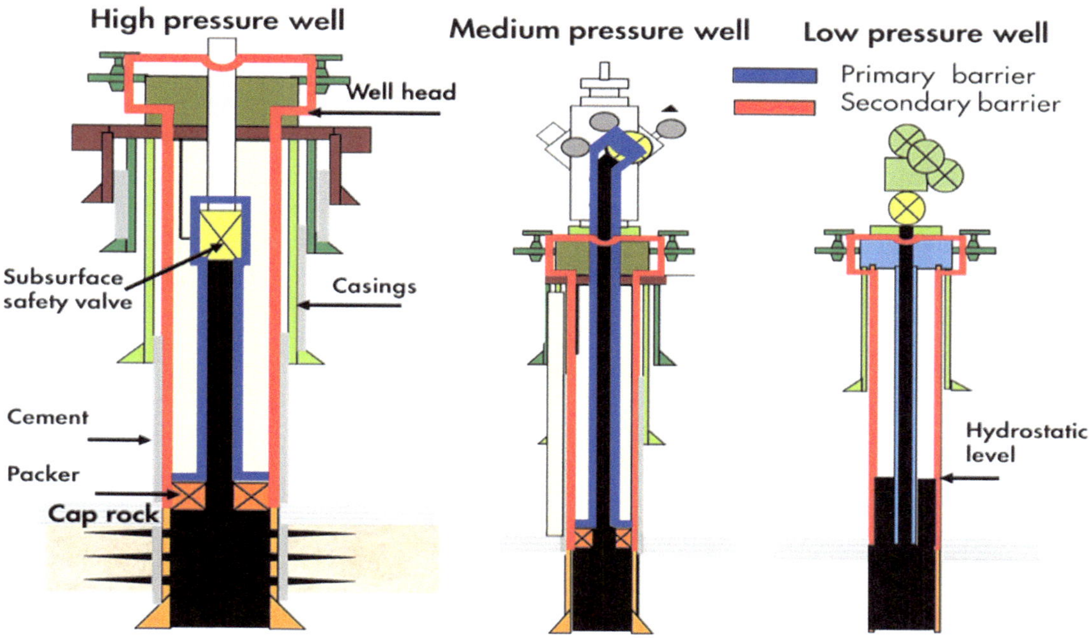

FIG. 8 Primary and secondary barriers of the HPHT wells (Shenold and Teodoriu, 2021).

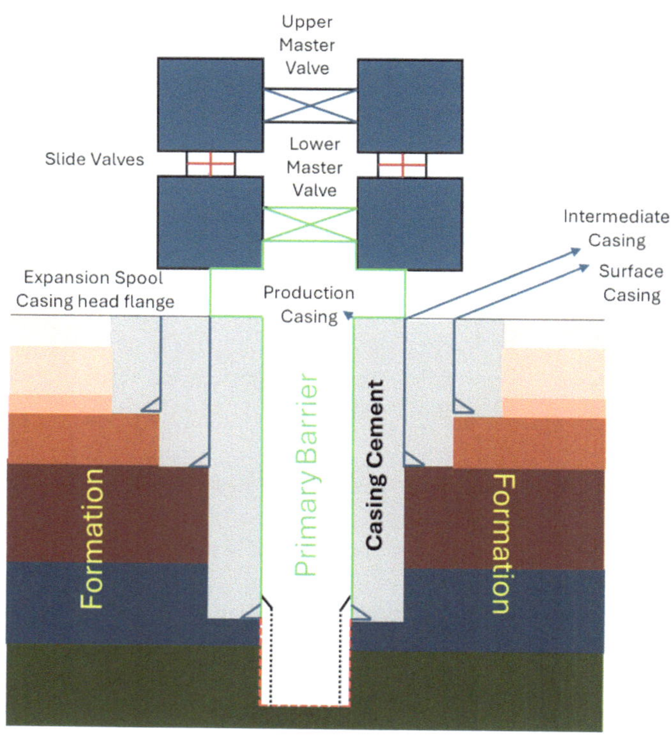

FIG. 9 Well barrier system of a geothermal well. *Modified from Lohne, H.P., Ford, E.P., Majoumerd, M.M., Randeberg, E., Aldaz, S., Reinsch, T., Wildenborg, T., Brunner, L.G., 2017. Barrier Definitions and Risk Assessment Tools for Geothermal Wells. International Research Institute of Stavanger.*

4 Well cement

Well cement is considered to be one of the most important parameters that control the integrity of the well, especially during the production phase (Teodoriu et al., 2019). Edwards et al. (1982) put forward the characteristics that well cement should have to be used in the geothermal well: (a) the strength of the cement should be sufficient enough to support the casing that might elongate due to exposure to high temperature; (b) cement slurry should displace the drilling mud when it is pumped into the well; and (c) the cement should be placed throughout the casing length to save it from the corrosive fluid. Moreover, cement also lowers the thermal fatigue in the casing caused by the contraction and expansion.

According to the American Petroleum Institute (API, 2005), well cement is divided into eight classes, from A to H. These classifications are based on depth, temperature, and sulfur resistance (Table 3). The mostly used well cement in the oil and gas industry is Classes G and H (Abid et al., 2020), whereas for the geothermal well, Classes A and G are used (Hole, 2008). Classes D, E, and F have become obsolete.

TABLE 3 API well cement classification (API, 2005; Drilling Manual, 2022).

Class	Depth (ft.)	Temperature (°F)	Purpose
A	0–6000	80–170	No special requirements are needed
B	0–6000	80–170	Moderate-to-high sulfate resistance
C	0–6000	80–170	Where early strength of the cement is required. Moderate-to-high sulfate resistance
D	6000–10,000	80–170	Used in HPHT conditions. Moderate-to-high sulfate resistance (not used anymore)
E	10,000–14,000	–	Used in HPHT conditions. Moderate-to-high sulfate resistance (not used anymore)
F	10,000–16,000	–	Used in HPHT conditions. Moderate-to-high sulfate resistance. Used in HPHT conditions. High sulfate resistance only (not used anymore)
G	–	–	Different additives are added to be used in all types of well with different ranges of pressure and temperature. Moderate-to-high sulfate resistance
H	–	–	Different additives are added to be used in all types of well with different ranges of pressure and temperature. Moderate sulfate resistance only

The zonal isolation is dependent upon the integrity of the cement sheath that is present in the annular space of the casing and the formation. Zhang and Eckert (2020) showed, through the finite element analysis, that tensile debonding on the cement can occur when the cement is exposed to thermal cyclic loading. Mueller et al. (2004) concluded, from their modeling, that the micro-annuli between the cement and casing/formation can occur due to the thermal loading of the cement. Carroll et al. (2016) presented different modes of failure (Fig. 10) that might occur in the cement sheath and can cause the flow of the subsurface fluid to the surface.

As high temperature exists in the geothermal environment, the cement sheath suffers thermal degradation. On the other hand, when the steel is exposed to high temperatures,

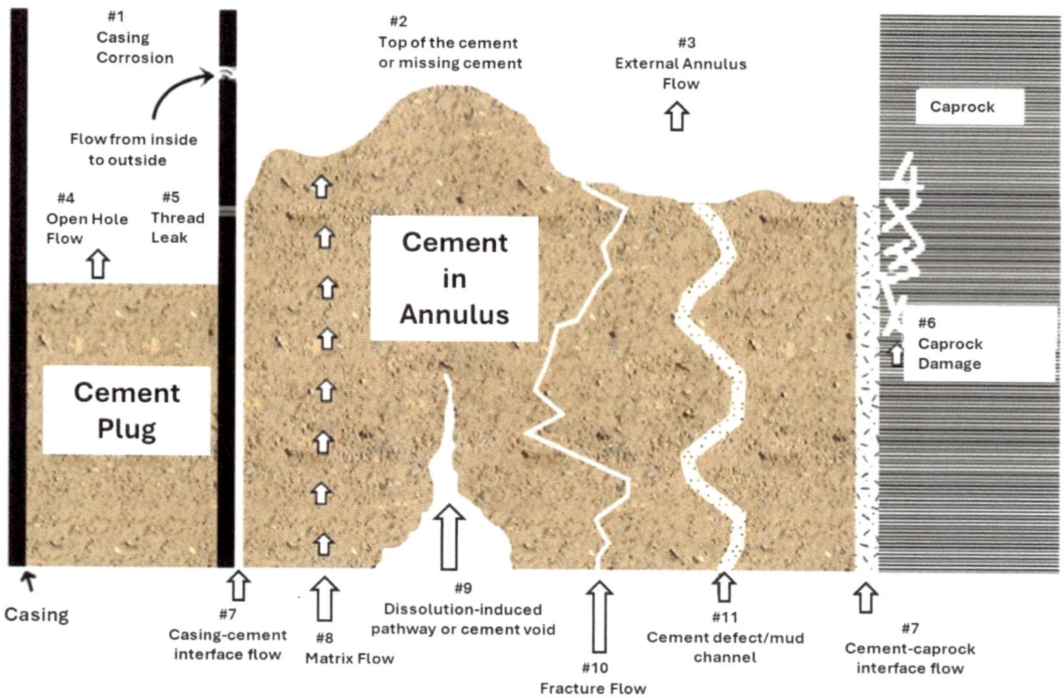

FIG. 10 Different modes of cement failure (Carroll et al., 2016).

expansion takes place, and when the production is stopped from the geothermal well or cold water is pumped down for some reason, then contraction of the casing takes place. As a result of the contraction and expansion of the casing, stresses are applied to the cement which can lead to the creation of micro-annuli and crack in the cement sheath (Nelson and Guillot, 2016). As cement and casing are the primary and only barriers in the geothermal well, failure of either one of them can be catastrophic and can severely risk the integrity of the well.

To understand the mechanical strength of the cement, with respect to the interfacial bonding shear strength (IBSS) and pure cement shear strength, Teodoriu et al. (2019) conducted laboratory-scale tests with Class H cement at room and high temperature (150 °F) on the samples that have been cured for 3, 7, and 14 days. It was shown that the IBSS is lower than the pure shear strength, which means that the cement in the well will be debonded first from the casing followed by a complete shear failure (Fig. 11). This scenario can happen when the cement will be exposed to cyclic loading in the geothermal well.

Kosinowski and Teodoriu (2012) presented the fatigue model on the cement casing contact pressure and the stress distribution using AnSys 2D. The purpose of using the fatigue parameter is to show that the continuous cyclic loading on the material can slowly lead to the failure of the material. In the initial stage, microcracks can be created, but over time, as more and more stress accumulates, these microcracks will be converted to big fissures or cracks which can cause the total failure of the material. Fig. 12 shows the result of cycling loading applied on the cement through casing expansion. It can be seen that, in the first stage, radial cracks appear on the sample. If the cycles and load on the cement are increased continuously, more cracks will propagate from the center of the cement sheath and in the direction that will be

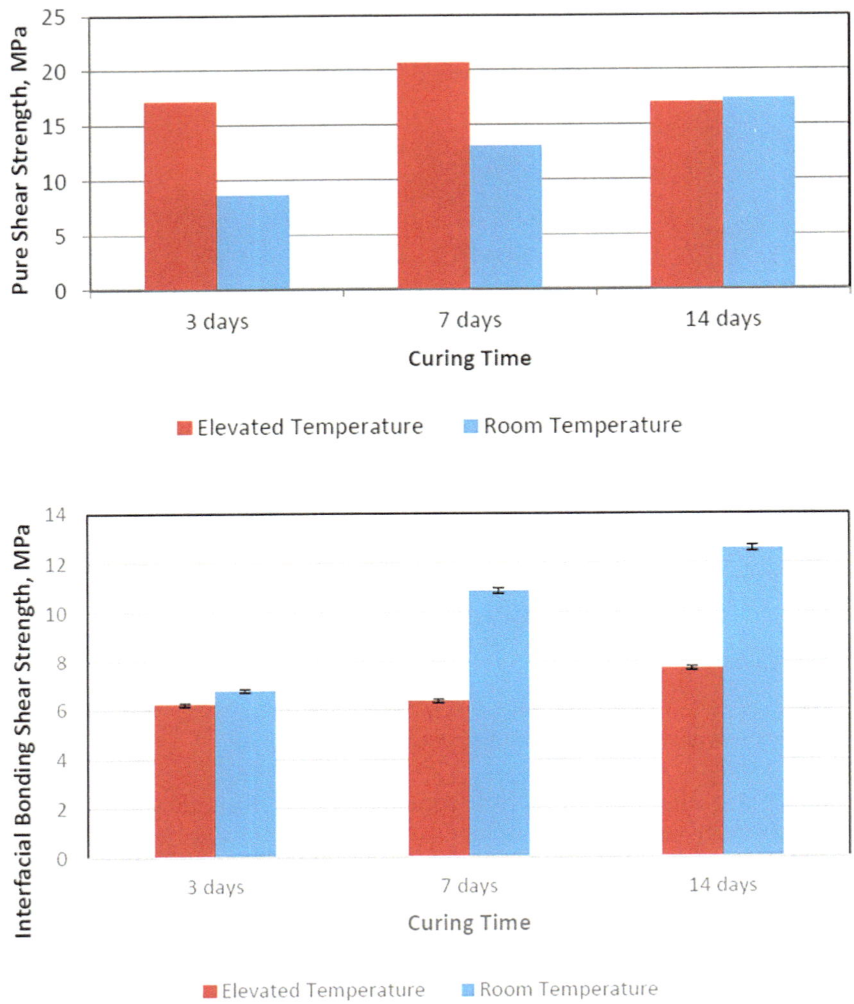

FIG. 11 Shear and interfacial bond shear strength for Class H cement (Teodoriu et al., 2019).

perpendicular to the initial crack and will continue to grow. When further load will be increased, more cracks will be developed to the point when further loading from the pipe to the cement sheath will not cause any more cracks. Hence, at this point, it can be said that cement is not in contact with the casing and has been completely debonded. If this sort of scenario arises in the geothermal well, it can give way to the formation fluids to flow up to the surface causing severe issues of well integrity.

While many studies have worked on the mechanical strength of the cement, it is also important to understand the thermal behavior of the cement. To find the thermal integrity of the well cement through modeling, thermal conductivity is an important parameter that has to be used. In that respect, Abid et al. (2023) did the test on the thermal conductivity of the cement, and it was observed that the mechanical properties of the cement evolves with time, in the

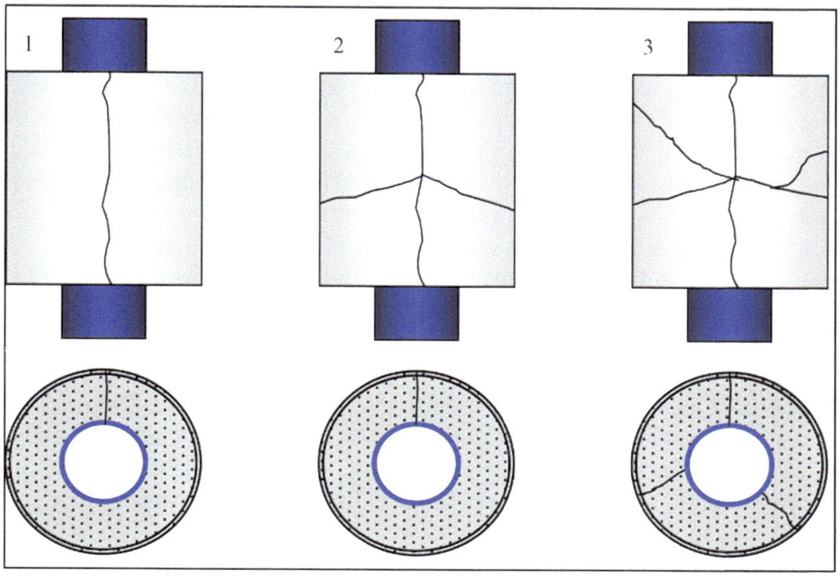

FIG. 12 Different modes of failure due to cyclic loading causing fatigue (Kosinowski and Teodoriu, 2012).

same way the thermal conductivity of the cement changes with the curing days and become constant after a certain period (Fig. 13). Therefore, it is important to use the value of the thermal conductivity carefully, as in the initial days of curing the value fluctuates, and if the value is used from the initial days of curing, then the modeling values can be over- or underestimated and may not be reliable.

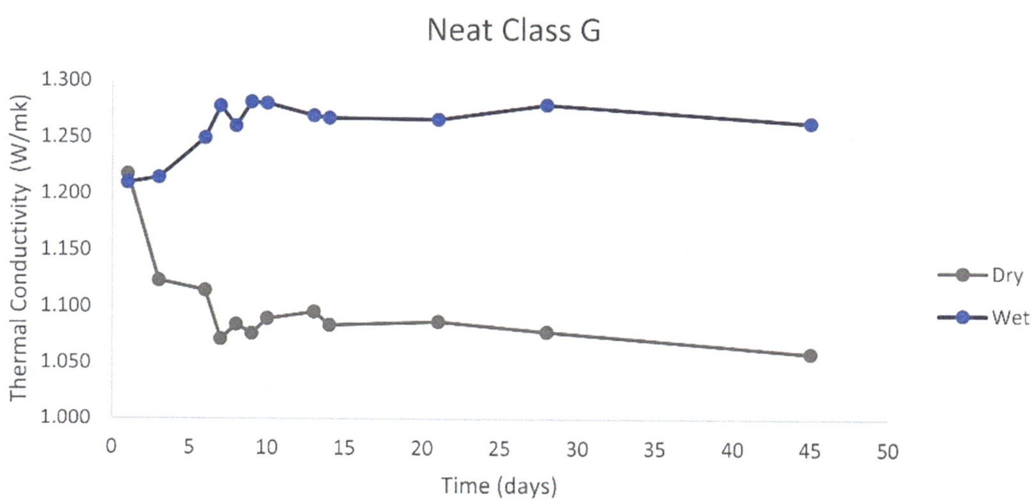

FIG. 13 Thermal conductivity of Class G cement during 45 days of curing well cement (Abid et al., 2023).

5 Well tubulars (oil country tubular goods, OCTG)

Though the condition to which the casing in the geothermal well is exposed is different from that in conventional oil and gas wells, still, the casing regulation that has been applied in the petroleum industry is used for geothermal wells. The casing acts as a conduit that serves as the connection between the bottom hole and the surface. Moreover, casing provides zonal isolation from the troublesome zone, prevents cross-flow to the shallow formation, protects freshwater aquifer, and gives access for the well control during a kick.

The difference between the casing design of geothermal and conventional oil and gas well is because of the following reasons (Teodoriu and Falcone, 2009): (a) the ID of the final (production) casing is much bigger in the geothermal well than that in the oil and gas well. The purpose of having a large ID in the geothermal well is to obtain a high flow rate to make the geothermal project economically feasible, and (b) for oil and gas wells, the production is taken from the tubing, whereas for geothermal wells, it is taken from the casing directly. Therefore, the life span of the production casing in geothermal wells should be longer than that of the conventional wells. Table 4 shows the comparison between the completion of gas, heavy oil, and geothermal wells.

TABLE 4 Completion comparison between gas, heavy oil, and geothermal well (Teodoriu and Falcone, 2009).

Parameter	Gas	Heavy oil	Geothermal
Temperature (°C)	60–50	60–350[a]	80–250
Production casing OD	9 5/8 in. with crossover to 7 in.	5 or 7 in.	9 5/8 in. and 7 in. liner
Depth (m)	3000–6000	300–1200	1000–5000 (or more)
Connection type	Premium	API long/buttress	API buttress

[a] *High temperature for the EOR method (steam injection).*

When the casing is exposed to high temperatures, radial, hoop, and axial stresses accumulate and lead to the failure of the casing integrity that will hamper the production and will require high workover cost. It is reported by Shenold and Teodoriu (2021) that the properties of the casing, especially burst and collapse pressure, changes drastically when the temperature goes above 200°C. Therefore, it is important to recalibrate the rating of the casing for geothermal applications. In the following section, different modes of casing failure will be explained.

6 Casing failure due to collapse

The casing collapse in the geothermal well occurs when the entrapped fluid present in the annular space (casing to casing) starts to expand due to high temperature. The collapse takes place within the body of the casing and not from the coupling. To avoid this problem, make sure that no fluid is trapped in the annular space. To achieve this goal, an external or tieback liner that assures proper cementing job in the annular space should be used. Moreover, the free water of cement that fills the annular space should be zero for all pumping, pressure, and temperature condition (Southon, 2005).

7 Casing failure due to compression

When the temperature is increased in the surrounding of the cemented casing, then expansion in the lateral direction of the casing is restricted. This causes the accumulation of compressive stresses in the casing, and its magnitude is dependent upon the temperature variation. If the temperature goes above 230°C, then the point comes when the compressive stresses become so high that plastic deformation in the casing occurs.

It is found that thread jumping in the buttress thread occurs when the casing is subjected to compressive loading. Maruyama et al. (1990) stated that the buttress thread loses its sealing ability when the temperature exceeds 200°C. Therefore, to avoid such scenarios, whenever a buttress thread connection is used, care should be given that the initial tensile stress in the casing is sufficient enough to counter the effect of compressive stresses when the temperature rises. Moreover, when two-stage cementing is used, pre-tension is also a possibility for tie-back and single string casing.

8 Casing failure due to degradation

The degradation of the casing in the geothermal condition occurs when the corrosion reduces the wall thickness which causes buckling. Fig. 14 shows the localized buckling that happened in the simulated corroded zone.

FIG. 14 Localized buckling of the casing caused by the corrosion (Teodoriu, 2015a).

The most corrosive gas present in the geothermal environment is CO_2, which attacks both nonmetal and metal components in the well, and when mixed with water/steam, it produces carbonic acid which is the main cause of carbon corrosion (Ueda and Takabe, 2001).

Corrosion can be through erosion or pitting. Pitting corrosion can create leakage problems while decreasing the fatigue life of the casing caused by the concentration of the localized stresses (Pohjanne et al., 2008), whereas erosion corrosion tends to remove the protective layer from the casing surface and gives way to more aggressive chemical corrosion. The cause of erosion corrosion in the geothermal well is the high flow rates that are required to make the project economically feasible.

9 Casing failure due to fatigue

The failure in the casing can also occur through fatigue, when the casing is continuously exposed to cyclic loading resulting in the accumulation of damages which can then lead to the ultimate failure of the casing (Teodoriu, 2015b). Fatigue can be of different types, and it depends on the downhole condition and mode of operation being performed on the well.

The dynamic loading in the casing string builds up when the casing is lowered in the well and is stopped multiple times during the trip-in process. The lowest casing will be exposed to a maximum dynamic load due to start/stop of operation, whereas during the drilling operation, fatigue is caused by the vibration and drilling of the subsequent section. Drilling in the hard formation can cause repeated interaction of the drilling string with the casing, which creates severe fatigue that can lead to casing failure.

In the geothermal environment, the casing will be exposed to an additional fatigue, which is due to the cyclic thermal loading. Large changes in the temperature along with the limited movement of the casing increase the level of stress, compromising the casing from its connection. Teodoriu and Falcone (2009) concluded that the fatigue resistance of the N80 buttress connection is dependent on temperature fluctuation. Moreover, based on the experiments, they suggested that within the temperature range of 100–250°C, the N80 buttress connection is prone to failure when subjected to fluctuation cycles between 10 and 110.

10 Risk assessmenet through feature, event, and process

In order to reuse the conventional oil and gas wells, a risk assessment has to be made to know whether or not the well can be repurposed for geothermal purposes. Srivastava et al. (2022) presented a risk assessment technique that was based on feature, event, and process (FEP). However, it is worth mentioning that there is no particular FEP for a geothermal well, but the authors used the generic FEPs from the carbon capture and sequestrations that is given in Quintessa (2013). The FEPs were initially designed for the disposal of the radioactive waste, and then, the concept was adopted to the CCS wells. The FEP gives the qualitative risk assessment and can be used as a checklist to ensure that all the scenarios related to FEP have been considered during the designing phase.

In order to convert a qualitative risk assessment to a quantitative one, the interaction matrix along with the incident potential matrix (IPM) has to be used, whereas to find the critical element in the given geothermal system that can be vulnerable to failure, cause-and-effect plot diagram has to be utilized. In their study, they presented the FEP along with the interaction matrix, IPM, and cause-and-effect plot diagram for the tubingless geothermal well completion, as this will be the most common completion if the conventional oil and gas wells have to be repurposed for geothermal purposes. Tables 5–7 and Figs. 15 and 16 show the interaction matrix along with the cause-and-effect plot. The critical elements used in their study were (a) casing, (b) cement sheath and plug, (c) water composition, and (d) temperature. However, more elements can be added which will make the interaction matrix bigger. For further explanation on this particular topic, go through the paper of Srivastava et al. (2022).

TABLE 5 Interaction matrix of tubingless geothermal well completion (Srivastava et al., 2022).

[1,1] Casing	[1,2] a) Contraction and expansion b) Debonding c) Micro annulus d) Cracks e) Tensile stresses	[1,3] a) Sorption	[1,4] a) Heat loss
[2,1] a) Contraction and expansion b) Corrosion e) Compression failure	**[2,2] Cement sheath and plug**	[2,3] a) Chemical equilibria b) Sorption	[2,4] Heat loss
[3,1] a) Corrosion b) Erosion c) Scaling	[3,2] a) Degradation b) Carbonation and bi carbonation c) Advection flow d) Precipitation e) Buoyancy flow f) Change in transfer properties	**[3,3] Water composition**	[3,4] Heat transfer
[4,1] a) Thermal stresses/loading b) Reduction in burst pressure d) Reduction in collapse pressure e) Reduction in yield strength f) Thermal expansion g) Plastic strain h) Thread jumping (API long)	[4,2] a) Circumferential cracking b) Axial cracking c) Debonding of the cement from formation or casing d) Shear failure e) Cement crushing b) Cyclical axial stress	[4,3] a) Dissolution of gas b) Precipitation and dissolution of minerals	**[4,4] Temperature**

TABLE 6 Cause-and-effect made from IPM and fitted in interaction matrix (Srivastava et al., 2022).

	1	2	3	4	Cause
	[1,1] **Casing**	[1,2] **E5** *4*	[1,3] **A2** *1*	[1,4] **E1** *3*	8
	[2,1] **E4** *4*	[2,2] **Cement sheath and plug**	[2,3] **A2** *1*	[2,4] **E1** *3*	8
	[3,1] **D4** *4*	[3,2] **B4** *3*	[3,3] **Water composition**	[3,4] **E1** *3*	10
	[4,1] **E4** *4*	[4,2] **E5** *4*	[4,3] **C1** *2*	[4,4] **Temperature**	10
Effect	12	11	4	9	

TABLE 7 Cause-and-effect of different elements (Srivastava et al., 2022).

Elements	Cause	Effect
Casing [1,1]	8	12
Cement sheet and plug [2,2]	8	11
Water composition [3,3]	10	4
Temperature [4,4]	10	9
Mean	9	9

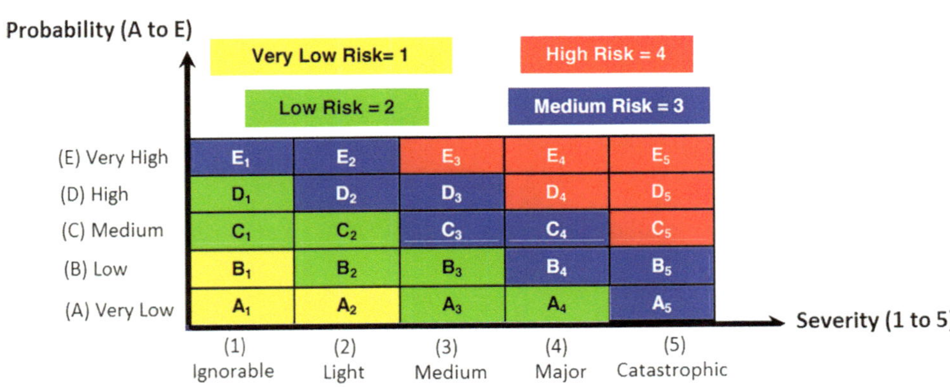

FIG. 15 Incident potential matrix for the risk assessment (Condor and Asghari, 2009).

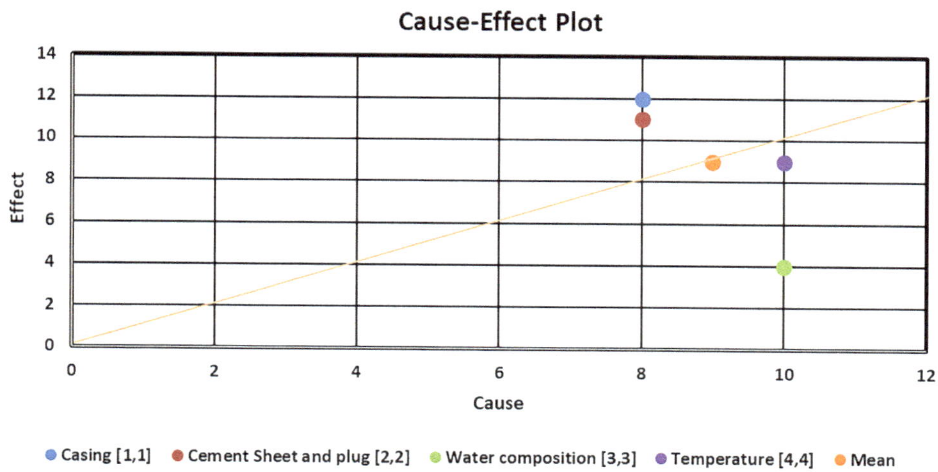

FIG. 16 Cause-and-effect plot diagram of tubingless casing (Srivastava et al., 2022).

From the figure, it can be seen that the most affected element in the geothermal well is the casing and then cement, while the elements that cause the most problem are the water composition and temperature. Therefore, care should be given to the casing when old/abandoned oil and gas wells have to be reused for geothermal purposes, as this is the most vulnerable element in the given geothermal system.

References

Abid, K., Raoof, G., Muktadir, G., 2020. A pozzolanic based methodology to reinforce Portland cement used for CO_2 storage sites. J. Nat. Gas Sci. Eng. 73, 103062.

Abid, K., Sharma, A., Ahmed, S., Srivastava, S., Velazco, A.T., Teodoriu, C., 2022. Review on geothermal energy and HPHT packers for geothermal applications. Energies 15, 7357.

Abid, K., Srivastava, S., Romero, M.L., Tellez, M.A., Teodoriu, C., 2023. Experimental and machine learning study of thermal conductivity of cement composites for geothermal wells. Geothermics 110, 102659.

API, 2005. API Speciation 10-A. API.

Bronicki, L.Y., 2017. Geothermal power conversion technology. In: Encyclopedia of Sustainability Science and Technology. Springer.

Budisulistyo, D., Krumdieck, S., 2015. Thermodynamic and economic analysis for the pre-feasibility study of a binary geothermal power plant. Energy Convers. Manag. 103, 639–649.

Carroll, S., William Carey, J., Dzombak, D., Huerta, N.J., Li, L., Richard, T., Um, W., Walsh, S.D.C., Zhang, L., 2016. Review: role of chemistry, mechanics, and transport on well integrity in CO_2 storage environments. Int. J. Greenhouse Gas Control 49, 149–160.

Condor, J., Asghari, K., 2009. An alternative theoretical methodology for monitoring the risks of CO_2 leakage from wellbores. Energy Procedia 1, 2599–2605.

DeMong, K., Rivenbark, M., 2003. Planning the well construction process for the use of solid expandable casing. In: SPE/IADC Middle East Drilling Technology Conference and Exhibition. SPE, Abu Dhabi.

DiPippo, R., 2016. Geothermal Power Generation (Development and Innovation). Woodhead.

DOE. 2022. A History of Geothermal Energy in America. Accessed July 14, 2022. https://www.energy.gov/eere/geothermal/history-geothermal-energy-america#:~:text=Archaeological%20evidence%20shows%20that%20the,as%20a%20source%20of%20healing.

Dudley, B., 2017. BP Energy Outlook, 2017 ed. British Petroleum.

Dupal, K., Naquln, C.J., Dalgle, C., Cook, L., York, P., 2000. Well Design with Expandable Tubulars Reduces Costs and Increases Success in Deepwater Applications. Deep Offshore Technology.

Edwards, L.M., Chilingar III, G.V., Rieke, and W H. Fertl., 1982. Handbook of Geothermal Energy. Gulf Publishing Co, Houston.

EIA, 2016. International Energy Outlook 2016. U.S. Energy Information Administration.

Fichter, C., Reinicke, K.M., Teodoriu, C., Wang, Y., Lotz, U., 2009. Influences and limiting parameters of casing design on the success of hydrogeothermal projects - a new approach for the design of geothermal wells. In: Der Geothermiekongress 2009, Bochum.

Goetzl, G., 2020. MUSE—Differences Between Deep and Shallow Geothermal Energy. pp. 7–22. Accessed July 24, 2023 https://geoera.eu/blog/muse-differences-between-deep-and-shallow-geothermal-energy/#:~:text=Moreover%2C%20deep%20geothermal%20energy%20allows,combined%20heat%20and%20power%20applications.

Hole, H., 2008. Geothermal well cementing. In: 26th Workshop—Petroleum Engineering Summer School, Dubrovnik.

ISO/TS 16530-2, 2014. Well Integrity—Part 2: Well Integrity for the Operational Phase. ISO/TS.

Kosinowski, C., Teodoriu, C., 2012. Study of class G cement fatigue using experimental investigations. In: SPE/EAGE European Unconventional Resources Conference and Exhibition. SPE, Vienna.

Kumar, N., Marker, R., Corneliussen, R., Kirkemo, E.G., Rønneberg, E., 2010. Mono-diameter expandable drilling liner applications in deepwater—ready or not? In: Drilling Contractor. https://drillingcontractor.org/mono-diameter-expandable-drilling-liner-applications-in-deepwater-%E2%80%93-ready-or-not-4565.

Drilling Manual, 2022. API Cement Classes in Oil and Gas. 10 06 https://www.drillingmanual.com/api-cement-classes-in-oil-and-gas/.

Marinovation, 2011. CoilFlat CT Liner. Technical University of Clausthal, Germany.

Maruyama, K., Tsuru, E., Ogasawara, M., Inoue, Y., Peters, E.J., 1990. An experimental study of casing performance under thermal cycling conditions. SPE Drill. Eng., 156–164.

Moya, D., Aldás, C., Kaparaju, P., 2018. Geothermal energy: power plant technology and direct heat applications. Renew. Sust. Energ. Rev. 94, 889–901.

Mueller, D.T., GoBoncan, V., Dillenbeck, R.L., Heinold, T., 2004. Characterizing casing-cement-formation interactions under stress conditions: impact on long-term zonal isolation. In: SPE Annual Technical Conference and Exhibition. SPE, Houston.

Nelson, E.B., Guillot, D., 2016. Well Cementing. Schlumberger, Houston.

NORSOK D-010, 2013. Well Integrity in Drilling and Well Operations. NORSOK.

Pohjanne, P., Carpén, L., Hakkarainen, T., Kinnunen, P., 2008. A method to predict pitting corrosion of stainless steels in evaporative conditions. J. Constr. Steel Res. 64, 1325–1331.

Quintessa, 2013. The Generic CO_2 Geological Storage FEP Database, Version 2.0.0. December. Accessed April 21, 2022 https://www.quintessa.org/co2fepdb/v2.0.0/#toc.

Shenold, C., Teodoriu, C., 2021. Systematic Well Integrity Approach in HTHP and Geothermal Applications. GRC Transactions. GRC.

Shortall, R., Davidsdottir, B., Axelsson, G., 2015. Geothermal energy for sustainable development: a review of sustainability impacts and assessment frameworks. Renew. Sust. Energ. Rev. 44, 391–406.

Siddiqui, A., York, P., 2012. Monobore solid expandable liners—redesigning Wells for a more economical and operational benefit. In: International Petroleum Technology Conference. SPE, Bangkok.

Southon, J.N.A., 2005. Geothermal well design, construction and failures. In: World Geothermal Congress. International Geothermal Association, Antalya.

Srivastava, S., Teodoriu, C., 2018. Gas well conversion to geothermal: a case study for Oklahoma. In: 43rd Workshop on Geothermal Reservoir Engineering Stanford University. Stanford University, Stanford.

Srivastava, S., Abid, K., Teodoriu, C., 2022. Re-visiting geothermal well integrity definitions. In: Geotherm. Resour. Council Trans., pp. 1850–1862.

Teodoriu, C., 2015a. Why and when does casing fail in geothermal wells: a surprising question? In: Proceedings World Geothermal Congress, Melbourne.

Teodoriu, C., 2015b. Dynamic casing shoe while drilling: a smart drilling concept for future monobore technologies. J. Nat. Gas Sci. Eng. 27, 1279–1286.

Teodoriu, C., Falcone, G., 2009. Comparing completion design in hydrocarbon and geothermal wells: the need to evaluate the integrity of casing connections subject to thermal stresses. Geothermics 38, 238–246.

Teodoriu, C., Lehr, J., Grimmer, H., 2012. A new method for casing shoe depth calculation with application to monobore well construction. In: DGMK/ÖGEW-Frühjahrstagung 2012, Fachbereich Aufsuchung und Gewinnung. DGMK, Celle.

Teodoriu, C., Yi, M.C., Salehi, S., 2019. A novel experimental investigation of cement mechanical properties with application to geothermal wells. Energies (MDPI) 12, 3426.

Ueda, M., Takabe, H., 2001. The formation behavior of corrosion protective films of low Cr bearing steels in CO_2 environments. In: CORROSION 2001. SPE, Houston.

Varahram, A., Bar, F., Hassel, T., Bach, F.-W., 2011. Cost-efficient Monobore well COnstruction for geothermal energy. In: 16th Annual Conference on Industrial Engineering Theory, Application & Practice. IJIE, Stuttgart.

Zhang, W., Eckert, A., 2020. Micro-annulus generation under downhole conditions: insights from three-dimensional staged finite element analysis of cement hardening and wellbore operations. J. Rock Mech. Geotech. Eng. 12, 1185–1200.

CHAPTER

6

High-temperature geothermal well cementing

Axel-Pierre Bois[a], *Simon James*[b], *and Anthony Badalamenti*[c]

[a]XGS Energy, Houston, TX, United States [b]CS8 Consulting, Le Plessis-Robinson, France
[c]CURISTEC, Neuville-sur-Saône, France

Abbreviations

APB	annular pressure buildup
BHCT	bottom hole circulating temperature
BHST	bottom hole static temperature
BWOC	by weight of cement
C/S	mole ratio of CaO to SiO_2
CAC	calcium aluminate cement
CAP	calcium aluminate phosphate
CAPEX	capital expenditure
CASH	calcium aluminum silicate hydrate
CCS	confined compressive strength
CHI	cement hydraulic integrity
CMI	cement mechanical integrity
DV	differential valve
ECD	equivalent circulation density
FC	forward circulation
OPEX	operational expenditure
PBR	polished bore receptacle
RC	reverse circulation
SAGD	steam-assisted gravity drainage
SRC	system response curve
SWOT	strengths, weaknesses, opportunities, threats
UCS	unconfined compressive strength

Geothermal Energy Engineering
https://doi.org/10.1016/B978-0-443-21662-6.00005-0

Copyright © 2025 Elsevier Inc. All rights are reserved, including those for text and data mining, AI training, and similar technologies.

High-temperature hydro-geothermal, enhanced geothermal, and advanced geothermal wells are different from oil and gas wells since, in geothermal wells (Fig. 1):

1. The deepest tubular shoe is placed high above the reservoir. The fluids are produced through a slotted liner instead of a perforated cemented tubular, to maximize flow rate since thermal power is a direct function of the flow rate.
2. All casing strings are cemented to the surface, with no open annular space. The rationale behind this design is to minimize the free axial expansion of the casing and the APB (annular pressure buildup) related to increases in temperature, which could lead to critical loading cycles. For example, an increase of temperature from 100°C to 200°C would submit a 1000-m free pipe to more than 1-m axial expansion, which would require special expansion spool completion wellhead designs (Putri et al., 2020). An increase of 100°C in the temperature of a trapped fluid could result in a pressure increase of about 180 MPa.
3. There are often loss zones above and in the reservoir. These zones may be naturally fractured formations, formations with low fracture gradients that are fractured while drilling, or formations that are thermally fractured. Lost circulation could impair the perfect placement of the cement slurry around the tubular, hence leading to local non-cemented tubular sections, where the casing could be submitted to corrosion or buckling. Such was encountered on the IDDP2 well in Iceland at about 2300-m depth, and this is probably the cause of the severe casing damage encountered at this depth (Hoang et al., 2022). Lost circulation can be addressed by the use of lightweight cement slurries (formations with low fracture gradients) and lost circulation treatments.
4. Cement placement often happens after having significantly reduced the formation temperature by pumping/circulating cold fluid (Friðleifsson et al., 2021).
5. Formation fluids can be very acidic and may degrade the cement sheath after it has set (Pyatina et al., 2023) and corrode the casing.
6. Cement sheaths are submitted to very high temperatures and very large temperature cycles; the temperatures could reach super critical conditions (Sugama and Pyatina, 2022; Pyatina and Sugama, 2023a,b).

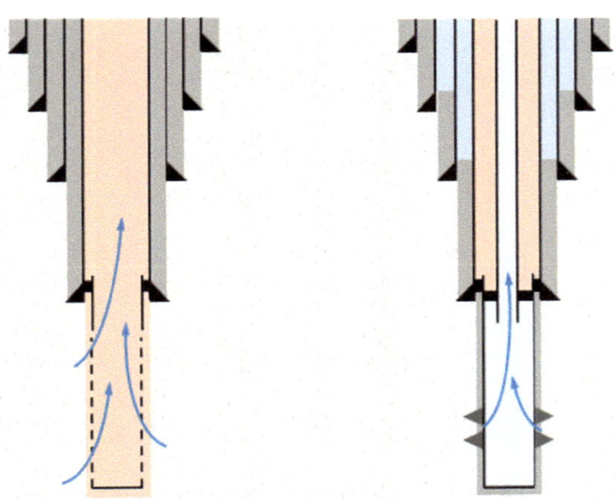

FIG. 1 Geothermal well (left) versus oil and gas well (right) architectures.

7. The objective of a producer well, being to bring heat to surface, the thermal conductivity of the set cement should be low for the cement sheath to behave as a thermal insulator, hence decreasing thermal losses (Pyatina and Sugama, 2023a,b).

1 Introduction

The 1960s marked the beginning of the design of cement systems dedicated to geothermal wells. Such work mainly focused on the long-term cement chemical stability when cement is exposed to high temperatures and acidic environments (Ostroot and Shyrock, 1964; Kukacka, 1977, 1981, 1982; Zeldin et al., 1979; Milestone et al., 1986). More recently, the importance of thermal cycling on the mechanical failure of cement systems has been identified (Kosinowski and Teodoriu, 2012; Yuan et al., 2013; Shadravan et al., 2015), and work has been engaged to modify the mechanical and thermal properties of cement systems (Sugama, 2006). However, a quick look at the available literature shows that there is still a lack of studies and data on the cement hydraulic, chemical, mechanical, and thermal behavior at temperatures above 200°C, as will be exposed in this book chapter.

The performance requirements for high-temperature geothermal well sealants are bounded by three extreme conditions:

- The first condition is when low-pH, chemically aggressive brines are encountered in the drilled formations. This can occur anywhere along the well. When these fluids contact the annular sealant, they may chemically degrade it, creating a path for these fluids to attain the casing, leading to casing corrosion and, possibly loss of the well.
- The second condition is for the shallow well depths where sealants are placed at low temperature and set at low temperature. During production, the upper well section is exposed to high temperature, and therefore, to significant thermal stress. When production is stopped, or cold water is injected, during a workover for example, the temperature tends to decrease, inducing thermal cycles during which the temperature remains higher than the temperatures at which cement has set.

  This extreme condition is also encountered for the deepest casing/liner strings when the well has been significantly cooled down prior to cementing. Production leads to increasing the downhole temperature and cold water injection to decreasing it.
- The third extreme condition is for the deepest casing/liner strings, where the sealants are placed at elevated BHCT and set at high temperature if no cooling has occurred prior to cementing. The sealant is exposed to cooling if cold water is injected in an injection well for example, or during a workover made on a producer well. This subjects the cement sheath to thermal stresses and possibly temperature cycles during which the temperature remains lower than the temperature at which the cement has set.

The structure of this chapter is based on the workflow used when designing a cement system:

1. Chemical design.
2. Hydraulic design.
3. Integrity design.

Generally, during the cement design process, one task is not engaged or validated if the previous one is not successful since, for example, having a cement with good mechanical properties is of little interest if the cement slurry is not stable and not efficiently placed in the annular space between the tubular and the wellbore.

2 Chemical design

There are four main parts to designing cement systems for high-temperature geothermal wells. Firstly, the chemistry of the cement system must be selected to ensure thermal stability. Secondly, the cement system must have suitable chemical resistance to any corrosive brines with which they will be in contact. The density of the cement slurry must also be adjusted to minimize the risk of fracturing weak formations leading to loss of cement slurry to the formation. Finally, the properties of the cement slurry must be controlled to ensure the correct placement and setting of a homogeneous fluid in the annulus.

3 Cement system chemistry—High temperatures

By far, the most common sealant systems used for oil and gas wells contain Portland cement as the main hydraulic binder. However, for high-temperature geothermal wells, alternate systems have been developed to overcome some of the perceived limitations of Portland cement systems when placed in very hot formations and in contact with acidic brines. The following sections give an overview of the different high-temperature resistant sealant systems used in geothermal wells.

3.1 Portland cement systems

It has been known, for many years, that at least 35% BWOC (by weight of cement) fine silica must be added to Portland cement to provide thermal stability (prevent the decrease in compressive strength and increase in permeability) at temperatures above about 110°C (Menzel, 1935; Kalousek, 1952). For geothermal well applications, preventing the increase in permeability is the most important as increased permeability may lead to more rapid corrosion of the casing.

Several authors have also demonstrated that for different-sized silica particles, 35% BWOC is a minimum concentration of silica, and they assume that the silica particles are sufficiently small that all the silica is consumed during the reactions. Patchen (1960) studied the effect of silica particle size and concentration on Portland cement systems aged for up to 30 days at (160°C) 320 °F. He concluded that a minimum of 35% BWOC silica was required to eliminate strength retrogression and high permeability. In one set of data, the effect of different silica particle sizes was shown. The larger size silica (100 mesh—149 μm) took longer to react but provided the same ultimate performance.

Eilers and Root (1974) studied the impact of silica particle size and concentration on set cement properties after aging at temperatures of (232°C) 450 °F and (316°C) 600 °F for up to 2 years. They evaluated 325 mesh powdered silica (particle sizes <44 μm) and 100 to 325 mesh silica sand (particle sizes between 44 μm and 149 μm). The powdered silica used would be classified today as silica flour. They found that 35 to 40% BWOC silica was sufficient to prevent strength retrogression of most Portland cement systems. This concentration of silica brings the C/S ratio (mole ratio CaO to SiO_2) to 1. If a minimum amount of silica is used, then it should be 325 mesh or finer. Systems with at least 35% BWOC silica formed tobermorite after 24 h, which transformed to xonotlite at longer times. The tests of Eilers and Root were performed at the saturated steam pressures: 2.8 MPa at 232°C and 10.5 MPa at 316°C.

The conversion temperature of tobermorite to xonotlite is significantly influenced by pressure. Heller and Taylor (1951) observed that tobermorite was formed hydrothermally at temperatures between 110°C and 150°C from lime silica mixtures with a C/S = 1 at saturated steam pressures. Above 160°C, xonotlite was produced. Speakman (1968) examined the stability range of tobermorite at pressures from 30 to 280 MPa and temperatures up to 398°C. At pressures above 69 MPa, tobermorite was stable at temperatures up to 285°C. Few tests were performed at lower pressures, but there was evidence to suggest that the stability temperature of tobermorite decreased with decreasing pressure.

Nelson and Eilers (1985) indicated that, for practical purposes, at least for thermally cycled wells, the temperature limit for Portland cement + silica systems is about 320°C. Much more recent work by Stiles (2006) has shown that systems based on Portland cement with at least 40% BWOC silica were stable when exposed to temperatures of 340°C for 2 years. The paper did not contain any mineralogical analysis of the formed phases, but did include values of Young's modulus, Poisson's ratio, and tensile strength in addition to the usual compressive strength measurements.

Roy et al. (1979) examined the minerals present in calcium silicate systems at temperatures from 250°C to 400°C. At 400°C, xonotlite dehydrated to wollastonite (β-$CaSiO_3$) and water. Systems containing wollastonite fibers were "more fragile" than those containing xonotlite. At 400°C, tests at different pressures showed that increasing pressures delayed the formation of wollastonite: after 10 days, only xonotlite was seen in systems cured at a pressure of 172 MPa, while, at lower pressures (68.9 and 17.2 MPa), wollastonite was observed. This is not surprising as the transition from xonotlite to wollastonite liberates water leading to an increase in overall volume. There was some indication that wollastonite was forming during tests performed at 350°C but the limited duration of tests did not demonstrate this convincingly. However, these results do not contradict the observations of Nelson and Eilers (1985) and Stiles (2006) that Portland cement and silica systems are stable up to 320°C.

Several authors (Stiles, 2006; Chougnet-Sirapian et al., 2011) have cured cement samples in water-filled cells containing silica sand, with the sand used to simulate the formation. They compared the properties of silica-stabilized Portland-cement systems cured in water and in water and silica sand at temperatures of 250°C and 344°C for 6 months. The systems cured in water and sand had significantly higher compressive strength, tensile strength, and Young's modulus than those cured in water. XRD analysis showed that xonotlite was the main phase formed under both aging conditions, and there were no signs of phases with lower C/S ratios in the systems cured in silica sand, with identical traces from both the center and exterior of the samples. For samples cured in silica sand there was a broad peak centered on a d-spacing of 4 which the authors attributed to baseline differences.

Da Silva et al. (2012) examined the interface between cement set inside rock samples, both with and without the presence of a bentonite filter-cake. The systems were initially cured at

90°C for 2 days and subsequently cured for 26 days at either 150°C or 300°C. They observed interactions at the interface between cement and rock, with the zone being wider if bentonite filter-cake was present. At 300°C, there was a distinct break at the interfaces, and this was attributed to molar volume changes, shrinkage, or differential thermal expansion. The latter is a key observation as it may be a laboratory-induced artifact (samples are cooled down for analysis), but it may also reproduce to some extent the thermal conditions the cement at shallow well depths may experience when production starts and stops.

As a summary, Portland cement stabilized with at least 35% BWOC silica has been shown to have long-term thermal stability at temperatures up to 340°C, where xonotlite will be the main binding phase present. Potential issues with chemical and mechanical resistance of systems based on Portland cement will be discussed later.

3.2 CASH cement systems

The use of CASH (calcium aluminum silicate hydrate) cement systems for very high-temperature applications was first examined in the late 70s. Roy et al. (1979) examined formulations of calcium aluminum silicate hydrate systems at temperatures up to 400°C using different starting materials with particular interest around the anorthite composition. The formation of cementitious products was dependent on the starting materials. Systems formed from calcium aluminate cement and fine silica formed cementitious systems containing hexagonal anorthite. However, systems formed from boron-stabilized di-calcium silicate, metakaolin, and silica produced triclinic anorthite, which were not cementitious. In a subsequent publication, difficulties in retarding CASH cement systems were mentioned (Roy et al., 1980).

Later work (Barlet-Gouédard et al., 2008) examined systems formed from class G cement and calcined metakaolin to supply the required aluminum ions. The inventors used different particle sizes to provide systems with suitable slurry properties for well applications. A thickening time of 3 h at 149°C was shown in an example. There has not been any subsequent publication to indicate that these types of systems have been used commercially.

There have been more recent studies on the chemistry and morphology of CASH cement systems cured above 200°C (Kyritsis et al., 2009; Meller et al., 2005, 2007, 2009), but little or no work has been published on controlling the slurry properties of these systems. The control of slurry properties is one of the key areas requiring investigation for many newer systems that have promising set material properties.

3.3 CAC systems

CAC (calcium aluminate cement) systems have generally found limited use in oil well cementing. However, due to their stability at very high temperatures, they have been used in fire floods, also called in-situ combustion wells. In these wells, in-situ combustion of high-viscosity hydrocarbons is used to enhance the production of said hydrocarbons. Cement slurries are placed at low temperatures but are exposed to very high temperatures (370°C–930°C) in the injection wells during combustion.

Nelson and Eilers (1985) refer to the use of calcium aluminate cements for wells with high-temperature fluctuations as they have better thermal shock resistance than Portland cement systems. They attributed this performance to the absence of calcium hydroxide in calcium aluminate cement systems.

When calcium aluminate cement hydrates at low temperatures, several metastable phases are formed (CAH_{10}, C_2AH_8, and C_4AH_{13}), which then convert to C_3AH_6. C_3AH_6 is the stable hydrated phase of calcium aluminate cement at temperatures up to about 225°C. The conversion of the metastable phases to C_3AH_6 leads to a reduction in compressive strength. This reduction in compressive strength is generally not an issue if the water-to-cement ratio is kept below about 0.4. Above 225°C, water is lost and $C_3AH_{1.5}$ is formed. At temperatures higher than 550°C, the $C_3AH_{1.5}$ decomposes with the liberation of CaO and the formation of $C_{12}A_7$. This transition would indicate that calcium aluminate cement systems may not be thermally shock resistant if heated to above 550°C.

Kulu et al. (2021) examined the thermal properties of calcium aluminate cement systems. They showed that, for pure calcium aluminate cement systems, significant shrinkage occurred from about 260°C to 380°C. This corresponded to the loss of water with the transition from C_3AH_6 to $C_3AH_{1.5}$. Systems with quartz sand showed significant expansion above about 550°C, which corresponded with the conversion of alpha to beta quartz.

One limitation with the application of calcium aluminate cements in high-temperature wells is control of the thickening time. Michaux (1999) examined a range of retarders for high alumina cement systems using isothermal calorimetry and thickening time tests. Citric acid at 0.2% BWOC gave a thickening time of 125 min at 50°C while the onset of the exothermic peak was 200 min using isothermal calorimetry at the same temperature. The longest thickening time achieved at 70°C (the highest temperature tested) was 4 h with 0.5% BWOC glucose. These retarders should be sufficient to allow placement of calcium aluminate cement systems in the shallow sections of geothermal wells.

3.4 CAP cement systems

A detailed discussion on the development and first applications of CAP (calcium aluminate phosphate) cement systems is given by Sugama (2006). The system was developed to withstand the chemical environments in high-temperature geothermal wells where cement may be exposed to:

1. Saline brines containing high concentrations of CO_2 (pH ~5) at temperatures up to 320°C (deep sections of the well).
2. Brines at $pH < 1.5$ containing H_2SO_4 and CO_2 at temperatures up to 200°C (shallow sections of the well).

The system developed and applied commercially consists of calcium aluminate cement, fly ash, and sodium hexametaphosphate. The hydrothermal chemical reactions of the system are described below.

Firstly, there is a rapid reaction between calcium ions from the calcium aluminate cement and the sodium polyphosphate to form calcium hydrogen phosphate. There is a further reaction with calcium aluminate to forming hydroxy apatite and aluminum oxide hydroxide (boehmite).

The reaction scheme for calcium monoaluminate:

Release of calcium ions:

$$CaO.Al_2O_3 + 4H_2O \rightarrow Ca^{2+} + 2Al(OH)_4^-$$

Acid-base reaction with phosphate ions:

$$Ca^{2+} + NaHPO_4^- \rightarrow Ca(HPO_4) + Na^+$$

$$Al(OH)_4^- + H^+ \rightarrow AlH(OH)_4$$

Hydrothermal hydration:

$$Ca(HPO_4) + 4Ca^{2+} + 2NaHPO_4^- + xH_2O \rightarrow Ca_5(PO_4)_3(OH) + 2Na^+ + xH^+ + yOH^-$$

$$AlH(OH)_4 \rightarrow \gamma - AlOOH + 2H_2O$$

$$3\,Ca^{2+} + 2Al(OH)_4^- + 4OH^- \rightarrow 3CaO.Al_2O_3.6H_2O$$

where:

$Ca_5(PO_4)_3(OH)$ phase is hydroxyapatite.

γ-ALOOH is boehmite.

$CaO.Al_2O_3.6H_2O$ is hydrogarnet.

Subsequently, the fly ash, a source of alumina silicate, reacts with the sodium ions from the polyphosphate at high temperatures to form analcime ($NaAlSi_2O_6{\cdot}H_2O$).

CAP cement systems have been used commercially for ~25 years (Sugama, 2006), although in a small number of applications. There is, therefore, far less information available on the chemistry and mechanisms of action of the slurry additives used in this system than for systems based on Portland cement. The information that is available has generally come from one Service Company and may therefore reflect the preferences of that company for specific chemistries.

As can be seen from the reaction scheme of the CAP cement system described above, the polyphosphate ions will play a significant effect in the retardation of the system compared with calcium aluminate cement systems. Indeed, the CAP cement system appears to be easier to retard at high temperatures.

The first retarder used for CAP cement systems was citric acid (Sugama, 2006). It was postulated that carboxylate groups on the citric acid complexed calcium ions from the calcium aluminate cement, forming a retarding barrier on the surface of the CAC grains. The performance of the retarders was examined using differential scanning calorimetry measurements at a heating rate of 10°C/min. The results indicated that increasing the concentration of citric acid increased the onset temperature of the exothermic reaction. Relatively high concentrations (up to 8% BWOC) of citric acid were used.

Combinations of tartaric acid and citric acid, each added at 2.1% BWOC, gave a thickening time of almost 6h at 118°C (Reddy et al., 2005).

Xu et al. (2014) demonstrated the effects of using combinations of retarders: zirconium lactate (CL-23), citric acid, and tartaric acid. Long thickening times were achieved at temperatures up to 149°C (300 °F), although in some cases, the sensitivity of thickening time to retarder concentration was quite high.

Combinations of low molecular weight polymers and citric acid have been shown to provide long thickening times at temperatures up to at least 177°C (Joseph et al., 2014). The low molecular weight polymers were sodium lignosulfonate and carboxymethyl cellulose. Using a combination of retarders, with different retarding mechanisms, for high-temperature applications, is common practice with Portland cement systems (Cadix and James, 2021).

In many of the slurry examples given in the patent literature cited above, very few formulations have contained a dispersant, possibly because retarders such as citric acid may also function as a dispersant. In one patent (Brothers, 2008), a sulfonated acetone formaldehyde condensation product was used as a dispersant. These dispersants are also used for Portland cement systems.

Benge and Dew (2006) discussed the field application of a CAP cement system for a well with a BHCT of 121°C. They stated that conventional fluid loss control additives are not effective in calcium aluminate cement systems and that an acceptable high-temperature fluid loss control additive had been developed for the application. The fluid loss control at a BHCT of 124°C was 79 mL. The fluid loss control additive was probably a cationic starch as described in reference (Reddy et al., 2005). In the patent example, an API fluid loss of 40 mL was obtained at 118°C.

4 Cement system chemistry—Corrosive fluids

The corrosion resistance of any material will depend on the conditions to which it is exposed, e.g., fluid type, duration, temperature, and static or flowing conditions. A given material may experience severe corrosion under one set of conditions but may be acceptable for application under a different set of conditions. This is why there are differing views as to whether Portland cement-based systems are suitable for application in environments with high concentrations of CO_2.

4.1 Portland cement systems

Portland cement-based systems are thermodynamically unstable in many acidic fluids including those containing high concentrations of CO_2. However, the rate of attack is slow as shown in many tests summarized in a report from Benedictus et al. (2009). Santra and Sweatman (2011) have discussed the applicability of Portland cement-based systems for CO_2 storage wells and highlighted issues with evaluating sealant performance and indicated ways to improve the performance of Portland-cement systems.

The two main concepts to improve system performance were:

- Reducing the permeability of the systems.
- Reducing the amount of Portlandite present by including pozzolanic materials.

Note that pozzolanic reactions also result in a reduction in permeability due to the formation of additional calcium silicate hydrates, so the benefit of adding pozzolans cannot be entirely attributed to the removal of Portlandite.

Koons et al. (1993) evaluated a range of Portland cement-based systems for geothermal wells, exposing the samples to brines saturated with CO_2 at a pressure of 20.7 MPa and a temperature of 288°C. One of their key findings was that the active solids:liquid weight ratio for geothermal slurries should be higher than 2:1. Active solids were described as those that contributed to the binder phases, so they would include cement, silica, and some pozzolans. This finding basically relates to the reduced permeability of low water content systems. Koons et al. recommended a Portland cement/class C fly ash blend (Bruckdorfer and Coleman, 1987) with an active solids-to-liquid ratio higher than 2 for applications in corrosive environments up to 204°C. An alternate recommendation was to use class G cement with 30% to 40% silica flour and up to 10% BWOC fume silica. The recommendation for low-density systems was to either foam the base slurry up to 30% quality, or to include glass bubbles in the designs.

Bruckdorfer and Coleman (1987) disclosed a sealant system with improved resistance to CO_2 that comprised class H cement and class C fly ash. The system was also designed at a high (compared with typical fly ash systems) density, thus providing a higher solids content and consequent lower permeability. The class C fly ash system showed better performance than the class F fly ash systems. This may be due to lime in the class C fly ash carbonating rapidly in the presence of CO_2 thus reducing the rate of penetration of CO_2 into the system.

Barlet-Gouédard et al. (2012) disclosed a Portland cement-based system with improved resistance to corrosion by CO_2. The system contained a high volume of solids to reduce permeability and a reactive pozzolan (metakaolin) to reduce the amount of Portlandite present and decrease the system permeability. The system also contained voids from air bubbles or hollow cenospheres where precipitation of calcite preferentially occurred. The precipitation of calcite in voids reduces the expansion of the system and reduces the risk of mechanical failure, at least in unconfined laboratory experiments.

Milestone and Aldridge (1990) evaluated the performance of cement systems based on Portland cements in fluids containing CO_2. They showed that carbonation occurs quickly in high-temperature Portland cement systems where the phases formed are tobermorite and xonotlite. Carbonation of these phases gave increased porosity systems. However, systems with a high C/S ratio when carbonated had high strength and low permeability. For high-temperature stabilized formulations (at least 35% BWOC silica), they found that high concentrations of CO_2 caused significant degradation of the cement but commented that at low concentrations of CO_2 carbonation may not be a problem. Their findings also support the discussion on evaluating cement chemical resistance in realistic conditions. Even if the formation fluid contains a high level of CO_2, the concentration will rapidly decrease at the cement/formation interface, and replenishment of the CO_2 concentration via diffusion will be an extremely slow process. This is the reason Bois et al. (2013), among others, described an integrated perspective to designing cement sheath in CO_2 environment, which consists in ensuring that no defect is induced in the cement sheath during the life of the well, such that CO_2 cannot penetrate deep in the cement sheath. Of course, this approach is only valid when looking at the process of cement axial degradation, and does not work for radial degradation since the cement sheath thickness is of the order of a few centimeters.

4.2 CAP cement systems

As discussed above, the main reason for the development of CAP cement systems was to develop a sealant system with improved resistance to acidic fluids containing CO_2 and

H_2SO_4. Studies (Sugama and Carciello, 1995; Sugama et al., 2002; Pyatina and Sugama, 2023a, b, Pyatina et al., 2023) have shown that CAP cement systems are less affected by fluids containing high concentrations of CO_2 than Portland cement-based systems. The CO_2 resistance of these systems has been attributed (Sugama, 2006) to the formation of very stable carbonated apatite and cancrinite (from analcime) and to the fact that boehmite is stable in the presence of CO_2.

However, in the presence of sulfuric acid containing brines, calcium sulfate (bassanite) was formed from the reaction with hydroxyapatite and hydrogarnet. The bassanite initially formed provided a protective layer against further attack, but spallation of the bassanite layers led to further attack. The chemical durability tests were performed with cylindrical cement samples (30 mm diameter, 70 mm long) and placed directly into the chemical solution 25 times the volume of the cement samples. As is discussed later in this chapter, these types of tests are extreme. In an actual field situation, confinement of the samples would not only reduce the contact between the brine and sample surface, but the rock would limit the spallation of the sample surface. These tests show that what is generally considered a chemically resistant sealant degrades under certain conditions.

5 Cement system chemistry—Chemical resistance testing

In geothermal wells, fluids corrosive to cement and casing are typically found at two depths (Sugama, 2006). The production zones at temperatures >300°C have fluids at moderate pH (~5) but with high (40,000 ppm) concentrations of CO_2 and concentrations of H_2S around 100 ppm. At shallow depths (<1000 m) formations can contain low pH (~1.5) fluids with sulfuric acid and 5000 ppm CO_2. These shallow formations may reach temperatures of 200°C during production.

Much of the early work on sealant systems for high-temperature geothermal wells concentrated on their chemical stability in the presence of corrosive fluids. For a given well, the importance of a chemical attack will depend on the location of the fluid that can chemically attack cement and corrode casing.

There are two main situations to consider (Fig. 2a): cement in front of a permeable zone containing a corrosive fluid (A) and in front of an impermeable formation (B).

The most critical situation is in front of zone A where chemical attack will be through the cement thickness (Fig. 2b). The presence of defects (radial or disking cracks) in the cement can provide a direct route for the brine to contact the outer casing surface and therefore accelerate corrosion of the casing.

In zone B, the cement will only be attacked by brine once the brine has migrated along the annulus from the permeable zone (Fig. 2c). This migration can occur through channels (poor drilling fluid displacement), cracks or microannuli in the cement sheath, or by attack of the cement in the annulus from the permeable zone upwards. As the length of the cemented annulus is much larger than the thickness of the cement sheath, corrosion of casing far from the zone containing the corrosive fluid is unlikely to occur if there are no defects (channels, microannuli) in the cemented annulus.

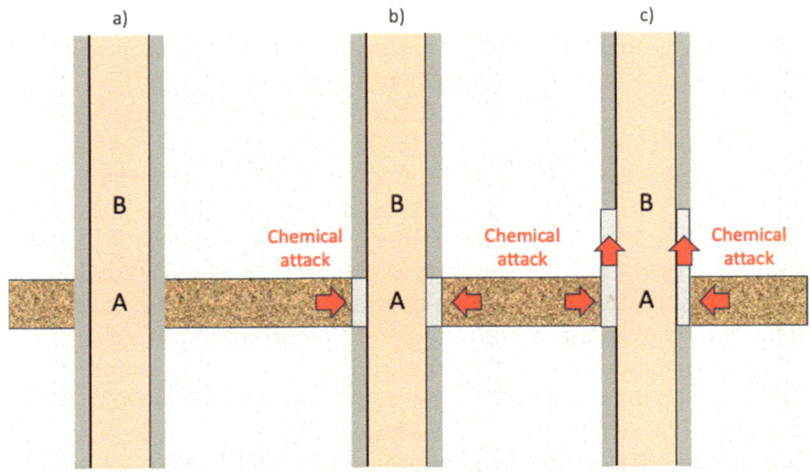

FIG. 2 Chemical attack of cement systems (a) Situations A and B; (b) Chemical attack for situation A; (c) Chemical attack for situation B.

5.1 Laboratory evaluation of chemical attack

In many laboratory tests to evaluate the chemical degradation of set cement, relatively small samples of cement are placed in large volumes of the reactive chemical in unconfined conditions. These conditions are not representative of field situations and may lead to pessimistic evaluations of sealant performance for a given situation and potential overdesign of the sealant system.

Since the early studies on the degradation of cement systems, significant work has been performed over the last 20 years related to the degradation of Portland cement systems by CO_2 for underground storage of CO_2. Although the chemical attack depends on the chemistry of the fluid and sealant and the exposure temperature, much can be learned from the work that has been performed. A technical report issued by the IEA Greenhouse Gas Programme (Benedictus et al., 2009) summarizes, in its Table 4.1, the results from many studies on the degradation of Portland cement systems in fluids containing CO_2 under different conditions.

The degradation rate was shown to be controlled by the diffusion of CO_2 into the pores of the cement system, with the diffusion rate being dependent on the conditions of the test. Even in the harshest conditions, the rate of attack along the length of the wellbore was extremely slow—12 m in 10^4 years in the worst case (wet CO_2 gas at 204°C). These results indicate that a chemical attack along the wellbore axis is not a significant issue unless there are voids (channels, microannuli) to allow the flow of brine along the axis.

In conventional tests (cement placed directly in a fluid), the rate of attack in many situations was sufficient to degrade the cement over its thickness (2–3 cm) in weeks to months. There was a rapid replenishment of ions at the surface of the cement system due to direct contact with the brine and a high brine-to-cement volume ratio. However, in actual conditions there is not an almost unlimited supply of the chemically aggressive ions at the cement surface. The ions are in the pores of the formation and need to diffuse through the formation pores to reach the cement sheath. Duguid et al. (2004) and Duguid (2008) presented two types of experiments demonstrating the importance of diffusion through rock. In the first case, the cement was placed directly in contact with carbonated brine, and in the second case, the cement was surrounded by permeable rock to represent a field situation more accurately. In both experiments, the brine was 0.5 mol NaCl and the temperature of exposure 50°C. In the first situation (direct contact with brine), the calculated time to degrade 25 mm of cement was between 60 and 100 years. In the second case, with cement surrounded by permeable rock, the degradation time was 28,000 years. With higher concentrations of CO_2 and higher temperatures the rate of attack will likely increase, but this will not change the fact that the presence of rock will significantly decrease the rate of attack.

Another issue with testing sealant samples in unconfined conditions is the impact of chemically induced mechanical failure. Sugama (2006) refers to this in relation to sulfate attack on CAP cements and Pyatina and Sugama (2020) show this in relation to sulfuric acid attack on Portland cement/silica systems. The attack by sulfate ions forms calcium sulfate on the surface of the samples, which should protect the sample from further attack. However, formation of calcium sulfate creates significant expansion of the cement system with consequent spalling of the surfaces and removal of the protective surface layer leading to further attack. In a confined test environment, spallation would not occur as the surface layer would be kept in place by the confining formation. On the contrary, in unconfined laboratory tests, this type of failure mode is likely to occur in situations where the chemical attack causes a significant increase in volume of solid components. Conversion of calcium hydroxide to calcium carbonate has a theoretical volume change of solids of 10%, whereas conversion of calcium hydroxide to calcium sulfate has a theoretical solid volume change of 120%. The relatively small volume change due to carbonation tends to plug the sealant porosity, reducing the rate of chemical degradation. However, the large volume changes of sulfate attack create mechanical failure.

Note that the ratio of acid volume to sample volume in the tests by Pyatina and Sugama (2020) was two and the acid was replaced every 3 days.

Milestone et al. (2012) also showed the spalling failure of laboratory-prepared samples exposed to CO_2-containing fluids in a well, with the outer carbonated layer just loosely held on the sample. The samples were aged in a well, and if the samples were brought to the surface too quickly (rapid depressurization), then the surface layer would detach. Milestone et al. also noted that the presence of carbonated layers has implications for the measurements of compressive strength and permeability and what the results meant for a well application. For example, brine could flow along the interface between the carbonated layer and the non-carbonated layer, whereas a permeability measurement is typically made with flow through the carbonated layer.

Laudet et al. (2011) conducted a coupled chemo-mechanical test at 90°C whereby lime-saturated water was injected initially into Portland cement plugs, followed by CO_2-saturated water, while applying radial and axial stresses equal to 3 and 6 MPa, the injection pressure being 2.5 MPa. They observed that adding CO_2 to the water injected throughout the mechanically stressed cement leads to ultimate clogging of the sample.

5.2 Field observations

An early example of severe corrosion of casing in geothermal wells was reported by Hedenquist and Stewart (1985). The wells were 12 to 16 years old and casing corrosion was clearly from the outside of the casing. The corrosive brine contained a high concentration of CO_2. Evaluation of the cement indicated that it had almost completely changed to calcite. However, there was no information given on the type of cement, how successfully it had been placed or whether it had failed mechanically.

A later paper (Shen and Pye, 1989) combined analysis of cement samples taken from geothermal wells and laboratory tests where Portland cement systems were placed directly in CO_2 laden fluids at 204°C and 232°C. The field analysis was quite revealing and, with today's knowledge, indicates a major contributor to CO_2 attack of cement and casing is mechanical failure of the cement sheath. Carbonated cement layers were found next to the cement-casing interface and along fractures in the cement. The carbonated layers were from 1 to 6 mm thick, so did not originate from carbonation through the cement from the outer formation but from the microannulus and fractures. The paper gave the size of fractures up to 500 μm although the size could have changed during recovery of the samples. However, the fact that carbonation was observed around the fractures indicates that the fractures were present downhole and that they were a significant contributor to cement carbonation and potential casing corrosion.

5.3 Conclusion

Although sealant systems may show significant degradation from chemical attack under typical laboratory test conditions, the consequences for field applications may not be severe. More realistic laboratory tests should be performed with the sealant surrounded by permeable rock to represent the depletion of ions more closely at the cement rock interface. The surrounding rock will also provide some mechanical support to the sealant, reducing the risk of mechanical failure caused by excessive expansion.

In a recent paper (Pyatina et al., 2023), samples of different cement systems were placed in the bottom of a non-producing geothermal well (temperature between 300°C and 350°C). The samples were cured for 24 h at 300°C in a laboratory autoclave before being placed in the well. Although the samples were aged in a well the geometry and exposure conditions were similar to those of a laboratory test: small samples were placed in a large volume of fluid, rather than being sandwiched between casing and formation. An additional complication of running exposure tests in a well is that samples must be recovered from the well slowly, to allow pore pressure to reduce.

6 Cement system chemistry—Mitigating loss circulation

To minimize the risk of fracturing weak formations with high hydrostatic pressure, lower density cement systems are used. There are three generic methods to reduce the slurry density of any of the systems discussed above, although there may be differences in details, e.g., chemistry of foaming surfactants or extenders, depending on the chemistry of the base system. The three methods are:

1. Adding more water.
2. Foaming with inert gas.
3. Adding lightweight particles.

Details on the chemical additive requirements and limitations of these methods for general oil and gas well applications have been described in detail in Nelson et al. (2006) and Cadix and James (2021), so they will not be reproduced here.

6.1 Water extended systems

Adding water to a system is the least efficient method of density reduction although it is generally the most cost-effective. The properties (e.g., compressive strength, permeability) of the base system may be significantly degraded, which may make the system unsuitable for high-performance applications. These systems would not be suitable in cases where the cement is required to isolate a formation containing a corrosive brine. The properties of different water-extended systems at different densities and curing conditions have been presented by Nelson and Eilers (1985).

Extended systems that contain extenders with high pozzolanic activity, such as natural pozzolans and fly ashes, can be used at temperatures up to about 232°C but, at higher temperatures, they lead to weaker systems, possibly due to the formation of reyerite (Eilers and Root, 1974).

6.2 Foamed systems

Foaming cement systems with inert gas such as nitrogen is an effective way to reduce the density of cement slurries. The thermal and chemical stability of the set foamed cement will be close to that of the base slurry, provided that the foam quality is less than 35%. For a conventional class G cement with 35% BWOC silica base fluid, this corresponds to a slurry density around $1200\,kg\,m^{-3}$.

One of the main challenges with foam cement systems is to ensure that the foam is stable during placement and up until initiation of setting. This may be particularly challenging at high temperatures, where stable foaming agents are required. Once a stable foam system has set, its thermal stability will be identical to that of the base slurry.

The stability of foam cement depends on the properties of the base slurry, the volume of gas, the surfactants used, and the method of preparing the foamed slurry. All these parameters impact the size of the gas bubbles and hence foam stability.

Sugama et al. (1986) used a sodium alpha-olefin sulfate surfactant and a cocodiethanolamide stabilizing agent to prepare foamed cements for evaluation at 300°C. They also used graphite fibers to prevent bubble separation in the slurry phase. There was no information on the shear conditions for foaming, which is key in determining the gas bubble size.

De Rozières and Ferrière (1991) examined the impact of pressure during the mixing of foamed cement. They observed that the higher the pressure (up to 7 MPa), the smaller the bubble size for a given foam quality. At a given pressure, increasing the foam quality increased the bubble size. De Rozières et Ferrière also linked their findings to observations made by Davies et al. (1981) that foamed cement generated under field conditions (high pressure and high shear) was much more stable than those made under laboratory conditions.

Typical foaming agents currently used are summarized in Table 1. There is little information available about their effectiveness in foam cement systems at very high temperatures.

Indeed, Heathman and Beck (2006) indicated that "foamed cement is not conducive to an environment that might reach 395°F during slurry placement." Foam cement has been used in an HTHP well in the North Sea (Ravi et al., 2003). The static temperature at the bottom of the section cemented with foam cement was 177°C, but no information was given about the BHCT.

TABLE 1 Typical foaming agents used to foam Portland cement systems (Cadix and James, 2021).

Type	Hydrophobic group	Hydrophilic group
Cationic	C8 to C16 alkyl chain	Quaternary ammonium
Anionic	C8 to C16 alkyl chain	Sulfonate
	C6 to C10 alkyl chain	Polyether sulfonate
Amphoteric	C10 to C18 alkyl chain	Betaine
	C10 to C18 alkyl chain	Dimethylamine oxide

6.3 Lightweight particles

There are two main types of hollow spheres that are regularly used in low-density slurries:

- Cenospheres (ceramic microspheres) are produced during the combustion of coal in power plants and contain mostly aluminosilicates.
- Hollow borosilicate glass bubbles.

The limitation with using hollow particles is that they will crush under hydrostatic pressure, with different particles crushing at different pressures. The hollow borosilicate glass bubbles are the most efficient at reducing slurry density as, for a given resistance to hydrostatic pressure, borosilicate glass bubbles have a lower density than cenospheres. When designing systems with lightweight particles, additional care should be taken to ensure that the slurries are stable, i.e., there is no creaming of the particles.

Systems containing cenospheres have been shown to degrade with time at elevated temperatures (Nelson and Barlet-Gouédard, 2006). This is the reason why Koons et al. (1993) recommended using cenospheres for applications up to 204°C and glass bubbles at higher temperatures. The limitation on hollow glass bubbles was economic and hollow glass bubbles can be used at low temperatures.

To maximize the benefit of using lightweight particles, the slurry should have low permeability, which is achieved by maximizing the solids-to-liquid weight ratio as recommended by Koons et al. (1993). Drochon and Garnier (2003) disclosed designs using cenospheres and glass bubbles that contained a high volume (>50%) of solids and with densities below 1300 $kg\,m^{-3}$. Compressive strengths higher than 7 MPa were achieved with a slurry density of 900 $kg\,m^{-3}$. The performance was achieved by optimizing the sizes of the different blend components to provide optimum packing. The optimized particle packing concept has been applied to design high-temperature stable (Barlet-Gouédard et al., 2008) and CO_2-resistant cement systems (Barlet-Gouédard et al., 2012).

Hollow glass microspheres can also be used in CAP cement systems (Kukacka and Sugama, 1994).

7 Cement system chemistry—Slurry properties control

The discussion in previous sections has concentrated on the properties of the set material. However, the slurry must be placed correctly in the annulus and therefore the slurry properties must be controlled. Methods for the determination of slurry properties are detailed in an API-recommended practice (API, 2013). A discussion of the more important methods, with images of typical laboratory equipment used, has been given by Dargaud and Boukhelifa (2006). The methods are usually used to evaluate the performance of Portland cement-based slurries, but in most cases, the methods can be used to evaluate the performance of other sealant systems.

For geothermal well applications the most important slurry parameters to control are:

- Density.
- Rheological properties.
- Slurry stability.
- Thickening time.

Although other parameters, such as fluid loss may require control.

The physical and chemical methods to control these properties for Portland cement-based systems have been discussed in detail by Cadix and James (2021). The chemistry of additives for Portland cement-based systems has also been reviewed by Nelson et al. (2006). There is significantly less information available on the chemistry of additives suitable to control the properties of alternate sealant systems such as CAP cement systems.

7.1 Density control

Methods to reduce the density of sealant systems have been discussed above. These are physical methods (addition of lightweight particles, nitrogen, water), which are effectively independent of the chemistry of the sealant system. Foaming surfactants that are used for Portland cement systems may also be used with CAP cement systems.

7.2 Rheological properties

Cadix and James (2021) identified four main parameters affecting the rheological properties of cement slurries:

1. The volume of solids.
2. The viscosity of the interstitial fluid.
3. The particle size distribution and shape of the solids.
4. Interactions between particles.

Items 1 and 3 are physical parameters and are independent of the chemistry of the sealant system. The viscosity of the interstitial fluid depends on the presence of soluble additives and polymers. The chemistry of these additives and polymers may be different for different sealant systems. Interactions between particles depend on the chemistry of the sealant systems and the chemistry of additives, such as dispersants. Hence, to control these interactions depends on the chemistry of the sealant system. Experimental methods to determine the rheological properties of cement slurries are detailed in clause 11 of API RP10B-2 (API, 2013). The documented measurement methods are currently limited to a maximum temperature of 88°C although, equipment is available to determine the rheological parameters at temperatures up to 300°C. However, it is essential that any sealant system tested have a setting time much longer than the test time so that the sealant is not set in the instrument.

7.3 Slurry stability

Slurry stability is related to the rheological properties of the sealant system at low shear rates and is therefore predominantly controlled by the rheological properties of the interstitial fluid and the interactions between particles. Stable slurries are generally obtained by using a combination of dispersants and antisettling agents (Nelson et al., 2006). Biopolymers such as Welan gum and Diutan gum, regularly used as antisettling agents in Portland cement systems, can also be used in CAC and CAP cement systems. At rest, their rigid-rod structure entangling gives rise to a yield stress, that limits particle settling. These biopolymers are generally limited to use at temperatures below about 150°C. At higher temperatures, bentonite may be used to provide a yield stress.

The experimental methods to determine the slurry stability are given in clause 12 of API RP10B-2 (API, 2013). The two key parameters are free fluid and sedimentation. Free fluid is syneresis where fluid is expelled to the surface of the cement slurry. This occurs if the slurry is insufficiently dispersed. The measurement is taken before the slurry has set.

Sedimentation is the settling of particles under gravity and is measured once the cement has set. Sedimentation occurs if the yield stress of the slurry is too low to prevent settling of particles. The sedimentation test should be run with a slurry formulated for the specific application, with the correct thickening time, and under temperature conditions that represent the field application. Cement is allowed to set in a mold (typically 150 mm high and 25 mm diameter) at temperature and pressure. The cylinder of set cement is then cut into slices, and the density of each slice is measured to give a density profile along the length of the set sample. There should not be any significant density gradient along the column.

7.4 Thickening time

The control of the thickening time of sealant systems is highly dependent on the chemistry of the system. A summary of the chemistry used for CAC and CAP cement systems is given in the previous sections. Retarders for Portland cement systems are discussed in detail by Cadix and James (2021). Equipment and procedures are available to allow thickening time to be determined at temperatures up to 300°C and pressures up to 275 MPa.

8 Cement hydraulics—Slurry placement

Cement placement design and operations have been studied for decades and syntheses of the state of the art have been presented by Arnold (1982), Smith (1990), Bittleston and Guillot (1991), Piot and Cuvillier (2006), Tardy et al. (2017), and Bois et al. (2023). The objective of this section is not to replace these reviews, but to focus on the special needs for high-temperature geothermal wells.

8.1 Failure mechanisms

Theoretically, placing a cement slurry into the annular space between the tubular and the wellbore is simple. The cement, chemical additives, and water are transported to the well site. Water, chemical additives, and cement are mixed into a cement slurry. Then, if the Forward Circulation (FC) placement method is selected, for example, the slurry is pumped into the well's casing, then pumped out of the casing into the annulus.

However, cement placement requires the successful integration of multiple complex engineering skills and the use of multiple types of equipment operated by people with various skills and expertise. As such there are hundreds (thousands) of tasks and processes that must be completed correctly to obtain a successful cement annular placement. Listed below are the most common areas where problems can arise in the case of a geothermal well:

- Engineering design errors due to incorrect well information and input data:
  - Well data mining due to scarce old available data.
  - Cementing slurry design due to testing equipment not rated for high temperatures.
  - Hydraulic and thermal simulations due to the uncertainties in the fluid physical and rheological parameters and in tubular standoff values.
- Cement slurry design errors:
  - Laboratory testing due to the testing equipment not rated to high temperatures.
  - Chemicals selection and chemical compatibility at high temperatures.
  - Sensitivity to temperature.
  - Laboratory testing not aligned with onsite mixing equipment and processes.
- Cement slurry not placed in the tubular/wellbore annulus as per design:
  - Cement slurry not achieving design height in the annulus due to loss circulation related to rock porosity, rock fracturing, annular bridging, or washouts.
  - Cement slurry not achieving complete radial placement due to poor tubular centralization or well fluids not correctly conditioned prior to pumping.
  - Premature slurry dehydration during pumping due to slurry not mixed to designed density, chemical failure, annular bridging, or mechanical equipment failure.
- Surface equipment mechanical failures during mixing and pumping of the cement slurry:
  - Unable to mix the cement slurry to designed density and rate.
  - Unable to transport the cement solid to the mixer at the designed rate.
  - Equipment failure due to maintenance.
  - Plug container failure.
- Casing equipment (such as centralizers, liner hanger, liner packer, float collar, float shoe, stage tool, and cementing plugs) failure in the well due to harsh environmental conditions.
- On site communication errors:
  - Non-coordination of multiple execution processes on rig site.
  - Lack of chain of communication structure and roles/responsibility.
  - Lack of contingency plans.

Regardless of the type of casing system used, geothermal cementing presents unique challenges in terms of cement slurry design and placement methods. Below is a summary of six of the most common technologies used for geothermal operations and a summary of eight steps that have proven effective in obtaining successful cement placement (see Fig. 3).

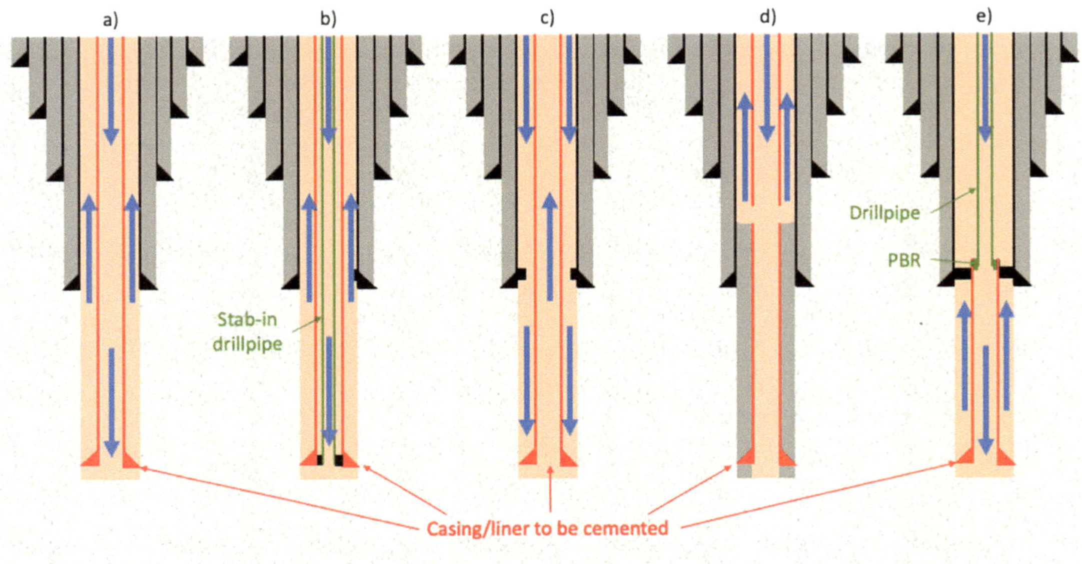

FIG. 3 Cement placement methods; (A) Forward circulation; (B) Stab-in; (C) Reverse circulation; (D) Staged cementing; (E) Liner.

8.2 Forward circulation placement

The conventional or forward circulation (FC) placement method involves mixing and pumping the cement slurry into the casing, then, it turns at the bottom of the casing into the annular space between the casing and the wellbore and moves up to its target. The FC placement method typically utilizes two casing plugs (bottom and top). The bottom plug is pumped into the casing directly ahead of the cement slurry and provides a mechanical separation between the wellbore drilling fluid and the cement slurry. The bottom plug also wipes the internal walls of the casing, removing any contaminants, which may have accumulated over time. The bottom plug is designed with a differential pressure rupture device that opens when an internal casing pressure differential of between 2.4 and 5.2 MPa is applied, therefore allowing the cement slurry to exit the casing and enter the casing-wellbore annulus. Once all the cement slurry has been pumped into the casing, a top plug (solid) is released into the casing, and the drilling fluid is pumped to force the cement slurry down the casing and into the well's annulus. The top plug provides mechanical separation between the cement slurry and drilling mud and provides a positive indication that all cement slurry is pumped out of the casing. A SWOT analysis, dedicated to high temperature geothermal specific needs is presented in Table 2.

TABLE 2 Forward circulation placement method—specific SWOT for hot geothermal projects.

Strengths	Opportunities
Proven slurry placement method Proven engineering tools and design simulators 2-plug method used by all cementing service providers	Circulate the well before pumping cement slurry Engineering, chemist, laboratory, and operations discussions for designing the cement slurry and placement processes
Weaknesses	**Threats**
Mechanical equipment failure at high temperature Loss circulation since gravity plays against pumping in the annular	Cement slurry performance at high temperatures Loss circulation, hole collapse

8.3 Inner string or stab-in placement

Inner string or stab-in cementing is a technique typically used with large-diameter casings in which drill pipe is placed inside the casing as the conduit for pumping fluids from the surface to the casing annulus. Typically, a stab-in stinger is placed on the bottom of the drill pipe; when the drill pipe reaches the float collar, the stinger is inserted into the float collar, which creates a pressure seal and prevents the cement slurry from entering the casing during cementing operations.

In general, performing a cementing job through drill pipe can substantially reduce the cement placement time, which translates into a reduction in the cement retarder concentration, which results in a reduction in waiting on cement time and a reduction in rig cost. Additionally, inner string cementing coupled with a casing pack-off system can prevent casing collapse due to annular pressures during and after cement placement. When the Inner string or stab-in cementing placement method is used, casing movement such as rotation and/or reciprocation cannot be used. A SWOT analysis, dedicated to high geothermal specific needs is presented in Table 3.

TABLE 3 Inner string or stab-in placement method—Specific SWOT for hot geothermal projects.

Strengths	Opportunities
Proven slurry placement method Proven engineering tools and design simulators	Circulate the well before pumping cement slurry Reduce cement placement duration Reduction of retarder concentration
Weaknesses	**Threats**
Casing rotation/reciprocation not feasible	Cement slurry performance at high temperatures Loss circulation, hole collapse

8.4 Reverse circulation placement

The reverse-circulation (RC) placement method involves pumping the cement slurry down the annular space between the casing and the wellbore, and displacing the drilling fluid back up through the casing. This method reduces the cement slurry placement time and ensures that the cement slurry does not encounter the very high temperatures that exist at the tubular shoe, therefore, facilitating the design of the cement slurry, especially its retarding. Finally, it also reduces the equivalent circulating density, hence minimizing the risk of loss circulation and not achieving the designed height of cement slurry in the annulus. The RC placement method requires special casing float equipment, modified well head assemblies and extensive engineering and operational planning. A SWOT analysis, dedicated to high geothermal specific needs is presented in Table 4.

TABLE 4 RC placement method—specific SWOT for hot geothermal projects.

Strengths	Opportunities
Proven slurry placement method Reduction in cement slurry placement time Reduction of risk of loss circulation	Circulate the well before pumping cement slurry Tailor chemical loading to decrease waiting on cement slurry hydration
Weaknesses	**Threats**
Limited experienced service providers Additional engineering modeling and design required Non-typical casing equipment and well heads required	Annular bridging Cement slurry not prevented from entering into the casing

8.5 Pipe rotation/reciprocation

Achieving complete cement annular fill is essential to providing the highest degree of cement placement and annular zonal isolation. Casing rotating or reciprocating has proven effective in enhancing the slurry placement into the annulus. As the casing is moved, the drag forces created over the circumference of the casing help to remove areas of gelled drilling fluid and provide improved cement placement in non-centralized casing annular sections. Casing rotation is more efficient than casing reciprocation since it helps to move the slurry from the wide side of the annulus into the narrow side when the casing is poorly centralized. Unfortunately, it is difficult to implement in the case of liner and/or the large casing sizes as encountered in geothermal wells, either due to lack of power at the rig, or the use of a stab-in approach to pump the slurry. A SWOT analysis, dedicated to high geothermal specific needs is presented in Table 5.

TABLE 5 Pipe rotation/reciprocation method—specific SWOT for hot geothermal projects.

Strengths	Opportunities
Tried and proven slurry placement method Proven engineering tools and design simulators	Circulate the well before pumping cement slurry Enhance cement slurry placement
Weaknesses	**Threats**
Special casing tools required such as centralizers Liner rotation difficult to implement Casing rotation not feasible when stab-in	Casing stuck Rig power not sufficient to move casing

8.6 Stage cementing with differential valve

Differential valve (DV) tools (also called stage tools) are used to cement multiple well sections behind the same casing string, or to cement a critical long section in multiple stages. Stage tools are installed at specific points in the casing string as the casing is being run into the hole. The first (or bottom) cement stage is pumped through the tool to the end of the casing and up the annulus to the calculated-fill volume (height) using the FC placement method. When this stage is completed, first stage shut-off plug is released into the casing, pumped down and it lands on the casing float collar. The first stage shut-off plug provides a mechanical seal to prevent additional fluids from being pumped out of the bottom of the casing. A free-falling plug or pump down dart is used to hydraulically set the stage tool and open the side ports, allowing the second cement stage (top stage) to be displaced above the tool. A closing plug is used to hydraulically close the sliding sleeve over the side ports at the end of the second stage and serves as a check valve to keep the cement from U-tubing above and back through the tool. A SWOT analysis, dedicated to high geothermal specific needs is presented in Table 6.

TABLE 6 Stage cementing—differential valve method—specific SWOT for hot geothermal projects.

Strengths	Opportunities
Reduced risk of loss circulation Method to ensure cement placement over critical areas	Reduces formation stress during cementing Eliminate use of low-density cement
Weaknesses	**Threats**
Special casing tools required DV tool port seal failure under high temperature Depth at which to place DV should be known before running the casing into the hole	PBR seal failures due to well temperatures Expensive remedial operations

8.7 Liner cementing

Liners are used to eliminate the need to run a full string of casing. Unlike a full string of casing which is installed back to the surface, liners terminate downhole at point in the last casing string. A liner is installed into the well using drill pipe and is secured inside the previous casing with a casing anchor commonly referred to as a "liner hanger." Typically the liner cement is designed using the same cement slurry design process as that of the full string of casing and the FC placement method is used. The cement slurry is mixed and pumped into the drill pipe and enters the liner, after the required cement volume has been pumped into the drill pipe, drill pipe plug is released and the cement slurry is displaced out of the drill pipe into the liner casing when the drill pipe plug arrives at the liner hanger it lands and locks into a liner wiper plug. Pressure is applied to release the wiper plug into the liner and it displaces the cement from the liner into the annulus. Liner cementing prohibits the use of the RC placement method and casing movement such as rotation and reciprocation cannot be used due to the liner hanger design. But it allows to avoid cementing up to the surface.

The liner casing can be converted to a full string of casing with the use of a casing tie-back and polished bore receptacle (PBR) tool. The value of converting the liner casing into a full string of casing increases the internal pressure rating of the well. The casing tieback and polished bore receptacle (seal assembly) is a mechanical system that enables the bottom of the casing to be stabbed and locked into the liner hanger and the PBR provides the pressure seal, thus providing a continuous high-pressure conduit for well fluids. A SWOT analysis, dedicated to high geothermal specific needs is presented in Table 7.

TABLE 7 Liner-tieback cementing method—specific SWOT for hot geothermal projects.

Strengths	**Opportunities**
Reduction in cement slurry placement time Reduction in initial casing cost	Tieback casing can be recovered
Weaknesses	**Threats**
Reduces use of casing movement during primary cementing Eliminated the use of reverse cementing methods	PBR seal failures due to well temperatures Mechanical tools failures

8.8 Technical hurdles related to geothermal cementing

Bois et al. (2023) presented a review of cement placement modeling for oil and gas wells and highlighted three main categories to consider for the model to be efficient:

- Modeling performed on the whole cement slurry placement process, both in the tubular and in the annulus at the same time:
  - Heat transfer simulations to calculate the temperatures in and in the near-vicinity of the wellbore.
  - Fluid dynamics simulations to calculate the equivalent circulation density (ECD).
  - U-tube simulations to account for vacuum effects.

- Modeling performed only on the cement slurry placement process in the tubular or at its shoe:
  - Stability at the fluid interfaces during pumping the fluid down the tubular.
  - Changes in pipe diameter (annular tolerances) such as liner hanger, casing shoe, and shoe track.
- Modeling performed only on the cement slurry placement process in the annulus:
  - Fluid stability at interfaces during pumping the fluid in the annulus.
  - Changes in annulus geometry due to under-gauge or over-gauge wellbore sections.
  - Changes in annulus geometry due to casing joints, centralizers, and scratchers.
  - Changes in tubular standoff ratio with space and time.

As mentioned in the introduction of this chapter, the main features that distinguish a high-temperature geothermal well from an oil & gas well are the higher temperatures, the presence of significant loss circulation zones, and the fact that all casing cement jobs are conducted to surface.

Therefore, design fields that deserve special attention are:

- The heat transfer simulations.
- The fluid dynamics simulations.
- The fluid interface stability simulations.

Not much has been published to describe and validate numerical models applied to downhole temperatures higher than 200°C.

8.9 Primary cementing best practices

The industry recognizes that primary cementing success is a direct driver for the safe and effective harvesting of a reservoir and the financial success of a well. Over the years, the oil and gas industry has dedicated millions of R&D dollars to the study of primary cement successes and failures and in developing technologies to increase primary cementing success. In July 2001, Crook et al. (2001) published a paper, which outlines eight proven steps that help achieve primary cementing success. The eight steps are listed below together with a short description and their potential use in geothermal wells. Very seldom do well conditions allow for the use of all eight of the below "best practices"; however, the use of as many of the eight steps as possible will amplify the ability to achieve a successful cement placement in the well.

1. Condition the drilling fluid

   The function of a drilling fluid is to remove drilled solids from the well bore, maintain well pressure control, and cool the drill bit. Drilling fluids are composed of three main components: liquid (water or hydrocarbon), weighting material (barite, hematite, sand), and chemicals (viscosifiers/dispersants), and are thixotropic. They begin to develop gel strength during casing installation in the well. Research has proven that a drilling fluid with high gel strengths is detrimental in obtaining successful cement slurry placement in the well.

   Due to the high well temperatures encountered in geothermal wells, the development of areas of high gel strengths is accelerated, so advanced cementing technologies must be used to help ensure that the drilling fluid is properly conditioned and that the well is prepared to receive cement. Listed below are two proven technologies to help prepare the well to receive cement.

- Well temperature simulators are used to help the engineer determine when the well has been cooled to the cement slurry design temperature. Prior to casing installation, the engineer uses a well temperature simulator to determine the change in well temperature as the well fluid is circulated at various rates over a given time. The engineer monitors the temperature of the drilling fluid exiting the well during conditioning and compares the actual to the simulated temperature values to determine the BHCT of the well. This technology was proposed by Beirute (1991) and has been studied frequently over the years and has evolved several times.
- Circulating hole pressure match was initially developed and introduced to the industry by Beirute et al. (1991) and has proven to be an excellent tool to help the engineer determine when the gelled drilling fluid has been removed from the well. In application, prior to casing installation, the engineer uses a cement hydraulic simulator to determine the surface pressure values when the well is circulated at various rates with a given fluid rheology. Once the casing is installed, the well is circulated with the treated (less viscous) drilling fluid; the engineer compares the theoretical to actual surface pressure values and determines how much of the gelled drilling fluid is begin removed.

2. *Use spacers and flushes*
Spacer and flushes provide a separation between incompatible well construction fluids such as drilling fluids and cementing slurries. Additionally, they help thin immobile or gelled drilling fluid therefore assisting in drilling fluid displacement from the wellbore. They are also used to change the wettability of the wellbore and casing from oil-wet to water-wet if the drilling fluid is a non-aqueous fluid.

In the past 60 years, the industry has accelerated R&D studies in fluid systems to enhance cement placement and developments in the area of spacers and flushes to improve primary cementing success. Examples of the new technologies developed are: 1970s: the use of reactive flushes (sodium silicate-based flushes) for cementing operations; 1980s: beginning of the inclusion of surfactants into the design of spacers and flushes to improve displacement of non-aqueous drilling fluids; 1990s: development of laboratory testing equipment and methods to optimize surfactant selection and loading and measure changes in wettability to enhance drilling mud removal and bond strength; 2000s: use of fluid system design using rheological hierarchy and subsequent fluid design simulators. So, what is the difference between a spacer and a flush?

- Spacers are typically weighted and viscosified fluid systems that provide a physical separation between the drilling fluid and cement slurry. The spacer can be designed for use with either water-based or oil-based drilling fluids. Spacers help to change the wettability of the wellbore and casing, and help maintain pressure control in the well.
- Flushes are used to separate incompatible fluids and are usually not weighted with solids as such they achieve turbulent flow regimes at low pump rates much easier than weighted spacers. Flushes can provide fluid thinning of gelled drilling-fluid and are used to separate drilling fluids and cementing slurries. Flushes can be designed for use with either water-based or oil-based drilling fluids.

3. *Move the pipe*
   As previously described, casing movement can help obtain a successful primary cementing job. Unfortunately, these technologies are difficult to apply to high-temperature geothermal wells due to the casing design.
4. *Centralize the casing*
   Casing centralization creates even fluid flow paths in the well annulus thereby enhancing drilling fluid removal and cement slurry placement. Casing centralization is measured by the term "Casing Standoff" which is the measure of the smallest distance between the casing outside diameter and the diameter of the wellbore. A casing that is perfectly centered in the wellbore has a casing standoff of 100%. If the casing is touching the wellbore, then the casing standoff is calculated to be 0%. The minimum industry standard for optimum cement placement is 70% casing standoff. Casing standoff can be determined by performing numerical simulations with inputs that are the casing description, the centralizer types, properties, number, and location along the tubular. An error on the inputs due to a special behavior of the centralizers at high-temperature, for example, or the failure of some centralizers may lead to lower casing standoff during the operations, and therefore, to poor cement placement.
5. *Maximize the displacement rate*
   Maximizing the cement annular placement rates (pump rate) is one key to achieving efficient drilling fluid removal and the displacement of one fluid by another one. Pump rates should be selected to obtain the highest cement placement velocity, which will allow for the safe placement of cement into the annulus. Such evaluation is generally made by running cement hydraulic placement simulators, which rely on the knowledge of the fluid properties versus time. However, no such model has been validated at the very high temperatures encountered in some hot geothermal wells, and the inputs to these models may suffer from a lot of uncertainties due to the lack of testing equipment to determine them.
6. *Design slurry for proper temperature*
   Bottom Hole Static Temperature (BHST) sets the basic criteria for cement slurry design, and this topic was previously analyzed in this chapter.
7. *Select and test cement compositions*
   This topic was previously analyzed in this chapter.
8. *Select a proper cementing system*
   This topic was previously analyzed in this chapter.

8.10 Conclusions

Southon (2005) used downhole video cameras to identify casing collapses in geothermal wells. He showed that, most of the time, the casing deformation is segmented to one side of the casing circumference and observed no evidence of a complete necking or symmetrical collapse of the casing. He concluded that this non-symmetric collapse tendency is due to the thermal expansion of trapped fluid in the casing-to-casing annulus. Furthermore, he observed that collapse is normally located in the body of the casing and not near the couplings, which he explained by the yielding of the coupling, which induces a loss of thread sealing capacity, and therefore, the pressure relief of the expanded trapped fluid.

The visual absence of alternative collapse behavior does not necessarily mean that collapse at couplings and collapse over the entire 360-degree circumference does not exist. The observations suggest that casing collapse is determined by the geometry of the trapped fluid.

Since all casing strings are usually cemented to surface in geothermal wells, sound slurry design is especially important. For example, the long cement columns may cause problems with borehole fracturing around the shoe, where reverse cementing may be a solution. Furthermore, proper retarder selection and slurry stabilization are important to avoid sedimentation before curing and avoid formation of pockets of free water in the cement. If formed, such water pockets may cause casing integrity problems due to high water pressures during subsequent super-hot steam production.

Obtaining a successful primary cementing job for geothermal operations requires the use of multiple engineering and science disciplines as well as extensive computer modeling, laboratory testing and the discussions with experienced well construction, service providers and rig operators. However, a lot of testing equipment and numerical simulators have not been extended to the high temperatures encountered in geothermal wells, and some key technologies for good cement placement cannot be implemented due to the size of the tubular.

9 Cement integrity—Experimental understanding

Geothermal wells present many unique technical challenges that are not encountered in classical hydrocarbon-producing wells. In terms of geothermal cementing, the cement sheath must perform at temperatures which are much higher than conventional high-temperature wells, additionally coupling the cyclic changes in well temperature and pressures and exposure to harsh/corrosive fluids create technical challenges that are beyond hydrocarbon. The study of geothermal well cementing began with what were the best chemicals and processes required to safely placement cement into a geothermal well. In the early 1990s, the industry began the investigation into why the frequency of failure of geothermal wells was greater than the hydrocarbon well average. The initial studies indicated that one area that needed further work was geothermal well cementing. Below is a summary of the research and development work that has been funded by both governments and industry in terms of advanced geothermal cementing.

9.1 First experiments

Goodwin and Crook (1992) and Jackson and Murphey (1993) were the first authors to publish the results of laboratory experiments showing that changes of wellbore pressure or temperature could induce the failure of a cement sheath. Their apparatus consisted of two concentric steel casings between which cement was pumped and cured.

Some of their experiments consisted of increasing the pressure inside the inner casing and then decreasing the pressure to the initial state, while the axial permeability of the cement sheath was monitored. These cycles were performed in steps from 13.9 MPa up to 69.0 MPa. These experiments indicated a decrease of cement sheath permeability when the inner-casing pressure was at its maximum and an increase of permeability when it was bled down to the initial value. Furthermore, the observation of the specimens sectioned after each test showed that radial cracks radiated directly from the inner to the outer casing surface. Hoop cracks tended to generate either within the cement system or at the casing/cement interface and parallel to the inner casing surface cracks. They sometimes radiated toward the outer casing surface (in-echelon cracks).

Other experiments consisted of curing the cement while the inner-casing pressure was at 69.4 MPa and then in decreasing the inner-casing pressure in stages to 55.6, 41.7, 27.8, 13.9, and 6.9 MPa, while applying an air pressure across the cement sheath length. These experiments showed that gas started to flow as soon as the source of air was fixed to the bottom of the annulus, and continued throughout the experiment, independently of inner-casing pressure. Little flow was detected until the inner-casing pressure was lowered below 20.8 MPa. Plots of hoop strains measured during the experiments showed that, at pressures lower than about 27.8 MPa, the strain in the outer pipe remained constant, hence proving that a microannulus was opened. Hence, it was necessary to decrease the inner-casing pressure by 41.7 MPa to open a microannulus in a cement sheath that had set under 6.9 MPa, which is equivalent to a vertical depth of about 400 m.

Goodwin and Cook and Jackson and Murphey set-ups were reproduced/adapted by Teodoriu et al. (2010a), Yang et al. (2012), Thérond et al. (2017), Luo et al. (2020), Moghadam et al. (2020), and Beltrán-Jiménez (2022).

Thiercelin et al. (1998) developed an instrumented annular experiment to measure cement shrinkage. The experimental set-up consisted of two concentric steel cylinders of 196 and 111 mm external diameters, 80 mm in height, and 1 mm thickness. Strain gauges were mounted on the cylinders to monitor their tangential deformation and their consequent radial displacement. The cylinders were fixed on a steel plate, and placed in an oven at 77°C. The cement was then poured into the annular space. Depending on the experiment, the slurry was in contact with air, or covered with a 2-cm water layer. The temperature of the slurry was measured with a thermocouple sensor.

The experiment with the slurry in contact with air showed a different behavior for the inner and outer strain gauges. Outer gauge readings stayed at a constant zero because a microannulus of about 300 µm had been created at the outer interface of the cement sheath. The deformation of the inner cylinder showed a contraction due to the shrinkage of the cement and some sudden drops in the applied pressure, the last one being first progressive, followed by an abrupt decrease, after which strain remained constant up to the end of the experiment. These drops could be related to the creation of the radial cracks that were observed at the end of the experiment. The second experiment with the slurry covered by water indicated minimal changes in the bulk volume during the hardening phase, hence highlighting the significant impact of the access of cement to external water.

However, these experiments were not truly representative of what occurs downhole for two reasons:

1. Cement was in contact with air in the first experiment, which allowed replacing consumed water by air, and creating capillary pressures that could be at the origin of cement shrinkage. This is the famous self-desiccation concept encountered in civil engineering (Coussy, 2010), which may only exist when cementing in front of gas reservoirs. Cement was in contact with water in the second experiment, which allowed replacing the consumed water by the external water cap. Such conditions may occur when the cement is

in front of aquifers, and if there is no mud cake and reservoir permeability is large. The true downhole condition, when cement is in front of a barrier, or between two tubulars, is little or no access to any fluid during hydration (Bois et al., 2012).
2. The experiments were performed under atmospheric pressure, meaning that cement consolidation, in the sense of Biot's theory (Coussy, 2010), could not occur, even if it is at the origin of cement shrinkage under downhole conditions.

Therefore, one may conclude that the magnitude of cement shrinkage in these experiments cannot be representative of what is expected downhole.

Boukhelifa et al. (2005) developed a laboratory experiment to simulate the impact of the contraction or expansion of the inner casing on cement-sheath damage. The equipment included two main components: (1) A central core that could be manually expanded or contracted to simulate the expansion and contraction of the inner casing string of a well. Displacement was measured by strain gauges placed on its outer diameter; and (2) Outer metal rings that simulated weak or stiff formations, depending on the ring thickness. The deformation of the outer rings was measured with strain gauges placed around the outer surface of the ring. The cement to be tested was poured directly in the annular space. A layer of water was placed on top of the cement surface, and a plastic film was placed on the top of the mold to prevent water evaporation and maintain water saturation of the cement. The cement system hydrated for 3 days while displacements of the central core and outer rings were recorded. After 3 days, a lateral-sealing system was installed to perform permeability measurements. Loading and unloading were done manually by gradually increasing or decreasing the central core diameter.

The experiments showed that:

1. Contraction of the central core can open a microannulus at the cement-sheath inner interface, and going back to the initial state may not close the micro annulus since the initial fit between the cement body and the inner metallic core could not be reestablished.
2. Expanding the central core can induce radial tensile cracks in the cement sheath. In the large unloading stages, these cracks closed and then the inner micro annulus opened.
3. A flexible cement system has a lower risk of being radially cracked when expanding the central core than a stiff one.
4. The use of expansive cement helps to prevent the opening of an inner microannulus or the creation of radial cracks in the cement.
5. Increasing formation stiffness helps to prevent the creation of radial cracks in the cement when expanding the central core. Moreover, tensile cracks generated in the cement have a lower impact on permeability with the stiff formation.
6. Increasing formation stiffness seems to have a contradictory impact on the risk of opening an inner microannulus when contracting the central core: for a low expansion sealant, the risk of creating a micro annulus would be increased, while it would be decreased for cements with large expansion.

Again experiments were not truly representative of what actually occur downhole and they suffered from most of the drawbacks experienced by Thiercelin et al., i.e., the cement was in contact with water and the experiments were performed under atmospheric pressure conditions.

9.2 Cement fatigue

Kosinowski and Teodoriu (2012) developed a specific set-up to investigate the resistance of cement when submitted to low cycles of fatigue. The mold was made of a Plexiglas tube with an inner diameter of 110 mm and a length of 160 mm, and a centered casing, with inner diameter of 66 mm, outer diameter of 75 mm, and length 200 mm. Cement was poured into the mold and cured under atmospheric conditions for 2 weeks. After unmolding, the specimens were placed under a hydraulic load frame, and a cyclic axial load was applied on the casing. The load was increased to a predetermined value, then released to zero, then increased to the same value and released to zero multiple times until the sample failed or a defined number of cycles were performed. If the cement was able to withstand the cyclic loading, the maximum force per cycle was increased and the testing was continued. In general, the first visible mode of cement failure was a radial crack. As the applied load was increased, or more cycles were executed, new cracks developed, including transversal cracks (disking).

These experiments were supposed to mimic what could be expected under cycling thermal loading, but this was not the case for the following reasons:

1. Mechanically loading a cement sheath is not the same as thermally loading it.
2. Axially loading the steel tube induced radial pressure to the cement sheath by Poisson effect, but also created shear stresses at the tube/cement interface and axial stresses in the cement that may not be representative of what is encountered downhole.
3. The existence of a formation behind the cement sheath is not simulated.
4. Testing is performed under ambient conditions, neglecting the effect of downhole stresses.
5. Cement is cured under ambient conditions.
6. The high cycling frequency used in the laboratory is much higher than what is encountered in the well.

Yuan et al. (2013) used the same set-up to perform tests on a class G cement system. Shadravan (2013) and Shadravan et al. (2015) modified it to allow pressure to be applied directly inside the tubing, meaning that a load frame was not required and shear stresses at the tubing/cement interface were annealed. Johns (2014) used this experiment set-up to study the fatigue of fibered cement systems. McDaniel et al. (2014) extended it by using an external PVC pipe to simulate an outer barrier.

9.3 Thermal cycling

Garnier et al. (2010) seem to be the first authors to publish the results of experiments dedicated to study the impact of thermal loading. They designed a specific experiment to evaluate how cement systems would perform during SAGD (steam-assisted gravity drainage) operations. It basically consisted of a cement sheath placed around a piece of steel casing of unknown diameter that was heated. No formation was included as its stiffness was supposed to be very low for unconsolidated formations. Due to the unconsolidated formation that the cement was being designed for, the cement slurry was poured between the casing and a cardboard or resin framework. The framework was removed after several days when the cement had set, and the casing with its cement sheath was placed in a water fill-container.

Low-toxicity mineral oil was poured inside the steel casing with a mechanical stirrer and an electrical heater.

The tests were run by heating the oil with a ramp fixed at 0.6°C/min until a temperature of 130°C was reached, and the temperature was maintained for 2h before the heating was stopped. Several sealants were tested and the results showed that a radial crack appeared at 74°C for a conventional class G cement system with silica, while no such crack was induced for a resilient cement system though the test was pursued up to 180°C.

These experiments suffer from three drawbacks:

1. The absence of formation.
2. The full access to water.
3. Curing under atmospheric pressure.

This set-up was also used by Patil et al. (2010) and Pershikova et al. (2010) and extended by Corina and Moghadam (2024). Hence, Corina and Moghadam's set-up includes an outer tubular to better mimic downhole conditions, and their tests were conducted at a maximum temperature of 80°C. However, these experiments suffered from the fact that a finite-thickness cylindrical pipe does not behave like an infinite medium in which a cylindrical hole has been drilled. When changes of temperatures occur, the inner diameter of an empty finite-thickness cylindrical pipe varies, while the inner diameter does not change if the thickness is infinite. Therefore, the test results cannot be used to evaluate what happens downhole.

Albawi (2013) and Albawi et al. (2014) developed an experiment to investigate the effect of thermal cycling on annular cement sheath integrity, which was supposed to account for the existence of the formation. The experimental set-up consisted of three parts:

1. A heating/cooling source to provide temperatures as high as 200°C, and as low as −50°C, with an adjustable ramp rate up to 30°C/min.
2. The specimen to be tested included a pipe, a cement sheath, and a rock. The dimensions were designed with respect to a typical production casing wellbore with a 12¼″ borehole and a 9⅝″ casing, scaled by about 1:4. Hence, the casing had a diameter of 60mm, a thickness of 3mm, and a length of 240mm, and the annular cement had a thickness of 10mm. The formation was made of Castlegate sandstone, with outer and inner diameters of 200 and 80mm, respectively, and a length of 200mm.
3. A heat-conducting medium was used to transfer heat from the thermal platform radially to the pipe.

Temperature variations were applied radially to the casing, and the effect of these variations on cement sheath integrity was observed through visual inspection, acoustic emission monitoring, and CT scanning.

Three specimens were tested: The first sample was not exposed to thermal cycles ("virgin" specimen) and was directly sent to CT scanning after cement curing. The second sample was cemented with a centralized casing and the third was cemented with casing 50% off position. The last two specimens were exposed to the same thermal cycle schedule. The sealant was made of class G cement with a water-to-cement ratio of 0.42, cured under atmospheric pressure and at a temperature of 66°C during 5 to 7 days. All experiments were performed in a dry state, without access to water. These tests showed that casing-cement debonding occurred before thermal cycling and that severe debonding between both cement-formation and casing-cement interfaces and radial cracking happened for the cycled specimens.

In summary, these tests indicated that the use of CT scanning to evaluate cement sheath integrity should be looked at as a major improvement; however, the testing protocol and capacity to mimic downhole conditions are not evident:

1. The specimens were prepared at 66°C, then cooled down to room conditions, before performing the initial state CT scan. Assuming that the room temperature was 21°C, the specimens were cooled by 45°C prior to CT scanning. This decrease in temperature could be at the origin of the initial damage observed on the specimens.
2. The outer boundary condition of the experiment is not representative of what exists in a downhole because heating a finite formation does not behave like heating an infinite formation.
3. The sandstone is a porous body, meaning that, depending on the stability of the cement slurry, some water may have leaked into the sandstone pores, altering cement properties.
4. Tests are performed under atmospheric conditions and not under downhole pressure.

de Andrade et al. (2015) modified the Albawi apparatus to allow curing and testing of cement under pressure. They used it to analyze the impact of the type of formation by performing experiments with Saltwash North sandstone and Mancos shale as formation rock. They also analyzed the impact of the casing outer surface, by considering three different pipe surfaces, mill varnished, sandblasted, and sandblasted with oil-based mud (OBM). These experiments showed that:

1. The pressurized set-up gives better bonding and less cracks/voids than the ambient set-up, showing the benefit of the initial applied pressure.
2. Thermal cycling increased cement-to-casing and cement-to-formation debonding. Radial and shear cracks were observed to propagate from the outer-casing surface toward the inner-rock surface.
3. Debonding and cracking due to thermal cycling initiated from already present debonded areas and cracks.
4. The casing surface condition did not significantly impact the cement sheath integrity during thermal cycling.

However, the set-up of Andrade et al. only partially resolves the atmospheric pressure condition of the Albawi set-up since the maximum pressure outside the rock cylinder was 3.45 MPa, which is insufficient to simulate what occurs downhole, in most situations. The initial cement state, after cooling the specimens, showed that most of the defects had occurred and further cycling seemed only to slightly propagate some of the defects. Therefore, because the initial stress state is not representative of what is encountered downhole, the results of the experiments may not be representative of what is encountered downhole.

This experimental set-up was used by Vrålstad et al. (2015) with a larger temperature range of 5° to 125°C using a class G cement system with 25% BWOC of silica flour mixed with tap water to 1.80 sg. Both Castlegate sandstone and Mancos shale were used as formation rock. After the cement was placed in the annulus, the pressure was increased to 3.5 MPa and the temperature was set at 125°C from inside the casing, and left to cure for 5 days. The samples were CT scanned and exposed to 20 thermal cycles, where one cycle consists of approximately

1.5h of heating, 4h holding time at maximum temperature, approximately 1.5h of cooling, and 4h holding time at minimum temperature. The experiments showed cracking and debonding for the sandstone sample, whereas the shale sample remained almost unaffected. Vrålstad et al. explained this as the effect of rock stiffness, while it could also be explained if water leaked from cement to sandstone but not to shale, especially as no fluid loss control additives had been used.

The equipment was also used by Todorovic et al. (2016), Roy et al. (2016a), Skorpa et al. (2019), Vrålstad and Skorpa (2020), Skorpa and Vrålstad (2022). Lavrov et al. (2014, 2015) and Roy et al. (2016b) used Abaqus to simulate these experiments, showing that radial tensile stresses contributing to annular cement debonding are likely to develop during cooling of such an assemblage.

9.4 Conclusions

This review shows that conducting experiments under thermal loadings is more difficult than performing them with pressure loadings, especially if the outer boundary conditions should reflect the boundary condition encountered downhole. Moreover, most laboratory experiments have been conducted under conditions encountered in oil & gas, and no such experiments have been conducted under the temperatures encountered in high-temperature geothermal wells. Alternative mechanisms have been studied, such as fatigue, but their use to design a cement sheath is not evident since their boundary conditions may not be representative of what happens downhole. Furthermore, most experiments were only conducted on Portland cement-based systems.

10 Cement integrity—Thermo-mechanical properties

Theoretically, CMI modeling requires an understanding of the cement's thermal-porous-chemical–mechanical behavior and how changes in temperature, pressure, and stress will affect the following parameters:

- Thermal: Specific heat, coefficients of thermal conduction, coefficient of thermal dilation.
- Porous: Biot's coefficient, Skempton's coefficient.
- Chemical: Cement hydration.
- Mechanical: Elastic, plastic, and failure properties.

This is a very broad domain of study and many times not all the required cement parameters are measured for a given project. One may only need to measure a limited number of these parameters if a simplified, but still precise, model is used or if a data base has been built, correlations have been calculated, and theoretical models have been developed to estimate the unmeasured parameters from the measured ones.

The discussion of set-cement properties measurement can be extensive. However, this section will only focus on the static measurement of the mechanical properties of set-cement under compression. No reference is made to the measurement of the dynamic elastic parameters because they do not compare with the static ones. The reader interested in the poro-mechanical behavior of set-cement is invited to read the papers by Ghabezloo et al. (2008, 2009) and Meng et al. (2020).

The current protocol used in the industry to measure the mechanical properties of set cement is the following (Thiercelin, 2006; Garnier et al., 2007; James and Boukhelifa, 2008; API, 2017):

1. The water, cement, and chemical additives are weighted and mixed according to API recommended practice (API, 2013).
2. The slurry is poured into molds. Two types of mold are used. The first type of molds have exactly the shape of the samples on which the tests are performed. The second type of molds are larger than the plugs on which the tests are performed, and they require additional cutting/coring work to obtain the required plugs.
3. The molds are placed in a curing chamber full of water and the curing chamber is closed. The temperature and pressure of the chamber are increased at a rate which enables the target values to be reached before the cement slurry becomes a porous solid.
4. The temperature and pressure remain constant during the curing of the cement.
5. The temperature and pressure are decreased to ambient conditions at a low rate so as not to damage the set-cement samples. During the curing process, the pressure is maintained high enough to ensure that the water remains in its liquid phase.
6. The curing chamber is opened, and the molds taken out.
7. The samples are removed from the molds and are shaped and/or surface ground depending on the type of test to perform and the type of mold used.
8. The cement plugs are placed under the load frame, and the test is performed.
9. The test data are analyzed and the required parameters evaluated.

If the objective is only to measure the unconfined compressive strength at ambient temperature, the crush testing protocol can be used. The cement samples are 5-cm cubes, they are rotated by 90°, then are placed under a conventional load frame, and the axial load is uniformly applied until the cement cube fails. This testing protocol can result in a high dispersion in the results (Garnier et al., 2007), which is why the use of the UCS testing protocol is highly recommended.

If the objective is to measure the unconfined compressive strength and the elastic properties at ambient temperature, the UCS testing protocol can be used. With the UCS testing protocol the cement plugs are solid cylinders with a length-to-diameter ratio of 2, they are axially placed under a conventional load frame with platens of the same diameter as the plugs. Sensors are installed to measure the axial and radial deformation of the plugs. Then the axial load is cycled and/or uniformly applied up to failure. Most of the time, the elastic properties can be obtained with more precision by using the CCS protocol.

If the objective is to measure the complete set-cement constitutive law parameters (unconfined compressive strength, friction angle, elastic, and plastic behavior), or to perform tests at temperature, the CCS (confined compressive strength) testing protocol must be used. The cement plugs are solid cylinders with length to diameter ratio of 2. The plug is placed in a special cell that allows an axial load and radial confining pressure to be applied to the sample. Sensors are installed to measure the axial and radial deformation of the plugs. An isotropic confining pressure is applied, and the radial confining pressure is kept constant while the axial load is cycled and/or uniformly increased up to failure.

Usually it is easy to determine the failure point but, since the elastic properties are determined as slopes measured on the stress–strain curves, their evaluation is often less precise. Furthermore, the elastic properties can be determined on a loading phase or on an unloading phase when cycles are performed. This means that a test can result in different values, depending on how the analysis is made.

We will focus on the measurement of the elastic and plastic properties, and failure in compression and will review some key papers. As early as 1965, Handin (1965) applied the CCS protocol to study the effect of testing temperature and confining and pore pressure on the strength of water-saturated and air-dried oilfield cements. He observed that conventional cement systems become ductile even under low effective confining pressures and that strain of up to 30% could be achieved before rupture. The testing temperature had little effect on the strength of cement but reduced the ductility, which may be an artifact of the testing protocol. The effect of the drainage conditions was very significant. Later, Thiercelin et al. (1998) used the same protocol on conventional 1.89-sg class G cement systems under different confining pressures. The deviatoric phases of the tests were performed at deformation rates of 7 or 2 μStrain/sec, showing no significant difference between the two rates. On the basis of this observation, the measured permeability of the sample, and the strain rate used, the authors considered that the tests were fully drained. Thiercelin et al. also showed that the conventional Class G cement system showed ductile behavior under relatively low confining pressures; deformation of more than 20 millistrains could be supported without failure with a confining pressure of 30 MPa. The cement Young's modulus was relatively independent of confining pressure.

The impact of the loading rate and of the associated drainage conditions is most of the time not considered when testing well cements. Hence, Ravi et al. (2007) performed CCS tests with cyclic loading phases between 50% and 90% of the UCS. Although not specifically detailed in the paper, the initial strain rate in the cycling phases was around 10 μStrain/sec, but this was increased to at least 80 μStrain/sec for most of the cycles for foam cement, and it was even higher for the non-foamed systems. This high-strain-rate cycling could make a significant difference in system behavior because the effectively undrained conditions would lead to pore pressure increase in the samples during testing. Iverson et al. (2010), among others, did not give the loading rate. However, too high loading rates lead to undrained behavior and, if no correction is performed, lead to an overestimation of the Young' modulus and Poisson's ratio of the set-cement, and to an underestimation of the strength. On the contrary, too low loading rates translate to creep and, therefore, to underestimation of all parameters (Ghabezloo et al., 2008; Vu, 2012; Vu et al., 2012; Bois et al., 2013).

The possibility for set-cement to behave plastically, with a collapsible behavior, as identified by Handin (1965) and Thiercelin et al. (1998) has been studied by Ghabezloo et al. (2008), Vu (2012), Vu et al. (2012), and Bois et al. (2013) who showed that collapsible plasticity can induce the opening of microannuli at a casing/cement interface due to compressive cycles: Loading the cement into compression induces its plastic compaction and unloading does not permit the cement to return to its initial state since the plastic strains are not recovered.

Vrålstad et al. (2021) cured cement plugs inside separate high-temperature autoclaves, which were placed inside an oven at 500°C. The total test duration of the curing was 8 weeks, with 2 weeks for temperature ramp up to 500°C, 4 weeks held constant at 500°C, and 2 weeks ramp down to room temperature. However, the limitation of this curing process is that the actual pressure inside the autoclaves during aging was unknown and visual observation after exposure showed that all cement plugs were completely dry. The UCS tests performed on Portland-based cement plugs showed that cement becomes stiffer after high-temperature exposure. On the contrary, the tests on CAC plugs showed that the cement becomes less stiff and more ductile after high-temperature exposure. These results are difficult to interpret since at least four mechanisms concur: increase of porosity, change of mineral phases, modification of capillary pressure, and plug damage during preparation, hence showing the difficulty that are encountered when curing at very high temperature. ter Heege et al. (2019) and Vu et al. (2021) performed similar tests, showing that curing should replicate downhole conditions for the tests to be representative.

These papers indicate there are many testing variables which can affect the determination of cement properties for high-temperature geothermal wells. These test variables include the plug preparation, the water saturation, the loading rate, the use of loading cycles, and the drainage status. Among these variables, the sample preparation deserves a special attention since the samples are generally cured with access to water, while many times cement hydrates under downhole conditions with little or no access to water. The process of reducing the pressure and temperature at the end of the curing phase to reach ambient conditions is critical as a too rapid decrease in temperature and/or pressure could damage the plugs. Precisely characterizing the mechanical behavior of a cement system at high temperature is challenging with the testing equipment available and the applied procedures. What is needed is new test equipment and related testing procedures, which would enable the cement slurry to be poured, cured, and tested under geothermal well conductions without removal from the test cell. This will minimize artifacts and therefore allow the impact of the test variables (temperature, pore pressure, loading rate…) to be studied, giving a better understanding of the behavior of cement (elasticity, plasticity, creep…) at high temperatures.

11 Cement integrity—Designing cement sheath

Cement sheath integrity is defined as the ability of the cement sheath not to leak during the life of the well. At the end of the 1990s, the industry started looking at cement integrity under the perspective of mechanical failure, i.e., understanding how cement stresses created during drilling, well completion and stimulation, and reservoir harvesting, could damage a cement sheath. Saint-Marc et al. (2008) identified five failure modes, debonding at the cement sheath's inner or outer interface, shear failure, radial cracking, and disking (Fig. 4).

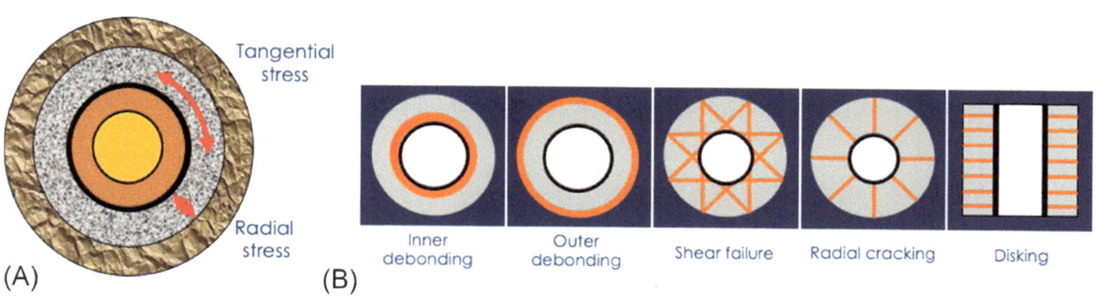

FIG. 4 Cement Mechanical integrity; (A) stresses in the cement sheath; (B) failure modes.

The hidden assumptions were that a mechanically failed cement sheath would lead to losses of hydraulic integrity and a casing surrounded by a damaged cement sheath would become in contact with formation fluids, which could corrode it. A lot of work has been published on cement mechanical integrity (CMI), some of the publications will be reviewed in this section. However, the hidden assumptions are quite pessimistic since mechanically damaged cement leads to leakage only if the defects are connected, which can be the case with a microannulus, but may not be the case for a cement sheath damaged in shear. Furthermore, the risk/importance of casing corrosion depends not only on the nature of the formation fluid, but also on where the reservoir is located along the well since axial corrosion is a slow process, while radial corrosion can be a fast one. This is why cement integrity should also be looked at under the perspective of cement hydraulic integrity (CHI), which is the field that studies how cement mechanical failure can translate to hydraulic leakages. CHI modeling is still in its infancy, as will be shown later in this section.

This section focuses on cement integrity models built to analyze how a cement sheath performs when it initially has no defects, since there is no practical interest in simulating how a damaged cement sheath can become more damaged. It does not focus on the method to solve the equations, be it analytical, semianalytical, or numerical, since this method is only a mathematical tool, and no solving method is bad if it has been well implemented.

11.1 Cement mechanical integrity

To our knowledge, the first paper published on CMI modeling was written by Charara and Nguyen-Minh (1992). They simulated the loading of a cased-hole by a change in either the mud pressure or a change in temperature, and assumed that cement is an incompressible, linear viscoelastic, non-porous body that fails according to Mohr-Coulomb criterion with zero tensile strength. The initial state of stress of the cement after it has set was not computed.

Then, Atkinson and Eftaxiopoulos (1996) developed a model of a cased-hole subjected to anisotropic stress variations in the formation, assuming that cement is a linear isotropic elastic, non-porous body. They considered this model as a representation of a cased hole after cement has set. However, this is not accurate since cement is pumped in a liquid state. Hence, their model only characterizes a cased hole subjected to anisotropic stress variations in the formation.

In the mid-1990s, Schlumberger, Halliburton, and Baker started working to understand how a cement sheath could become mechanically damaged when submitted to stress variations which occurs during the life of a well. The results of these works were published in Thiercelin et al. (1999), Bosma et al. (1999, 2000a, b), Di Lullo and Rae (2000), Ravi et al. (2002a, b), and Thiercelin (2006). Even if these models used different tools to solve the equations, they have six features in common:

1. They consider 2-dimensional cross-sections of the well.
2. They assume that cement is pumped as a liquid and simulate cement loading history phase by phase.
3. The formation and tubular behave according to linear elasticity.
4. Cement failure is analyzed based on Mohr-Coulomb and tensile criteria.

5. The cement sheath behaves according to linear elasticity until it becomes damaged. The cement's porous nature and compacting behavior (pore collapse) were not considered.
6. Cement state of stress after it has set is calculated using simple well hydrostatics assumptions.

Feature 1 is a standard for CMI models, which has been removed by some models. It does not impair the precision of model. Features 2–4 are just common sense, but they have not always been implemented in other models (Atkinson and Eftaxiopoulos, 1996; Fleckenstein et al., 2005, Shahri et al., 2005; Li et al., 2010; Nabipour et al., 2010; Bui and Tutuncu, 2013; Fang et al., 2015; Wang et al., 2016; Zhang and Yao, 2016; Liu et al., 2018). Features 5 and 6 are simplifications that alter the models' predictions. They were discussed in the 2000s. Hence, Fourmaintraux et al. (2004) used the SRC (System Response curve) method to develop a model, which highlighted the importance of simulating the cement hydration processes to evaluate the state of stress in the cement after the cement has set. Saint-Marc et al. (2008) presented the results of simulations performed with a model, which includes the possibility of modeling cement hydration. Bois et al. (2011), 2012) developed mechanistic models to highlight the consequences of correctly calculating the initial state of stress and modeling cement poro-mechanical behavior, including its pore-collapsing behavior. Indeed, this last behavior is at the origin of opening microannuli during compressive loading cycles.

11.2 Cement initial state of stress

The models published by Schlumberger, Halliburton, and Baker (Thiercelin et al., 1998; Bosma et al., 1999, 2000a, b; Di Lullo and Rae, 2000; Ravi et al., 2002a, b) used a very simple approach to calculate the initial state of stress in the cement after it has set, which ranged from zero stress (DeBruijn et al., 2008) to simple fixed value based on the shrinkage/expansion behavior of cement during its hydration, without considering what takes place during cement hydration. Such approach has been followed by many models. Hence the models published by Philippacopoulos and Berndt (2002a,b), Myers et al. (2005), Gray et al. (2009), Teodoriu et al. (2010b), Giasuddin et al. (2012), Xu et al. (2015), Xu et al. (2018a, b), Patel and Salehi (2019), Patel et al. (2019), and Ahmed et al. (2021) considered that the initial state of stress in the cement is zero, following Thiercelin et al. (1998). This is a very pessimistic hypothesis, which increases the risk of concluding that cement will become damaged even if this is not the case. Furthermore, Ferla et al. (2009), Jo and Gray (2010), Nygaard et al. (2011), Arias (2013), Weideman and Nygaard (2014), Guo et al. (2015), Schreppers (2015), De Simone et al. (2017), Zhang et al. (2017), Orlic (2018), Frash and Carey (2018), and Niu et al. (2019) assumed that the initial state of stress in the cement is equal to the hydrostatic pressure of the cement slurry, following Bosma et al. (1999, 2000a, b) and Ravi et al. (2002a, b) when they assumed that there is no shrinkage. Finally, Philippacopoulos and Berndt (2002a,b), Gholami et al. (2016), Li et al. (2016), Hu et al. (2019), Wu et al. (2020, 2021) calculated the initial state of stress in the cement by applying volumetric variations resulting from the shrinkage/expansion behavior of cement during hydration to the set-cement sheath instead of doing it while cement hydrates, again following Bosma et al. (1999, 2000a, b) and Ravi et al. (2002a, b). This method was a direct improvement over the previous approaches, but it tends to over-estimate the effect of cement shrinkage because it is applied while the cement is hard.

Following the papers published by Fourmaintraux et al. (2004) and Saint-Marc et al. (2008), other work has been conducted to correctly model cement hydration and calculate the cement initial state of stress after it has set (Lecampion, 2013; Petersen and Ulm, 2016; Thérond et al., 2017), Onaisi et al., 2017; Ramos and Sánchez Camus, 2017; Zhang and Eckert, 2018, 2020; Gheibi et al., 2020), but it is not clear how these models perform with latest advanced cement systems, which are not just made of Portland cement and water. To alleviate this issue, Luo et al. (2020) and Meng et al. (2021), among others, followed another approach, which consists of performing experiments in the laboratory, and then back analyzing the data to obtain the required cement hydration parameters. Unfortunately, this approach could be cumbersome and expensive if one has to proceed in this manner for each cement sheath configuration to design.

11.3 Cement poro-mechanical behavior

Many consider that there is no need to simulate set-cement as a poro-mechanical material since its permeability is so low that its pore pressure variations are relatively slow. This contradicts the poro-mechanics theory as developed by Biot (1941, 1956) and structured by Rice and Cleary (1976) and Coussy (2010). If this were true, there would be no need to model clays and shales as poro-mechanical materials, there would be no need to distinguish drained from undrained behaviors, and effective stress from total stresses. In fact, wellbore cement is not only a porous material, but it is also a high-porosity material, for which the saturation degree defines, in a large part, its mechanical behavior.

Such poro-mechanical behavior should be considered not only after cement has set, but also while hydrating. For example, Bois et al. (2012) defined the following stages to simulate cement hydration:

1. At a hydration degree lower than the percolation threshold, cement behaves according to fluid mechanics.
2. At a hydration degree higher than the percolation threshold, cement behaves according to poro-mechanics.
3. Cement hydration leads to an increase in grain volume and to a decrease in fluid volume in the cement pores.
4. Cement hydration leads to a change in cement-material physical and mechanical properties.
5. Cement-volume variation during hydration occurs while cement properties evolve.

With the exception of Saint-Marc et al. (2008), Garnier et al. (2010), Lecampion (2013), Petersen and Ulm (2016), Ramos and Sánchez Camus (2017), Thérond et al. (2017), Onaisi et al. (2017), Zhang et al. (2017), Xia et al. (2017), Zhang and Eckert (2018, 2020), Li et al. (2018), Sasaki et al. (2018), Gheibi et al. (2020), and Meng et al. (2021), most authors do not consider the poro-mechanical behavior of the cement, which the authors feel is not the correct approach.

11.4 Cement hydraulic integrity

The main approach followed to simulate CHI consists of assuming a given hydraulic path geometry, and performing hydraulic simulations, the same manner sustained casing pressure is simulated (Huerta et al., 2009; Deremble et al., 2010, 2011; Lewis et al., 2012; Carey et al., 2013;

Kelkar et al., 2014). More advanced models are available where the hydraulics is coupled with the mechanics (Lecampion et al., 2011, 2013; Wang and Dahi Taleghani, 2014, 2017; Dahi Taleghani and Wang, 2016; Feng et al., 2016, 2017; Kim et al., 2016; Gao, and abd Cheng, Y., 2018; Xu et al., 2020). However, these models are still conceptual since the description of the cement sheath before propagating defects is an input parameter, which is not fully understood. To our knowledge, the only model where CMI and CHI have been coupled was presented by Bois et al. (2019) to study the integrity of cement plug. It was validated based on downhole data. No such model has been presented and validated for the annular cement sheath. Furthermore, even for the simplest models, no extension to geothermal conditions has been found in the literature. This is why CHI is a field still in its infancy.

11.5 Requirements for geothermal wells

The previous sections have shown that predicting cement integrity requires having a model that considers cement as a collapsible pro-mechanical solids and that simulates cement hydration to calculate set-cement initial state of stress. However, not much has been published on the application of CMI modeling for hot geothermal wells. The papers found in the literature show that nothing was done to consider the specificities of such wells (Bour and Hernandez, 2003; McCulloch et al., 2003; Petty et al., 2003; Ravi et al., 2008).

Only papers by Vu et al. (2020, 2021) tried to specifically evaluate what would be the consequences of producing fluids with temperatures higher than 200°C. They focused on modeling the increase in pore pressures, the opening of microannuli at the cement sheath's boundaries, and the phase changes which take place in the cement when it is heated to high-temperature values. This work showed that heating cement up to 250°C can result in pore pressures higher than 100 MPa unless the pore pressures can be released. Additionally, it identified that there are three mechanisms that can lead to such release of pore pressures:

1. During cement hydration, due to the water consumption by the chemical reactions.
2. When a microannulus opens due to the large pore pressures, therefore allowing venting the pressures to the surface or to a downhole reservoir.
3. When a change of phase occurs in the cement, as this leads to the creation of additional porosity in the cement.

All this means that the cement sheath should not be simulated as a closed system, but rather as an open thermo-hydro-chemo-mechanics. However, much work remains to be done to validate these concepts and take them into account when designing cement integrity of geothermal wells.

12 Conclusions

Reliable annular isolation is an essential element to minimize maintenance costs and extend the commercial life of high-temperature geothermal wells. The key steps in the selection and implementation of a suitable annular sealant are:

- Chemical design
  The chemical design should consider both the properties of the set material (thermal and chemical resistance) and the slurry properties that are critical for correct placement in the annulus. Although Portland cement-based systems are the most widely used, several

alternate sealant systems with improved chemical resistance have been investigated. However, the current limitation with these alternate sealant systems is that little work has been performed to develop additives to allow them to be placed and set uniformly in the annulus at high temperatures.

The most important slurry properties to control are thickening time, slurry stability and viscosity. Within the oil and gas industry equipment is available to allow these properties to be measured at temperatures up to 300°C.

- Hydraulic design
A good hydraulic design is essential to ensure that the sealant is placed where it is required in the annulus. Good practices from the oil and gas well industry should be applied where possible. However, cementing geothermal wells presents two specific challenges which can lead to poor cement placement in the annulus: (1) cementing to surface and (2) not being able to rotate the casing. Therefore, it is highly recommended to ensure that the casing standoff is above 70%–80%. Otherwise mud pockets may exist, which could lead to casing collapse when the temperature is increased, and/or mud channels may remain in the annulus, offering a path for acidic fluids to migrate faster along the cement sheath.

One area where additional studies would be beneficial is the rheological behavior of well construction fluids (spacers, cement slurries) at high temperatures.

- Integrity design
The high temperatures and large temperature fluctuations of geothermal wells pose challenges to the mechanical and hydraulic integrity of the casing/cement/formation assembly. Recent simulation work has identified potential failure mechanisms for geothermal wells, where the very high temperatures may cause large increases in the pore pressure of the sealant unless the pressure can be relieved. However, experimental work is still required to validate these concepts. Furthermore, the industry has not a rigorous process to measure cement thermo-mechanical properties and to perform cement integrity simulations. The industry together with the academic world should urgently engage in R&D studies to overcome this weakness, which can threaten the capacity to design geothermal wells that do not fail. Sometimes, it is better to increase the CAPEX slightly if this allows to decrease the OPEX significantly.

This review was dedicated to high-temperature hydro-geothermal and enhanced geothermal wells. What about new advanced geothermal wells, such as closed-loop geothermal wells? Closed loop wells have the advantage that little to no formation fluid enters the wellbore. Coaxial closed loop wells have another advantage since the fluid is pumped down the annulus. This means that, in such wells, the upper cement sheaths experience lower increases in temperature than in hydro-geothermal and enhanced geothermal-producing wells. However, such advanced geothermal wells require that the cement sheath be thermally conductive to maximize the conductive transfer of heat from the formation to the harvesting fluid (Ichim et al., 2018; Song et al., 2019; Abid et al., 2023). This is a new field of investigation since what is needed is not cement samples with high thermal conductivity in the laboratory, but a high thermally conductive cement sheath downhole. More work is required to move from R&D to operations.

References

Abid, K., Srivastava, S., Romero Tellez, M.L., Amani, M., Teodoriu, C., 2023. Experimental and machine learning study of thermal conductivity of cement composites for geothermal wells. Geothermics 110 (102659).

Ahmed, S., Patel, H., Salehi, S., 2021. Effects of wait on cement, setting depth, pipe material, and pressure on performance of liner cement. J. Pet. Sci. Eng. 196 (15), 108008.

Albawi, A., 2013. Influence of Thermal Cycling on Cement Sheath Integrity. Master of Science Thesis, Norwegian University of Science and Technology (NTNU), Trondheim, Norway.

Albawi, A., de Andrade, J., Torsæter, M., Opedal, N., Stroisz, A., Vrålstad, T., 2014. Experimental set-up for testing cement sheath integrity in Arctic wells. In: OTC Arctic Technology Conference, Houston, Texas, USA, 10-12 February. OTC-24587-MS.

API, 2013. Recommended Practice for the Testing of Well Cements, second ed. API RP 10B-2. April.

API, 2017. Mechanical Behavior of Cement, first ed. API TR 10TR7. December.

Arias, H., 2013. Use of Finite-Element Analysis to Improve Well Cementing in HPHT Conditions. PhD dissertation, Texas A&M University, College Station, Texas, USA, p. 243.

Arnold, E.S., 1982. Cementing: bridging the gap from laboratory results to field operations. J. Pet. Technol. 34 (12), 2843–2852 (SPE-9708-PA).

Atkinson, C., Eftaxiopoulos, D.A., 1996. A plane model for the stress field around an inclined, cased and cemented wellbore. Int. J. Numer. Anal. Methods Geomech. 20, 549–569.

Barlet-Gouédard, V., Danican, S., Nelson, E., Goffé, B., Cambus, C., 2008. Cement Compositions for High Temperature Applications. (US patent 7,459,019 B2).

Barlet-Gouédard, V., James, S., Drochon, B., Piot, B., Caritey, J.-P., 2012. Cement Composition for Carbon Dioxide Supercritical Environment. (US patent 8,091,642 B2).

Beirute, R.M., 1991. A circulating and shut-in well-temperature-profile simulator. J. Pet. Technol. 43 (9), 1140–1146. SPE-17591-PA.

Beirute, R.M., Sabins, F.L., Ravi, K.M., 1991. Large-scale experiments show proper hole conditions: a critical requirement for successful cementing operations. In: SPE Annual Technical Conference and Exhibition, Dallas, Texas, USA, 6-9 October 1991. SPE-22774-MS.

Beltrán-Jiménez, K., 2022. Characterization of Annulus Well Cement Quality to Estimate the Leakage Potential in Cement Well Barriers. Ph.D. dissertation, University of Stavanger, Norway.

Benedictus, T., van der Kuip, M.D.C., Kronimus, R.A., Huibregtse, J.N., Yavuz, F., Remmelts, G., Remmelts, G., 2009. Long-Term Integrity of CO2 Storage—Well Abandonment. Technical report 2009/08, International Energy Agency, Greenhouse Gas R&D Programme.

Benge, G., Dew, E.G., 2006. Meeting the challenges in completion liner design and execution for two high-rate acid-gas injection wells. SPE Drill. Compl. 21 (3), 180–184. SPE-91861-PA.

Biot, M.A., 1941. General theory of three-dimensional consolidation. J. Appl. Phys. 12 (2), 155–164.

Biot, M.A., 1956. General solutions of the equations of elasticity and consolidation for a porous material. J. Appl. Mech. 27 (3), 91–96.

Bittleston, S.H., Guillot, D., 1991. Mud removal: research improves traditional cementing guidelines. In: Oilfield Review April, pp. 44–55.

Bois, A.-P., Garnier, A., Rodot, F., Saint-Marc, N., Aimard, N., 2011. How to prevent loss of zonal isolation through a comprehensive analysis of microannulus formation. SPE Drill. Compl. 26 (1), 13–31. SPE-124719-PA.

Bois, A.-P., Garnier, A., Galdiolo, G., Laudet, J.-B., 2012. Use of mechanistic model to forecast cement-sheath integrity. SPE Drill. Compl. 27 (02), 303–314 (SPE-139668-PA).

Bois, A.-P., Vu, M.-H., Ghabezloo, S., Sulem, J., Garnier, A., Laudet, J.-B., 2013. Cement sheath integrity for CO2 storage—an integrated perspective. Energy Procedia 37, 5628–5641.

Bois, A.-P., Vu, M.-H., Noël, K., Badalamenti, A., Delabroy, L., Thérond, E., Hansen, K., 2019. Evaluating cement-plug mechanical and hydraulic integrity. SPE Drill. Compl. 34 (2), 92–102. SPE-191335-PA.

Bois, A.-P., Zhao, H., Wen, D., Luo, Y., Li, Y., Badalamenti, A.M., Song, M., Calvo, C., Reñe, J., Liang, H., 2023. Cement placement modeling—a review. SPE Drill. Compl. 38 (2), 342–357. SPE-214331-PA.

Bosma, M., Ravi, K., van Diel, W., et al., 1999. Design approach to sealant selection for the life of the well. In: SPE Annual Technical Conference and Exhibition, Houston, Texas, USA, 3–6 October. SPE-56536-MS.

Bosma, M.G.R., Cornelissen, E.K., Schwing, A., 2000a. Improved experimental characterization of cement/rubber zonal isolation materials. In: SPE International Oil and Gas Conference and Exhibition in China, Beijing, China, 7-10 November. SPE-64762-MS.

Bosma, M.G.R., Cornelissen, E.K., Schwing, A., 2000b. Improved experimental characterization of cement/rubber zonal isolation materials. In: SPE Asia Pacific Oil and Gas Conference and Exhibition, Brisbane, Australia, 26-28 February. SPE-74497-MS.

Boukhelifa, L., Moroni, N., James, S.G., Le Roy-Delage, S., Thiercelin, M.J., Lemaire, G., 2005. Evaluation of cement systems for oil- and gas-well zonal isolation in a full-scale annular geometry. SPE Drill. Compl. 20 (1), 44–53. SPE-87195-PA.

Bour, D.L., Hernandez, R., 2003. CO2 resistance, improved mechanical durability, and successful placement in a problematic lost-circulation interval achieved: reverse circulation of foamed calcium aluminate cement in a geothermal well. GRC Trans. 27, 163–167.

Brothers, L.E., 2008. Cement Compositions with Improved Mechanical Properties and Methods of Cementing in Subterranean Formations. (US patent 7,351,279 B2).

Bruckdorfer, R.A., Coleman, S.E., 1987. CO_2-Enhanced Oil Recovery with Corrosion Resistant Cement. (US patent 4,635,724 A).

Bui, B.T., Tutuncu, A.N., 2013. Modeling the failure of cement sheath in anisotropic stress field. In: SPE Unconventional Resources Conference-Canada, Calgary, Alberta, Canada, 5-7 November. SPE-167178-MS.

Cadix, A., James, S., 2021. Cementing additives. In: Wang, Q. (Ed.), Fluid Chemistry, Drilling and Completion. Elsevier, pp. 187–254.

Carey, J.W., Lewis, K., Kelkar, S., Zyvoloski, G.A., 2013. Geomechanical behavior of wells in geologic sequestration. Energy Procedia 37, 5642–5652.

Charara, I., Nguyen-Minh, D., 1992. Mechanical effects of pressure and temperature variations on annular cement in oil wells. In: 33rd US Rock Mechanics Symposium, Santa Fe, New Mexico, USA, 3-5 June. ARMA-92-0429.

Chougnet-Sirapian, A., Pershikova, E., Loiseau, A., 2011. New steam resilient cement: evaluation of long-term properties under extreme conditions. In: SPE International Symposium on Oilfield Chemistry, the Woodlands, Texas, USA, 11-13 April. SPE-141202-MS.

Corina, A.N., Moghadam, A., 2024. The sealing performance of cement sheaths under thermal cycles for low-enthalpy geothermal wells. Energies 17 (239), 12.

Coussy, O., 2010. Mechanics and Physics of Porous Solids. John Wiley and Sons, p. 281.

Crook, R.J., Benge, G., Faul, R., Jones, R.R., 2001. Eight steps ensure successful cement jobs. Oil Gas J., 37–43. 2 July.

Da Silva, J.R.M.C., Milestone, N.B., Johnston, J.H., 2012. The effect of drilling fluid and temperature on the cement-rock interaction in geothermal wells. GRC Trans. 36, 277–285.

Dahi Taleghani, A., Wang, W., 2016. Emergence of delamination fractures around the casing and its stability. J. Energy Resour. Technol. 139 (012904), 11.

Dargaud, B., Boukhelifa, L., 2006. Laboratory testing, evaluation, and analysis of well cements. In: Nelson, E.B., Guillot, D. (Eds.), Well Cementing, second ed, pp. 627–657. Schlumberger.

Davies, D.R., Hartog, J.J., Cobbett, J.S., 1981. Foamed cement—a cement with many applications. In: SPE Middle East Oil Technical Conference, Manama, Bahrain, 9–12 March. SPE-9598-MS.

de Andrade, J., Sangesland, S., Skorpa, R., Toropovic, J., Vrålstad, T., 2015. Cement sheath integrity during thermal cycling: experimental tests on wellbore sections. In: SPE Bergen One Day Seminar, Bergen, Norway, 22 April. SPE-173871-MS.

De Rozières, J., Ferrière, R., 1991. Foamed-cement characterization under downhole conditions and its impact on job design. SPE Prod. Eng. 6 (3), 297–304. SPE-19935-PA.

De Simone, M., Pereira, F.L.G., Roehl, D.M., 2017. Analytical methodology for wellbore integrity assessment considering casing-cement-formation interaction. Int. J. Rock Mech. Min. Sci. 94 (4), 112–122.

DeBruijn, G., Siso, C., Reinheimer, D., Whitton, S., Redekopp, D., 2008. Flexible cement improves wellbore integrity for steam assisted gravity drainage (SAGD) wells. In: SPE/PS/CHOA International Thermal Operations and Heavy Oil Symposium, Calgary, Alberta, Canada, 20-23 October. SPE-117859-MS.

Deremble, L., Loizzo, M., Lecampion, B., Quesada, D., 2010. Assessment of leakage pathways along a cemented annulus. In: SPE International Conference on CO2 Capture, Storage, and Utilization, New Orleans, Louisiana, USA, 10–12 November. SPE-139693-MS.

Deremble, L., Loizzo, M., Huet, B., Lecampion, B., Quesada, D., 2011. Stability of a leakage pathway in a cemented annulus. Energy Procedia 4, 5283–5290.

Di Lullo, G., Rae, P., 2000. Cements for long term isolation—design optimization by computer modelling and prediction. In: IADC/SPE Asia Drilling Technology Conference and Exhibition, Kuala Lumpur, Malaysia, 11-13 September. SPE-62745-MS.

Drochon, B., Garnier, A., 2003. Low Density and Low Porosity Cement Slurries for Oil Wells and the Like. (US Patent 6626991 B2).

Duguid, A., 2008. An estimate of the time to degrade the cement sheath in a well exposed to carbonated brine. Energy Procedia 1 (1), 3181–3188.

Duguid, A., Radonjic, M., Bruant, R., Mandecki, T., Scherer, G., Celia, M., 2004. The effect of CO2 sequestration on oil well cements. In: 7th International Conference on Greenhouse Gas Control Technologies, Vancouver, British Columbia, Canada, 5-9 September.

Eilers, L.H., Root, R.L., 1974. Long-term effects of high temperature on strength retrogression of cements. In: 49th Annual Fall Meeting of the Society of Petroleum Engineers of AIME, Houston, Texas, USA, 6-9 October. SPE-5028-MS.

Fang, J., Wang, Y., Gao, D., 2015. On the collapse resistance of multilayer cemented casing in directional well under anisotropic formation. J. Nat. Gas Sci. Eng. 26, 409–418.

Feng, Y., Podnos, E., Gray, K.E., 2016. Well integrity analysis: 3D numerical modeling of cement interface debonding. In: 50th US Rock Mechanics/Geomechanics Symposium, Houston, Texas, USA, 26–29 June. ARMA-16-246.

Feng, Y., Li, X., Gray, K.E., 2017. Development of a 3D numerical model for quantifying fluid-driven interface debonding of an injector well. Int. J. Greenhouse Gas Control 62, 76–90.

Ferla, A., Lavrov, A., Fjaer, E., 2009. Finite-element analysis of thermal-induced stresses around a cased injection well. J. Phys. Conf. Ser. 181 (012051), 8.

Fleckenstein, W.W., Eustes, A.W.I., Rodriguez, W.J., et al., 2005. Cemented casing: the true stress picture. In: AADE National Technical Conference and exhibition, Houston, Texas, USA, 5-7 April. AADE-05-NTCE-14.

Fourmaintraux, D., Bois, A.-P., Franco, C., Fraboulet, B., Brossollet, P., 2004. Efficient wellbore cement sheath design using the SRC (system response curve) method. In: 14th Europec Biennal Conference, Madrid, Spain, 13-16 June. SPE-94176-MS.

Frash, L.P., Carey, J.W., 2018. Engineering prediction of axial wellbore shear failure caused by reservoir uplift and subsidence. SPE J. 23 (4), 1039–1066. SPE-189981-PA.

Friðleifsson, G.Ó., Pálsson, B., Stefánsson, B., Albertsson, A., Gíslason, Þ., Gunnlaugsson, E., Thorsteinsson, H.H., Ketilsson, J., Sæther, S., Sörlie, C., Elders, W.A., Zierenberg, R.A., 2021. The IDDP success story—highlights. In: World Geothermal Congress 2020+1, Reykjavik, Iceland, April–October 2021.

Gao, Q., abd Cheng, Y., 2018. Numerical investigation of fluid-driven debonding fracture propagation along wellbore interfaces during hydraulic fracturing. In: Energy Sources, Part A: Recovery, Utilization, and Environmental Effects, p. 11.

Garnier, A., Fraboulet, B., Saint-Marc, J., Bois, A.-P., 2007. Characterization of cement systems to ensure cement sheath integrity. In: Offshore Technology Conference, Houston, Texas, USA, 30 April-3 may. OTC-18754.

Garnier, A., Saint-Marc, J., Bois, A.-P., Kermanac'h, Y., 2010. An innovative methodology for designing cement-sheath integrity exposed to steam stimulation. SPE Drill. Compl. 25 (1), 58–69. SPE-117709-PA.

Ghabezloo, S., Sulem, J., Guédon, S., Martineau, F., Saint-Marc, J., 2008. Poromechanical behavior of hardenend cement paste under isotropic loading. Cem. Concr. Res. 38 (11), 1424–1437.

Ghabezloo, S., Sulem, J., Saint-Marc, J., 2009. The effect of undrained heating on a fluid-saturated hardened cement paste. Cem. Concr. Res. 39 (1), 54–64.

Gheibi, S., Agofack, N., Sangesland, S., 2020. Modified discrete element method (MDEM) as a numerical tool for cement sheath integrity in wells. J. Pet. Sci. Eng. 196 (107720), 15.

Gholami, R., Aadnoy, B.S., Fakhari, N., 2016. A thermo-poroelastic analytical approach to evaluate cement sheath integrity in deep vertical wells. J. Pet. Sci. Eng. 147, 536–546.

Giasuddin, H.M., Sanjayan, J.G., Ranjith, P.G., 2012. Modeling of a wellbore composite cylinder system for cement sheath stress analysis in geological sequestration of CO2. In: 46th US Rock Mechanics/Geomechanics Symposium, Chicago, Illinois, USA, 24–27 June. ARMA-2012-369.

Goodwin, K.J., Crook, R.J., 1992. Cement sheath stress failure. SPE Drill. Eng. 7 (4), 291–296. SPE-20453-PA.

Gray, K.E., Podnos, E., Becker, E., 2009. Finite element studies of near-wellbore region during cementing operations: part 1. SPE Drill. Compl. 24 (1), 127–136. SPE-106998-PA.

Guo, S., Bu, Y., Yan, X., 2015. Cement sheath integrity under two different formation conditions in steam stimulation well. In: 25th International Ocean and Polar Engineering Conference, Kona, Big Island, Hawaii, USA, 21-26 June.

Handin, J., 1965. Strength of oil well cements at downhole pressure-temperature conditions. SPE J. 5 (4), 341–347. SPE-1300-PA.

Heathman, J., Beck, F.E., 2006. Finite element analysis couples casing and cement designs for HP/HT wells in East Texas. In: SPE/IADC Drilling Conference, Miami, Florida, USA, 21-23 February. SPE-98869-MS.

Hedenquist, J.W., Stewart, M.K., 1985. Natural CO_2-rich, steam-heated, waters in the Broadlands Ohaaki geothermal system, New Zealand: their chemistry, distribution, and corrosive nature. GRC Trans. 9, 245–250.

Heller, L., Taylor, H.F.W., 1951. Hydrated calcium silicates. Part II. Hydrothermal reactions: lime: silica ratio 1: 1. J. Chem. Soc., 2397–2401.

Hoang, H.N., Bois, A.-P., Naumann, M., Sæther, S., Coudert, T., 2022. Effect of cement behaviour on casing integrity in superhot geothermal wells: a numerical study. In: 56th US Rock Mechanics/Geomechanics Symposium, Santa Fe, New Mexico, USA, 26–29 June 2022. ARMA 22–738.

Hu, J., Wang, L., Feng, J., et al., 2019. Full-life-cycle analysis of cement sheath integrity. Math. Probl. Eng. 2019 (8279435), 11.

Huerta, N.J., Checkai, D., Bryant, S.L., 2009. Utilizing sustained casing pressure analog to provide parameters to study CO2 leakage rates along a wellbore. In: SPE International Conference on CO2 Capture, Storage, and Utilization, San Diego, Califormia, USA, 2-4 November. SPE-126700-MS.

Ichim, A., Teodoriu, C., Falcone, G., 2018. Estimation of cement thermal properties through the three-phase model with application to geothermal wells. Energies 11 (2839), 11.

Iverson, B.J., Maxson, J., Bour, D.L., 2010. Strength retrogression in cements under high-temperature conditions. In: 35th Workshop on Geothermal Reservoir Engineering, Stanford University, California, USA, 1–3 February.

Jackson, P.B., Murphey, C.E., 1993. Effect of casing pressure on gas flow through a sheath of set cement. In: SPE Drilling Conference, Amsterdam, the Netherlands, 23–25 February. SPE-25698-MS.

James, S., Boukhelifa, L., 2008. Zonal isolation modeling and measurements—past myths and today's realities. SPE Drill. Compl. 23 (1), 68–75. SPE-101310-PA.

Jo, H., Gray, K.E., 2010. Mechanical behavior of concentric casing, cement, and formation using analytical and numerical methods. In: 44th US Rock Mechanics Symposium and 5th US-Canada Rock Mechanics Symposium. Salt Lake City, Utah, USA, 27-30 June. ARMA-10-142.

Johns, A.W., 2014. Cyclic Loading of fiber-Containing Cement Sheaths in HPHT Conditions. Texas A&M University, College Station, Texas, USA.

Joseph, T., Chakraborty, P.P., Melbouci, M., 2014. Calcium Aluminate Cement Composition Containing a Set Retarder of an Organic Acid and a Polymeric Mixture. (US patent 8,720,563 B2).

Kalousek, G.L., 1952. The reactions of cement hydration at elevated temperatures. In: 3rd International Symposium on the Chemistry of Cement, London, UK, 15-20 September, pp. 334–354.

Kelkar, S., Carey, J.W., Dempsey, D., Lewis, K., 2014. Integrity of pre-existing wellbores in geological sequestration of CO2—assessment using a coupled geomechanics-fluid flow model. Energy Procedia 63, 5737–5748.

Kim, J., Moridis, G.J., Martinez, E.R., 2016. Investigation of possible wellbore cement failures during hydraulic fracturing operations. J. Pet. Sci. Eng. 139, 254–263.

Koons, B.E., Free, D.L., Frederick, A.F., 1993. New design guidelines for geothermal cement slurries. GRC Trans. 17, 43–51.

Kosinowski, C., Teodoriu, C., 2012. Study of class G cement fatigue using experimental investigations. In: SPE/EAGE European Unconventional Resources Conference and Exhibition, Vienna, Austria, 20-22 March. SPE-153008-MS.

Kukacka, L.E., 1977. The applicability of concrete polymer materials for use in geothermal environments. In: SPE International Symposium on Oilfield and Geothermal Chemistry, San Diego, California, USA, 27–29 June. SPE-6611-MS.

Kukacka, L.E., 1981. Current status of geothermal well cement development. In: International Geothermal Drilling and Completions Technology Conference, Albuquerque, New Mexico, USA, 21 January.

Kukacka, L.E., 1982. Geothermal well cements; current status of R and D and downhole testing. In: International Conference on Geothermal Energy. Florence, Italy, 11–14 May.

Kukacka, L.E., Sugama, T., 1994. Lightweight carbon dioxide resistant cements for geothermal well completions. In: The Geothermal Program Review XII, pp. 119–126.

Kulu, P., Goljandin, D., Traksmaa, R., Kaljuvee, T., Gregor, A., 2021. Thermal properties of calcium-aluminate based materials. IOP Conf. Ser.: Mater. Sci. Eng. 1140 (012028), 6.

Kyritsis, K., Meller, N., Hall, C., 2009. Chemistry and morphology of hydrogarnets formed in cement-based CASH hydroceramics cured at 200 to 350°C. J. Am. Ceram. Soc. 92 (5), 110–1111.

Laudet, J.B., Garnier, A., Neuville, N., Le Guen, Y., Fourmaintraux, D., Rafai, N., Burlion, N., Shao, J.F., 2011. The behavior of oil well cement at downhole CO2 storage conditions: static and dynamic laboratory experiments. Energy Procedia 4, 5251–5258.

Lavrov, A.V., Torsaeter, M., Albawi, A., Todorovic, J., Opeda, N., Cerasi, P., 2014. Near-well integrity and thermal effects: a computational road from laboratory to field scale. In: 48th US Rock Mechanics / Geomechanics Symposium, Minneapolis, Minnesota, USA, 1–4 June. ARMA-14-7109.

Lavrov, A., Todorovic, J., Torsæter, M., 2015. Numerical study of tensile thermal stresses in a casing-cement-rock system with heterogeneities. In: 49th US Rock Mechanics / Geomechanics Symposium, San Francisco, California, 28 June-1 July. ARMA-15-110.

Lecampion, B., 2013. A macroscopic poromechanical model of cement hydration. Eur. J. Environ. Civ. Eng. 17 (3), 176–201.

Lecampion, B., Quesada, D., Loizzo, M., Bunger, A., Kear, J., Deremble, L., Desroches, J., 2011. Interface debonding as a controlling mechanism for loss of well integrity: importance for CO2 injector wells. Energy Procedia 4, 5219–5226.

Lecampion, B., Bunger, A., Kear, J., Quesada, D., 2013. Interface debonding driven by fluid injection in a cased and cemented wellbore: modeling and experiments. Int. J. Greenh. Gas Control 18, 208–223.

Lewis, K., Zyvoloski, G.A., Kelkar, S., Carey, J.W., 2012. Coupled stress and flow along interfaces in the wellbore environment in relation to CO_2 sequestration. In: 46th US Rock Mechanics / Geomechanics Symposium, Chicago, Illinois, USA, 24–27 June. ARMA-12–437.

Li, Y., Liu, S., Wang, Z., Yuan, J., Qi, F., 2010. Analysis of cement sheath coupling effects of temperature and pressure in non-uniform in situ stress field. In: CPS/SPE International Oil & Gas Conference and Exhibition in China, Beijing, China, 8-10 June. SPE-131878-MS.

Li, W., Chen, M., Jin, Y., et al., 2016. The application of cement sheath failure criterion in determining the wellbore internal pressure window. In: 50th US Rock Mechanics / Geomechanics Symposium, Houston, Texas, USA, 26–29 June. ARMA-2016-627.

Li, W., Chen, M., Jin, Y., et al., 2018. The porothermoelastic response of casing-cement-stratum. In: 52nd US Rock Mechanics / Geomechanics Symposium, Seattle, Washington, USA, 17–20 June. ARMA-2018-1164.

Liu, K., Gao, D., Dahi-Taleghani, A., 2018. Analysis on integrity of cement sheath in the vertical section of wells during hydraulic fracturing. J. Pet. Sci. Eng. 168, 370–379.

Luo, Y., Shan, Y., Liu, H., Liu, W., Cao, W., Wang, Y., 2020. Influence of the initial stress state of the cement sheath on the casing-cement sheath-formation system stress distribution. In: 54th US Rock Mechanics/Geomechanics Symposium, Golden, Colorado, USA, 28 June-1 July 2020. ARMA-20-1868.

McCulloch, J., Gastineau, J., Bour, D.L., Ravi, K.M., 2003. Life cycle modeling of wellbore cement systems used for enhanced geothermal system development. GRC Trans. 27, 147–154.

McDaniel, J., Watters, L.T., Shadravan, A., 2014. Cement sheath durability: Increasing cement sheath integrity to reduce gas migration in the Marcellus shale play. In: SPE Hydraulic Fracturing Technology Conference, the Woodlands, Texas, USA, (4-6 February. SPE-168650-MS).

Meller, N., Hall, C., Phipps, J., 2005. A new phase diagram for the $CaOAl_2O_3SiO_2H_2O$ hydroceramic system at 200°C. Mater. Res. Bull. 40 (5), 715–723.

Meller, N., Hall, C., Kyritsis, K., Giriat, G., 2007. Synthesis of cement based $CaO–Al_2O_3–SiO_2–H_2O$ (CASH) hydroceramics at 200 and 250°C: ex-situ and in-situ diffraction. Cem. Concr. Res. 37 (6), 823–833.

Meller, N., Kyritsis, K., Hall, C., 2009. The mineralogy of the $CaO–Al_2O_3–SiO_2–H_2O$ (CASH) hydroceramic system from 200 to 350 °C. Cem. Concr. Res. 39 (1), 45–53.

Meng, M., Frash, L.P., Carey, J.W., Li, W., Welch, N.J., 2020. Rapid measurement of Biot effective stress coefficient for class-G oil well cements. In: 54th US Rock Mechanics/Geomechanics Symposium, Golden, Colorado, USA, 28 June-1 July. ARMA-20-1745.

Meng, M., Frash, L.P., Carey, J.W., Niu, Z., Zhang, W., Guy, N., Lei, Z., Li, W., Welch, N., 2021. Predicting cement sheath integrity with consideration of initial state of stress and thermoporoelastic effects. SPE J. 26 (6), 3505–3528 (SPE 205344-PA).

Menzel, C.A., 1935. Studies of high pressure steam curing of tamped hollow concrete block. J. Am. Concr. Inst. 32 (9), 51–64.

Michaux, M., 1999. Controlling Setting in High-Alumina Cement. Patent WO 99/33763 A1.

Milestone, N.B., Aldridge, L.P., 1990. Corrosion of cement grouts in aggressive geothermal environments. GRC Trans. 14, 423–429.

Milestone, N.B., Kukacka, L.E., Carciello, N., 1986. Effects of carbon dioxide attack on geothermal cement grouts. In: GRC Annual Meeting, Palm Springs, California, USA. 29 September-1 October.

Milestone, N.B., Bigley, C.H., Durant, A.T., Sharp, M.D.W., da Silva, J.R.M.C., 2012. Chemical reactions in geothermal cements. In: New Zealand Geothermal Workshop, Auckland, New Zealand, 19–21 November.

Moghadam, A., Castelein, K., Ter Heege, J., van der Valk, K., Orlic, B., Wollenweber, J., 2020. Large-scale laboratory investigation of the microannulus behavior in the casing-cement interface. In: 54th U.S. Rock Mechanics/ Geomechanics Symposium, Physical Event Cancelled, 26 June-1 July. ARMA-20-1325.

Myers, S., El Shaari, N., Dillenbeck III, R.L., 2005. A new method to evaluate cement systems design requirements for cyclic steam wells. In: SPE Western Regional Meeting, Irvine, California, USA, 30 March-1 April. SPE-93909-MS.

Nabipour, A., Joodi, B., Sarmadivaleh, M., 2010. Finite element of downhole stresses in deep gas wells cements. In: SPE Deep Gas Conference and Exhibition, Manama, Bahrain, 24-26 June. SPE-132156-MS.

Nelson, E.B., Barlet-Gouédard, V., 2006. Thermal cements. In: Nelson, E.B., Guillot, D. (Eds.), Well Cementing, second ed, pp. 319–341. Schlumberger.

Nelson, E.B., Eilers, L.H., 1985. Cementing steamflood and fireflood wells—slurry design. The journal of. Can. Pet. 24 (5), 58–63. PETSOC-85-05-06.

Nelson, E.B., Michaux, M., Drochon, B., 2006. Cement additives and mechanisms of action. In: Nelson, E.B., Guillot, D. (Eds.), Well Cementing, second ed, pp. 49–92. Schlumberger.

Niu, Z., Shen, J., Wang, L., et al., 2019. Thermo-poroelastic modelling of cement sheath: pore pressure response, thermal effect and thermo-osmotic effect. Eur. J. Environ. Civ. Eng. 26 (2), 657–682.

Nygaard, R., Salehi, S., Lavoie, R., 2011. Effect of dynamic loading on wellbore leakage for the Wabamun area CO2 sequestration project. In: Canadian Unconventional Resources Conference, Calgary, Alberta, Canada, 15-17 November. SPE-146640-MS.

Onaisi, A., Urbanczyk, C., Lansot, J.-Y., et al., 2017. Impact of poro-chemo-mechanical parameters on pre-stress and cement sheath integrity under well pressure and thermal loadings. In: Abu Dhabi International Petroleum Exhibition & Conference, Abu Dhabi, UAE, 13-16 November. SPE-188761-MS.

Orlic, B., 2018. Numerical investigations of cement interface debonding for assessing well integrity risks. In: 52nd US Rock Mechanics / Geomechanics Symposium, Seattle, Washington, USA, 17–20 June. ARMA-2018-1298.

Ostroot, G.W., Shyrock, S., 1964. Cementing geothermal steam wells. J. Pet. Technol. 16 (12), 1425–1429. SPE-904-PA.

Patchen, F.D., 1960. Reaction and properties of silica-Portland cement mixtures cured at elevated temperatures. Petrol. Trans. AIME 219, 281–287. SPE-1499-G.

Patel, H., Salehi, S., 2019. Development of an advanced finite element model and parametric study to evaluate cement sheath barrier. J. Energy Resour. Technol. 141 (9), 8. 092902.

Patel, H., Salehi, S., Teodoriu, C., 2019. Assessing mechanical integrity of expanding cement. In: SPE Oklahoma City oil and gas symposium, Oklahoma City, Oklahoma, USA, 9-10 April. SPE-195225-MS.

Patil, R., Garnier, A., Galdiolo, G., Patil, S., Ravi, K., Ferreira, L., 2010. Designing and testing cement system for SAGD application. In: SPE Annual Technical Conference and Exhibition, Florence, Italy, 19-22 September. SPE-134169-MS.

Pershikova, E.M., Chougnet-Sirapian, A., Loiseau, A., Khater, W., Garnier, A., 2010. Evaluation of specialized cement system for long-term steam injection well integrity. In: Canadian Unconventional Resources & International Petroleum Conference, Calgary, Alberta, Canada, Oct. 19-21. SPE-137710-MS.

Petersen, T.A., Ulm, F.-J., 2016. Early-age stress and pressure developments in a wellbore cement liner: application to eccentric geometries. J. Appl. Mech. 83 (9), 091012.

Petty, S., Gastineau, J., Bour, D.L., Ravi, K.M., 2003. Life cycle modeling of wellbore cement systems used for enhanced geothermal system development. In: 28th Workshop on Geothermal Reservoir Engineering, Stanford University, Stanford, California, USA, 27–29 January.

Philippacopoulos, A.J., Berndt, M.L., 2002a. Mechanical response and characterization of well cements. In: SPE Annual Conference and Exhibition. SPE. San Antonio, Texas, USA, 29 September-2 October. SPE-77755-MS.

Philippacopoulos, A.J., Berndt, M.L., 2002b. Structural analysis of geothermal well cements. Geothermics 31 (6), 657–676.

Piot, B., Cuvillier, G., 2006. Primary cementing techniques. In: Nelson, E.B., Guillot, D. (Eds.), Well Cementing, second ed, pp. 459–501. Schlumberger.

Putri, F.R., Alamsyah, D., Adityatama, D., Putri, R., Nainggolan, S., Burnama, A., Krishamidan, A., 2020. Comparison of wellhead completion designs for geothermal operations in Indenosia: expansion spool and casing head completions. In: New Zealand Geothermal Workshop, Auckland, New Zealand, 15–17 November.

Pyatina, T., Sugama, T., 2020. Cements with supplementary cementitious materials for high-temperature geothermal wells. Geothermics 86 (101840), 22.

Pyatina, T., Sugama, T., 2023a. Cements for supercritical wells at 400°C. GRC Trans. 47, 447–467.

Pyatina, T., Sugama, T., 2023b. Hydrophobic, thermal shock-resistant latex-modified lightweight class G cement composites in reservoir thermal energy storage (RTES) systems. In: 48th Workshop on Geothermal Reservoir Engineering, Stanford University, Stanford, California, USA, 6–8 February.

Pyatina, T., Sugama, T., Garrison, G., Bour, D., Petty, S., 2023. Results of high temperature cement blends exposure in Newberry well Oregon. GRC Trans. 47, 468–488.

Ramos, R.M., Sánchez Camus, A., 2017. Borehole cement sheath integrity - numerical simulation under reservoir conditions. Mecánica Computacional 35, 193–225.

Ravi, K., Bosma, M., Gastbled, O., 2002a. Improve the economics of oil and gas wells by reducing the risk of cement failure. In: IADC/SPE Drilling Conference, Dallas, Texas, USA, 26-28 February. SPE-74497-MS.

Ravi, K., Bosma, M., Gastbled, O., 2002b. Safe and economic gas wells through cement design for life of the well. In: SPE Gas Technology Symposium, Calgary, Alberta, Canada, 30 April-2 may. SPE-75700-MS.

Ravi, K., Bosma, M., Hunter, L., 2003. Optimizing the cement sheath design in HPHT shearwater field. In: SPE/IADC Drilling Conference, Amsterdam, the Netherlands, 19-21 February. SPE-79905-MS.

Ravi, K., McMechan, D.E., Reddy, B.R., Crook, R., 2007. A comparative study of mechanical properties of density-reduced cement compositions. SPE Drill. Compl. 22 (2), 119–126. SPE-90068-PA.

Ravi, K., Fukuzawa, M., Hunter, W.J., Noerdin, A.I., 2008. Advanced cement systems used to improve geothermal well reliability in Java. In: SPE Annual Technical conference and Exhibition, Denver, Colorado, USA, 21–24 September. SPE-115638-MS.

Reddy, B.R., Brothers, L.E., Palmer, A.V., 2005. Methods of Cementing High Temperature Wells and Compositions Therefor. (US patent 6,846,357 B2).

Rice, J.R., Cleary, M.P., 1976. Some basic stress diffusion solutions for fluid-saturated elastic porous media with compressible constituents. Rev. Geophys. Space Phys. 14 (2), 227–241.

Roy, D.M., White, E.L., Langton, C.A., Grutzeck, M.W., 1979. Potential new high temperature cements for geothermal wells. In: SPE International Symposium on Oilfield and Geothermal Chemistry. Houston, Texas, USA, 22–24 January. SPE-7677-MS.

Roy, D.M., Langton, C.A., Grutzeck, M.W., White, E.L., 1980. Hydrothermal high temperature cements for potential geothermal applications. In: 5th SPE International Symposium on Oilfield and Geothermal Chemistry, Stanford, California, USA, 28–30 May. SPE-8994-MS.

Roy, P., Morris, J.P., Walsh, S.D.C., Iyer, J., Carroll, S., Todorovic, J., Gawel, K., Torsæter, M., 2016a. Assessment of thermal stress on well integrity as a function of size and material properties. Energy Procedia 114, 5241–5248.

Roy, P., Walsh, S.D.C., Morris, J.P., Iyer, J., Hao, Y., Carroll, S., Gawel, K., Todorovic, J., Torsæter, M., 2016b. Studying the impact of thermal cycling on wellbore integrity during CO2 injection. In: 50th US Rock Mechanics/ Geomechanics Symposium. Houston, Texas, USA, 26–29 June. ARMA-16-0668.

Saint-Marc, J., Garnier, A., Bois, A.-P., 2008. Initial state of stress: the key to achieving long-term cement-sheath integrity. In: SPE Annual Technical Conference and Exhibition, Denver, Colorado, USA, 21-24 September. SPE-116651-MS.

Santra, A., Sweatman, R., 2011. Understanding the long-term chemical and mechanical integrity of cement in a CCS environment. Energy Procedia 4, 5243–5250.

Sasaki, T., Soga, K., Abuhaikal, M., 2018. Water absorption and shrinkage behaviour of early-age cement in wellbore annulus. J. Pet. Sci. Eng. 169, 205–219.

Schreppers, G.J., 2015. A framework for wellbore cement integrity analysis. In: 49th US Rock Mechanics / Geomechanics Symposium. San Francisco, California, USA, 28 June-1 July. ARMA-2015-349.

Shadravan, A., 2013. A Method for Cement Integrity Evaluation in Unconventional Wells. Master of Science Thesis. Texas A&M University, College Station, Texas, USA.

Shadravan, A., Schubert, J.J., Amani, M., Teodoriu, C., 2015. Using fatigue-failure envelope for cement-sheath-integrity evaluation. SPE Drill. Compl. 30 (1), 68–75. SPE-168321-PA.

Shahri, M.A., Schubert, J.J., Amani, M., 2005. Detecting and modeling cement failure in high-pressure/high-temperature (HP/HT) wells, using finite element method (FEM). In: International Petroleum Technology Conference, Doha, Qatar, 21-23 November. IPTC-10961-MS.

Shen, J.C., Pye, D.S., 1989. Effects of CO_2 attack on cement in high-temperature applications. In: SPE/IADC Drilling Conference, New Orleans, Louisiana, USA, 28 February - 3 March. SPE-18618-MS.

Skorpa, R., Vrålstad, T., 2022. Leakages through radial cracks in cement sheaths: effect of geometry, viscosity, and aperture. J. Energy Resour. Technol. 144 (013006), 10.

Skorpa, R., Werner, B., Vrålstad, T., 2019. Effect of rock on cement sheath integrity: shale vs. sandstone. In: 38th International Conference On Ocean, Offshore and Artic Engineering. Glasgow, Scotland, 9-14 June. OMAE2019-96738.

Smith, T.R., 1990. Cementing displacement practices - field applications. J. Pet. Technol. 42 (5), 564–566. 628–629. SPE-18617-PA.

Song, X., Zheng, R., Li, R., Li, G., Sun, B., Shi, Y., Wang, G., Zhou, S., 2019. Study on thermal conductivity of cement with thermal conductive materials in geothermal well. Geothermics 81, 1–11.

Southon, J.N.A., 2005. Geothermal well design, construction and failures. In: World Geothermal Congress, Antalya, Turkey, 24–29 April.

Speakman, K., 1968. The stability of tobermorite in the system CaO-SiO_2-H2O at elevated temperatures and pressures. Mineral. Mag. J. Mineral. Soc. 36 (284), 1090–1103.

Stiles, D.A., 2006. Effects of long-term exposure to ultra-high temperature on the mechanical parameters of cement. In: IADC/SPE Drilling Conference, Miami, Florida, USA, 21-23 February. SPE-98896-MS.

Sugama, T., 2006. Advanced cements for geothermal wells. In: Brookhaven National Laboratory Report (BNL-77901-2007-IR).

Sugama, T., Carciello, N.R., 1995. Sodium phosphate-derived calcium phosphate cements. Cem. Concr. Res. 25 (1), 91–101.

Sugama, T., Pyatina, T., 2022. Cement formulations for super-critical geothermal wells. In: 47th Workshop on Geothermal Reservoir Engineering, Stanford University, Stanford, California, USA, 7–9 February 2022.

Sugama, T., Kukacka, L.E., Galen, B.G., 1986. Advanced high-temperature lightweight foamed cements for geothermal well completions. Trans. - Geotherm. Resour. Counc., 87–92.

Sugama, T., Brothers, L.E., Weber, L., 2002. Calcium aluminate cements in fly ash/calcium aluminate blend phosphate cement systems: their role in inhibiting carbonation and acid corrosion at a low hydrothermal temperature of 90°C. J. Mater. Sci. 37, 3163–3173.

Tardy, P.M.J., Flamant, N., Lac, E., Parry, A., Sri Sutama, C., Baggini Almagro, S.P., 2017. New generation 3D simulator predicts realistic mud displacement in highly deviated and horizontal wells. In: SPE/IADC Drilling Conference and Exhibition, the Hague, the Netherlands, 14-16 march. SPE-184677-MS.

Teodoriu, C., Reinicke, K.M., Fichter, C., Wehling, P., 2010a. Investigations on casing-cement interaction with application to gas and CO_2 storage wells. In: SPE EUROPEC/EAGE Annual Conference and Exhibition, Barcelona, Spain, June 14–17. SPE-131336-MS.

Teodoriu, C., Ugwu, I., Schubert, J., 2010b. Estimation of casing-cement-formation interaction using a new analytical model. In: SPE Europec/Eage Annual Conference and Exhibition, Barcelona, Spain, 14–17 June. SPE-131335-MS.

ter Heege, J.H., Wollenweber, J., Naumann, M., Pipilikaki, P., Vercauteren, F., 2019. Well integrity of high temperature wells: Effect of mineralogical changes on mechanical properties of well cement. In: 53rd US Rock Mechanics/Geomechanics Symposium Held in New York, New York, USA, 23–26 June. ARMA-19-1981.

Thérond, E., Bois, A.-P., Whaley, K., Murillo, R., 2017. Large-scale testing and modeling for cement zonal isolation in water-injection wells. SPE Drill. Compl. 32 (4), 290–300. SPE-181428-PA.

Thiercelin, M.J., 2006. Mechanical properties of well cements. In: Nelson, E.B., Guillot, D. (Eds.), Well Cementing, second ed, pp. 269–288. Schlumberger.

Thiercelin, M.J., Baumgarte, C., Guillot, D., 1998. A soil mechanics approach to predict cement sheath behavior. In: SPE/ISRM Rock Mechanics in Petroleum Engineering (Eurock'98), Trondheim, Norway, 8-10 July. SPE-47375-MS.

Thiercelin, M.J., Dargaud, B., Baret, J.-J., Rodriguez, W.J., 1999. Cement design based on cement mechanical response. SPE Drill. Compl. 13 (4), 266–273. SPE-52890-PA.

Todorovic, J., Gawel, K., Lavrov, A., Torsaeter, M., 2016. Integrity of downscaled well models subject to cooling. In: SPE Bergen One Day Seminar, Bergen, Norway, 20 April. SPE-180052-MS.

Vrålstad, T., Skorpa, R., 2020. Digital cement integrity: 3D visualization of cracks and microannuli in well cement. Sustain. For. 12 (10), 13.

Vrålstad, T., Skorpa, R., Opedal, N., de Andrade, J., 2015. Effect of thermal cycling on cement sheath integrity: realistic experimental tests and simulation of resulting leakages. In: SPE Thermal Well Integrity and Design Symposium, Banff, Alberta, Canada, 23-25 November. SPE-178467-MS.

Vrålstad, T., Skorpa, R., Opedal, N., Todorovic, J., Agofack, N., Hoang, N.-H., 2021. Cement sheath integrity during high temperature geothermal well operations. In: 40th International Conference on Ocean, Offshore and Arctic Engineering, 21–30 June. OMAE2021–65116.

Vu, M.-H., 2012. Effect and Stress and Temperature on Cement Integrity of Oil Wells. PhD dissertation, Université, Paris Est, France.

Vu, M.-H., Bois, A.-P., Badalamenti, A., 2020. Cement mechanical integrity for geothermal wells. GRC Trans. 44, 129–142.

Vu, M.-H., Sulem, J., Ghabezloo, S., Laudet, J.-B., Garnier, A., Guédon, S., 2012. Time-dependent behaviour of hardened cement paste under isotropic loading. Cem. Concr. Res. 42 (6), 789–797.

Vu, M.-H., Bouhours, A., Bouhours, J., Bouchair, R., Bois, A.-P., Badalamenti, A., 2021. Advanced cement mechanical integrity for thermal wells. In: SPE Annual Technical Conference and Exhibition, Dubai, UAE, 21-23 September. SPE-206144-MS.

Wang, W., Dahi Taleghani, A., 2014. Three-dimensional analysis of cement sheath integrity around wellbores. J. Pet. Sci. Eng. 121, 38–51.

Wang, W., Dahi Taleghani, A., 2017. Impact of hydraulic fracturing on cement sheath integrity; a modelling approach. J. Nat. Gas Sci. Eng. 44, 265–277.

Wang, Y.B., Gao, D.L., Fang, J., 2016. Assessment of wellbore integrity of offshore drilling in well testing and production. J. Eng. Mech. 142 (6), 9.

Weideman, B., Nygaard, R., 2014. How cement operations affect your cement sheath short and long term integrity. In: AADE Fluids Technical Conference and Exhibition. Houston, Texas, USA, 15-16 April. AADE-14-TTCE-20.

Wu, Y., Patel, H., Salehi, S., 2020. Thermal considerations of cement integrity in geothermal wells. In: 45th Workshop on Geothermal Reservoir Engineering, Stanford University, Stanford, California, USA, 10–12 February.

Wu, Y., Patel, H., Salehi, S., 2021. Parametric study of mechanical stresses within cement sheath in geothermal wells. Geothermics 90 (102000), 13.

Xia, Y., Jin, Y., Chen, M., et al., 2017. Dynamic analysis of a cylindrical casing-cement structure in a poroelastic stratum. Int. J. Numer. Anal. Methods Geomech. 41 (12), 1362–1389.

Xu, W., Reddy, B.R., Morgan, R., Brothers, L., 2014. Wellbore Servicing Compositions and Methods of Using Same. (US patent 8,685,901 B2).

Xu, H., Zhang, Z., Shi, T., et al., 2015. Influence of the WHCP on cement sheath stress and integrity in HTHP gas well. J. Pet. Sci. Eng. 126, 174–180.

Xu, H., Ma, T., Peng, N., et al., 2018a. Influences of fracturing fluid injection on mechanical integrity of cement sheath under four failure modes. Energies 11 (3534), 18.

Xu, H., Peng, N., Ma, T., et al., 2018b. Investigation of thermal stress of cement sheath for geothermal wells during fracturing. Energies 11 (2581), 22.

Xu, Y., Yan, Y., Xu, S., Guan, Z., 2020. Numerical simulation study on propagation of initial microcracks in cement sheath body during hydraulic fracturing process. Energies 13 (1260), 18.

Yang, Z., Meng, Q., Chen, X., Gao, L., 2012. The theoretical radial stress solution and its experimental verification of expansion cement at annulus interfaces. Pet. Explor. Dev. 39 (5), 644–651.

Yuan, Z., Teodoriu, T., Schubert, J.J., 2013. Low cycle cement fatigue experimental study and the effect on HPHT well integrity. J. Pet. Sci. Eng. 105, 84–90.

Zeldin, A., Kukacka, L.E., Fontana, J., Carciello, N., 1979. Organosiloxane polymer concrete for geothermal environments. In: 108th AIME Annual Meeting held in New Orleans, Louisiana, USA, 18–22 February.

Zhang, W., Eckert, A., 2018. Numerical investigation of the influence of cement failure to micro annuli generation. In: 52nd US Rock Mechanics / Geomechanics Symposium, Seattle, Washington, USA, 17–20 June. ARMA-2018–224.

Zhang, W., Eckert, A., 2020. Microannulus generation under downhole conditions: insights from three-dimensional staged finite element analysis of cement hardening and wellbore operations. J. Rock Mech. Geotech. Eng. 12 (6), 1185–1200.

Zhang, X., Yao, A., 2016. Stress analysis of cement sheath in thermal production well: effects of high temperature changes. Int. J. Earth Sci. Eng. 9 (5), 1933–1937.

Zhang, W., Eckert, A., Liu, X., 2017. Numerical simulation of microannuli generation by thermal cycling. In: 51st US Rock Mechanics / Geomechanics Symposium, San Francisco, California, USA, 25–28 June. ARMA-2017-0354.

CHAPTER

7

Geothermal production, injection, and storage engineering

Gioia Falcone, Christopher S. Brown, Isa Kolo, David Banks, and Orestis Angelidis

University of Glasgow, Glasgow, United Kingdom

Abbreviations

AGS advanced geothermal systems
ASHP air source heat pump
ATES aquifer thermal energy storage
BHEs borehole heat exchangers
BTES borehole thermal energy storage
DBHEs deep borehole heat exchangers
DC district cooling
DH district heating
DHC district heating and cooling
EGS enhanced or engineered geothermal system
GSHP ground source heat pump
HDPE high-density polyethylene
HSA hot sedimentary aquifer
MDPE medium-density polyethylene
MTES mine water thermal energy storage
PEX cross-linked polyethylene

1 Introduction

Similar to other geo-energy sectors, production engineering, in the geothermal context, is a specialized field that focuses on designing, monitoring, and optimizing the necessary equipment to extract, store, and/or convert thermal energy, through processes that link the subsurface environment with the ultimate point of sale or use. This section aims to provide an

Geothermal Energy Engineering
https://doi.org/10.1016/B978-0-443-21662-6.00012-8

Copyright © 2025 Elsevier Inc. All rights are reserved, including those for text and data mining, AI training, and similar technologies.

overview of the different types of geothermal systems that can be engineered to deliver and store thermal energy, together with their typical installation parameters and ranges of performance.

1.1 System classification

Geothermal engineering revolves around the coupling of a given geological setting and the surface demand end-uses of the thermal energy. Geological settings are derived from the types of geothermal plays. Although the definition of geothermal play is not straightforward, and several fit-for-purpose classifications are used worldwide (Breede et al., 2015), Moeck (2014) gives an example of a classification based on the combination of geological reservoir characteristics and exploitation strategy, which distinguishes between hydrothermal and petrothermal systems. According to Moeck (2014), a play type can be defined by the heat source and the geological controls on heat transport and thermal energy storage (TES) capacity. Within this chapter, the geological settings are simplified as:

1. Near-surface sediments and rocks (dry or saturated)
2. Aquifers and hot sedimentary aquifers (HSAs)
3. Disused, flooded mine workings
4. Crystalline basement
5. Magmatic and supercritical
6. Geopressured sedimentary rocks

Items 5 and 6 are regarded as unconventional settings.

The surface demand end-uses of the thermal energy resource can be summarized as:

1. **Direct use** of hot water directly from springs or reservoirs.
2. **Ground source heat pumps (GSHPs)**, relying on stable ground or water temperatures from the subsurface to deliver controlled temperatures aboveground.
3. **District thermal networks** to distribute heat and cooling from a central source to multiple buildings via a network of underground pipes (can be in combination with GSHPs).
4. **Electricity generation** via a power plant generally built where the geothermal field is located.
5. **Combined heat and power (CHP)** for concurrent production of electricity and useful thermal energy (heating and/or cooling) from a single source.
6. **Underground thermal energy storage (UTES)** where thermal energy is alternatively produced and rejected to the subsurface and also injected from renewable sources, waste heat, or via electrical network balancing.

Different exploitation methods can be implemented to couple the varying geological settings to end-user demand via either open-loop or closed-loop systems. Open-loop systems involve the circulation of a geological fluid, which acts as a heat transfer fluid that exchanges heat between the reservoir and surface end-user (i.e., via either a heat pump or heat exchanger; Brown and Howell, 2023). These types of systems can carry higher geological risk due to their dependence on subsurface permeability, heat storage, and the presence of fluid in the reservoir. In contrast, closed-loop systems rely on an artificial heat transfer fluid with no hydraulic interaction with the reservoir. They usually exchange heat via conduction through the borehole wall to the heat transfer fluid in the subsurface before being advected to the surface (e.g., Brown et al., 2023a).

Examples are illustrated in Figs. 1–3, providing indicative ranges of the project's depth and corresponding temperature (Eavor Technologies—Eavor, 2022; Eden Geothermal, 2024). At shallower depths, closed-loop borehole heat exchangers (BHEs), which utilize coaxial, single U-tube or double U-tube configurations, are often employed (García Gil et al., 2022), although other versions exist (such as helical exchangers—see Zarrella et al., 2013). The coaxial configuration in vertical BHEs can be used at depth and could play an important role in repurposed infrastructure such as oil and gas wells or ex-geothermal exploration wells (Caulk and Tomac, 2017). Closed-loop systems can be developed at depth as advanced geothermal systems (AGS) that operate in a similar manner to shallower systems, exchanging heat through a closed-loop circuit after the heat transfer fluid extracts heat via conduction through an impermeable zone or pipe wall (e.g., Malek et al., 2022). In contrast to shallower systems, they consist of more elaborate "radiator-type" structures (e.g., Kelly and McDermott, 2022).

Open-loop systems operate at almost any depth, targeting hydrothermal systems relying on natural reservoirs in the subsurface. Petrothermal systems may be used but generally require a level of stimulation, creating enhanced or engineered geothermal systems (EGSs) which rely on a fracture network (Breede et al., 2013, 2015). Flooded mines could be developed using a heat pump for heating and cooling, as they have a relatively stable temperature and can use high flow rates (Ramos and Falcone, 2013). Open-loop mine water projects have been progressing in Europe, for instance, at Heerlen, Netherlands (Verhoeven et al., 2014), and at Gateshead and Lanchester Wines, United Kingdom (Banks et al., 2022). It should also be noted that there are a variety of methods in developing geothermal mine water systems including closed-loop (Walls et al., 2021).

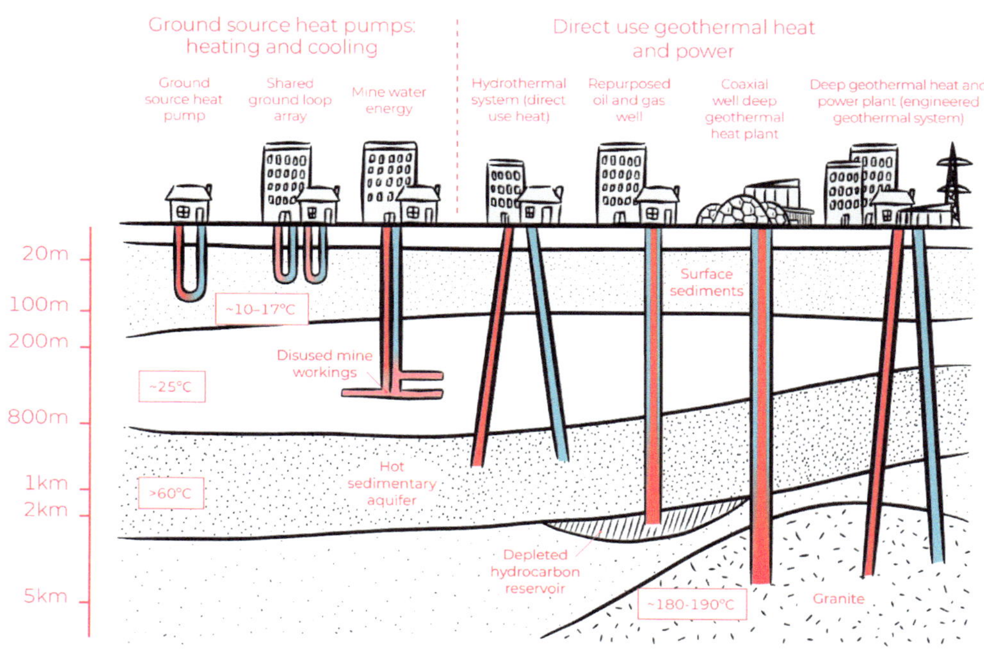

FIG. 1 Indicative depth ranges and corresponding temperature for various open-loop and closed-loop geothermal systems. ©*Eden Geothermal Ltd., 2023.*

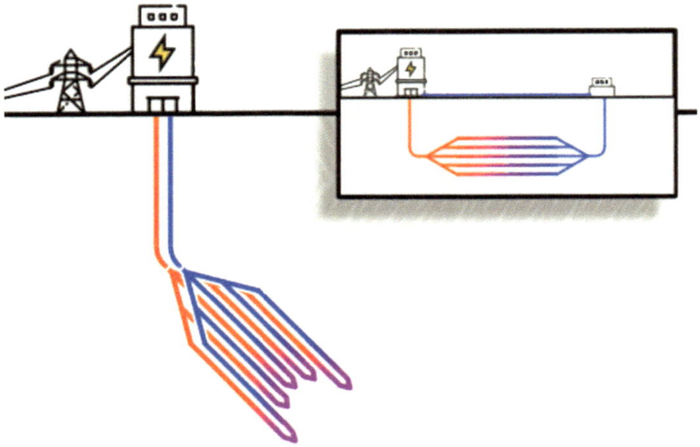

FIG. 2 Schematic of a doublet well advanced geothermal system (AGS) concept: one in a deviated forked configuration and the other in a multipronged horizontal configuration (Beard and Jones, 2023). *Adapted from Eavor Technologies—Eavor, 2022. Closed-Loop Geothermal Technology for a 24/7 Carbon-Free Future. Retrieved November 11, 2022, from https://www.eavor.com/technology/*

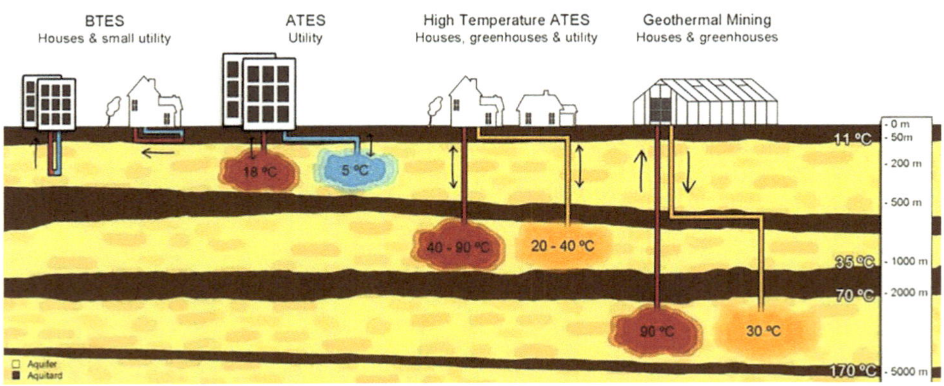

FIG. 3 Underground thermal energy storage (UTES) options, with indicative depth ranges and corresponding temperature (MMB, 2012).

The threshold between shallow and deep systems is often loosely defined in the literature. It has been set at 300 m in the United Kingdom, based on the Infrastructure Act (DECC, 2016), 400 m based on the German Association of Engineers (VDI) Guideline 4640i (Sanner et al., 1997), or 500 m based on the past UK Renewable Heat Incentive subsidies (OFGEM, 2020; Watson et al., 2020). Others have also defined a middle-deep interval from 400 m to 1 km (Breede et al., 2015; Homuth et al., 2016).

1.2 Typical system performance range

Examples of performance indicators and key engineering parameters for various open- and closed-loop systems are given in Table 1 (after Mégel, 2014; Abesser et al., 2020; Causeway Energies, 2023). These highlight the variable engineering operational range and required surface components/methods of heat transmission. For all systems, the cutoffs for depths can vary depending on the classification, as highlighted in Section 1.1.

Typically, for shallow closed-loop BHEs, they operate at flow rates <0.5 L/s/borehole (to minimize hydraulic losses while ensuring turbulent flows—further discussed in Section 3) and rely on relatively low temperatures in the subsurface, which are climate and seasonally dependent (e.g., Sofyan et al., 2020). Deep borehole heat exchangers (DBHEs) exploit greater temperatures at depth and require higher flow rates to minimize the impact of thermal short-circuiting, which occurs when there is thermal interaction between the up- and down-flowing heat transfer fluid. Short-circuiting can also be minimized through insulation of the central pipe and modification of the geometrical design (Le Lous et al., 2015; Morchio and Fossa, 2019). Both systems typically require a heat pump to supply space heating, cooling, or TES to the end-user. There are deeper closed-loop AGSs, which are designed for power production, but often have high initial costs. Therefore, they must produce fluid at temperatures that can sustain efficient power generation (i.e., over 100°C to ensure conversion efficiency >10%). It should be noted that advances in drilling technology are required for these schemes to be commercial (Malek et al., 2022); however, there are several experimental/prototype examples, such as the Eavor-Lite loop in Alberta, Canada (Toews and Holmes, 2021), and in Xi'an (Li et al., 2018). The first commercial-scale Eavor-Lite loop is currently under construction in Germany (EIB et al., 2024).

Open-loop systems can tap into shallow aquifers for heating, cooling, or TES, using GSHPs with low-to-high flow rates. Flow rates are usually within the range as indicated in Table 1, but they can be over 50 L/s (Abesser, 2010). Shallow systems can range in capacity from a few kilowatts, for domestic application, to tens of megawatts. Deeper aquifers can be exploited for direct heat use (i.e., via a heat exchanger) or for power generation, usually using binary power plants if fluid is produced (Zarrouk and Moon, 2014). Open-loop HSAs usually rely on natural permeability, but can be stimulated via hydraulic, chemical, or thermal processes (e.g., Huenges et al., 2007; Sutra et al., 2017). In contrast, open-loop EGSs often use an artificial reservoir created from stimulation. Initial permeability is usually limited to a targeted fault or fracture (e.g., Reinecker et al., 2021). They exploit higher-temperature petrothermal systems and are primarily used for power generation, requiring high flow rates.

Aquifer thermal energy storage (ATES) occurs when heat is intentionally injected into the ground via an open-loop system into an aquifer or geothermal reservoir. When using a doublet system, bi-directional flow is employed—one well extracts heat during winter, while the other injects coolth. In summer, flow is reversed, and heat is injected into the warm well, while coolth is extracted. At present, most ATES systems are employed at shallow depths (<500 m) in combination with heat pumps or heat exchangers; however, they can be deployed up to 1.25 km depths as is the case in Neubrandenburg, since 2004 (Holstenkamp et al., 2017; Kallesøe et al., 2020). The system is still in operation, injecting fluid at 85–90°C into a reservoir of ~55°C natural temperature. As reported by Adams (1982), one of the first tests recorded, evaluating the storage potential in 1982, occurred in Cambridge, targeting the Lower Greensand aquifer of 13 m of thickness. Four boreholes were drilled (one active and three observational), and a 280-day charge–discharge cycle test was undertaken. During the test, heat at 57°C was injected at a flow rate of 0.23 L/s, and following extraction, a recovery factor of 32% was achieved. In recent times, there are several more commercial-scale systems (Kallesøe et al., 2020). A further, highly regarded example of low-temperature ambient ATES is the system installed at Arlanda Airport in Stockholm, which stores surplus summer heat (from air conditioning of the airport terminal) and surplus winter coolth (from de-icing of aircraft aprons and pre-heating of ventilation air) in separate portions of a shallow sand/gravel aquifer (Andersson, 2009).

Lastly, mine water systems target flooded worked mine seams for heating or cooling and can sustain very high flow rates. They are becoming increasingly common, usually utilizing heat pumps, and reviews are published by Banks et al. (2003, 2004, 2017), Peralta Ramos et al. (2015), Chu et al. (2021), and Walls et al. (2021). The potential of flooded mines for TES is also considered by, e.g., Hahn et al. (2018). Although other operational arrangements do exist (Banks et al., 2017), the two main models for large-scale mine water geothermal exploitation are as follows.

Open loop, with discharge. A shaft or borehole into a mine complex is pumped, or an existing gravity discharge from a flooded mine is captured. The mine water passes through a heat exchanger, coupled to a heat pump, which extracts the heat. The water is then discharged to a surface recipient: a river, an estuary, or a sea. Because such mine water is often of poor quality (e.g., rich in iron or manganese), some form of treatment is usually required before the mine water can be responsibly discharged (Banks, 2003). The cost of this treatment (and of pumping) is like establishing a new purpose-built geothermal scheme that is often prohibitively expensive. Thus, such schemes are often established where there is an existing need to pump and treat the water for other purposes—e.g., mine dewatering or regional control of mine water levels (i.e., to prevent mine water levels from rising and contaminating overlying aquifers or breaking out at the surface—Younger, 1993; Neymeyer et al., 2007). An excellent international example of such a scheme is that at Barredo coal mine shaft, Mieres, northern Spain (Banks, 2017; Loredo et al., 2017), which provides heating and cooling to a local hospital and university.

Open loop, with reinjection. A pair (or more) of boreholes is drilled into flooded mine workings. One borehole is used for pumped abstraction of mine water to a surface heat exchanger, usually coupled to a heat pump, while the other is used for reinjection of thermally spent mine water. Because the water is reinjected into the ground, the necessity to treat water prior to discharge is avoided. Moreover, the lack of net abstraction means that large mine water drawdowns do not occur, keeping pumping costs within reasonable bounds. These schemes can be subject to other risks, however. If the flow path between injection and abstraction boreholes is too short and rapid, then there is a risk of feedback of thermally spent water to the abstraction borehole. For this reason, abstraction and reinjection boreholes should either be located at large distances from each other or be completed in different stratigraphic levels of a mine system to achieve "vertical separation." Moreover, problematic water chemistry can still affect boreholes, heat exchangers, filters, and pipeline components—via either corrosion (saline, H_2S-rich, reducing, warm water) or scaling (e.g., barite, calcite, or ferric oxyhydroxide, if iron-containing water is aerated). A highly regarded international example of such a scheme is that at Heerlen, Netherlands, where widely spaced abstraction and reinjection boreholes are sunk to different levels of a coal mine system and are used to support district heating and cooling (DHC) (Verhoeven et al., 2014).

TABLE 1 Typical ranges of performance indicators and key engineering parameters for various open-and closed-loop geothermal systems.

	Shallow BHE	Deep BHE	Conventional open-loop doublet/triplet	EGS	AGS	Mine water geothermal systems
System	Closed	Closed	Open	Open	Closed	Open/closed
Depth (m)	<500 m	>500 m	Any (typically, <500 m is classified as shallow GSHP)	>3 km	>3 km (for power generation)	Any, but typically <1000 m
Production temperature (°C)	0–10	>4	>5	>150	>100	10–40
Flow rate	<0.5 L/s/ borehole	<15 L/s	5–70 L/s	20–60 L/s	Unproven at scale	A few liters per second up to 200–300 L/s
Applications	Heating/ cooling/ BTES	Heating/ cooling	Heating/cooling/ ATES	Power/ heating	Power/ heating	Heating/cooling/ MTES
Energy delivery to end-user	Heat pump	Heat pump	Heat pump/direct heat	Power plant/direct heat	Direct heat	Heat pump/direct heat
Rock permeability of the rock mass	n/a	n/a	Natural (in the aquifer)/can still be stimulated	Artificially created/ stimulated	n/a	Artificially created by worked seams/ voids

Modified from Mégel, T., 2014. Introduction to geothermal energy technologies. The National Geothermal Resource Assessment for Lebanon, Mövenpick Hotel, Beirut, 24 March 2014.

2 Open-loop borehole systems: Engineering and completion aspects

In this section, key geothermal production and injection engineering aspects of open-loop borehole systems are discussed, including well productivity and injectivity in the near-wellbore region of the subsurface and well design, as they have a direct impact on the overall performance of a geothermal project over its lifetime.

2.1 Well productivity and injectivity

Depending on the type of system, whether in open- or closed-loop mode, a production, injection, and/or thermal storage component is included. For open-loop systems, productivity and injectivity of the near-wellbore region contribute to the overall performance of the system. Often, the productivity and injectivity indices are used to evaluate the success of a well.

Assuming single-phase (liquid) conditions in a production well, its productivity can be captured by the productivity index, which is defined as the ratio between the produced flow rate and the difference between the current reservoir and flowing bottomhole pressures in the well.

Similarly, for an injection well, its injectivity can be captured by the injectivity index, defined as the injection flow rate divided by the difference between flowing bottomhole and reservoir pressures.

Both the productivity and the injectivity indices are typically expressed in liters/second/bar and quantify the ability of a well to be poor or good producer/injector. Typical relevant injectivity or productivity values for an industrial geothermal project are above 2L/s/bar (Baujard et al., 2017).

For shallow, groundwater-based open-loop systems, the productivity/injectivity indices are essentially analogous to specific capacity. Specific capacity is the ratio of discharge/reinjection rate (m^3/day) to drawdown/upconing in the production/reinjection well (m). The units of specific capacity are thus m^2/day; specific capacity is related to aquifer transmissivity.

Both the productivity index and the injectivity index are a function of the permeability of the formation in the near-wellbore region. In well doublet systems, there needs to be compatibility between the productivity index of the production well and the injectivity index of the injection well; the same is applied to well triplets, with two production wells and one reinjection well, for example. When the natural permeability is not sufficient, hydraulic, chemical, or thermal well stimulation options might be possible, depending on the geological settings (sedimentary formations vs. crystalline basement, depth, temperature, and geochemistry), as well as any applicable health, safety, and environmental regulations.

In closed-loop systems, where the produced/injected/stored fluid does not get into physical contact with the natural formation, the productivity and injectivity indices play no role in the overall performance of the system. In some AGS, a chemical sealant is used to enable open-hole completion of single- or multi-branch sections, to reduce the permeability of the near-wellbore region by several orders of magnitude (Zatonski and Brown, 2023).

2.2 Well design

Given the significant variety of geothermal exploitation methods, there is no universal way of designing and completing a well. Currently, there is a limited number of international standards and guidelines dedicated to geothermal drilling and completions. For open-loop systems, the following are key references: Geothermie Nederland (2021) and CIBSE (2019), African Union (2016), Standards New Zealand (2015), and Finger and Blankenship (2010). Other standards and best practices relevant to open-loop geothermal systems can be leveraged from the petroleum sector (e.g., API standards and recommended practices) and the water sector.

2.2.1 *Example of a mid-deep open-loop well design: A typical Dutch geothermal well, the Netherlands*

Fig. 4 illustrates an example of a design of a geothermal well for the producer of a typical doublet in aquifers in the Netherlands, where the reservoir pressure is typically subhydrostatic, so formation water will not flow to the surface without an artificial lift (e.g., an installed electrical submersible pump (ESP) to lift the hot brine to a surface heat exchanger). After heat extraction at the surface, a paired injection well pumps the cooled water back to the source reservoir.

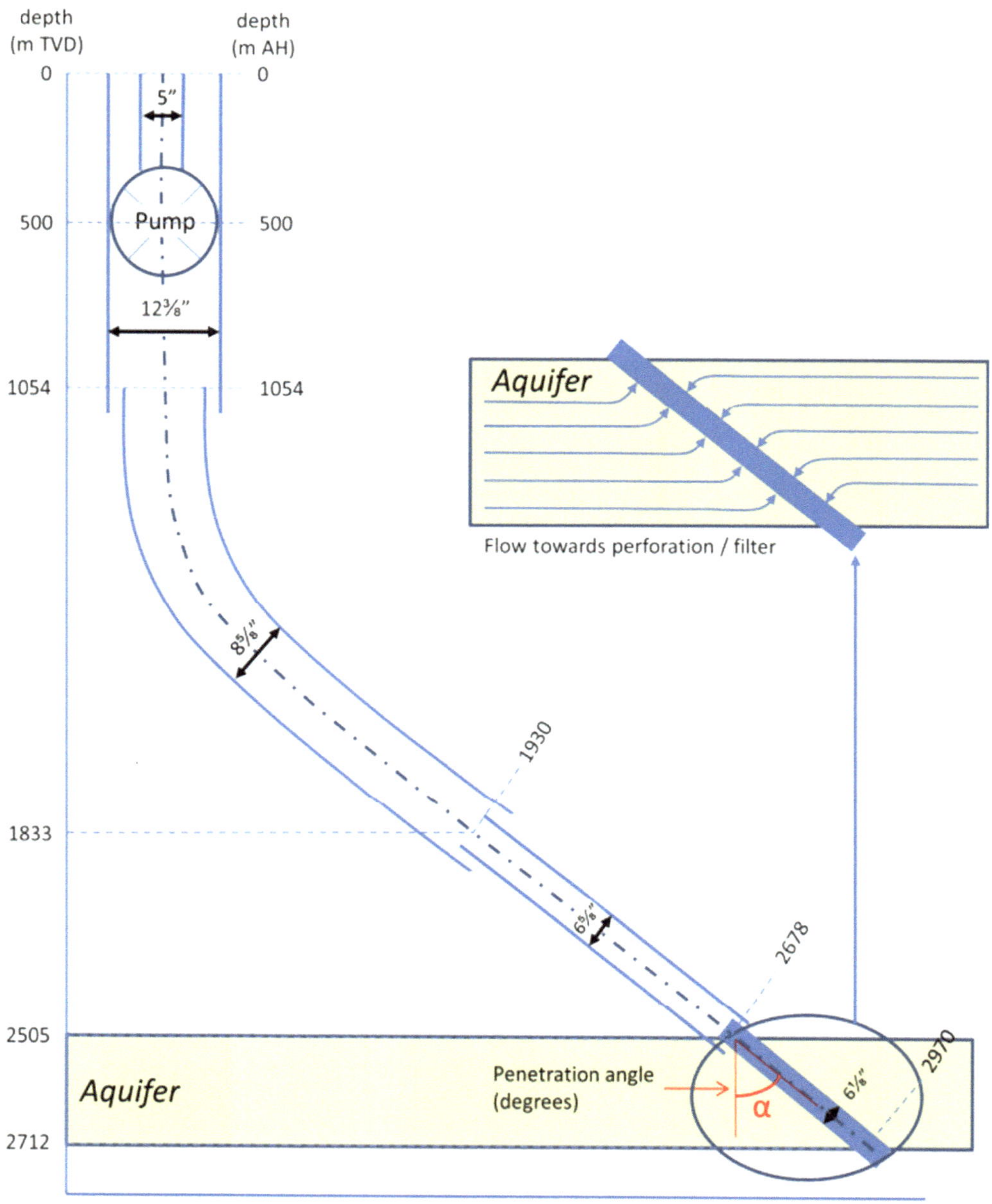

FIG. 4 Diagram of a typical production well of geothermal doublet in aquifers in the Netherlands (Mijnlieff et al., 2014).

2.2.2 Example of a shallow open-loop well design: UK Geoenergy Observatory, Glasgow, United Kingdom

The UK Geoenergy Observatory in Glasgow was drilled in 2020 to provide a research facility that could be used for practical testing of shallow mine water geothermal systems. It is composed of 12 boreholes, targeting coal mine workings, with the deepest borehole reaching 199 m, located 1.5 km away from the Cuningar Loop (Monaghan et al., 2022). An example of an installed borehole is GGA02, highlighted in Fig. 5, with a drill length of

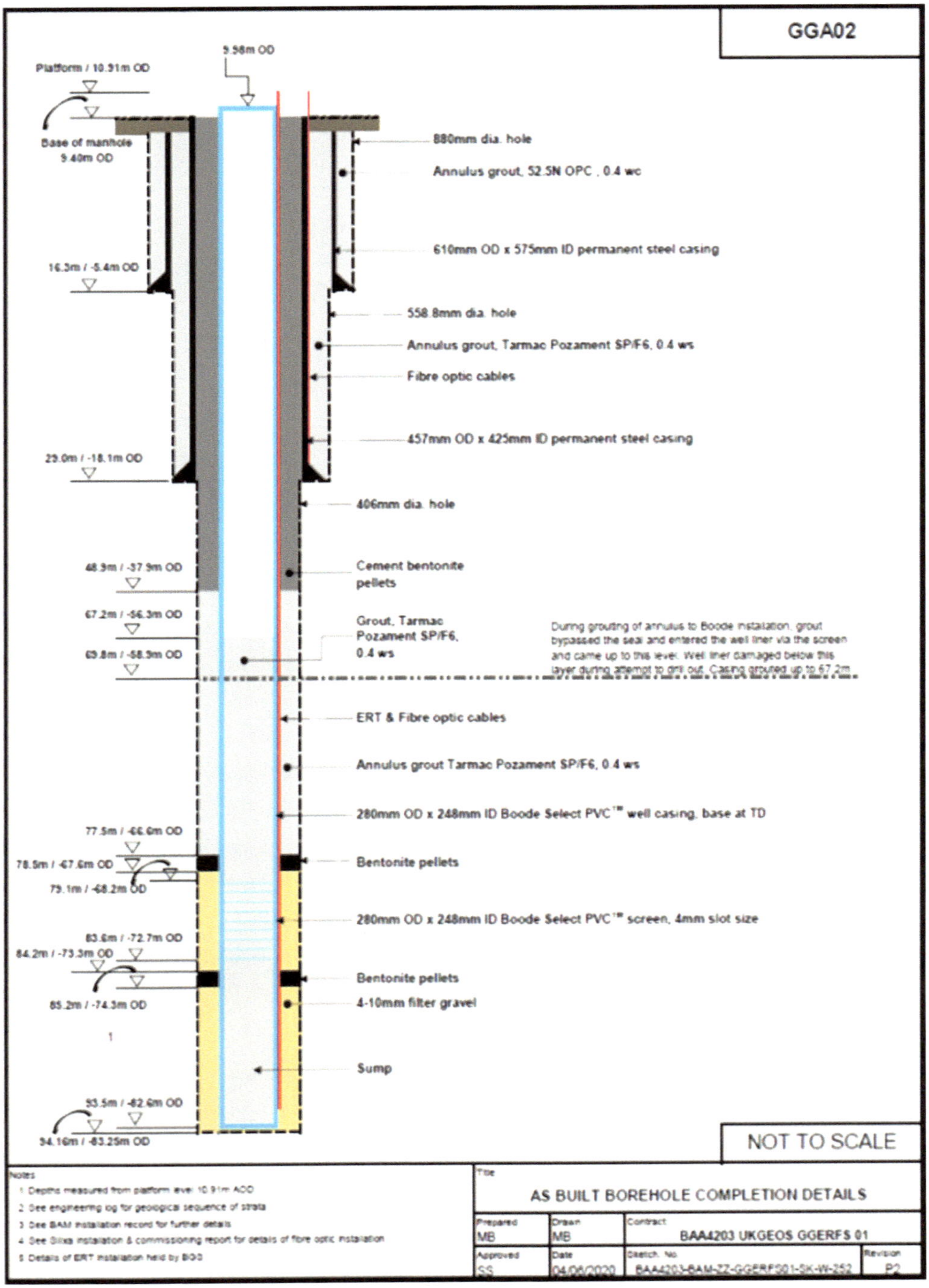

FIG. 5 Example of GGA02 well design. British Geological Survey © UKRI 2020 (BGS permit no. CP24/007). All rights reserved. *Reproduced from Monaghan, A.A., Starcher, V., Barron, H.F., Shorter, K.M., Walker-Verkuil, K., 2020. Borehole GGA02, UK Geoenergy Observatory, British Geological Survey, Glasgow, Nottingham, UK, p. 31 (OR/20/022) (Unpublished)*

94.16 m (below the drilling platform). The borehole is cased to 67.2 m, with varying hole diameters of 8.8–4.06 cm. The screened interval consists of a Boode select PVC screen with 4 mm slot size. This is 280 mm OD × 248 mm ID (Monaghan et al., 2020). As with deeper systems, shallow drilled wells often require a series of casing strings, although it is possible to complete via a single casing string (especially in lithified aquifers, where a single casing string, grouted to below rockhead, is simply required to exclude shallow superficial or weathered material).

2.2.3 *Example of a deep open-loop well design: United Downs Project, Cornwall, United Kingdom*

Two wells have been drilled at the United Downs Project, which aim to be the first power-producing geothermal system in the United Kingdom. Reinecker et al. (2021) report that the two wells (UD-1 and UD-2) were drilled to 5058 m TVD and 2214 m TVD in 2018 and 2019, respectively. They are both deviated wells aiming to target a fault zone with an 8.5 in. open-hole section (Table 2 and Fig. 6). The borehole diameter for UD-1 was set at 24 in. at the surface, while for the shallower UD-2, 17.5 in. (Ledingham and Cotton, 2021).

2.3 Flow assurance and long-term sustainability issues

Coined by the company Petrobras in the early 1990s, "Garantia de Fluxo" stands for "Guarantee of Flow." Flow assurance refers to ensuring the performance of the system is (safely and economically) maintained throughout the life of the project. It encompasses thermohydraulic analysis, operability, blockages, fluid rheology, mechanical integrity, and mitigation. Examples of flow assurance issues in the context of open-loop geothermal systems are:

- Pipeline blockage by, e.g., scale, sand, and salt.
- Mechanical integrity—corrosion (CO_2, H_2S) and erosion (sand).
- Compromised safety—safety valves that are scaled open.
- H_2S—toxic and highly corrosive.
- In open-loop geothermal doublets, formation, gravel pack and well-screen damage is common at the injection well due to (Ungemach, 2003):
  - Chemical incompatibility between injected and formation (native) fluids
  - Microbiological effects
  - Water sensitivity of sandstones
  - Suspended solids (fines, corrosion products, and scale)
  - Fines migration within the injected formation
  - Trapped gases
  - Air contamination
  - Incompatible chemical additives and inhibitors
  - Thermodynamic changes (pressures and temperatures) induced by the injection process, injection flow rates, and inadequate completions of the well.

Each type of open-loop geothermal system presents its own potential flow assurance issues, depending on the geological settings, the geochemistry involved, the engineering

TABLE 2 Completion summaries of UD-1 and UD-2.

UD-1 production well			UD-2 injection well		
Drilled diameter	Depth (MD)	Casing size	Drilled diameter	Depth (MD)	Casing size
24 in.	245 m	18 5/8 in.			
17.5 in.	900 m	13 5/8 in.	17.5 in.	804 m	13 5/8 in.
12.25 in.	4000 m	9 5/8 in.	12.25 in.	1820 m	9 5/8 in.
8.5 in.	5275 m		8.5 in.	2393 m	

From Ledingham, P., Cotton, L., 2021. The united downs deep geothermal power project. In Proceedings of the World Geothermal Congress, Reykavik, Iceland.

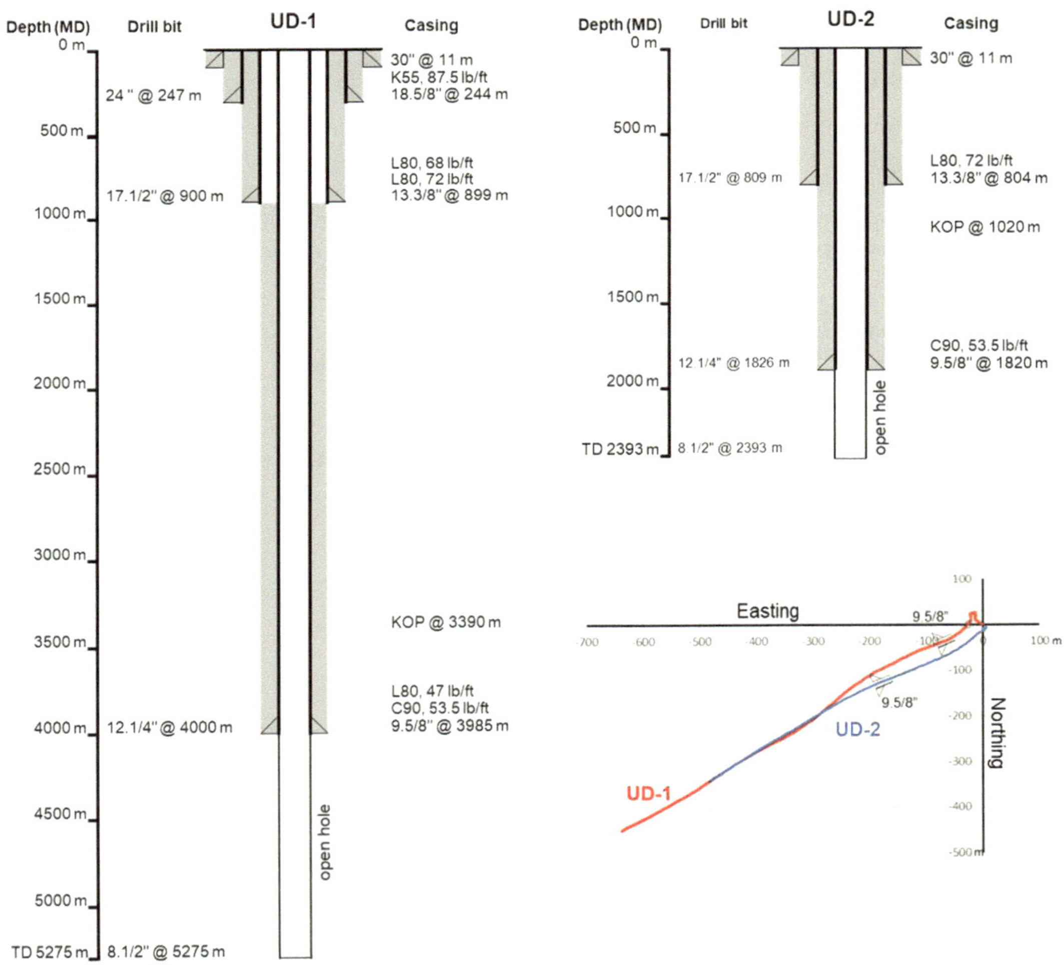

FIG. 6 Example of UD-1 and UD-2 well design. *Reproduced from Reinecker, J., Gutmanis, J., Foxford, A., Cotton, L., Dalby, C., Law, R., 2021. Geothermal exploration and reservoir modelling of the United Downs deep geothermal project, Cornwall (UK). Geothermics 97, 102226.*

design, and the operating conditions. If not prevented or mitigated, these issues can negatively affect the long-term performance of the system and hence the quantity of recoverable or storable thermal energy associated to the project, which in turn increases the financial risk to the investor.

3 Closed-loop borehole systems: Engineering and completion aspects

Conventional geothermal systems that produce warm fluids rely on convective heat transport (subsurface fluid movement to access heat from a large subsurface volume) and are heavily dependent on adequate reservoir permeability to allow fluid production and induce subsurface fluid flow. The warm reservoir fluid is abstracted to the surface, where it typically passes through a heat exchanger. In geological settings without adequate permeability at the appropriate depth, closed-loop boreholes allow the possibility of extracting heat, primarily by conduction, through the surrounding rocks and borehole wall to a heat exchange system that is installed within the borehole itself. Fig. 7 illustrates a selection of possible closed-loop borehole completions.

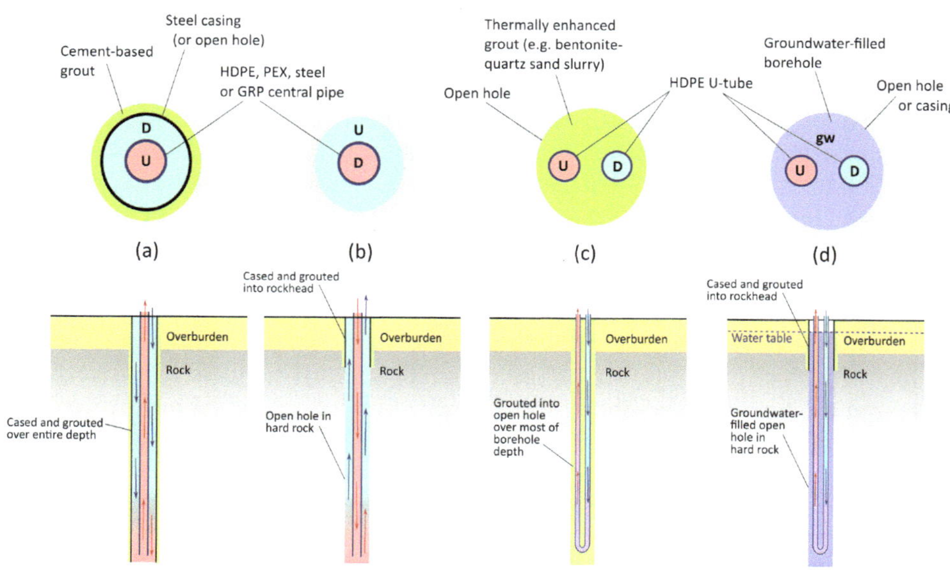

FIG. 7 A selection of possible closed-loop borehole completions. In all diagrams, *U* and *D* are the upflow and downflow heat transfer fluid, respectively; gw, groundwater-filled borehole; green shading, grout; pink, warm fluid; and pale blue, cold fluid. (A) CXA coaxial heat exchanger in fully cased borehole, used for heat extraction. (B) CXC coaxial heat exchanger in open borehole, used for heat rejection (cooling). (C) U-tube grouted into the borehole with thermally enhanced grout and used for heat extraction. (D) U-tube suspended in groundwater-filled drilled well ("Scandinavian" concept) and used for heat extraction. *Reproduced with permission of © David Banks (2024).*

The concept of a closed-loop system is decades old. In the late 1940s, Robert C. Webber of Indianapolis, USA, pioneered a closed-loop, direct expansion GSHP system, inspired by experiments with his domestic freezer. He buried copper, refrigerant-filled evaporator tubing in the ground outside his house (Zogg, 2008). By the 1950s, Miriam Griffith (1948, 1957, 1974) was researching closed-loop heat pump systems in the United Kingdom with her colleague John Sumner, by burying heat exchangers made of copper pipe in the subsoil, filled with a brine-based heat transfer fluid, and using heat pumps to extract ground source heat from the brine. In fact, by 1947, the main configurations for coupling a heat pump to the ground or groundwater had already been fully identified by E.N. Kemler (Sanner, 2005).

Although copper was initially favored as a choice for subsurface heat exchange elements, it suffers several drawbacks, including high cost and susceptibility to corrosion under certain geochemical settings. Thus, most shallow closed-loop GSHP systems today are based on high-density polyethylene (HDPE) or cross-linked polyethylene (PEX) pipe—or sometimes, in very shallow systems, medium-density polyethylene (MDPE). In fact, in the overall series of combined subsurface thermal resistances, the component represented by the choice of PE pipe (lower thermal conductivity, but cheap and robust) over copper pipe is not decisive. A sufficient length of PE heat exchanger pipe is installed, either in shallow trenches (horizontal GSHP system) or in vertical boreholes (vertical GSHP system or BHE). The circulating heat transfer fluid is usually a water-based solution of ethylene or propylene glycol, or an alcohol such as ethanol, with additives such as biocides and corrosion/toxicity inhibitors. The use of inorganic salts (true "brines") as heat transfer fluids is now largely superseded, although organic salts such as formates remain candidates.

In BHEs to depths of down to c.300 m, a common choice of downhole heat exchanger is the "U-tube"—the technology is comprehensively described by Sanner (2005), Sanner et al. (2007), Geotrainet (2011a, b), and Banks (2012). This "U"-shaped loop of HDPE (typically at least SDR11 wall thickness, PN16 grade) comprises two straight "shanks" of pipe, fused to a prefabricated "U" configuration end-piece. Depending on the drilling method, the borehole into which the loop is installed may be filled with drilling mud or water. The U-tube should be factory-new, prefabricated when delivered, pressure-tested pre-installation, and should be installed from a large-diameter spool to minimize the risk of pipe damage and "kinks." The number of joints in the pipe should be minimal, and any joints (e.g., between pipe and end-piece or at the top of the borehole) should be by quality-controlled heat fusion (e.g., controlled electrofusion) techniques. The end-piece will typically be weighted, and the U-tube filled with fluid and pressurized during installation to ensure that it is neither buoyant nor excessively "heavy" and that it does not collapse under fluid pressure in the borehole. The two shanks of the loop may be separated at a constant distance by prefabricated "spacers," to ensure that they do not come into close contact, creating a thermal short-circuit between upflowing and downflowing heat transfer fluid.

If the borehole is stable and groundwater-filled to near the surface (as is, e.g., the case in the wet, hard-rock terrain of Scandinavia), the U-loop can simply be left suspended in the groundwater column. In this case, the groundwater forms the thermal contact between the loop and the borehole wall. Any convection effects that establish themselves in the borehole groundwater column (convection cells, thermosiphon, and forced convection due to natural vertical head gradients) will typically serve to enhance heat transfer. If required by a national regulator, if the ground is unstable, or if the water table is deep, then the borehole may be backfilled, following loop installation, by a grout material, from the base of the borehole upward, via a tremie pipe. As the grout is fluid when installed, a calculation must be made of bottomhole grout pressure, and this should be counteracted by adequately pressurizing the loop during grouting. The grout should be selected to ensure a low hydraulic conductivity ($<10^{-9}$ m/s), yet a reasonably high thermal conductivity (the author suggests aiming for >1.5 W/m/K). The grout may be cement-based or bentonite-based; additives to enhance thermal conductivity can include graphite or silica (quartz). Indeed, a popular choice for thermal grout is simply a slurry of bentonite and fine silica sand (Mahmoud et al., 2021).

After installation, the pipework should be purged to remove any air pockets or bubbles (MCS, 2021a). The U-tube installation should be fully pressure-tested, according to recommended and documented procedure, before and after installation (EA, 2011; GSHPA, 2023).

The diameter of the pipes of the U-tube will be selected to minimize hydraulic resistance and thus parasitic pumping power in circulating the heat transfer fluid. For shallow BHE, one would normally aim for no more than a few tens of kilopascal pressure loss in the U-tube. The consequence of this is that, for shallow (several tens of meters deep) BHE, 32 mm outer diameter (OD) SDR11 HDPE pipe may be an appropriate selection; around and above 100 m, 40 mm OD SDR11 pipe will be required; and for progressively deeper boreholes, 45 mm or 50 mm OD pipe may be selected (Olsson, 2018a; Gehlin et al., 2020). Indeed, in Scandinavia, U-tube BHE to depths of 250–300 m are becoming increasingly common (Gehlin et al., 2016, 2020; Wirtén, 2017; Korhonen et al., 2018; Brown et al., 2024).

Installation of >50 mm OD U-tube in a deep borehole will often be prohibitively expensive, due to the requirements for borehole diameter. In deeper BHE, one may install "double" or even "triple" U-tubes of 32–40 mm OD PE pipe: the former comprising a cluster of four pipes (two U-tubes connected in parallel) and the latter six pipes (three U-tubes in parallel), to maximize heat transfer area and minimize hydraulic resistance.

A combination of heat transfer fluid, circulation flow rate, and pipe diameter should be selected to:

- Achieve a Reynolds number > 2500 in the U-tube, thus ensuring transient–turbulent fluid flow, which optimizes heat transfer (MCS, 2021b). Rifled pipe is available to assist in achieving transient–turbulent flow.
- Minimize total pressure losses in the U-tube. Combined losses in U-tube, header pipes, and surface heat pump/heat exchanger should ideally remain below 80 kPa (8 m equivalent water head; MCS, 2021b).
- Ensure that circulation pumping power around the hydraulic circuit remains low (MCS (2021b) recommend that it should be <2.5% of the total thermal output of the closed-loop system).

A number of standards are available regarding the installation of shallow closed-loop GSHP systems, including those of the German VDI (VDI standard 4640 *"Thermische Nutzung des Untergrunds"*), the UK-based Ground Source Heat Pump Association (GSHPA, 2023), and the USA-based IGSHPA (2017).

Below a certain depth (typically >c.300 m), U-tube heat exchangers become unattractive: mostly because of the hydraulic pressure losses (and parasitic pumping power) in circulating a heat transfer fluid around a narrow, increasingly long, fluid circuit and because installing a wider-diameter U-tube would entail expensive deep drilling at large diameter.

The preferred option at depths greater >300–500 m is usually a coaxial BHE (Olsson, 2018b, c; SiL, 2024), comprising a single pipe down the center of the completed borehole, which may be fully or partially cased or which may be open-hole in stable (e.g., hard rock) lithologies. Alternatively, a thin polymer liner (or "sock") can be installed against an open-hole rock wall, to prevent fluid exchange with the formation (Mazzotti et al., 2018). A heat transfer fluid, which is usually water (either natural groundwater or an introduced water of controlled chemical composition), is circulated down the annular space between the coaxial pipe and the borehole wall to the base of the borehole. At the base, the water enters the coaxial pipe and is pumped back to the surface (with the loading of heat it has acquired from the borehole wall on its downward journey), where heat is extracted via a heat exchanger or heat pump, before being recirculated down the borehole annulus. A helpful review of worldwide deep applications of coaxial BHE is given by Sapinska-Sliwa et al. (2015).

Closed-loop BHE can also be used to reject surplus heat *to* the ground from space cooling or dehumidification. Where this is the objective of a coaxial BHE, it can be demonstrated that it is beneficial to circulate the warm water *down* the coaxial pipe and *up* the annulus (i.e., the opposite to heat extraction mode). This phenomenon is analogous to the efficiencies of "parallel flow" and "counterflow" heat exchangers (Banks, 2022).

Because coaxial BHE need only accommodate a single pipe, it can be of a larger diameter than a U-tube, allowing hydraulically efficient flow within the annulus and up the coaxial pipe, without unacceptable parasitic power costs. However, because the coaxial tube may be deep and long, and because the difference in temperature between upflow and downhole fluid may be large (at least, if the heat yield is being extracted efficiently), there is great potential for "thermal short-circuiting"—heat exchange through the wall of the coaxial pipe between the upflow and downflow fluid streams (Wirtén, 2018).

The coaxial pipe can be constructed of polyethylene, and a high wall thickness will contribute toward insulating the upflow and downflow streams. However, issues of rigidity, strength, thermal tolerance, buoyancy (especially in saline groundwater), and pressure tolerance can restrict the use of plastics such as HDPE and PEX at high temperatures in very deep coaxial BHE. The buoyancy of polyethylene can also be problematic in deep boreholes, especially in saline groundwater, and can require significant weighting. Alternative choices of material include glass fiber–reinforced plastics or steel. Steel, which has a high thermal conductivity, is particularly ineffective in preventing thermal short-circuiting. High fluid flow velocities will reduce thermal short-circuiting (a strategy recommended by Holmberg, 2016), both by reducing contact time between upflow and downflow fluids and by reducing the temperature differential for a given heat extraction rate, but this will be achieved at the expense of higher parasitic power costs in circulation pumping. The hydraulic design of an optimal BHE will often be a "balancing act" (Lazzarotto et al., 2021; Brown et al., 2024) between achieving transient–turbulent fluid flow at the borehole wall, minimizing thermal short-circuiting, and maintaining circulation pumping costs (i.e., hydraulic resistance) within acceptable limits.

Several researchers and developers have considered the benefits of trying to fit insulation to deep coaxial pipes to reduce thermal short-circuiting, although high downhole pressures tend to compress and reduce the efficiency of conventional insulation materials (foams, etc.; Wirtén, 2018). Double-skinned coaxial pipe constructions, containing either an insulating fluid (e.g., air) layer or a vacuum, have also been proposed (Śliwa et al., 2017; Wirtén,

2018). An early example of a coaxial, vacuum-insulated pipe was installed at Weggis on Lake Lucerne (Rybach and Hopkirk, 1995).

It should be mentioned that deep coaxial BHE systems have been proposed for retrospective installation in unsuccessful geothermal exploration wells or in spent, failed, or abandoned hydrocarbon wells. Where installation is to take place in an existing well, particular attention should be paid to:

- Completed diameters (is the diameter sufficient to allow the hydraulically efficient installation of a coaxial pipe arrangement?).
- Any equipment that may remain in-hole and restrict access for fluid circulation.
- Fluid chemistry in hole: is it corrosive or scaling, can it be used for circulation during heat recovery, or should it be replaced by a controlled fluid?
- Is the well still open to a reservoir? Is there a risk of fugitive hydrocarbon (or other) gases being entrained with the water flow and escaping (suffocation, explosion, and odor and greenhouse gas hazards)?

In the United Kingdom, testing has recently been completed on a coaxial closed-loop installation in the Kirby Misperton KM-8 hydrocarbon exploration borehole in Yorkshire (Cariaga, 2023).

Any closed-loop system, whether shallow or deep and U-tube or coaxial, should be subject to some routine monitoring (the extent of which will be related to the size of the scheme):

- Fluid flow and return temperatures
- Fluid flow rate (which allows calculation of thermal output)
- Circulation pump electricity consumption
- Pressure (wellhead flow and return; heat exchanger or heat pump flow and return)
- Heat exchange efficiency or heat pump coefficient of performance
- Occasional sampling of circulating fluid to identify risk of scaling or corrosion; degradation of any anti-freeze

4 Case studies

This section aims to introduce a range of geothermal engineering case studies across the various geological settings, demand profiles, and exploitation techniques outlined in the chapter. These include conventional doublet/triplet systems, mine water exploitation, deep BHEs, and borehole TES. These case studies are designed to provide guidance on how to approach key production engineering questions, as well as results and lessons learned from a geothermal perspective.

4.1 Open-loop

4.1.1 Conventional doublet/triplet system: The Pullach geothermal field, Germany

Background

The Pullach geothermal field is located in the Molasse Basin in Southern Germany, targeting the Malm aquifer (Baumann et al., 2017), which forms a hydrothermal system (Breede et al., 2015). The Malm aquifer is an Upper Jurassic carbonate reservoir, with high secondary porosity and permeability due to karstification, and constitutes the main target for geothermal development (Schintgen and Moeck, 2021). It dips southward and reaches a maximum thickness of 600 m (Prestel et al., 1991). The original doublet was drilled to 3389 m TVD (TH1a) and 3443 m TVD (TH2) between 2004 and 2005, spaced 3.1 km apart (Fig. 8; Baumann et al., 2017).

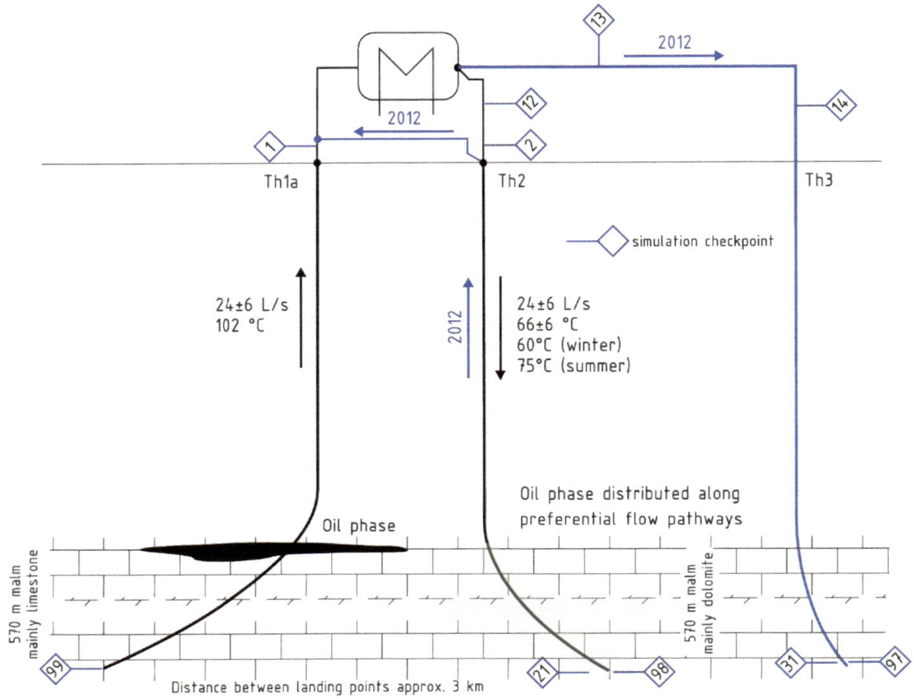

FIG. 8 Schematic of the Pullach field. Flow after conversion of Th2 in blue (Baumann et al., 2017). *License obtained from Elsevier for reproduction (5734210516505).*

Conversion to triplet

The Pullach geothermal plant was commissioned in 2005 as a doublet system (Dussel et al., 2016); however, it has since been extended into a triplet by reversing an injection well to be used for production (Agemar et al., 2014). The third well (TH3) was developed as an injection well in 2011, and well TH2 was then used for continuous production from 2012 (Baumann et al., 2017). The third well was drilled to expand the geothermal plant, doubling the capacity (Erdwerk, 2024). The system sustains a maximum flow rate of 79 L/s with production temperatures up to 104°C, sustaining a maximum capacity of 11.5 MW_{th} (Weber et al., 2015). Three-dimensional, thermal–hydraulic–mechanical modeling of the reservoir, calibrated by data, has highlighted that an operation could be sustained for at least 50 years with unlikely thermal breakthrough (Egert et al., 2022).

4.1.2 *Mine water geothermal: North-east England, United Kingdom*

Seaham Garden Village

An example of an "open-loop with discharge" mine water system is a planned district heating system for Seaham Garden Village, County Durham, based on the mine water that was already being pumped from the shaft of Dawdon Colliery for purposes of regional control of mine water levels. Following heat exchange, the mine water is treated and then discharged into the North Sea. A pilot-scale system has been operated for some time (Bailey et al., 2013), but the full-scale system has yet to be completed (Wood and Crooks, 2020; Coal Authority, 2020).

Gateshead: Geological setting

The United Kingdom's first large-scale mine water heating scheme is that operated by Lanchester Wines at Felling, Gateshead, and reported by Banks et al. (2022; of which the following overview is a summary). It comprises two closely spaced abstraction–reinjection heat pump systems, providing heating for warehouse facilities at Abbotsford Road and Nest Road. A hydrogeological conceptual model of the sites is shown in Fig. 9 and documented by Banks et al. (2022), who argue that there are three aquifer systems in play:

1. The Upper Aquifer System (UAS), which comprises unmined permeable sandstone units within the Carboniferous Coal Measure sequence and contains relatively fresh, iron-containing water.
2. The High Main Aquifer System (HMAS), which comprises extensive workings in the High Main coal seam and the suprajacent (probably mining-fractured) High Main Post sandstone. The water within this aquifer is brackish and reducing. Boreholes intercepting workings in the HMAS often support high yields with little drawdown. A relatively flat drawdown response to pumping may indicate some degree of head-induced recharge from the nearby Tyne Estuary (workings extend under the estuary) or simply good connectivity into a regional network of flooded workings.
3. The Deep Mined Aquifer System (DMAS), which comprises workings in several deeper coal mine seams (including the Maudlin, Brass Thill, Hutton, and Harvey-Beaumont seams) that were typically worked later than the High Main. While worked seams in the DMAS enjoy some degree of connectivity with each other, connectivity with the HMAS is poor. The water in the DMAS is saline and reducing.

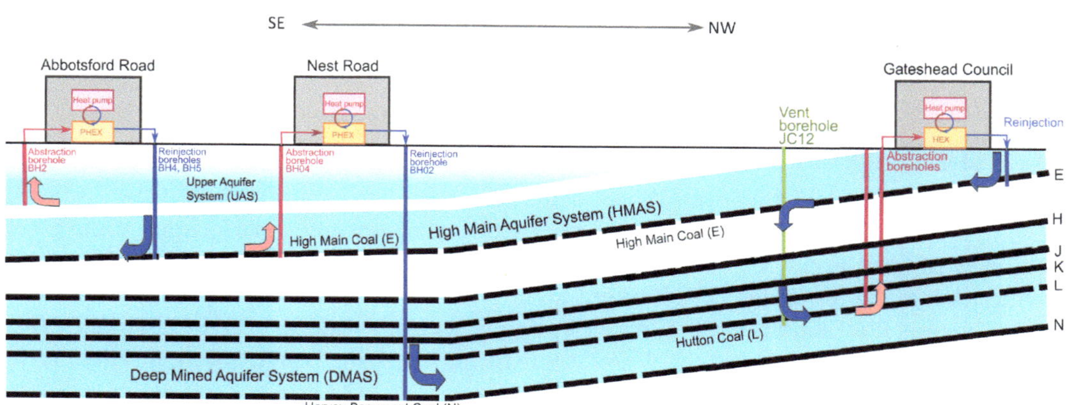

FIG. 9 Highly conceptual cross-section of the Gateshead mine water systems. *E–N* are the designated coal seams. Pink is abstracted water/boreholes and deep blue is the reinjected cold water. Not to scale. PHEX, plate heat exchanger. Abbotsford Road and Nest Road represent situation as of 2022. *Modified from Banks, D., Steven, J., Black, A., Naismith, J., 2022. Conceptual modelling of two large-scale mine water geothermal energy schemes: felling, Gateshead, UK. Int. J. Environ. Res. Public Health 19(3), 1643. Reproduced with permission from © David Banks (2024).*

Abbotsford Road

The Abbotsford Road site (as it existed in 2021–2022) abstracted water from a 110 m borehole in the UAS, at rates of up to 30 L/s. It was passed through a plate heat exchanger, coupled to a heat pump, before reinjection to two 155 m boreholes to the HMAS (a third borehole was occasionally used but had inferior injectivity to the others). The capacity of the system was limited by the yield of the production well. Moreover, the iron-rich nature of the produced water led to clogging of the production well and the heat exchanger. Indeed, the heat exchanger requires regular cleaning by circulation of solutions of citric or oxalic acid; and the borehole requires occasional rehabilitation (Banks et al., 2022).

It is widely recognized that dissolved iron will tend to precipitate out of solution if exposed to oxygen. It is thus usually recommended that the abstraction–heat exchange–reinjection line is sealed and pressurized, preventing contact with, or ingress of, air. Even in the iron-rich water of the Dawdon shaft, Bailey et al. (2013) found that exclusion of air was effective in preventing ferric oxyhydroxide precipitation. In some systems, however, exclusion of air is not wholly possible, due to the shallow nature of the aquifer or borehole (as seems to be the case at Abbotsford Road), or because oxidation of iron already seems to commence within the mine workings (as was the case at another UK colliery, Caphouse; Burnside et al., 2016; Banks et al., 2017).

In short, the Abbotsford Road site was able to deliver usable heat, but its capacity was limited by the yield of the production well, and it had a regular maintenance requirement in respect of ferric oxyhydroxide clogging. Further work is underway at the site to explore options for increasing mine water and thermal yield (Banks et al., 2022).

Nest Road

At the Nest Road site, some 700 m NW of Abbotsford Road, four boreholes were drilled, of which two were eventually completed as (i) a 131 m production borehole from the HMAS and (ii) a 280 m reinjection borehole to the DMAS. Yields of up to 60 L/s were proven from the production well, with relatively low drawdown. During testing of the reinjection borehole, the DMAS appeared to act almost as a "sealed tank." Heads rose quasilinearly during injection and only dissipated very slowly after reinjection stopped. The subsequent operation of the site appears to be relatively unproblematic, with respect to ferric oxyhydroxide scaling, although elevated H_2S has been reported in the produced HMAS water, as have some issues of corrosion of steel instruments (Banks et al., 2022).

Gateshead Council District heating scheme

Around the time the Lanchester Wines schemes commenced operating, the Coal Authority and Gateshead Council started prospecting the same set of workings in a location some 1.9 km to the NW of Nest Road, with a view to a public sector c.6 MW mine water district heating scheme (Adams et al., 2023a; Triple Point Heat Networks, 2023; Gateshead Council, 2024). At this scheme, it was intended that deeper, slightly warmer water would be abstracted from the DMAS (Hutton Seam at around 150 m depth) and, following heat extraction (via titanium heat exchangers, specified due to the corrosive nature of the DMAS mine water—Pashby, 2024), would be reinjected to the HMAS at around 40 m depth. This consortium recognized that the old mine plans of the workings (often pillar and stall) would be challenging to adequately georeference (Adams et al., 2023a). Narrow-diameter exploration boreholes were drilled to locate mine voids, and thereafter, full-scale production and reinjection boreholes were constructed. When the boreholes were tested, the HMAS and DMAS aquifers were found to behave hydraulically similar to Lanchester Wines—i.e., the DMAS experienced little connectivity and did not dissipate pumping-induced changes in head. The Gateshead Council consortium converted one of their boreholes to a "hydraulic vent," some 120 m NE of the main abstraction/reinjection wells, deliberately connecting the DMAS to the HMAS (Adams et al., 2023b). This thus allows flow from the HMAS to the DMAS (BGS, 2024). This source of recharge to the DMAS changed its hydraulic behavior and allowed it to be used for sustainable abstraction. Such a vent does, however, pose at least a theoretical pathway for thermal feedback within the mine system. The scheme has recently been commissioned.

Interaction between systems

As regards interaction between the schemes, a relatively limited degree of (beneficial) hydraulic interference between the Abbotsford Road reinjection and the Nest Road abstraction in the HMAS was demonstrated. According to Banks et al. (2022), no clear indication of thermal interference had been observed. Test pumping indicated a degree of (beneficial) hydraulic interference between the Gateshead Council and Lanchester Wines Nest Road schemes in the HMAS and (especially) the DMAS. Hydraulic interference does not necessarily imply a high risk of thermal interference, however. The Coal Authority (2022) is establishing a "Living Mine Water Laboratory," with a network of monitoring boreholes in the area around and between the Gateshead Council and Lanchester Wines schemes. This should give early warning of any significant hydraulic or thermal interference between the schemes.

Key lessons

The combined experiences of the two schemes suggest the following general lessons for mine water geothermal:

i. Old mine plans can be challenging to satisfactorily georeference, leading to uncertainties in targeting mine void spaces and a potentially significant borehole failure rate. A program of narrow-diameter pilot drilling to confirm mine voids, prior to production hole construction, can be beneficial.
ii. The hydraulic behavior of differing mined seams can be radically different, depending on the mutual regional interconnectivity of seams, and their interaction with adjacent permeable aquifers and surface water features. In highly confined mined seams, hydraulic interaction can propagate rapidly over large distances.
iii. The hydrochemistry of differing mined seams can vary significantly with depth at a given location. Shallow worked coal seams may contain fresh, relatively oxidizing water. Deeper seams may contain highly corrosive, H_2S-rich, saline waters. System components should be specified that are compatible with the given water chemistry: for corrosive waters, this may include high-strength u-PVC or stainless steel well casings, high-grade stainless steel, or even titanium heat exchangers.

iv. The high dissolved iron content of most coal mine (and many metal mine) waters can lead to potential precipitation of ferric (and other metal) oxyhydroxides in contact with oxygen or with pH change due to degassing of CO_2. The pressure in the sealed abstraction–heat exchange–reinjection pipe circuit should be carefully managed to prevent degassing and contact with oxygen. In some cases, it may not be fully possible to prevent oxidation of dissolved iron; in such cases, the resulting ferric oxyhydroxide scaling may require a significant routine maintenance commitment.

4.2 Closed-loop

4.2.1 Deep borehole heat exchangers: Newcastle Science Central Deep Geothermal Borehole, United Kingdom

Background

The Newcastle Science Central Deep Geothermal Borehole (NSCDGB) was drilled to 1821 m in Newcastle, United Kingdom, between 2011 and 2014, targeting the Mississippian Fell Sandstone Formation as a conventional geothermal resource (Younger et al., 2016). Unfortunately, the well was not productive due to limited hydraulic conductivity at the reservoir interval; thus (as of 2024), it is in the process of being repurposed as a pilot DBHE. This will enable testing of a DBHE to evaluate the potential of such a system to be used as a low-risk method of exploiting geothermal energy. DBHEs are often coaxial in design to limit pressure drop, and parasitic losses in the circulation pump, while maximizing the thermal yield due to the lower borehole thermal resistance in contrast to single and double U-tube configurations (Brown et al., 2024). A concentric pipe is fitted within the borehole, cold fluid is circulated down the annular space, warming with depth via conduction through the borehole wall, before being advected to the surface via the central pipe.

The NSCDGB provides an ideal scenario to test the implementation and capacity of DBHE technology due to the high heat flows and significant geothermal resource associated to the North Pennine Batholith (Howell et al., 2021; Brown, 2022). It is planned that the borehole will be repurposed to c.920 m due to limitations caused by a narrow-diameter liner inserted below this point. Otherwise, it is likely that there could be hydraulic issues (i.e., increased pressure drop of the heat transfer fluid; Brown et al., 2023b), leading to greater parasitic losses.

Modeling approach and wellbore design

Modeling of the NSCDGB has been undertaken to give preliminary indications of the performance over a 20-year lifetime. Brown et al. (2023c) investigated the thermal and hydraulic performance of a DBHE completed to 920 m, assuming a homogenous design of the subsurface (i.e., isotropic geology). The anticipated completion of the well is highlighted in Fig. 10, with the outer diameter of the central pipe set to 0.1005 m (~4 in.). All other parameters are listed in Table 3. Numerical modeling was undertaken using OpenGeoSys (OGS) software, which utilizes the "dual-continuum" method for evaluating the wellbore components and has been extensively validated (Chen et al., 2019; Cai et al., 2021; Brown et al., 2024; Brown and Falcone, 2024).

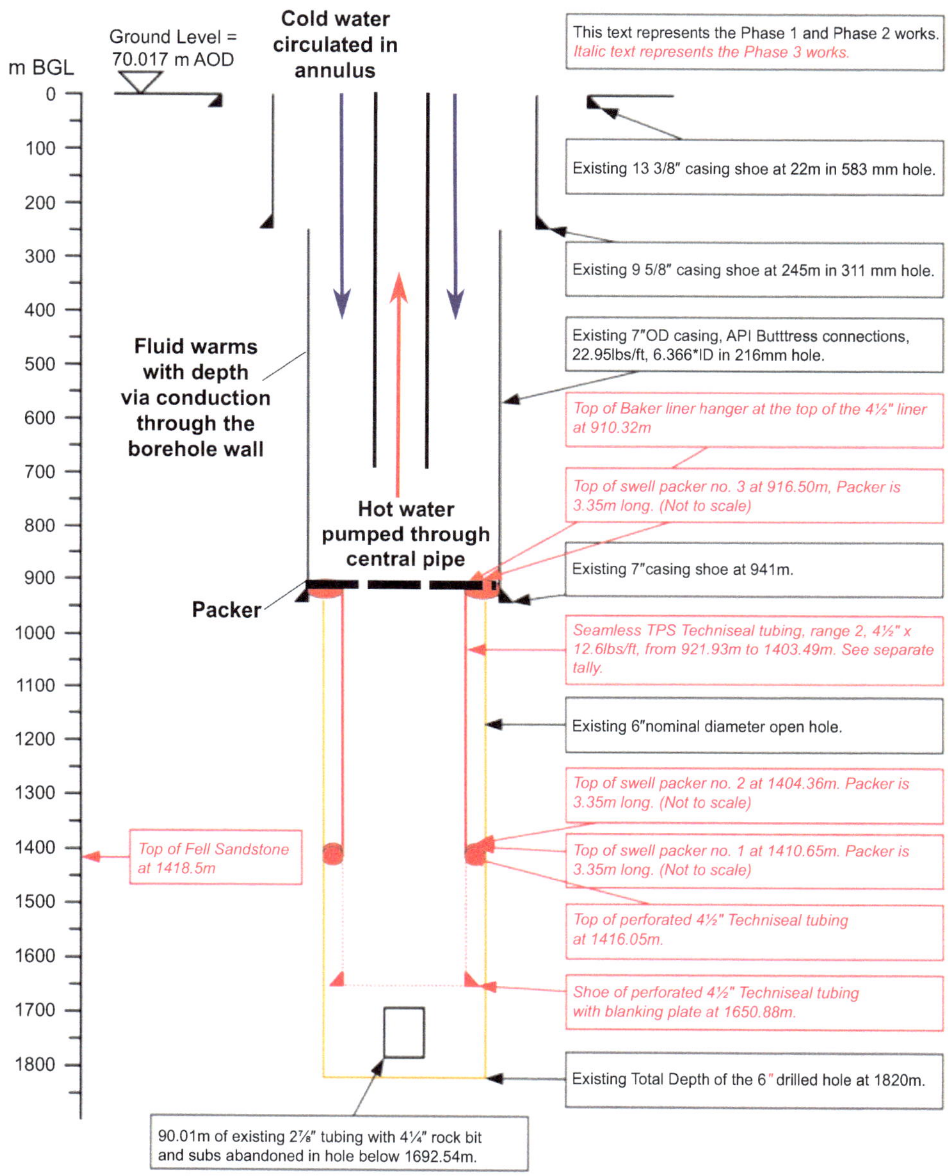

FIG. 10 Newcastle Science Central Deep Geothermal Borehole well construction. *Modified from Younger, P.L., Manning, D.A., Millward, D., Busby, J.P., Jones, C.R., Gluyas, J.G., 2016. Geothermal exploration in the Fell Sandstone Formation (Mississippian) beneath the city centre of Newcastle upon Tyne, UK: the Newcastle Science Central deep geothermal borehole. Q. J. Eng. Geol. Hydrogeol. 49(4), 350–363, to include DBHE location).*

TABLE 3 Key parameters in the study of Brown et al. (2023b).

Parameter	Value	Units
Borehole depth [1]	920	m
Borehole diameter	0.216	m
Outer diameter of inner pipe	0.1005	m
Thickness of inner pipe	0.00688	m
Thickness of outer pipe	0.0081	m
Thickness of grout	0.01905	m
Thermal conductivity of polyethylene inner pipe	0.45	W/(mK)
Thermal conductivity of steel outer pipe	52.7	W/(mK)
Density of rock	2480	kg/m^3
Thermal conductivity of rock	2.55	W/(mK)
Specific heat capacity of rock	950	J/kgK
Density of grout	995	kg/m^3
Thermal conductivity of grout	1.05	W/(mK)
Specific heat capacity of grout	1200	J/kgK
Density of fluid	998	kg/m^3
Thermal conductivity of fluid	0.59	W/(mK)
Specific heat capacity of fluid	4179	J/kgK
Heat load injected	50	kW
Surface temperature	9	°C
Geothermal gradient	33.4	°C/km
Volumetric flow rate	0.005	m^3/s
Circulation pump efficiency	70	%

Results

The key results from Brown et al. (2023c) highlighted that, over the lifetime of the system, a thermal yield of 50kW could be imposed on the borehole, corresponding to a constant baseload of 65kW to be supplied to a building in combination with a heat pump. This would be insufficient to meet the full demand of the building, and further sources of energy or a DBHE array would be required. The coefficient of system performance (i.e., the ratio of heat supplied to the building in contrast to parasitic losses from the heat pump and circulation pump) varied between 4.1 and 5.6 for the duration of the simulation. The study also highlighted that, by allowing periods of rest (i.e., 6months during summer when demand is likely to be low), the system could extract a greater thermal yield from the ground with warmer inlet and outlet temperatures from the DBHE (Fig. 11). It should be noted that deep closed-loop systems can produce a far lower thermal yield in contrast to open-loop equivalents.

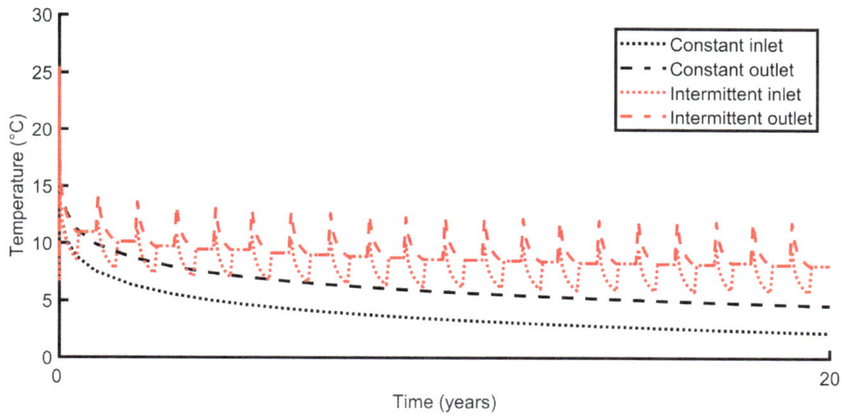

FIG. 11 Inlet and outlet temperatures for a single DBHE for constant and intermittent operation (with 50 kW applied as both constant and intermittent load) (Brown et al., 2023c).

4.2.2 *Borehole thermal energy storage: University of Glasgow, United Kingdom*

Background

It is common to use surface heat sources, such as industrial waste heat or solar thermal heat, in a BTES system (Gehlin, 2016). Here, a notional case study is investigated using the James Watt Building at the University of Glasgow, United Kingdom (Kolo et al., 2023). The building has a total area of 3193.53 m^2 and an annual heat demand of 3011 MWh with 1981 MWh winter heat demand. Roof-mounted solar thermal collectors are assumed to be installed on the building. The harvested solar heat is used to satisfy the building's summer heat demand, and any short-term surplus is directed to an intermediate storage tank to smooth out demand and supply peaks. Long-term surplus heat is subsequently directed to a BTES store in the summer and then extracted for use during winter. After a conceptual presentation of the system, the system modeling has been decoupled into surface (solar thermal) and subsurface (BTES) modeling. This facilitates detailed analysis of each system component.

The solar–BTES coupled system

A conceptual representation of the system operation is illustrated in Fig. 12. In summer (solid lines), solar heat is used to meet the building's heat demand. Surplus solar heat is stored in a buffer storage tank with a storage capacity of 18 days. Any subsequent long-term surplus is then directed to the underground BTES storage. During winter (dashed lines), the stored heat is gradually extracted as a contribution to the heat supply of the building. The storage is seasonal, since surplus solar heat is predominantly available in the summer but to be extracted in the winter.

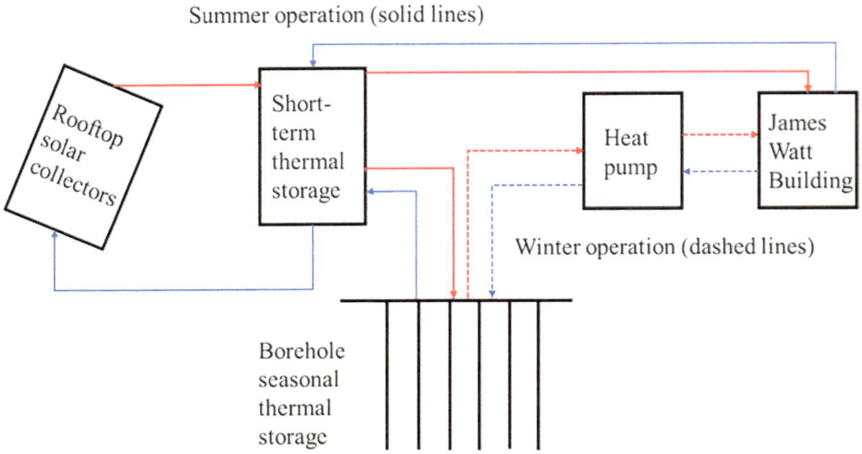

FIG. 12 Conceptual illustration of the coupled solar–BTES system for James Watt Building, University of Glasgow (Kolo et al., 2023).

Solar thermal modeling

The heat output from the solar thermal collectors was modeled using the open-source python package Atlite (Hofmann et al., 2021). The total area of the solar panels used was 4790.3 m^2, 1.5 times the roof size, assuming that additional area can be secured from neighboring areas. Atlite calculates the heat output based on the average of the actual hourly heat radiation on the solar collector (sourced from Hersbach et al., 2020) and the difference between the temperature of the connected BTES store (assumed to be 80°C) and hourly (average) ambient air temperature. For the solar collector technology, an azimuth of 180° and a slope of 40° were assumed. Efficiency coefficients C0 and C1 were assumed to be 0.8 and 3.0, respectively. Fig. 13A shows the available solar thermal heat over 1 year, from January, along with the heat demand; their difference—surplus solar—is also plotted. As expected, surplus solar is only available in the summer period (between days 100 and 250, i.e., April and September). To simulate short-term buffer storage, this is plotted over an 18-day moving average as shown in Fig. 13B. The moving average shows a peak charge of 7.7 MWh/day (322 kW) and a cumulative peak charge of 139 MWh over 18 days. The cumulative charge over the year adds up to 318 MWh with a peak charge rate of 322 kW. This represents the available heat that is directed to the underground BTES storage.

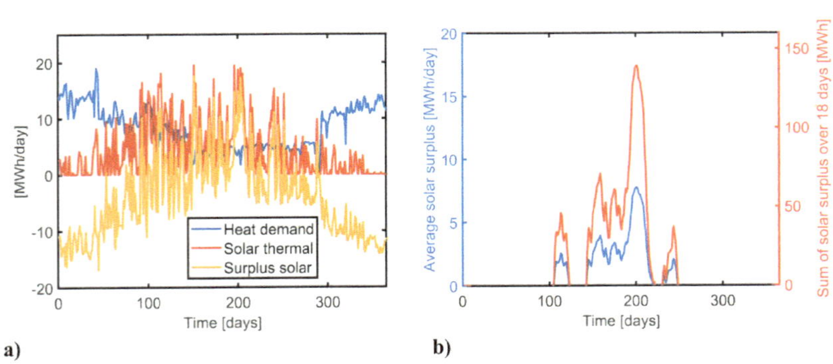

FIG. 13 Available surplus solar heat after meeting heat demand (Kolo et al., 2023). (A) Surplus solar heat—difference between solar thermal heat and heat demand. (B) 18-day moving average of surplus solar heat.

Design of BTES array

A heat of 318 MWh or 1144.8 GJ is to be stored during the summer (April–September), and the same amount will be extracted for use in winter. Discharge is distributed equally over 4 months from November to February; i.e., 79.48 MWh is discharged each month, with a peak discharge rate of 200 kW. October and March are taken to be recovery (rest) periods. The subsurface volume $V = H/(\rho c \Delta T)$ required to store a heat load, $H = 1144.8$ GJ, is 34,690 m^3 for a volumetric heat capacity, $\rho C = 2.2$ MJ/(m^3 K), and average temperature rise, $\Delta T = 15$ K. For an equidimensional cylinder (the same diameter and height), this volume implies a depth and diameter of 36 m and a surface area of 1018 m^2. Using the maximum heat exchange rate of 322 kW and a specific heat extraction rate of 40 W/m, the number of 36 m—deep boreholes required is 224 with an approximate inter-borehole spacing of 2.13 m, assuming square packing.

Subsurface BTES modeling

The subsurface BHE heat transport module of the finite element open-source software OGS (Kolditz et al., 2012) coupled with Thermal Energy Systems in Python (TESPy; Witte and Tuschy, 2020) is adopted for BTES modeling. While OGS models the detailed heat transfer between the surrounding rock, the grout, and the BHE pipe inlet and outlet, TESPy handles the surface components including the pipe network and cooling/heating load. Fig. 14 illustrates the TESPy pipe network where pipes are connected in parallel with a splitter that evenly distributes the flowrate among BHEs and a merger to combine flows from all BHEs.

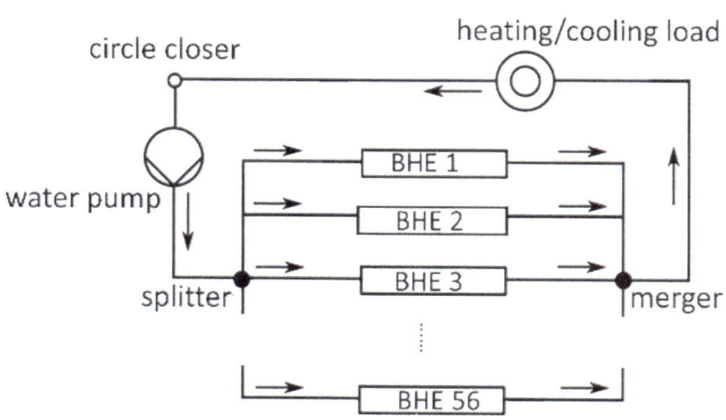

FIG. 14 Illustration of pipe network in TESPy. *Modified from Chen, S., Shao, H., Witte F., 2024. HEAT_TRANSPORT_BHE PipeNetwork Feature. https://www.opengeosys.org/docs/processes/heat-transport/heat_transport_bhe_pipelinenetwork/ (Accessed on 27 May 2024).*

OGS–TESPy coupling has been used in several works after verification and validation (Chen et al., 2020; Cai et al., 2021). Due to symmetry, only a quarter of the subsurface 224 boreholes with square packing is modeled in OGS. The heat load input is based on the previous section with 6 months charge, 1 month rest, 4 months discharge, and 1 month rest (see Fig. 15A). The 1 year simulation starts with the charge period, i.e., April–March. A coaxial BHE is assumed to have an outer pipe diameter of 0.127 m and an inner pipe diameter of 0.087 m, installed in a hole of 0.152 m diameter. Monoethylene glycol at 23.5 wt% with a flow rate of 0.112 L/s (per borehole) is used as the heat transfer fluid because it is an anti-freeze. It has a density of 1032 kg/m^3, a thermal conductivity of 0.482 W/m/K, and a specific heat capacity of 3842 J/kg/K at 10°C (Melinder, 2010). In areas around the University of Glasgow, a surface ground temperature of 10.17°C (Met Office, 2022) and a geothermal gradient of 0.03592°C/m (Browne et al., 1987) were assumed. Fig. 15B shows the computational domain which is 30 m × 30 m × 72 m (x, y, z). Zero heat flux boundary condition is imposed on all boundary faces. A quarter of the heat load shown in Fig. 15B is imposed.

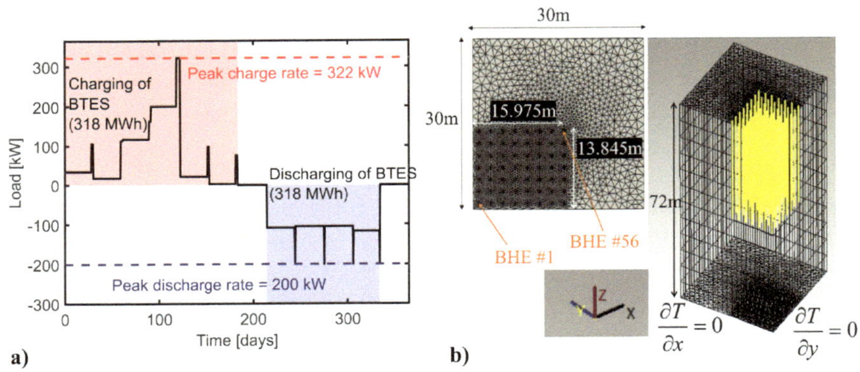

FIG. 15 Simulation setup and boundary conditions (Kolo et al., 2023). (A) Annual heat load profile starting from April. (B) Top and side views of the finite element mesh.

Results

The results obtained are shown in Fig. 16. Fig. 16A shows the fluid inlet (T_{in}) and outlet (T_{out}) temperatures for the central BHE, BHE 1, and the outer BHE, BHE 56. The temperature limits are between 33.33°C (at peak charge) and 2.19°C (at peak discharge), which are suitable operation conditions for the considered heat transfer fluid. Hence, the system can operate as desired for charge and discharge. While the inlet temperatures for all BHEs are the same, BHE 1 has a slightly higher T_{out} compared with BHE 56 during charge, because there are more heat losses at the boundaries close to BHE 56. This is clearly illustrated in Fig. 16B, which shows heat loss to the surroundings of BHE 56 extending over a region of thermal influence.

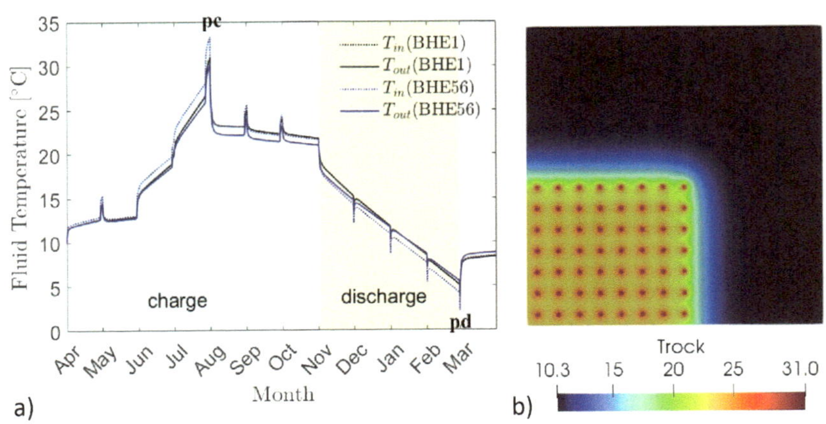

FIG. 16 Fluid and rock temperatures (Kolo et al., 2023). (A) Fluid temperatures over the year, indicating peak charge (pc) and peak discharge (pd). (B) Rock temperature (Trock) distribution at peak charge (pc).

The analysis presented here shows that, with a solar heat–sourced BTES system subject to the stated assumptions, 318 MWh representing 16% winter heat demand of James Watt Building can be obtained using a 224 borehole array BTES system. Even though the analysis shown here is only for 1 year, the performance of the BTES system is expected to improve over time.

Different approaches have been adopted here for modeling the solar thermal system on one hand and BTES system on another, which makes it possible to analyze the details of each borehole and the surrounding rock. The disadvantage of this numerical approach is that it is computationally very expensive. Using a whole-systems approach, dynamic modeling of the whole system can be undertaken with minimal computational time using tools like TRNSYS and Dymola. TRNSYS has been used recently, in a first, to study a BTES system with curtailed wind as fluctuating heat source (Brown et al., 2023d).

5 Integration of geothermal systems with district thermal energy networks

Depending on their temperature, not all geothermal resources are suitable for electric power generation. However, they can play a key role toward the decarbonization of heating and cooling.

Thermal energy networks offer a flexible approach to meeting the needs for heating and cooling, allowing for the harmonization of different aspects of thermal systems, including thermal sources, energy transformation units, and storage. They can therefore manage some aspects centrally (e.g., the geothermal source), while keeping energy transformation units or storage at a district/building/flat level. Such systems offer a unique opportunity to leverage geothermal resources for increased system efficiency, contributing to meeting the energy supply trilemma of low-cost heating/cooling, low emissions, and high security of supply.

5.1 Impact of centralizing thermal energy supply units and sources

Centralization enables economies of scale, diverse mix of energy transformation equipment, diversification of peak thermal power, and utilization of waste heat (Frederiksen and Werner, 2013). Centralized thermal sources can more effectively leverage low-zero carbon energy transformation technologies. Depending on the temperature of the thermal network and the depth of the geothermal well, a direct heat exchanger can be used. Alternatively, a GSHP is required to upgrade the heat/coolth, introducing electricity requirements (Tester et al., 2016). There is a balance between the capital costs associated with the depth of the well and the fuel cost savings from having a GSHP, which is bespoke to each project. In both cases, centralization allows for high full-load equivalent operational hours, which is the operation of each energy transformation unit/geothermal source at design capacity for extended time periods (Averfalk et al., 2021), due to demand diversification. In general, energy supply units/sources need to meet the peak power demand for the system other than the annual energy demand. However, demands throughout the year are significantly lower than the peak demand, with the load duration curve capturing this demand occurrence (CIBSE, 2020). For decentralized systems, frequent start–stop cycles tend to occur. Especially when considering energy transformation units with high turn-down ratios (ratio of minimum and design capacity), it is possible to oversize the system, resulting in low efficiencies, high maintenance costs, and compromised thermal comfort for the end-user (Lund et al., 2014). Centralization results in flatter load duration curves, due to the superimposition of complementary (rather than coinciding) individual demands. Optimizing the sizing and operation of the energy transformation unit and geothermal abstraction point requires a well-designed plan, considering the role of daily energy storage provided by tank TES and backup energy supply equipment such as boilers or chillers.

5.2 Geothermal vs alternative supply options

To further highlight the importance of geothermal sources for a cost-effective operation of thermal networks, the difference to counterfactual supply options is presented. Typically, when geothermal sources are not available, a mix of boilers or chillers would be employed. However, in the efforts to decarbonize the energy system and minimize energy use, heat pumps are seen as the predominant solution. Air source heat pumps (ASHPs) are identified as the key counterfactual supply option. As aforementioned, direct heat exchangers allow for minimum electricity use (only for pumping), while they also offer a robust and noncomplex solution. However, when the temperature of the well is not high enough for the end-user, a GSHP is required.

The benefits of using GSHPs vs ASHPs lie on their coefficient of performance (COP) differences. Heat pump's COP depends on multiple aspects, such as the type of refrigerant used and the temperature difference across the evaporator and condenser (Vering et al., 2022). The temperature difference between the source and the sink is often the most critical factor, as it directly affects the amount of electricity required to raise the source temperature to the desired level for the sink (thermal network, building, etc.). The environmental heat source can range from ambient air to geothermal or surface water. For ASHPs, there are typically large temperature differences between the ambient air and the desired sink temperature, while in geothermal systems, a more consistent temperature profile throughout the year is present (Singh and Sharma, 2017). Furthermore, low-temperature geothermal sources can be "charged" by raising their temperature during low-demand periods. If, for example, a thermal energy network is using a technology mix including solar thermal and GSHPs using a borehole ground heat exchanger array, the ground can be charged during times of high solar irradiance if low demand is present in the network (Pauschinger, 2016). By charging the ground, the COP of the GSHP during its operation will be higher. It effectively allows decoupling the demand from the supply, facilitating a smarter system operation. Also, the operation of GSHPs can be more easily shaped around times of low electricity prices (Sveinbjörnsson et al., 2019).

Geothermal systems can also function as thermal batteries, providing a critical solution for grid management by facilitating the utilization of curtailed wind energy and balancing grid voltage and frequency variations (Klyapovskiy et al., 2019). They allow for a truly smart energy system relying on energy synergies and sector coupling, seeing various utilities as complementary rather than individual problems (Angelidis et al., 2023). This way of viewing the thermal system is clearly captured in the evolution of thermal energy systems as shown in Fig. 17.

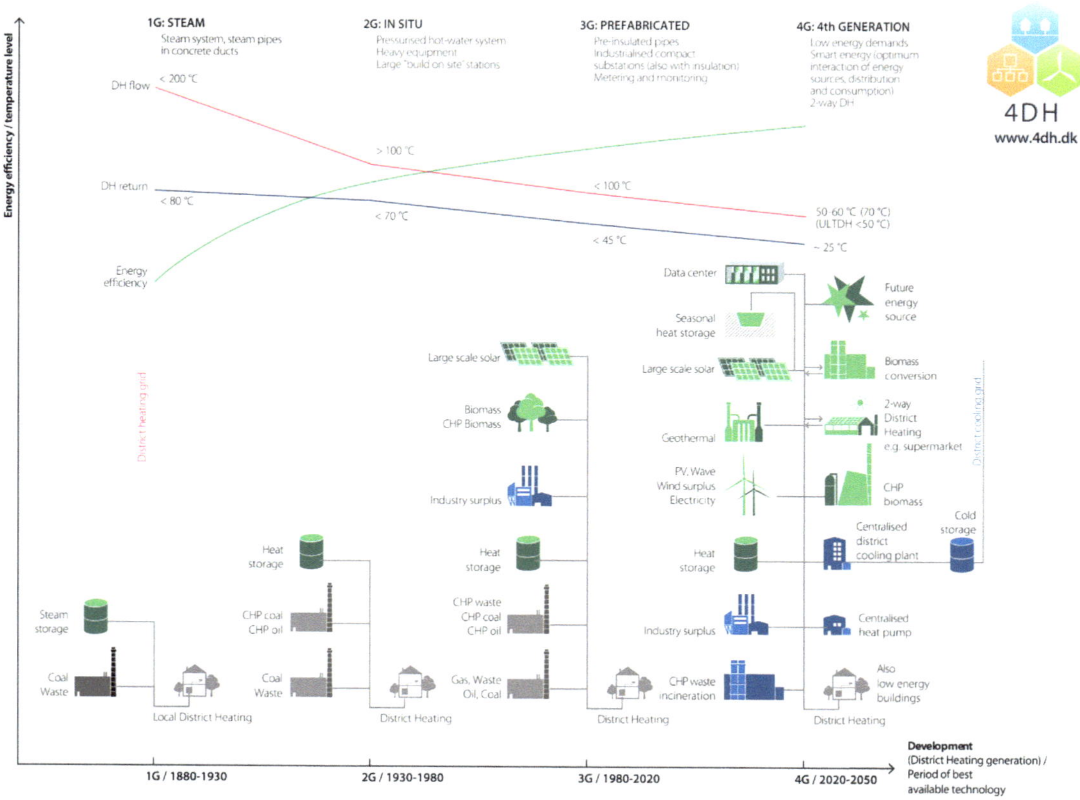

FIG. 17 Progression of district heating—1st–4th generation (Thorsen et al., 2018).

5.3 Types of thermal energy Networks

District thermal energy networks include district heating (DH), district cooling (DC), and DHC networks. DH and DC networks conventionally comprise energy center(s) with centralized energy transformation units and storage, insulated pipe network delivering heat or cooling to consumers, and connection units such as substations and heat/cooling interface units. DHC utilizes an ambient network to meet both heating and cooling demands in an area, featuring decentralizing energy transformation units but—typically—a centralized thermal source.

All these systems focus on synergizing with available waste heat opportunities and seasonal storages in both urban and rural areas. As can be seen from Fig. 17, there is a trend to reduce temperatures in the networks, mainly due to reduced temperature requirements for the end-users deriving from better building insulation.

For DHC, where one ambient temperature network is present, it can be easier to utilize distributed low-grade thermal resources and introduce "energy trading," since heating and cooling are met by the same system, making each energy substation a prosumer (producer and consumer; Lund et al., 2014; Revesz et al., 2020; Wirtz et al., 2019). However, this approach in thermal systems may result in lower economic and environmental performance than that in conventional DH and DC systems, since it heavily depends on heating and cooling synergies. According to Gudmundsson et al. (2022), when only heating demand is present, the system performs worse than the traditional DH, while Wirtz et al. (2020) specify that a seasonal demand co-occurrence of heating and cooling of 30% is required for the economics to start becoming comparable. Zhang et al. (2023) argue that due to demand seasonality (heating requirements during winter and cooling requirements during summer), only 0.1% of the building stock in Europe has a demand co-occurrence >30%. Geothermal wells can provide a solution to this lack of demand co-occurrence, by acting as seasonal thermal stores. This way, heat stored during summer can be used during winter and vice versa. Due to the ambient water temperatures used in DHC, ATES and flooded mines offer a unique opportunity; however, a detailed hydrogeological survey is needed to ensure that energy dissipation allows for seasonal storage of the temperature differences involved (Buffa et al., 2019).

Combination of an ambient DHC with decentralized low-temperature sources (e.g., shallow groundwater, mine water, or closed-loop BHEs) implies that each individual end-user typically has its own heat exchangers and heat pumps that can extract heat from (or reject surplus heat to) the ambient thermal transfer network. Such systems function best when heat rejection (cooling) and heat extraction loads are approximately balanced. When they are not, heat exchange with the ground is used to satisfy the imbalance, providing the excess heat or cooling.

There can be advantages of such decentralized ambient systems, in that each end-user has its own heat pumps, and so heating or cooling can be provided at a temperature controlled by the end-user. In addition, the electricity used by the heat pumps is paid directly by the end-users, who are responsible for the efficient operation of their respective heat pump. However, individual heat pumps on the network "see" individual end-user peak loads, potentially leading to higher total heat pump capacity being installed, compared with centralized systems where statistical probability of demand co-occurrence leads to smaller peaks seen by the source/unit. Decentralized, ambient systems can be large or small scale: the mine water system at Heerlen, Netherlands (Verhoeven et al., 2014; Banks, 2017; Brummer and Bongers, 2019) is an example of a system that is evolving to support a wide variety of thermal loads, using the subsurface mine water reservoir to absorb imbalances in heating and cooling demand. At a smaller scale, the concept of the shared communal ground loop has been proposed in the United Kingdom through the "Heat the Streets" project in Stithians, Cornwall (The Engineer, 2023). Instead of each property on a street having its own BHE, all the properties' heat pumps are coupled to a "shared ground loop," in turn extracting heat from shared BHEs.

Altogether, integrating geothermal sources with thermal energy networks for energy production and storage can facilitate effective sector coupling and a shift to a holistic approach to more sustainable energy system design.

6 Conclusions and future outlook

This chapter highlights that there are a range of geothermal shallow-to-deep, closed- to open-loop systems over a range of geological settings to meet a wide variety of domestic to nondomestic applications, including via coupling with thermal energy networks. The key conclusions of the chapter are:

- There are a variety of geothermal energy demands, which have various thermal load profiles, and subsurface exploitation techniques, which can supply varying amounts of heat at different grades, as well as TES. It is thus, essential to engineer the system vis-à-vis the end-user demand.
- Conventional open-loop systems can utilize traditional methods for calculating productivity and injectivity indices. However, closed-loop systems do not have any hydraulic interactions with the reservoir, and thus, the thermal interactions through the borehole wall must be assessed to determine the specific thermal yield. The pressure drop encountered by the circulating fluid is also of importance as it is essential to minimize parasitic losses.
- As there exists a great variety of methods to engineer geothermal systems, there are no universal well designs that are equally applicable to all situations. Case studies can provide suitable analogs to highlight the key lessons learned from past developments and methodology on how to engineer a geothermal system.
- While not all geothermal resources lend themselves to power generation, they offer a great opportunity for coupling with thermal energy networks toward decarbonization of heating and cooling, also providing strategic large-scale energy storage.

References

Abesser, C., 2010. Open-Loop Ground Source Heat Pumps and Groundwater Systems: A Literature Review of Current Applications, Regulations and Problems. Report, OR/10/045. British Geological Survey.

Abesser, C., Busby, J.P., Pharaoh, T.C., Bloodworth, A.J., Ward, R.S., 2020. Unlocking the Potential of Geothermal Energy in the UK. Open Report, OR/20/049. British Geological Survey. p. 22.

Adams, B., 1982. An Aquifer Thermal Energy Storage Experiment in the Lower Greensand at the IGS Research Site. Reach, Cambridgeshire.

Adams, C., Gordon, J., Parker, K., 2023a. The Gateshead mine water heat scheme. In: Presentation to the Society of Petroleum Engineers, Geothermal Seminar 2023 "Realising the Ambition", February 2023, Aberdeen, UK. https://www.spe-aberdeen.org/uploads/1115-SPE-Aberdeen-Feb23-Adams.pdf.

Adams, C., Parker, K., Gillon, J., 2023b. Developing the Gateshead mine heat scheme. In: Presentation to the 10th UK Geothermal Symposium, 20th–22nd November 2023, Geological Society, London, UK. https://www.geolsoc.org.uk/~/media/shared/documents/events/Energy%20Group/Geothermal/10th%20Geothermal%20Symposium%20-%20Abstract%20Book%202023.pdf.

African Union, 2016. The African Union Code of Practice for Geothermal Drilling. https://www.bgr.bund.de/EN/Themen/Zusammenarbeit/TechnZusammenarb/Downloads/2029_CodeofPractice_en.pdf;jsessionid=68C7B005CFD4A8321B8AB7A976DA41A7.internet001?__blob=publicationFile&v=3. Accessed on 28 February 2024.

Agemar, T., Weber, J., Schulz, R., 2014. Deep geothermal energy production in Germany. Energies 7 (7), 4397–4416.

Andersson, O., 2009. The ATES Project at Stockholm Arlanda Airport. In: Proceedings of Effstock, the 11th International Conference on Underground Thermal Energy Storage, June 2009, Stockholm, Sweden.

Angelidis, O., Ioannou, A., Friedrich, D., Thomson, A., Falcone, G., 2023. District heating and cooling networks with decentralised energy substations: opportunities and barriers for holistic energy system decarbonisation. Energy 269, 126740. https://doi.org/10.1016/j.energy.2023.126740.

Averfalk, H., Benakopoulos, T., Best, I., Dammel, F., Engel, C., Geyer, R., Gudmundsson, O., Lygnerud, K., Nord, N., Oltmanns, J., Ponweiser, K., Schmidt, D., Schrammel, H., Østergaard, D.S., Svendsen, S., Tunzi, M., Werner, S., 2021. Low-temperature district heating implementation guidebook: final report of IEA DHC Annex TS2 implementation of low-temperature district heating systems (Report No. 978-3-8396-1745–8). In: Low-Temperature District Heating Implementation Guidebook. IEA DHC.

Bailey, M.T., Moorhouse, A.M.L., Watson, I.A., 2013. Heat extraction from hypersaline mine water at the Dawdon Mine Water Treatment Site. In: "Mine Closure 2013", Proceedings of the Eighth International Seminar on Mine Closure. Australian Centre for Geomechanics, Perth, Australia, pp. 559–570, https://doi.org/10.36487/ACG_rep/1352_47_Bailey.

Banks, S.B., 2003. The UK coal Authority minewater-treatment scheme programme: performance of operational systems. Water Environ. 17 (2), 117–122. https://doi.org/10.1111/j.1747-6593.2003.tb00444.x.

Banks, D., 2012. An Introduction to Thermogeology: Ground Source Heating and Cooling, second ed. John Wiley & Sons, Chichester, ISBN: 978-0-470-67034-7, p. 526.

Banks, D., 2017. Integration of cooling into mine water heat pump systems (v.F1.1). In: Report by University of Glasgow for EU Research Fund for Coal and Steel, Project Low-Carbon After-Life (LoCAL): Sustainable Use of Flooded Coal Mine Voids as a Thermal Energy Source. Funded by European research Fund for Coal and Steel, https://doi.org/10.13140/RG.2.2.27377.48486. March 2017.

Banks, D., 2022. Examining issues of circulation asymmetry in closed loop borehole heat exchangers. In: University of Glasgow Report for the EPSRC INTEGRATE and NetZero-GEORDIE Projects., https://doi.org/10.13140/RG.2.2.15735.29608. Final version F1.2, 6th April 2022.

Banks, D., Skarphagen, H., Wiltshire, R., Jessop, C., 2003. Mine water as a resource: space heating and cooling via use of heat pumps. Land Contamin. Reclamat. 11 (2), 191–198. https://doi.org/10.2462/09670513.814.

Banks, D., Skarphagen, H., Wiltshire, R., Jessop, C., 2004. Heat pumps as a tool for energy recovery from mining wastes. In: Gieré, R., Stille, P. (Eds.), Energy, Waste and the Environment: A Geochemical Perspective. Vol. 236. Geological Society Special Publication, pp. 499–513, https://doi.org/10.1144/GSL.SP.2004.236.01.27.

Banks, D., Athresh, A., Al-Habaibeh, A., Burnside, N., 2017. Water from abandoned mines as a heat source: practical experiences of open- and closed-loop strategies, United Kingdom. Sustain. Water Resour. Manag. 5, 29–50. https://doi.org/10.1007/s40899-017-0094-7.

Banks, D., Steven, J., Black, A., Naismith, J., 2022. Conceptual modelling of two large-scale mine water geothermal energy schemes: felling, Gateshead, UK. Int. J. Environ. Res. Public Health 19 (3), 1643.

Baujard, C., Genter, A., Dalmais, E., Maurer, V., Hehn, R., Rosillette, R., Vidal, J., Schmittbuhl, J., 2017. Hydrothermal characterization of wells GRT-1 and GRT-2 in Rittershoffen, France: implications on the understanding of natural flow systems in the Rhine Graben. Geothermics 65, 255–268.

Baumann, T., Bartels, J., Lafogler, M., Wenderoth, F., 2017. Assessment of heat mining and hydrogeochemical reactions with data from a former geothermal injection well in the Malm Aquifer, Bavarian Molasse Basin, Germany. Geothermics 66, 50–60.

Beard, J.C., Jones, B.A. (Eds.), 2023. The Future of Geothermal in Texas the Coming Century of Growth & Prosperity in the Lone Star State. George Mitchell Foundation.

BGS, 2024. Borehole JC12, Saltmeadows, Albany road. In: BGS Borehole Records 21361255. British Geological Survey. reference NZ26SE3245. GeoIndex website https://api.bgs.ac.uk/sobi-scans/v1/borehole/scans/items/21361255.

Breede, K., Dzebisashvili, K., Liu, X., Falcone, G., 2013. A systematic review of enhanced (or engineered) geothermal systems: past, present and future. Geotherm. Energy 1, 1–27.

Breede, K., Dzebisashvili, K., Falcone, G., 2015. Overcoming challenges in the classification of deep geothermal potential. Geotherm. Energy Sci. 3 (1), 19–39.

Brown, C.S., 2022. Regional geothermal resource assessment of hot dry rocks in Northern England using 3D geological and thermal models. Geothermics 105, 102503.

Brown, C.S., Falcone, G., 2024. Investigating heat transmission in a wellbore for low-temperature, open-loop geothermal systems. Therm. Sci. Eng. Prog. 48, 102352.

Brown, C.S., Howell, L., 2023. Unlocking deep geothermal energy in the UK using borehole heat exchangers. Geol. Today 39 (2), 67–71.

Brown, C.S., Doran, H., Kolo, I., Banks, D., Falcone, G., 2023a. Investigating the influence of groundwater flow and charge cycle duration on deep borehole heat exchangers for heat extraction and borehole thermal energy storage. Energies 16 (6), 2677.

Brown, C.S., Kolo, I., Falcone, G., Banks, D., 2023b. Repurposing a deep geothermal exploration well for borehole thermal energy storage: implications from statistical modelling and sensitivity analysis. Appl. Therm. Eng. 220, 119701.

Brown, C.S., Kolo, I., Falcone, G., Banks, D., 2023c. Investigating scalability of deep borehole heat exchangers: numerical modelling of arrays with varied modes of operation. Renew. Energy 202, 442–452.

Brown, C.S., Desguers, T., Lyden, A., Kolo, I., Friedrich, D., Falcone, G., 2023d. Modelling borehole thermal energy storage using curtailed wind energy as a fluctuating source of charge. In: Proceedings of the 48th Workshop on Geothermal Reservoir Engineering, Stanford, CA, USA, pp. 6–8.

Brown, C.S., Kolo, I., Banks, D., Falcone, G., 2024. Comparison of the thermal and hydraulic performance of single U-tube, double U-tube and coaxial medium-to-deep borehole heat exchangers. Geothermics 117, 102888. https://doi.org/10.1016/j.geothermics.2023.102888.

Browne, M.A.E., Robins, N.S., Evans, R.B., Monro, S.K., Robson, P.G., 1987. The upper Devonian and carboniferous sandstones of the Midland Valley of Scotland. In: Investigation of the Geothermal Potential of the UK. British Geological Survey, HMSO, Keyworth.

Brummer, N., Bongers, J., 2019. Mijnwater Heerlen: roadmap to 2040. In: District Heating and Cooling Grid Parkstad, Limburg. EU InterReg NW Europe Report, p. 33. Available at https://guidetodistrictheating.eu/wp-content/uploads/2019/12/HeatNetNWE_Heerlen-Transition-Roadmap_District-Heating.pdf.

Buffa, S., Cozzini, M., D'Antoni, M., Baratieri, M., Fedrizzi, R., 2019. 5th generation district heating and cooling systems: a review of existing cases in Europe. Renew. Sust. Energ. Rev. 104, 504–522. https://doi.org/10.1016/j.rser.2018.12.059.

Burnside, N., Banks, D., Boyce, A., 2016. Sustainability of thermal energy production at the flooded mine workings of the former Caphouse Colliery, Yorkshire, United Kingdom. Int. J. Coal Geol. 164, 85–91. https://doi.org/10.1016/j.coal.2016.03.006.

Cai, W., Wang, F., Chen, S., Chen, C., Liu, J., Deng, J., Kolditz, O., Shao, H., 2021. Analysis of heat extraction performance and long-term sustainability for multiple deep borehole heat exchanger array: a project-based study. Appl. Energy 289, 116590.

Cariaga, C., 2023. CeraPhi Holds Geothermal Demonstrator Project at Former Fracking Site in the UK. Think Geoenergy. website, 31st August 2023 https://www.thinkgeoenergy.com/ceraphi-holds-geothermal-demonstrator-project-in-former-fracking-site-in-the-uk/. Accessed on 15 February 24.

Caulk, R.A., Tomac, I., 2017. Reuse of abandoned oil and gas wells for geothermal energy production. Renew. Energy 112, 388–397.

Causeway Energies, 2023. Geothermal System Archetypes: Classification, Applications and Technology Development Opportunities. 1st February 2023 https://causewaygt.com/2023/02/geothermal-system-archetypes-classification-applications-and-technology-development-opportunities/. Accessed on 28 February 2024.

Chartered Institution of Building Services Engineers (CIBSE), 2020. CP1 Heat Networks: Code of Practice for the UK (2020), second ed. Association for Decentralised Energy (ADE), London.

Chen, C., Shao, H., Naumov, D., Kong, Y., Tu, K., Kolditz, O., 2019. Numerical investigation on the performance, sustainability, and efficiency of the deep borehole heat exchanger system for building heating. Geotherm. Energy 7, 1–26.

Chen, S., Witte, F., Kolditz, O., Shao, H., 2020. Shifted thermal extraction rates in large borehole heat exchanger array—a numerical experiment. Appl. Therm. Eng. 167, 114750.

Chu, Z., Dong, K., Gao, P., Wang, Y., Sun, Q., 2021. Mine-oriented low-enthalpy geothermal exploitation: a review from spatio-temporal perspective. Energy Convers. Manag. 237. https://doi.org/10.1016/j.enconman.2021.114123. article 114123.

CIBSE, 2019. CP3 Open-Loop Groundwater Source Heat Pumps: Code of Practice for the UK (2019). Chartered Institution of Building Services Engineers, ISBN: 9781912034444, p. 156.

Coal Authority, 2020. Seaham Garden Village Mine Energy District Heating Scheme. Coal Authority Information Brochure, Mansfield, UK. https://www2.groundstability.com/wp-content/uploads/2020/01/CA_Seaham_Brochure.290120.pdf. Accessed on 14 February 2024.

Coal Authority, 2022. Mine Heat "Living Lab"—Gateshead Multiple Monitoring Borehole Construction. Tender notice 2022/S 000-020543, dated 27 July 2022, Coal Authority, Mansfield, UK. https://www.find-tender.service.gov.uk/Notice/020543-2022/PDF. Accessed on 21 February 2024.

DECC, 2016. Energy Bill 2015–2016. Infrastructure Act. Department of Energy and Climate Change. https://assets.publishing.service.gov.uk/government/uploads/system/uploads/attachment_data/file/490997/Infrastructure_Act_2015__Energy_Bill_2015-16_Keeling_Schedule_.pdf. Accessed on 30 September 2022.

Dussel, M., Lüschen, E., Thomas, R., Agemar, T., Fritzer, T., Sieblitz, S., Huber, B., Birner, J., Schulz, R., 2016. Forecast for thermal water use from upper Jurassic carbonates in the Munich region (South German Molasse Basin). Geothermics 60, 13–30.

EA, 2011. Environmental Good Practice Guide for Ground Source Heating and Cooling: Version 3. Environment Agency, Bristol, UK. Report GEHO0311BTPA-E-E https://mail.gshp.org.uk/pdf/EA_GSHC_Good_Practice_Guide.pdf.

Eavor Technologies—Eavor, 2022. Closed-Loop Geothermal Technology for a 24/7 Carbon-Free Future. Retrieved November 11, 2022, from https://www.eavor.com/technology/.

Eden Geothermal, 2024. Types of Geothermal Energy Technologies. https://www.edengeothermal.com/about/geothermal-energy/why-geothermal-energy/. Accessed on 28 February 2024.

Egert, R., Gaucher, E., Savvatis, A., Goblirsch, P., Kohl, T., 2022. Numerical determination of long-term alterations of THM characteristics of a Malm geothermal reservoir during continuous exploitation. In: Proceeding of the European Geothermal Congress, Berlin, Germany.

Erdwerk, 2024. Pullach, Germany. Accessed on 28 February 24 https://www.erdwerk.com/en/pullach-germany.

European Investment Bank (EIB), European Commission, Eavor, 2024. EIB and EU Innovation Fund Support Eavor's Innovative Geothermal Technology in Bavaria. Press Release. 2nd May 2024 https://eavor.de/blog/eib-and-eu-innovation-fund-support-eavors-innovative-geothermal-technology-in-bavaria/. Accessed on 31 May 2024.

Finger, J., Blankenship, D., 2010. Handbook of Best Practices for Geothermal Drilling. Sandia National Laboratories. SANDIA report SAND2010-6048. December 2010.

Frederiksen, S., Werner, S., 2013. District Heating and Cooling. Studentlitteratur, Sweden.

García Gil, A., Garrido Schneider, E.A., Mejías Moreno, M., Santamarta Cerezal, J.C., 2022. Shallow geothermal systems with closed-loop geothermal heat exchangers. In: Shallow Geothermal Energy: Theory and Application. Springer International Publishing, Cham, pp. 121–180.

Gateshead Council, 2024. District Energy Scheme Benefits. Web page https://www.gateshead.gov.uk/article/2994/District-Energy-Scheme-benefits. Accessed on 14 February 2024.

Gehlin, S., 2016. Borehole thermal energy storage. In: Advances in Ground-Source Heat Pump Systems. Woodhead Publishing, pp. 295–327.

Gehlin, S.E.A., Spitler, J., Hellström, G., 2016. Deep boreholes for ground source heat pump systems—Scandinavian experience and future prospects. In: Proc. ASHRAE Winter Meeting, Orlando, Florida, January 23–27. https://media.geoenergicentrum.se/2016/02/Gehlin_et_al_2016.pdf.

Gehlin, S., Andersson, O., Rosberg, J.E., 2020. Country update for Sweden 2020. In: Proc. World Geothermal Congress 2020+1, Reykjavik, Iceland, April–October 2021. https://www.geothermal-energy.org/pdf/IGAstandard/WGC/2020/01040.pdf.

Geothermie Netherland, 2021. Industriestandaard Duurzaam Putontwerp voor aardwarmteputten. https://geothermie.nl/downloads/Industriestandaard_Duurzaam_Putontwerp_GNL.pdf. Accessed 28 February 2024.

Geotrainet, 2011a. Geotrainet Training Manual for Designers of Shallow Geothermal Systems. European Federation of Geologists, Brussels, Belgium. https://geotrainet.eu/wp-content/uploads/2015/10/Geotrainet-Manual-for-Designers-on-Shallow-Geothemal.compressed.pdf.

Geotrainet, 2011b. Geotrainet Training Manual for Drillers of Shallow Geothermal Systems. European Federation of Geologists, Brussels, Belgium. https://geotrainet.eu/wp-content/uploads/2015/11/Drillers-Manual-Final-V2-10.11.pdf.

Griffith, M.V., 1948. The heat pump and its economic possibilities. Woman Eng. (J. Women's Eng. Soc.) 6 (14), 250–257. https://twej.theiet.org/twej/WES_Vol_6.html.

Griffith, M.V., 1957. Some aspects of heat pump operation in Great Britain. Proc. IEE A Power Eng. 104 (15), 262–271. https://doi.org/10.1049/pi-a.1957.0068.

Griffiths, M., 1974. 'After profligacy'-3: heat pump potential. Built Environ. 3 (7), 360–362.

GSHPA, 2023. Vertical Borehole Standard. Issue 3.0. Ground Source Heat Pump Association (UK). March 2023.

Gudmundsson, O., Schmidt, R.-R., Dyrelund, A., Thorsen, J.E., 2022. Economic comparison of 4GDH and 5GDH systems—using a case study. Energy 238, 121613. https://doi.org/10.1016/j.energy.2021.121613.

Hahn, F., Bussmann, G., Jagert, F., Ignacy, R., Bracke, R., Seidel, T., 2018. Reutilization of mine water as a heat storage medium in abandoned mines. In: Proc. 11th ICARD/IMWA/MWD Conference—"Risk to Opportunity". International Mine Water Association. article 1057 https://www.imwa.info/docs/imwa_2018/IMWA2018_Hahn_1057.pdf. Accessed on 21 February 2024.

Hersbach, H., Bell, B., Berrisford, P., Hirahara, S., Horányi, A., Muñoz-Sabater, J., Simmons, A., 2020. The ERA5 global reanalysis. Q. J. R. Meteorol. Soc. 146, 1999–2049.

Hofmann, F., Hampp, J., Neumann, F., Brown, T., Hörsch, J., 2021. Atlite: a lightweight Python package for calculating renewable power potentials and time series. J. Open Sour. Softw. 6 (62), 3294.

Holmberg, H., 2016. Transient Heat Transfer in Boreholes with Application to Non-Grouted Borehole Heat Exchangers and Closed Loop Engineered Geothermal Systems. Norwegian University of Science and Technology (NTNU), Department of Energy and Process Engineering, p. 76. Doctoral thesis 2016 https://ntnuopen.ntnu.no/ntnu-xmlui/bitstream/handle/11250/2385716/Henrik_Holmberg_PhD.pdf.

Holstenkamp, L., Meisel, M., Neidig, P., Opel, O., Steffahn, J., Strodel, N., Lauer, J.J., Vogel, M., Degenhart, H., Michalzik, D., Schomerus, T., 2017. Interdisciplinary review of medium-deep aquifer thermal energy storage in North Germany. Energy Procedia 135, 327–336.

Homuth, S., Hornich, W., Krenn, H., Sass, I., Spahn, T., 2016. Down-the-hole water-powered hammer drilling method for medium-deep Geothermal energy drilling. Oil Gas-Eur. Mag. 42 (1), 39–41.

Howell, L., Brown, C.S., Egan, S.S., 2021. Deep geothermal energy in northern England: insights from 3D finite difference temperature modelling. Comput. Geosci. 147, 104661.

Huenges, E., Moeck, I., Geothermal Project Group, 2007. Directional drilling and stimulation of a deep sedimentary geothermal reservoir. Sci. Drill. 5, 47–49.

IGSHPA, 2017. Closed-Loop/Geothermal Heat Pump Systems—Design and Installation Standards 2017. International Ground Source Heat Pump Association/Oklahoma State University.

Kallesøe, A.J., Vangkilde-Pedersen, T., Guglielmetti, L., 2020. HEATSTORE—underground thermal energy storage (UTES)—state of the art, example cases and lessons learned. In: Proceedings World Geothermal Congress, p. 1.

Kelly, J.J., McDermott, C.I., 2022. Numerical modelling of a deep closed-loop geothermal system: evaluating the Eavor-loop. AIMS Geosci. 8 (2), 175–213.

Klyapovskiy, S., You, S., Michiorri, A., Kariniotakis, G., Bindner, H., 2019. Incorporating flexibility options into distribution grid reinforcement planning: a techno-economic framework approach. Appl. Energy 254, 113662. https://doi.org/10.1016/j.apenergy.2019.113662.

Kolditz, O., Bauer, S., Bilke, L., Böttcher, N., Delfs, J.O., Fischer, T., Görke, U.J., Kalbacher, T., Kosakowski, G., McDermott, C.I., Park, C.H., 2012. OpenGeoSys: an open-source initiative for numerical simulation of thermo-hydro-mechanical/chemical (THM/C) processes in porous media. Environ. Earth Sci. 67, 589–599.

Kolo, I., Brown, C., Lyden, A., Watson, S., Banks, D., Falcone, G., Friedrich, D., 2023. A modelling study of seasonal borehole thermal energy storage in Scotland: integrating surface demand and the subsurface heat store. In: World Geothermal Congress 2023, Beijing, China, 4–6 September, 2023.

Korhonen, K., Leppäharju, H.P., Arola, T., 2018. Simulated temperature evolution of large BTES—case study from Finland. In: Proc. 2018 IGSHPA Research Conference, Sweden., https://doi.org/10.22488/okstate.18.000033.

Lazzarotto, A., Mazzotti Pallard, W., Richert, T., Vautrin, A., 2021. Improved Borehole Technology for Geothermal Heat Pumps Development. Project Report 43692-1, Stockholm, Sweden, KTH and Energimydigheten. https://www.liveinlab.kth.se/polopoly_fs/1.1049873.1613474011!/Report_Improved_Borehole_Tech_final.pdf.

Le Lous, M., Larroque, F., Dupuy, A., Moignard, A., 2015. Thermal performance of a deep borehole heat exchanger: insights from a synthetic coupled heat and flow model. Geothermics 57, 157–172.

Ledingham, P., Cotton, L., 2021. The united downs deep geothermal power project. In: Proceedings of the World Geothermal Congress, Reykavik, Iceland.

Li, C., Guan, Y., Wang, X., 2018. Study on reasonable selection of insulation depth of the outlet section of vertical deep-buried U-bend tube heat exchanger. Energy Build 167, 231–239. https://doi.org/10.1016/j.enbuild.2018.02.047. https://www.sciencedirect.com/science/article/pii/S0378778817328293.

Loredo, C., Ordóñez, A., Garcia-Ordiales, E., Álvarez, R., Roqueñí, N., Cienfuegos, P., Peña, A., Burnside, N.M., 2017. Hydrochemical characterization of a mine water geothermal energy resource in NW Spain. Sci. Total Environ. 576, 59–69. https://doi.org/10.1016/j.scitotenv.2016.10.084.

Lund, H., Werner, S., Wiltshire, R., Svendsen, S., Thorsen, J.E., Hvelplund, F., Mathiesen, B.V., 2014. 4th Generation District Heating (4GDH). Energy 68, 1–11. https://doi.org/10.1016/j.energy.2014.02.089.

Mahmoud, M., Ramadan, M., Pullen, K., Abdelkareem, M.A., Wilberforce, T., Olabi, A.-G., Naher, S.A., 2021. Review of grout materials in geothermal energy applications. Int. J. Thermofluids 10, 100070. https://doi.org/10.1016/j.ijft.2021.100070.

Malek, A.E., Adams, B.M., Rossi, E., Schiegg, H.O., Saar, M.O., 2022. Techno-economic analysis of advanced geothermal systems (AGS). Renew. Energy 186, 927–943.

Mazzotti, W., Acuña, J., Lazzarotto, A., Palm, B., 2018. Deep Boreholes for Ground Source Heat Pump. Swedish Energy Agency Project 40934-1 Report, Energimyndigheten. http://www.diva-portal.org/smash/get/diva2:1269108/FULLTEXT01.pdf.

MCS, 2021a. The Heat Pump Standard: Installation. MCS Standards Document MIS3005-I Issue 1.0, Microgeneration Certification Scheme (MCS) Charitable Foundation, Daresbury, Cheshire, UK. https://mcscertified.com/wp-content/uploads/2021/12/MIS-3005-I-Heat-Pump-Installation-Issue-1.0.pdf.

MCS, 2021b. GSHP hydraulics design guide. Chapter 3, In: Heat Pump Reference Information and Tools for Installers, Certification Bodies and Manufacturers. Microgeneration Certification Scheme (MCS) Charitable Foundation, Daresbury, Cheshire, UK. MCS Guidance Document MGD 007 Issue 1.0 https://mcscertified.com/wp-content/uploads/2021/10/MGD-007-Reference-Information-and-Tools-Issue-1.0.pdf.

Meer met Bodemenergie (MMB), 2012. Aquifer Thermal EnergyStorage in the Netherlands, a research programme (2010–2012). In: Extended English Summary of a Report by the Dutchresearch programme MMB. Meer Met Bodemenergie.

Mégel, T., 2014. Introduction to geothermal energy technologies. In: The National Geothermal Resource Assessment for Lebanon, Mövenpick Hotel, Beirut, 24th March 2014.

Melinder, Å., 2010. Properties of Secondary Work Fluids for Indirect Systems: Secondary Refrigerants or Coolants. International Institute of Refrigeration, Paris, France.

Met Office, 2022. Historic Station Data: Paisley. [Online]. Available at: https://www.metoffice.gov.uk/pub/data/weather/uk/climate/stationdata/paisleydata.txt. Accessed on 2 October 2022.

Mijnlieff, H.F., Obdam, A.N.M., van Wees, J.D.A.M., Pluymaekers, M.P.D., Veldkamp, J.G., 2014. TNO 2014 R11396 DoubletCalc 1.4 Manual English Version for DoubletCalc 1.4.3. https://www.nlog.nl/sites/default/files/6ab98fc3-1ca1-4bbe-b0a2-c5a9658a3597_doubletcalc%20v143%20manual.pdf. Accessed on 30 May 2024.

Moeck, I.S., 2014. Catalog of geothermal play types based on geologic controls. Renew. Sust. Energ. Rev. 37, 867–882.

Monaghan, A.A., Starcher, V., Barron, H.F., Shorter, K.M., Walker-Verkuil, K., 2020. Borehole GGA02, UK Geoenergy Observatory. British Geological Survey, Glasgow, Nottingham, UK, p. 31 (OR/20/022) (Unpublished).

Monaghan, A.A., Starcher, V., Barron, H.F., Shorter, K., Walker-Verkuil, K., Elsome, J., Kearsey, T., Arkley, S., Hannis, S., Callaghan, E., 2022. Drilling into mines for heat: geological synthesis of the UK Geoenergy Observatory in Glasgow and implications for mine water heat resources. Q. J. Eng. Geol. Hydrogeol. 55 (1), 1–19.

Morchio, S., Fossa, M., 2019. Thermal modeling of deep borehole heat exchangers for geothermal applications in densely populated urban areas. Therm. Sci. Eng. Prog. 13, 100363.

Neymeyer, A., Williams, R.T., Younger, P.L., 2007. Migration of polluted mine water in a public supply aquifer. Q. J. Eng. Geol. Hydrogeol. 40, 75–84. https://doi.org/10.1144/1470-9236/05-060.

OFGEM, 2020. Non-Domestic Renewable Heat Incentive (RHI). Office of Gas and Electricity Markets, London, UK. Available online: https://www.ofgem.gov.uk/environmental-programmes/non-domestic-rhi. Accessed on 25 June 2020.

Olsson, J., 2018a. Möjligheter och utmaningar med djupare borrhål [Possibilities and challenges with deeper boreholes—in Swedish]. Svensk Geoenergi 2018 (2), 14–16. http://media.geoenergicentrum.se/2018/11/svgeo_2018_02.pdf.

Olsson, J., 2018b. Koaxial en vinnare när Schweiz går på djupet [Coaxial a winner as Switzerland goes deep—in Swedish]. Svensk Geoenergi 2018 (2), 17–18. http://media.geoenergicentrum.se/2018/11/svgeo_2018_02.pdf.

Olsson, J., 2018c. Norge: djupare borrning kräver teknikskifte [Norway: deeper drilling requires different technology—in Swedish]. Svensk Geoenergi 2018 (2), 20–21. http://media.geoenergicentrum.se/2018/11/svgeo_2018_02.pdf.

Pashby, T., 2024. Water in disused coal mine could become low carbon heat source across Great Britain. In: New Civil Engineer. 25th January 2024 https://www.newcivilengineer.com/latest/water-in-disused-coal-mines-could-become-low-carbon-heat-source-across-great-britain-25-01-2024/.

Pauschinger, T., 2016. Solar thermal energy for district heating. In: Advanced District Heating and Cooling (DHC) Systems. Elsevier, pp. 99–120, https://doi.org/10.1016/B978-1-78242-374-4.00005-7.

Peralta Ramos, E., Breede, K., Falcone, G., 2015. Geothermal heat recovery from abandoned mines: a systematic review of projects implemented worldwide and a methodology for screening new projects. Environ. Earth Sci. 73, 6783–6795. https://doi.org/10.1007/s12665-015-4285-y.

Prestel, R., Stier, P., Jobmann, M., Schulz, R., Strayle, G., Wendebourg, J., Werner, J., Wolf, M., Eichinger, L., Salvamoser, J., Weise, S., Fritz, P., 1991. Hydrogeothermische Energiebilanz und Grundwasserhaushalt des Malmkarstes im süddeutschen Molassebecken. Bayer LfW und LGRB, München, Freiburg.

Ramos, E.P., Falcone, G., 2013. Recovery of the geothermal energy stored in abandoned mines. In: Clean Energy Systems in the Subsurface: Production, Storage and Conversion: Proceedings of the 3rd Sino-German Conference "Underground Storage of CO2 and Energy", Goslar, Germany, 21–23 May 2013. Springer, Berlin Heidelberg, pp. 143–155.

Reinecker, J., Gutmanis, J., Foxford, A., Cotton, L., Dalby, C., Law, R., 2021. Geothermal exploration and reservoir modelling of the united downs deep geothermal project, Cornwall (UK). Geothermics 97, 102226.

Revesz, A., Jones, P., Dunham, C., Davies, G., Marques, C., Matabuena, R., Scott, J., Maidment, G., 2020. Developing novel 5th generation district energy networks. Energy 201, 117389. https://doi.org/10.1016/j.energy.2020.117389.

Rybach, L., Hopkirk, R.J., 1995. Shallow and deep borehole heat exchangers – Achievements and prospects. In: Proc. World Geothermal Congress 1995, Italy, pp. 2133–2138. https://www.geothermal-energy.org/pdf/IGAstandard/WGC/1995/3-rybach.pdf.

Sanner, B., 2005. 60 Jahre erdgekoppelte Wärmepumpe. Geotherm. Energie 49, 30–33. https://sanner-online.de/media/5de5640de29e9c71ffff8019ffffff1.pdf.

Sanner, B., Hahne, E., Konstantinidou, E., 1997. VDI Guideline 4640: Harnessing of Underground Thermal Energy; VDI-Richtlinie 4640 Thermische Nutzung des Untergrundes. Verein Deutscher Ingenieure e.V., Düsseldorf, Germany.

Sanner, B., Karytsas, K., Abry, M., Coelho, L., Goldbrunner, J., Mendrinos, D., 2007. GROUNDHIT—advancement in ground source heat pumps through EU support. In: Proceedings European Geothermal Congress 2007, Unterhaching, Germany, 30th May—1st June 2007. Article 121 https://www.geothermal-energy.org/pdf/IGAstandard/EGC/2007/121.pdf.

Sapinska-Sliwa, A., Rosen, M.A., Gonet, A., Sliwa, T., 2015. Deep borehole heat exchangers—a conceptual review. In: Proc. World Geothermal Congress 2015, Melbourne, Australia, Paper 28045. https://www.geothermal-energy.org/pdf/IGAstandard/WGC/2015/28045.pdf.

Schintgen, T.V., Moeck, I.S., 2021. The interplay of Malm carbonate permeability, gravity-driven groundwater flow, and paleoclimate—implications for the geothermal field and potential in the Molasse Basin (southern Germany), a foreland-basin play. Geotherm. Energy 9, 1–43.

SiL, 2024. Ecoquartier des Plaines-du-Loup. Web Page of City of Lausanne. Services Industriels Lausanne. https://www.lausanne.ch/vie-pratique/energies-et-eau/services-industriels/a-propos-sil/nos-activites/chauffage-a-distance/production-energie-renouvelable/geothermie-moyenne-profondeur.html. Accessed on 14 February 2024.

Singh, R.K., Sharma, R.V., 2017. Numerical analysis for ground temperature variation. Geotherm. Energy 5, 22. https://doi.org/10.1186/s40517-017-0082-z.

Śliwa, T., Kruszewski, M., Assadi, M., Sapińska-Śliwa, A., 2017. The application of vacuum insulated tubing in deep borehole heat exchangers. Akad. Górniczo-Hutnicza Drill. Oil Gas 34 (2), 597–617. https://doi.org/10.7494/drill.2017.34.2.597.

Sofyan, S.E., Hu, E., Kotousov, A., Riayatsyah, T.M.I., 2020. A new approach to modelling of seasonal soil temperature fluctuations and their impact on the performance of a shallow borehole heat exchanger. Case Stud. Therm. Eng. 22, 100781.

Standards New Zealand, 2015. NZS 2403:2015—Code of Practice for Deep Geothermal Wells. Published 20th April 2015.

Sutra, E., Spada, M., Burgherr, P., 2017. Chemicals usage in stimulation processes for shale gas and deep geothermal systems: a comprehensive review and comparison. Renew. Sust. Energ. Rev. 77, 1–11.

Sveinbjörnsson, D., Jensen, L.L., Trier, D., Bava, F., Hassine, I.B., 2019. Large Storage Systems for DHC Networks (No. 2.3), FLEXYNETS Project. European Commission—Research & Innovation.

Tester, J.W., Reber, T.J., Beckers, K.F., Lukawski, M.Z., 2016. Deep geothermal energy for district heating. In: Advanced District Heating and Cooling (DHC) Systems. Elsevier, pp. 75–98, https://doi.org/10.1016/B978-1-78242-374-4.00004-5.

The Engineer, 2023. In-Road Ground Source Heat Pump Network Launches. The Engineer. 24 Mar 2023 https://www.theengineer.co.uk/content/news/in-road-ground-source-heat-pump-network-launches/.

Thorsen, J.E., Lund, H., Mathiesen, B.V., 2018. Progression of District Heating—1st to 4th Generation. Aalborg University, Denmark.

Toews, M., Holmes, M., 2021. Eavor-lite performance update and extrapolation to commercial projects. GRC Trans. 45, 86–109.

Triple Point Heat Networks, 2023. Gateshead District Energy Scheme. Information brochure https://tp-heatnetworks.org/wp-content/uploads/2023/08/gateshead-district-energy.pdf. Accessed on 14 February 2024.

Ungemach, P., 2003. Reinjection of cooled geothermal brines into sandstone reservoirs. Geothermics 32, 743–761. https://doi.org/10.1016/S0375-6505(03)00074-9.

Verhoeven, R., Willems, E., Harcouët-Menou, V., De Boever, E., Hiddes, L., Op't Veld, P., Demollin, E., 2014. Minewater 2.0 project in Heerlen the Netherlands: transformation of a geothermal mine water pilot project into a full scale hybrid sustainable energy infrastructure for heating and cooling. Energy Procedia 46, 58–67.

Vering, C., Maier, L., Breuer, K., Krützfeldt, H., Streblow, R., Müller, D., 2022. Evaluating heat pump system design methods towards a sustainable heat supply in residential buildings. Appl. Energy 308, 118204. https://doi.org/10.1016/j.apenergy.2021.118204.

Walls, D.B., Banks, D., Boyce, A.J., Burnside, N.M., 2021. A review of the performance of minewater heating and cooling systems. Energies 14 (19), 6215.

Watson, S.M., Falcone, G., Westaway, R., 2020. Repurposing hydrocarbon wells for geothermal use in the UK: the onshore fields with the greatest potential. Energies 13 (14), 3541.

Weber, J., Ganz, B., Schellschmidt, R., Sanner, B., Schulz, R., 2015, April. Geothermal energy use in Germany. In: Proceedings World Geothermal Congress, pp. 19–24.

Wirtén, L., 2017. Geoenergidagen lyfter potentialen [GeoEnergy day raises potential—in Swedish]. Svensk Geoenergi 2017 (2), 24–26. http://media.geoenergicentrum.se/2017/11/SvGeo_2_2017-web.pdf.

Wirtén, L., 2018. Hög termisk kortslutning I djupa borrhål [High thermal short-circuiting in deep boreholes—in Swedish]. Svensk Geoenergi 2018 (2), 18–19. http://media.geoenergicentrum.se/2018/11/svgeo_2018_02.pdf.

Wirtz, M., Kivilip, L., Remmen, P., Müller, D., 2019. Bidirectional low temperature networks design methodology based on mathematical optimisation. J. Phys.: Conf. Ser. 1343, 012111.

Wirtz, M., Kivilip, L., Remmen, P., Müller, D., 2020. Quantifying demand balancing in bidirectional low temperature Networks. Energy Build. 224, 110245. https://doi.org/10.1016/j.enbuild.2020.110245.

Witte, F., Tuschy, I., 2020. TESPy: thermal engineering systems in python. J. Open Sour. Softw. 5 (49), 2178.

Wood, J., Crooks, J., 2020. Mine energy at Seaham Garden Village. In: Presented at the Seaham Garden Village Supply Chain Engagement Day. 29 January 2020 https://www2.groundstability.com/wp-content/uploads/2020/02/02-03-Tolent-and-Coal-Authority.pdf. Accessed on 14 February 2024.

Younger, P.L., 1993. Possible environmental impact of the closure of two collieries in county Durham. Water Environ. 7 (5), 521–531. https://doi.org/10.1111/j.1747-6593.1993.tb00881.x.

Younger, P.L., Manning, D.A., Millward, D., Busby, J.P., Jones, C.R., Gluyas, J.G., 2016. Geothermal exploration in the Fell Sandstone Formation (Mississippian) beneath the city centre of Newcastle upon Tyne, UK: the Newcastle Science Central deep geothermal borehole. Q. J. Eng. Geol. Hydrogeol. 49 (4), 350–363.

Zarrella, A., Capozza, A., De Carli, M., 2013. Analysis of short helical and double U-tube borehole heat exchangers: a simulation-based comparison. Appl. Energy 112, 358–370.

Zarrouk, S.J., Moon, H., 2014. Efficiency of geothermal power plants: a worldwide review. Geothermics 51, 142–153.

Zatonski, V., Brown, C., 2023. Eavor-Lite update after four years of operation. GRC Trans. 47, 2023.

Zhang, Y., Johansson, P., Kalagasidis, A.S., 2023. Quantification of overlapping heating and cooling demand for the feasibility assessment of bi-directional systems over Europe. Energy Build. 294, 113244. https://doi.org/10.1016/j.enbuild.2023.113244.

Zogg, M., 2008. History of Heat Pumps: Swiss Contributions and International Milestones, Process and Energy Engineering. Oberburg/Swiss Federal Office of Energy, Bern, Switzerland, p. 114. https://www.osti.gov/etdeweb/servlets/purl/21381633.

CHAPTER

8

Stimulation in enhanced geothermal systems

Carlos A. Fernandez[a], *Zihao Li*[a], *Chao Zeng*[a], *Guoqing Jian*[b], *and Cheng Chen*[c]

[a]Pacific Northwest National Laboratory, Richland, WA, United States [b]Aramco Americas, Houston, TX, United States [c]Stevens Institute of Technology, Hoboken, NJ, United States

1 Introduction

In the United States alone, the total heat generated from enhanced geothermal systems (EGS) can be a viable source of energy, with a scale of about 1.4×10^{25} J (Duan et al., 2016). As described in the previous chapters, the main goal in EGS projects is to harness the heat stowed in deep, hot, and exceptionally low-permeability rocks. The concept of EGS was patented and developed in the early 1970s at the Los Alamos National Laboratory (LANL), creating the world's first installation in Fenton Hill. The practical HDR resource, as defined by LANL scientists, is the heat contained in the vast regions of the earth's crust, likely in granitic-type crystalline rocks, where water is not naturally present (this is now known not to be the case), other than the ones trapped in the highly impermeable formation (Brown, 2009). The concept of LANL was developed for the purpose of producing electricity. Nowadays, EGS include all geothermal reservoirs that require stimulation or permeability enhancement (Duan et al., 2016).

The economic success of EGS operations depends fundamentally on the opening of existing natural fractures and/or the creation and propagation of new fractures in formation rock at temperatures >150°C. In this fashion, a natural radiator/heat exchanger is created, which allows circulation of fluid (working fluid) with flow rates >5 m^3/min (85 kg/s) and heat exchange over 40 or more years (Clauser, 2006). To achieve such flow rates, the host rock is subjected to stimulation operations, to connect drilled wells. Stimulation, more commonly referred to as fracturing operations, can be performed on a single well followed by drilling a second (or third) well to intersect the stimulated volume of rock with or without additional

Geothermal Energy Engineering
https://doi.org/10.1016/B978-0-443-21662-6.00006-2

Copyright © 2025 Elsevier Inc. All rights are reserved, including those for text and data mining, AI training, and similar technologies.

stimulation in the new wells (Xing et al., 2021). A triplet configuration, consisting of two production wellbores and one injector, is often considered effective. However, the best setup will vary based on various factors, including reservoir characteristics, target temperature gradients, and economic considerations (Duan et al., 2016; Grant, 2011). Furthermore, it is critical to keep from extending the size of an EGS reservoir too much through stimulation to avoid significant losses of working fluid (typically water) that compromises the sustainability of the EGS project (Duan et al., 2016; Brown, 2009). An important factor to consider when developing an EGS project is to reduce the geothermal resource sustainability risk, i.e., the potential challenges and uncertainties associated with maintaining the long-term viability and productivity of geothermal resources. To do so, a combination of technical strategies, careful planning, monitoring, and regulatory compliance, specific to stimulation operations, include the following (McClure and Horne, 2013; Baisch et al., 2015; Maury et al., 2022): (1) the use of controlled hydraulic fracturing methods to create a network of fractures that enhance the permeability while minimizing the risk of induced seismicity. Techniques like multistage fracturing and low-rate fluid injection can help manage fracture propagation. (2) Implementation of real-time seismic monitoring during stimulation to detect and respond to induced seismic events quickly. This can involve adjusting injection rates or volumes to mitigate seismic risks. This also extends to the preparation of emergency response plans to address potential seismic events or other incidents, including ensuring that all personnel are trained and that the equipment is readily available. (3) Environmental protection measures which include (a) water management, i.e., the use sustainable water sourcing practices, such as using nonpotable water or treated wastewater and executing measures to prevent contamination of surface and groundwater resources; (b) waste handling, i.e., properly handle and dispose of any waste materials, such as flowback fluids, in accordance with environmental regulations; and (c) ensuring compliance with all relevant local, regional, and national regulations, obtaining the necessary permits, and conducting stimulation activities within regulatory guidelines.

Stimulation technologies have been derived from unconventional (tight) oil and gas recovery. Tensile hydraulic fracturing has not historically been considered the predominant method in geothermal. Recently, it has seen renewed interest following a prolonged period during which shear stimulation was the dominant concept. To design a stimulation plan, operators need to have in hand the tectonic and hydrogeological conditions of the field as well as a detailed analysis of several parameters including mechanical properties, petrographics, and petrophysics of the formation rock. Stimulation operations in EGS require a number of steps, including transportation of fluid components, fracturing rig mobilization to the well, fracturing fluid mixing, injection of fracturing fluid into the wellbore, produced water handling, and wastewater disposal and reuse. First, water, chemicals, and proppants are transported and stored at the wellsite. Simultaneously, fracturing equipment is transported and set up at the wellsite. Then, the fracturing fluid is mixed onsite and pumped into the well. The pumping rates and pressure need to reach values that are high enough, so that inducted stresses in the formation rock will exceed the critical stresses in reopening existing natural fractures or initiating and propagating new fractures. To avoid fracture closure, proppants are added in the frac mix, so the increase in hydraulic conductivity between the reservoir and the wellbore is maintained post-stimulation (Economides and Martin, 2007; Wasantha et al., 2021). The use of proppants in EGS is not common but was agreed to be necessary for long-term energy production. Nevertheless, further pumping leads to fracture propagation deeper in the reservoir in vertical and horizontal directions (Rafiee et al., 2012). The fracturing job is performed in several stages (typically 10–15 stages) to ensure adequate coverage of the EGS reservoir. Usually, stimulation occurs at the point of injection to complete the flow of the geofluid from the injection well to a production well. Slickwater is used without proppants in the first and last (flush) stage, while, in all the intermediate stages, proppants with two different sizes (40/70 and 100 mesh) are used, following a similar approach in stimulation operations in unconventional (tight) oil reservoirs. After these stages, and to enable complete circulation of a working fluid, additional wells are constructed within the well field. Additional stimulation of these wells may also be necessary to attain economic flow rates (>5 m^3/min).

The process of fracturing ends by stopping the pumping of fluid, which causes a slow pressure drop due to the filtration of fluid through the porous matrix of the formation rock. Then, the proppant that is distributed in the fracture network prevents the closure of the fractures, as the formation stresses begin to outweigh the inducted pressure. This highly permeable veins will then communicate the wellbore with other drilled wells, provided they intersect the propped fracture network. Once stimulation is over, pumps and the rest of the equipment are removed from the wellsite, and flow-testing equipment is installed. The ultimate technological goal/challenge during stimulation operations is the prediction of the geometry of the fractures and the assessment of the permeability of the fractures and of the propagation direction between wells. In this fashion, high hydraulic conductivity fractures are attained while simultaneously avoiding thermal short-circuiting, i.e., a decrease in temperature of the received liquid because of working fluid flowing through a single, major, fracture (McClure and Horne, 2014).

Typical water volumes, flow rates, and length of time for these steps are shown in Table 1 (Duan et al., 2016; Zimmermann et al., 2010; Asanuma et al., 2004; Michelet and Toksoz, 2006). The volume of water required, including circulation tests, for all stimulation activities in a single well range between 40,000 and 60,000 m^3, assuming no water is reused. These volumes are up to five times greater than in the US hydraulic fracturing operations in unconventional oil and gas recovery (7570 and 15,140 m^3) (Carpenter, 2013). Operators can argue with reason that these volumes of water are significant; however, compared with water loss in the formation associated to circulating water as a working fluid during geothermal plant operations, the volumes for hydraulic fracturing represent only a fraction of working fluid losses over the lifetime of the reservoir.

Another important factor that is considered when it comes to EGS stimulation operations and geothermal energy production is induced seismicity. Furthermore, a barrier to the development of geothermal energy is associated to injection-induced seismicity. This is because of the difficulty to assess, predict, and control induced seismicity during both stimulation and, later, geothermal energy extraction due to the complexity of fault behaviors and the lack of understanding of coupled thermo-hydro-mechanical-chemical (THMC) processes in EGS. Injection-induced seismicity in EGS generates public concern and remains a common topic of discussion in the scientific, political, and social spheres (Ellsworth, 2013; Elsworth et al., 2016). Although most of the seismic events associated with geothermal energy development are micro-seismic (moment magnitude $Mw < 3$), large magnitude-induced earthquakes have occasionally been recorded (Kim et al., 2018; Knoblauch and Trutnevyte, 2018). Understanding the THMC processes is critical in evaluating the probability and effect of injection-induced seismicity. Cyclic soft stimulation, chemical stimulation with permeability enhancement based on mineral dissolution, and feedback from borehole seismometers have been successfully used to mitigate the risks associated with fluid injection. However, occasionally, the activation of nearby critically stressed faults is uncontrollable, independently of volumes of fluid injected.

The reader is encouraged to access reference Gaucher et al. (2015) for a complete overview of the current approaches applied to injection-induced seismicity in EGS, and reference Evans et al. (2012), for a comprehensive review of over 40 European case histories of injection-induced seismicity in geothermal reservoirs.

TABLE 1 Typical volumes of water required for stimulation per well (McClure and Horne, 2013; Baisch et al., 2015; Economides and Martin, 2007; Wasantha et al., 2021; Rafiee et al., 2012).

Step	Flow rate (kg/s)	Volume (m^3)	Length of time (days)
Pre-stimulation test	5–7	400–600	1
Main stimulation	30–70	13,000–58,000	1–6.4
Post-stimulation	7–50	7200	2.5
Short-term circulation	20	2600–3600	21
Long-term circulation	50–100	4000–13,000	21

There are several differences when it comes to stimulation in an EGS reservoir as compared with stimulation in an unconventional oil (or gas) reservoir. EGS stimulation occurs at greater depths and at higher temperatures, requires a larger magnitude of stresses, and takes place in crystalline, igneous rocks, rocks that will become a heat exchanger post-stimulation (McClure and Horne, 2013).

In addition to water-based hydraulic fracturing, other stimulation methods derived from stimulation in tight oil and gas reservoirs, include (a) chemical stimulation (Nami et al., 2008a; Portier et al., 2009), which uses a range of chemicals, including acids and chelating agents, to dissolve minerals and, as a result, increase permeability, and (b) thermal stimulation based on the generation of new fractures by cold water–induced thermal shock. These methods will be described in detail in Section 2. In addition to these methods, state-of-the-art stimulation approaches have been proposed and studied at laboratory- and pilot-scale and include (c) foam stimulation, (d) stimulation based on CO_2-expanding polymeric (aqueous) fluids, and waterless stimulation methods such as (e) supercritical CO_2, (f) liquid nitrogen, (g) propellant-based stimulation, and (h) electrical pulse stimulation. Section 3 will discuss these stimulation approaches.

Tracers, naturally occurring or artificially synthesized materials, are injected into the reservoir to assess the connectivity and flow rates between wellbores in EGS during and after stimulation operations. Different types of tracers and approaches for their deployment and data analysis can be found in the open literature (Portier et al., 2009; Radilla et al., 2010; Wu et al., 2021a; Hawkins et al., 2018; Egert et al., 2020; Mattson et al., 2019; Sanjuan et al., 2015; Plummer et al., 2016). Tracers have been deployed at multiple EGS sites, including Hijiori (Japan) (Matsunaga et al., 2002, 2001; Yanagisawa et al., 2001), Fenton Hill (United States)

(Callahan, 1996; Duchane and Winchester, 1993; Rodrigues, 1993), Habanero (Australia) (Yanagisawa et al., 2011; Bendall et al., 2014; Yanagisawa, 2009), and Soultz-sous-Forêts (France) (Sanjuan et al., 2015, 2004). These cases have all demonstrated that tracer tests represent reliable methods for interrogating EGS reservoirs and assessing inter-well connectivity and fracture volume. A specific combination of tracers, e.g., the combination of an absorbing tracer and an inert tracer, can not only determine the fracture aperture distribution in a reservoir but also predict the thermal drawdown showing that heat transfer surface area evaluation is also possible using tracer technology (Wu et al., 2021b).

Prior to and after stimulation operations, it is very important to consider in situ stress and fracture characterization to better constrain the geometry and relative permeability of natural and/or artificially generated fractures (McClure and Horne, 2013). This should be followed by modelling flow and transport along the fractures of the 3D fracture network (Grant, 2011; Sausse et al., 2010; Dezayes et al., 2010; Karvounis and Jenny, 2011) based on the accurate knowledge of the fracture network which is obtained from seismic records during stimulation and/or production tests (Concha et al., 2010; Hébert et al., 2010, 2011; Ledésert et al., 2009). The ultimate goal of all these efforts is to anticipate the performance of the thermal exchanger and ensure its financial viability. Nevertheless, when a stimulated geothermal reservoir is operated, the viability of the EGS depends on thorough reservoir engineering and monitoring (Satman, 2011).

The following sections will describe the current advances in stimulation fluids and stimulation operations for the development of EGS.

2 Water-based hydraulic fracturing fluids

EGS are dedicated to the exploitation of the heat present in deep hot rocks that have limited permeability. The formation rock in EGS reservoirs could be composed of shale, basaltic, metamorphic, and/or granite rocks, with reservoir permeability values ranging from 10^{-19} to 10^{-17} m^2 (or 0.001–0.1 mD). Arguably, the economical and sustainable extraction of heat from EGS is only possible if the reservoir permeability is higher than 10^{-16} to 10^{-14} m^2 (0.1–10 mD), to ensure fluid circulation between the injection and production wells (Ghassemi, 2012; Tester et al., 2006a). Therefore, EGS wells must oftentimes be stimulated to increase the productivity of the well. Various stimulation techniques can be employed for this purpose, including hydraulic (Action, E. C, 2008), chemical (Tester et al., 2006b), thermal (Parker, 1999), and even electrical (Tenzer et al., 2004) stimulation, with hydraulic stimulation using water as fracturing fluid being, to the best of our knowledge, the only stimulation method applied in geothermal fields. In the first three stimulation techniques, stimulation fluid is pumped into the wells to expand preexisting fracture or to induce new fractures in the target horizon. For deep geothermal energy systems, primarily, hydraulic stimulation is applied. For this purpose, fresh water (near-surface groundwater or surface water) or previously produced formation fluid is used as stimulation fluid containing specialized chemical additives. In some cases, the cold injectate is pumped into the target reservoir to induce tensile fracture near the injection wells. This thermal stimulation provides more efficient and economic ways to increase permeability since the stimulation fluids (most often is cold water) do not need to be pressurized to high values. In other cases, the acidizing fluids are injected to dissolve sealed minerals in closed fractures and avoid clogging of the circulation pathways. The stimulation techniques for EGS projects are extensively discussed and documented in the literature, case by case. However, compositions of stimulation fluids are very limited in the literature. There is a paucity of published literature on stimulation fluids, highlighting the need to gather dispersed data from various EGS projects. The current section mainly focusses on the water-based stimulation fluids used in the field experiments in EGS. The water-based stimulation fluids are discussed based on their field application, with limited laboratory experiments and simulation work. Other waterless stimulation fluids and fluids that use less water will be covered in the next section. Foams and CO_2-expanding fracturing fluids (i.e., StimuFrac) are a few examples of the latter. Due to the water availability challenges with development of EGS, those waterless stimulation fluids show increasing potential to supplement or even replace current water-based stimulation fluids. The completed and ongoing EGS projects are summarized in Table 2. Note that most of the main EGS (or hot dry rock) projects around the world, as listed in Table 2, were taken in the last five decades. It is not meant to be inclusive, since more EGS projects are startups and some old oil wells are converted to geothermal sites. However, the chronicled summary infers the dominant stimulation techniques and stimulation fluids in field application. For water-based stimulation fluids, water is the main solvent with extra additives for specific needs. Therefore, the additives in water are highlighted based on their needs in the following analysis.

TABLE 2 Geothermal projects and the stimulation fluids in chronicled order.

No.	Country	Project name	Startup year	Stimulation method	Rock type	Stimulation fluids	Reference
1	Italy	Lardarello	1970	Hydraulic/ thermal	Metamorphic rock	Water; cold water	Cornet (1987)
2	USA	Fenton Hill	1974	Hydraulic	Crystalline rock	water	Walters and Dobson (2013)
3	Germany	Falkenberg	1977	Hydraulic	Granite	Water	
4	UK	Rosemanowes	1977	Hydraulic	Granite	Water; high-viscosity gel (700 cP) with bauxite	Gischig et al. (2020)
5	Germany	Bad Urach	1977	Hydraulic	Gneiss	Water	Hahne and Thomas (2014)
6	France	Le Mayet	1978	Hydraulic	Granite	Gel with sand; gel; water	Jupe et al. (1992)
7	USA	Northwest Geysers	1980	Thermal	Metasedimentary rock	Cold water	Portier et al. (2006)
8	Germany	Bruchsal	1983	Hydraulic	sandstone	Water	Portier et al. (2009)
9	Germany	Neustadt-Glewe	1984	Hydraulic	sandstone	Water	Kaieda et al. (2010)

TABLE 2 Geothermal projects and the stimulation fluids in chronicled order—cont'd

No.	Country	Project name	Startup year	Stimulation method	Rock type	Stimulation fluids	Reference
10	Sweden	Fjällbacka	1984	Hydraulic/chemical	Granite	Water; viscous gel (70, 80, and 150 cP) with proppant; HCl–HF with proppant	Nami et al. (2008b), Nami et al. (2007), and Pemecker (1999)
11	Japan	Hijiori	1985	Hydraulic	Granodiorite	Water	Gómez-Díaz et al. (2022)
12	France	Soultz	1987	Hydraulic/chemical	Granite	Brine/freshwater; HCl, RMA, NTA, OCA	Tester et al. (2006b), Nami et al. (2007), and Bertini et al. (2006)
13	Australia	Altheim	1989	Hydraulic/chemical	Limestone	Water; HF	Cornet (1987), Ladner and Häring (2009)
14	Japan	Ogachi	1989	Hydraulic	Granodiorite	Water	Gómez-Díaz et al. (2022)
15	Iceland	Nesjavellir	1990	unknown	Volcanic rocks	Unknown	Burns et al. (2000)
16	France	Bouillante	1996	Thermal	Volcanic lavas and tuffs	Cold water	Zimmermann et al. (2003)
17	Switzerland	Basel	1996	Hydraulic	Granite	Fresh water	Zimmermann et al. (2011))
18	Australia	Hunter Valley	1999	Hydraulic	Granite	Water	Portier et al. (2007)
19	Germany	Groβ Schönebeck	2000	Hydraulic/thermal/chemical	Volcanic rocks	High viscosity gel with proppant; cold water; 7.5% HF	Rose (2013) and Chabora et al. (2012)
20	El Salvador	Berlín	2001	Hydraulic/chemical	Volcanic rock	Water; HCl-HF	Schindler et al. (2010)
21	USA	Coso	2002	Hydraulic/thermal/chemical	Diorite, granodiorite Granite	Water; cold water; HCl and NTA	Pemecker (1999) and Zimmermann et al. (2009)
22	USA	Desert Peak	2002	Hydraulic/chemical	Volcanic and metamorphic rock	Water; SPA, RMA	Blöcher et al. (2016)
23	Germany	Landau	2003	Hydraulic/Chemical	Granite	33% inhibited HCl; fresh water	Riffault et al. (2018)
24	Australia	Cooper Basin	2003	Hydraulic	Granite	Water	Gómez-Díaz et al. (2022)

Continued

TABLE 2 Geothermal projects and the stimulation fluids in chronicled order—cont'd

No.	Country	Project name	Startup year	Stimulation method	Rock type	Stimulation fluids	Reference
25	Germany	Genesys Horstberg	2003	Hydraulic	Sedimentary rock	Water	Breede et al. (2013)
26	Germany	Unterhaching	2004	Chemical	Limestone	HCl	Jia et al. (2022)
27	Australia	Paralana	2005	Hydraulic/Chemical	Metasediments, granite	Water; HCl	Lutz et al. (2011) and Moeck et al. (2015)
28	Germany	Insheim	2007	Hydraulic	Sandstone, granite	Water	Law et al. (2019)
29	USA	Southeast Geysers	2008	Hydraulic	Greywacke	Water	Walters and Dobson (2013)
30	USA	Bradys Hot Spring	2008	Hydraulic/chemical	Rhyolite, tuff	Water	Cladouhos et al. (2012)
31	Germany	Genesys Hannover	2009	Hydraulic	Sandstone	Water	Breede et al. (2013)
32	Switzerland	St. Gallen	2009	Hydraulic/chemical	Limestone	Fresh water; soft acid	Bradford et al. (2017)
33	UK	Redruth	2009	Hydraulic	Granite	Water	Eden (2023)
34	USA	Newberry	2010	Hydraulic	Volcanic rock	Water	Lummer et al. (2015)
35	USA	Raft river	2010	Hydraulic/thermal	Quartz monzonite, schist, quartzite, and siltstone	Water; cold injectate	Schrage et al. (2012)
36	UK	Eden	2010	Electrical/chemical	Sandstone, granite	Biodegradable organic acid	Tenzer et al. (2004) and Baujard et al. (2017)
37	Germany	Mauerstetten	2011	Hydraulic/Chemical	Limestone	Water	Zhong et al. (2022)
38	France	Rittershoffen	2011	Hydraulic	Granite	Water	Wang et al. (2018)
39	China	Gonghe	2013	Hydraulic	Granite	Water	Allis et al. (2016)
40	China	Yilan	2014	Unknown	Metamorphic rock	Unknown	Park et al. (2017)
41	USA	Milford	2015	Hydraulic	Granite, gneiss	Water	Koroncz et al. (2022)
42	South Korea	Pohong	2016	Hydraulic	Granodiorite, gneiss	Water	Xiaofei et al. (2020)

TABLE 2 Geothermal projects and the stimulation fluids in chronicled order—cont'd

No.	Country	Project name	Startup year	Stimulation method	Rock type	Stimulation fluids	Reference
43	Hungary	Szeged	2016	Hydraulic/ chemical	Sandstone	Water; unknown	Toews et al. (2020)
44	China	Matouying	2019	Hydraulic	Granite	Water	Sharma et al. (2004)
45	Canada	Eavor-Lite	2019	Hydraulic	Sandstone	Water	Axelsson et al. (2006)

RMA, regular mud acid; NTA, nitrilotriacetic acid (chelating agent); OCA, organic clay acid; SPA, sulfophthalate (chelating acidic sodium).

2.1 Hydraulic stimulation fluids

Hydraulic stimulation fluids are injected at high flow rates and pressure buildup to open or create fractures in a reservoir. The injected fluids can be freshwater or previously produced formation fluids. The affecting distance for hydraulic stimulation can be over a length scale of several hundred meters around the injection well. Stimulation can either open preexisting fracture or create new tensile fractures. Opening of natural fractures by shearing is the dominant mechanism that provides fluid pathways, in comparison with tensile mechanism of new fracture. In natural fractures, the roughness of fracture planes can resist closure by the self-propping effect. However, for the new tensile fractures, exterior proppants are required to provide support after the release of pressure. Therefore, the viscosity of fluids has a special requirement to carry the heavy proppant in stimulation. High-viscous gels (polymers) are used to transport the proppant agents accurately into the fracture without prior sedimentation. So cross-linker is added to the gel to achieve the required high viscosity. The subsequent reduction of viscosity to enable a flow back of the remaining gel after placement of the proppant agents is achieved by a further additive, breaker. This additive activates automatically after a pre-set time. Both cross-linker and breaker are added to the stimulation water–proppant mixture before injection into the reservoir. The additives with specific purpose are listed in Table 3.

TABLE 3 Additives used in hydraulic stimulation fluids (Jia et al., 2022).

Additive	Definition/purpose	Example
Proppant	Ceramic spherule to keep induced fracture open	SiO_2, Al_2O_3
Gel	Increase viscosity of the stimulation fluids	Guar gum, polymer
Cross linker	Stabilize gel	Borate salts
Cross breaker	Destabilize gel	Ammonium persulfate
Biocide	Prevent activity of microorganisms	Glutaraldehyde
Scale inhibitor	Prevent formation of certain scales	Ethylene glycol, $CaSO_4$
Salt	Prevent clay swelling in the formation	KCl
Oxygen scavenger	Binds dissolved O_2	Ammonium bisulfite
Iron control	Prevent iron precipitation	Citric acid

2.2 Chemical stimulation fluids

Chemical stimulation techniques consist mainly of acid injection into the formation at low fracturing pressures with the aim of removing near-wellbore permeability damage and material deposited in fractures through transport and precipitation processes. The chemical stimulation techniques have been developed maturely in oil and gas industry to improve the productivity of oil and gas wells. These techniques are adopted and applied in geothermal systems.

Various chemical stimulations have been performed in geothermal wells over the last 30 years (Strawn, 1980; Batra et al., 1984; Barelli et al., 1985; Barrios et al., 2002; Aksoy and Serpen, 2010). The most widely used treatment in geothermal wells is injection of mixture of HF and HCl, which is the standard in sandstone-acidizing operations. HCl is selected to treat limestone, dolomite, and calcareous zones (Portier and Vuataz, 2010), whereas HF is used to dissolve clay minerals and silica. In geothermal wells, the strongest indication of acid-removable formation damage is a sharp drop of production rate. Nearly all geothermal wells that are acidizing candidates have been damaged by Ghassemi (2012) drilling mud solids and drill cuttings lost to the formation fractures and (Tester et al., 2006b) scale (calcium carbonate, silica, calcium sulfate, and mixture) (Schindler et al., 2010). A very successful method of acidizing geothermal wells has been a basic and high-rate operation. High-acid concentrations have been shown to be effective in geothermal wells produced from natural fractures that do not contain separate, large carbonate zones, such as the Soultz-Sous-Forets EGS sites in France.

Compared with the conventional chemical stimulation in oil industry, the high-temperature condition and single-phase fluid make some components in the fluid compositions unnecessary. The only acid additives necessary in a geothermal stimulation are (Ghassemi, 2012) corrosion inhibitor and inhibitor intensifier and (Tester et al., 2006b) high-temperature iron-control agent. High temperature in geothermal stimulation reduces the efficiency of corrosion inhibitors and increase their cost, as well as increasing the acid/rock reaction rate. The high-acid rock reaction rate requires the use of a retarded acid system to ensure that acid will not all be spent immediately next to the wellbore but will penetrate deeper into the formation. The standard operation of acidizing stimulation has been established in geothermal systems based on the experimental trials in last three decades. There are three main steps:

(1) A pre-flush, usually with hydrochloric acid (10%). The objective of this pre-flush is to displace the formulation brine and to remove calcium and carbonate materials in the formation. The pre-flush acid minimizes the possibility of insoluble precipitates.

(2) A main flush with hydrochloric–hydrofluoric acid mixture. A mixture of 10% HCl–5% HF (also called regular mud acid, RMA) is generally prepared by dissolving ammonium bifluoride (NH_4HF_2) in HCl. A mixture of 1% HCl and 56 kg of NH_4HF_2 will generate 1% HF solution. Regular mud acid (12% HCl–3% HF) is made from 15% HCl, where 3% HCl is used to hydrolyze the fluoride salts.

(3) A post-flush/over-flush usually by either HCl, KCl, NH_4Cl, or freshwater.

This treatment by HCl–HF mixture has been applied in more than ten geothermal fields (Schindler et al., 2010). The channel connectivity has been significantly improved by more than four times after chemical treatment, based on a statistical summary of geothermal field database. The concentration ratio of HCl and HF can be varied based on the formation mineralogy. Their effect on the precipitation reduction is shown in Fig. 1. Due to reservoir heterogeneities, it is unlikely that the acid pre-flush will remove all calcite. It is sufficient to reduce calcite <6% to avoid precipitation. HCl–HF mixture dissolves clays originating from drilling mud and reacts with most constituents of the sandstone formation. It results from enlargement and interconnections of the pores in the matrix, facilitating the fluid flow. Some other slow-reacting HF acid, such as fluoroboric, fluoroaluminic, and hexa-fluoro-phosphonic, are also used to avoid the drawback from fast reaction of HCl–HF mixture. For example, a new acid system based on phosphonic acid complex has been developed to hydrolyze NH_4HF_2 rather than HCl (Malate et al., 1998). The new acid system can penetrate deeper into the formation away from the borehole.

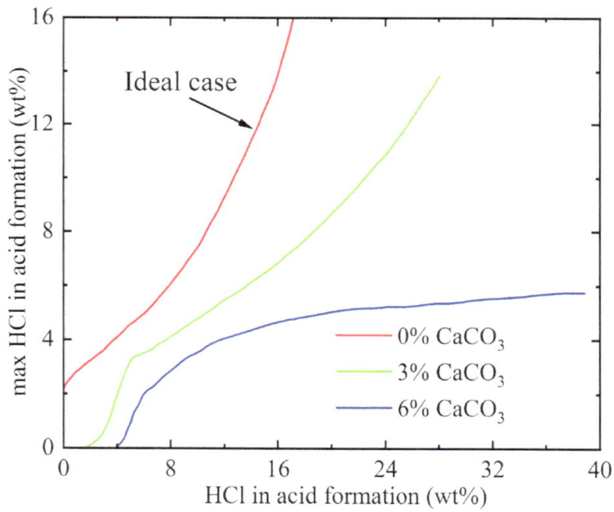

FIG. 1 HCl/HF ratio to avoid precipitation, based on AlF_3 and CaF_3 precipitation (Xu et al., 2021).

One issue for the HCl–HF mixture is the strong corrosion and fast reaction. It would corrode steel tubulars and also cause formation damage by creating wore holes. After strong acid stimulation, another mild stimulation is usually applied to remediate the damage and reach far away from the borehole. The mild acids are chelating agent, such as NTA, DTPA, and DTPA. For example, in the Coso EGS field in the United States, where calcite is the main mineral in the reservoir, 10% NTA solution was injected to form Ca–NTA complex. After alkaline stimulation, the productivity efficiency increased by 66% (Yang et al., 2012). The mud acid (mixture of HCl and HF) has been replaced by mild formic or acetic acid to implement stimulation by core-flooding experiments (Cao et al., 2016). The secondary precipitates have been significantly reduced by this operation.

Some strongly alkaline chemical reagents, such as 10% NaOH or Na_2CO_3, are also used for chemical stimulation of geothermal reservoir. Alkaline chemical reagents have a strong dissolution effect on quartz and calcite and other mineral particles. Its reaction rate is slower than RMA, resulting in a larger transformation area in the reservoir. Some chelating agents are also added to avoid the secondary precipitates. The alkaline reagent can be combined with a chelating agent to dissolve some minerals, such as feldspar and quartz.

Besides laboratory and field experiments, the numerical simulation has also been performed to simulate the effect of acidizing treatment. The chemical stimulation by mud acid (12 wt% HCl + 3 wt% HF) in Soultz EGS site has been simulated by self-developed code Frachem (Portier and Vuataz, 2010; Cao et al., 2016). The mineral dissolution rate of multiple minerals has been predicted as the function of distance from the injection well. Other widely used software to simulate dissolution and precipitation in acidizing treatment are PHREEQC, EQ3/6, WATEQ4F, and TOUGH-REACT (Yang et al., 2012). Table 4 shows some additional chemical stimulation fluids used.

TABLE 4 Chemical stimulants used in the geothermal acidization.

Stimulant	Purpose	Example
Strong alkaline	Quartz-dominated formation	Na_2CO_3, NaOH
Chelating agent	Reduce secondary precipitates	EDTA, NTA, DTPA, PESA, ATMP, GLDA
Strong acid	Carbonate formation	HCl, HF
Weak acid	Reduce secondary precipitates	CO_2, formic acid, acetic acid

EDTA, ethylenediaminetetraacetic acid; NTA, nitrilotriacetic acid; DTPA, diethylenetriaminepentaacetic acid; GLDA, glutamic acid.

2.3 Thermal stimulation fluids

For thermal stimulation of a geothermal well, a fluid (e.g., produced water, seawater, aquifer water, or surface water), considerably colder than the receiving hot formation, is injected. Injection of the cooler water leads to thermal contraction of the reservoir rock in the region near the injection well and reduces the stresses. The reservoir can be fractured at a significantly lower pressure than the original, in situ stress would indicate, when there is a large temperature contrast between the injection water and the formation. Since most hydraulic stimulations are also performed using water, which is colder than the reservoir rock, the boundary between the two methods may not be strictly defined. From the aspect of stimulation fluids, the composition of cold water is less than that in hydraulic stimulation with less additive. Thermal stimulation is not as common as hydraulic or chemical stimulations, since it only influences the area on the scale of several meters around the injection wells. The basic procedures for thermal stimulation is (Parker, 1999):

(1) Injection of cold water (often at elevated pressure): one period of injection of cold water at the well head or more targeted using packers.
(2) Intermittent cold water injection and well heating-up (thermal shocking): cycles of injection of cold water and letting the well heat up again area repeated.

Besides field tests on thermal stimulation, extensive theoretical and laboratory efforts have been made to understand the thermal effect on fracture initiation and propagation (Zeng et al., 2020, 2019). Based on the prior laboratory experiment (Dong et al., 2019), the injected water lags behind the fracture tip in the fracturing process, and there is an opening zone between the water tip and fracture tip. Consequently, the water transport in newly created fracture is slower than the fracture propagation. Therefore, the cooling effect is slower than the hydraulic fracturing if the water injection pressure is higher than the fracture initiation pressure. In this occasion, cold water transports in hydraulic fractures in certain thermal stimulation. The surface of new fracture is cooled by the fluid and results in thermal contraction and increase in fracture width, which in turn cause fracture propagation or formation of secondary cracks. The secondary cracks penetrate the rock matrix perpendicular to the main fracture and increase the permeability of the geothermal reservoir, as shown in Fig. 2. Such secondary cracks can be particularly important to enhance the heat exchange area or increase fluid loss in EGS. For a summary of related work on thermally induced fracture in the context of EGS, please refer to Jahan Bakhsh (2017). In summary, the thermally induced fractures are complimentary to the hydraulically induced fractures in terms of fracture length and distribution.

A successful case of thermal stimulation in a geothermal well in Guadeloupe in France has been reported by Tulinius et al. (2000). A 253°C reservoir was stimulated using seawater mixed with an inhibitor to prevent anhydrite scaling. Production results showed an output increase of 50% compared with the original production flow rate (Flores et al., 2005). It is demonstrated that thermal stimulation is very attractive option compared with the others, in cases where flow can be restored or enhanced by the generation of a relatively near-wellbore fracture network that will reconnect to a main reservoir flow system. More demonstration can be found in the review work (Grant et al., 2013). Two distinctive demonstrations are presented in Fig. 3. The relation of flow rate, wellhead pressure, and temperature difference are shown in Fig. 3A. The permeability enhancement in Fig. 3B is attributed to both hydraulic fracturing and thermal contraction. According to the analysis by Grant et al. (2013), the contribution of the thermal contraction is much greater than those due to pressure changes.

The prominent advantage of thermal stimulation lies in the reduced fracturing pressure and benign fracturing fluids, as compared with other two stimulation methods. In particular, the fluids in thermal stimulation are characterized by (1) benign compared with aggressive acids; (2) easy-to-prepare fluid with simple chemistry, especially when compared with a fully formulated, high-temperature, cross-linked fracturing fluids; (3) required mobilization of a minimum of equipment; (4) high pump pressure not normally required; (5) treatment fluids present minimal health, safety, and environmental issues; and (6) low cost (Flores et al., 2005).

However, thermal stimulation is not always a technically suitable solution. For example, stimulation is required to dissolve the material blocking the flow of steam, and thermal stimulation cannot overcome this. Another concern is the closure of fracture after thermal fracturing due to lack of proppant. Therefore, thermal stimulation presents peculiar advantage in certain conditions and complement the fracture network by hydraulic stimulation.

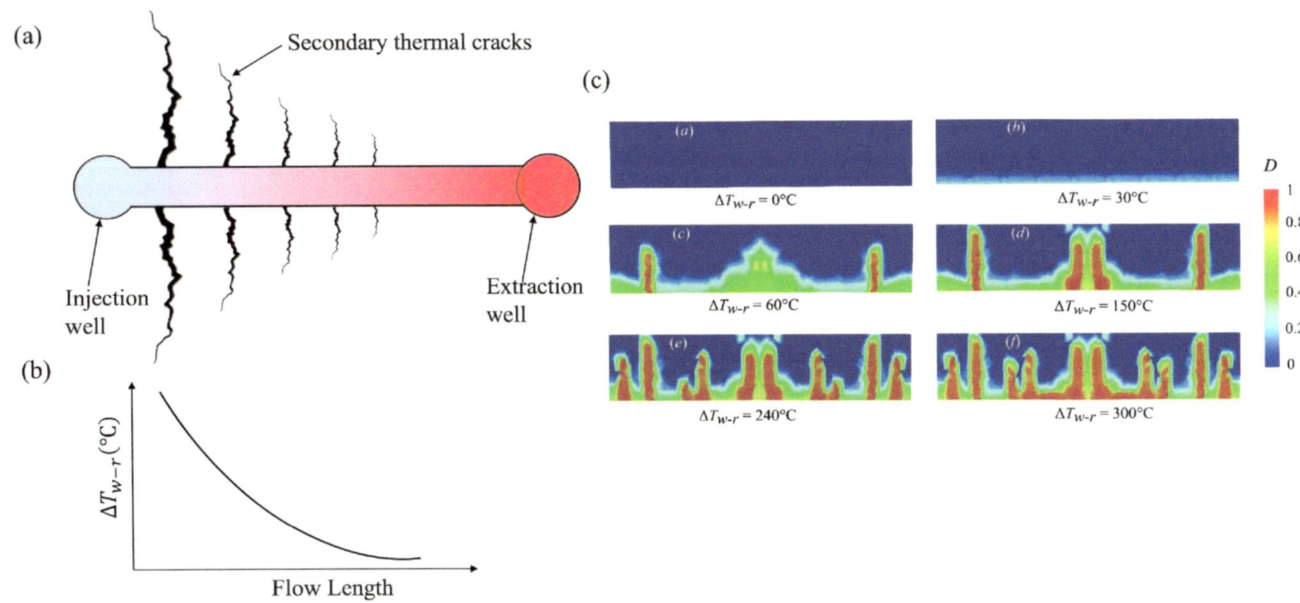

FIG. 2 Thermal cracking: (A) Schematic showing the secondary thermal cracks perpendicular to the main fracture surface, the secondary cracks is exaggerated in sketch. (B) The associated water-rock temperature difference along the flow direction, showing sharply drop. (C) The cooling simulation of half fracture surface, $D = 1$ indicates damage and $D = 0$ indicates intact status. *From Zeng, C., Deng, W., Wu, C., Insall, M., 2020. Thermal effect of cold fluid injection on fracture integrity of host rock: a microscale analysis of single asperity. Geothermics 87, 101872.*

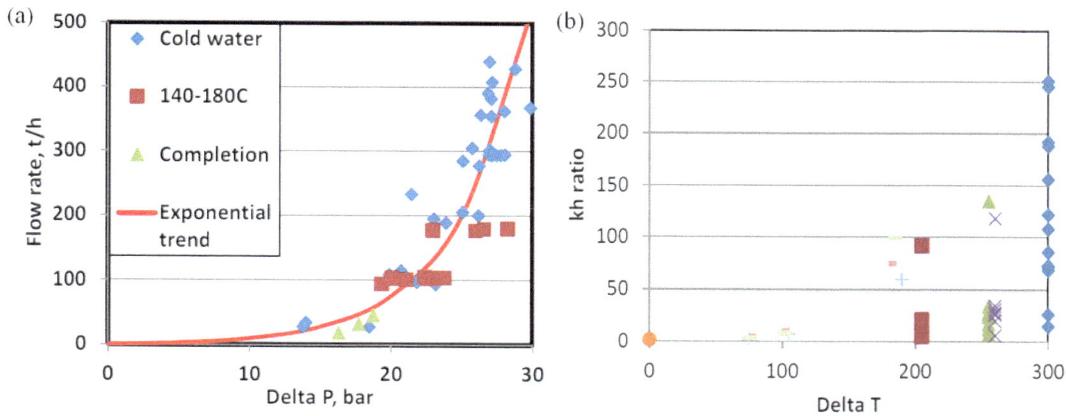

FIG. 3 (A) The relation of injection flow rate and wellhead pressure difference at different temperature differences and (B) the permeability enhancement of geothermal reservoir against temperature difference. Cold water is at 20°C, and completion fluid is also at 20°C. The data from a large number of wells were gathered by Grant et al. (2013).

3 Waterless hydraulic fracturing fluids

The emergence of waterless fracturing techniques signifies a revolutionary leap forward in hydraulic fracturing methodologies, challenging the conventional reliance on large water volumes and ushering in environmentally conscious alternatives. Techniques such as liquid nitrogen (LN_2) and carbon dioxide (CO_2) fracturing, CO_2 foam, and CO_2-responsive fluid fracturing, explosive and propellant fracturing, and electro-hydraulic fracturing collectively contribute to a paradigm shift in the industry. By negating the need for extensive water usage, these methods effectively address concerns related to water scarcity and align with increasingly stringent environmental regulations.

A pivotal advantage of these waterless fracturing approaches lies in their probable capacity to minimize the risk of formation damage near the wellbore while allowing for controlled fracture lengths. This dual benefit not only enhances reservoir performance but also contributes to the sustainable extraction of subsurface resources. However, despite these advantages, challenges persist, requiring a nuanced understanding and ongoing innovation.

Temperature and pressure considerations, limitations in proppant transport, and fluid phase transitions represent persistent challenges within the realm of waterless fracturing. Acknowledging these obstacles, the industry has witnessed promising innovations, including CO_2 foam fracturing. This method introduces a foam nature to the fracturing fluid, optimizing fluid mobility, minimizing loss to the formation, and showcasing the potential for improved fracture initiation and propagation. StimuFrac, an aqueous polymer–CO_2 blend, represents another noteworthy solution, offering a more efficient fluid for permeability enhancement based on lab-scale stimulation experiments (see Section 3.2) and a potentially less water-dependent alternative. This innovative blend demonstrates the commitment to overcome challenges and enhance the efficacy of waterless fracturing techniques. Furthermore, the integration of electro-hydraulic fracturing introduces a new frontier in this domain, leveraging electrical discharges to precisely control fracture initiation and propagation, thereby enhancing overall efficiency.

To fully optimize the effectiveness of these waterless fracturing techniques, a comprehensive understanding of fluid properties, stability, and rheology is paramount. Ongoing research endeavors and technological advancements play a pivotal role in addressing existing challenges and refining these innovative methods. This collective industry shift toward more sustainable and efficient practices not only underscores a commitment to environmental responsibility but also signifies a strategic pursuit of targeted reservoir stimulation methods. As waterless fracturing continues to evolve, it stands poised to redefine the landscape of hydraulic fracturing, fostering a more sustainable and environmentally friendly future for the industry, especially for enhanced geothermal systems.

3.1 Waterless fracturing fluids: N_2/CO_2

N_2/CO_2 fracturing fluids emerge as an environmentally conscious alternative, addressing the urgent need to reduce water consumption, particularly in regions facing water scarcity or stringent environmental regulations. Employing N_2/CO_2 as the fracturing fluid avoids introducing substances that could compromise the permeability of the reservoir, ultimately enhancing the overall productivity of the well. LN_2 stands out as a potential water-free fracturing fluid. Due to the low temperature of LN_2, it can induce significant thermal stress (Huang et al., 2020), leading to the formation of complex fractures. Yang et al. (2023) explored the use of both liquid nitrogen (LN_2) and no liquid nitrogen ($NoLN_2$) treatments for fracturing granite rock at temperatures ranging from 100 to 300°C under triaxial-confining stresses of 10 MPa. The breakdown pressure experienced a reduction of 10.1%–18.7% when employing LN_2 cooling compared with air-cooling treatment. Despite these advantages, the commercial application of LN_2 is limited, mainly due to the specialized demands of injection units and pumping systems.

Recent years have seen a notable increase in attention toward CO_2 fracturing as a waterless technique. In hydraulic fracturing tests conducted by Li et al. (2019), various fluids were tested, including gel, slick water, liquid CO_2, and supercritical CO_2 ($scCO_2$) with slick water, using a true triaxial fracturing simulation system. Tight sandstone samples from the Ordos Basin in China were utilized for the test. The breakdown pressure of slick water was ~12.3 MPa, while for gel, it was 19.9 MPa. Liquid CO_2 exhibited a breakdown pressure of around 8.8 MPa, and for supercritical CO_2, it ranged from 9.6 to 11.9 MPa. Topography results indicated that $scCO_2$ could form extensive fracture networks with multiple opened bedding planes, in contrast to water and gel fracturing, which only produced incompletely propagated transverse fractures. On the flip side, the width of fractures when using CO_2-based fluid was narrower than in other fluid systems, and a significant number of microfractures were observed. Jian et al. (2021). Performed laboratory experiments investigating the use of CO_2 in granite samples at a temperature of 200°C. The confining pressure was maintained at 9.7 MPa, while the axial pressure was set at 7.6 MPa. The findings indicate that, following stimulation in hot dry rock, CO_2 can produce larger fractures compared with water.

Song et al. (2019) further contributed to the understanding of CO_2 fracturing by revealing its ability to generate intricate fractures at lower breakdown pressures compared with water or thickened-water fracturing. The lab test, conducted with a confining pressure of 20 MPa and an axial load of 2 MPa on a cylindrical sample, demonstrated a breakdown pressure of 28 MPa for CO_2 fracturing and 37 MPa for water fracturing, with a constant pressure ramping at 500 kPa/min. In large-scale experiment using a 300 mm^3 sample from the Silurian Longmaxi Formation in Changning County, Sichuan Province, China, CO_2 and water were used as stimulation fluids. The study found that the surface area of fractures using CO_2 fracturing was much higher than that of water fracturing, with an increase of 81.9%–95.4% from the base case, while water fracturing only increased the surface area by 7.4%. Shear fractures accounted for 90% of the fractures generated, while tensile fractures were similar to the water fracturing process. Simulation results shown in Fig. 4 indicate the significant difference of fracture complexity using water or CO_2 as the stimulation fluid. The authors also reported a field pilot test using CO_2 as the stimulation fluid in the Jilin oilfield in China, where a significant increase in oil production, ranging from 4 to 20 times, was observed after stimulation. However, the absence of water, the conventional carrier of proppants in hydraulic fracturing, introduces challenges in the efficient transport and placement of proppants within fractures when employing waterless fluids.

3.2 CO_2 foam and CO_2 responsive fluid fracturing

Research conducted by Khan et al. (2023) has delved into CO_2 foam fracturing as a strategy to address the substantial issue of high leak-off linked to CO_2. While CO_2 foam is not entirely waterless, its liquid fraction can be as low as 5% or $<2\%$. Furthermore, foam has exhibited the capability to improve the transport of proppants to deep reservoirs. Nevertheless, the inherent thermodynamic instability of foams, especially under high-temperature conditions encountered in EGS, poses a significant challenge for their practical application in fracturing, as emphasized by Thakore et al. (2024).

In response to these challenges, Pacific Northwest National Laboratory (PNNL) has developed StimuFrac, an innovative fracturing fluid tailored for EGS. Consisting of an aqueous polymer solution of polyallylamine (PAA) combined with CO_2, StimuFrac provides a promising alternative that is both effective in enhancing reservoir permeability and less dependent on water. The cross-linking reactions, and associated volume expansion, and viscosity increase of PAA–CO_2 at elevated temperatures, as documented by Fernandez et al. (2019), Jung et al. (2015)), and Shao et al. (2015)), contribute to the generation of substantial fractures in hot dry rock and water-saturated rock systems, as elucidated in studies by Jian et al. (2021, 2023).

The research findings indicate that PAA–CO_2 is capable of generating large fractures in both dry and water-saturated granite rock systems with ultralow permeability of around 0.5–1 mD, making it a viable and versatile solution for addressing the challenges associated with traditional CO_2 foam fracturing in high-temperature EGS environments. The test was conducted at 200°C using a 6-in. granite sample, with horizontal stress equal to 1400 psi and vertical stress equal to 1100 psi. The PAA–CO_2 fluid system consistently showed much higher permeability enhancement after the stimulation process as compared with water, waterless fluids (CO_2), and combinations of the two, as shown in Fig. 5. StimuFrac showed that at relatively low temperatures (200°C), the main mechanism for the significantly large permeability enhancement, compared with other fracturing fluids, is the CO_2-activated and reversible viscosity increase (Jian et al., 2021, 2023). This viscosity increase reduces the mobility of the CO_2 phase in the porous matrix enabling the development of a larger pressure envelope with pressure values above the minimum principal stress, and, as a result, CO_2 then creates larger fractures as compared with H_2O, CO_2/H_2O, and CO_2 stimulation fluids. Since the CO_2-triggered expansion in volume of StimuFrac doubles at temperatures above 300°C

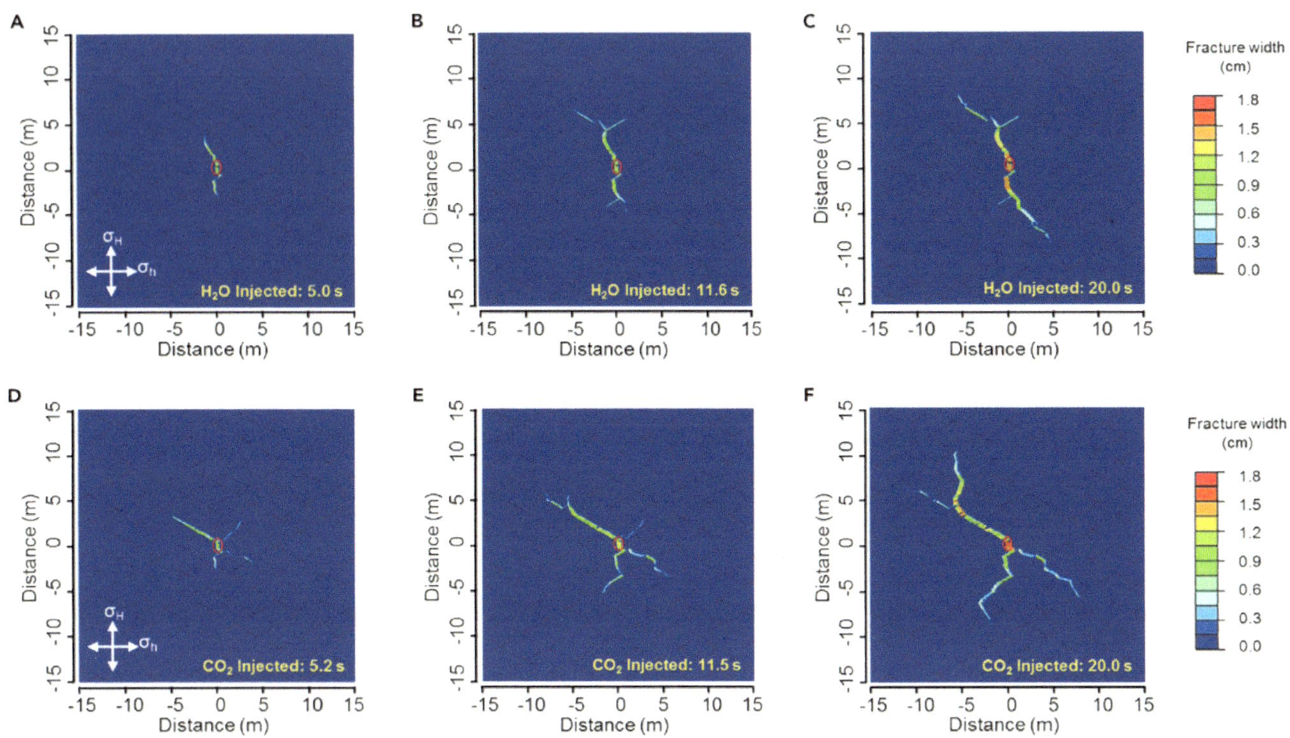

FIG. 4 Fracturing processes simulated by ABAQUS. Schematic fractures were magnified by 50 times for clarity; the red circles indicate injection points. (A–C) Fracture network during water fracturing with injection time of 5.0s (A), 11.6s (B), and 20.0s (C). (D–F) Fracture network during CO_2 fracturing with injection time of 5.2s (D), 11.5s (E), and 20.0s (F). *Reproduced with permission from Khan, A.M., BinZiad, A., Alshammari, F.O., Glaznev, I., 2023. Fiber assisted enhanced CO2 foam fracturing and proppant placement. In SPE Annual Technical Conference and Exhibition, Society of Petroleum Engineer, October 2023. D011S006R005.*

(Jung et al., 2015; Shao et al., 2015), the team predicts that the increase of both viscosity and volume could play a role in permeability enhancement at higher temperatures.

In summary, CO_2 foam fracturing offers several advantages in hydraulic fracturing operations. The foam nature of the fracturing fluid reduces fluid mobility and minimizes fluid loss to the formation. This method also improves proppant transport, contributing to enhanced conductivity and long-term well productivity. However, the complex design of the foam, including considerations of stability, surfactant selection, and injection conditions, poses challenges. Moreover, the sensitivity of foam stability to changes in temperature and pressure may limit its effectiveness in certain reservoir conditions. Economic considerations, including the costs associated with surfactants and additives (Thakore et al., 2024), further warrant careful assessment. On the other hand, CO_2-responsive fluid fracturing introduces unique benefits. These fluids can be tailored to undergo phase changes or viscosity alterations in response to specific reservoir conditions, enabling selective fracturing in desired zones.

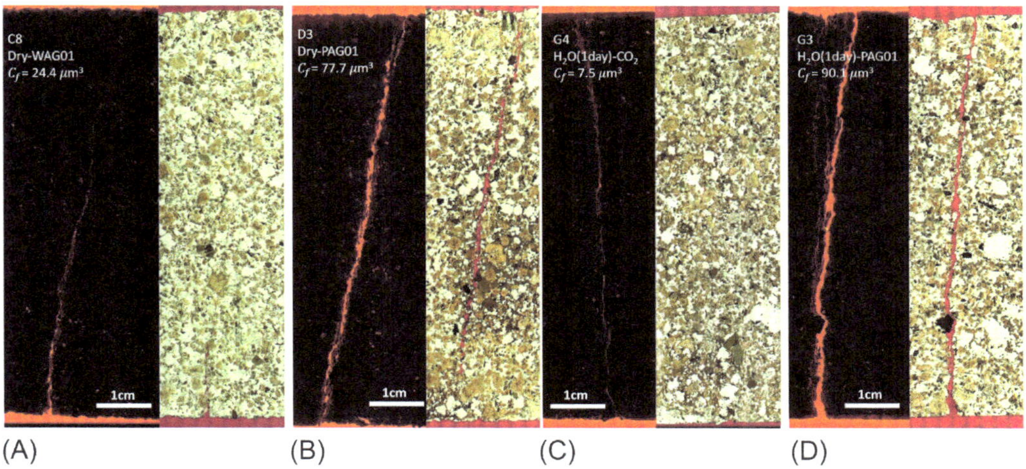

FIG. 5 Large fractures in dry and water-saturated systems generated by PAA–CO_2 [(B) and (C) images] as compared to low permeability fractures generated by stimulating with water-CO_2 [(A) and (C) images] (Jian et al., 2023).

3.3 Explosive and propellant fracturing

An alternative waterless fracturing method involves the use of explosive chemicals, such as nitroglycerin and TNT. Five wells were subjected to a shale oil formation test, conducted by Miller and Johansen (1976) in 1976, at depths of 60–100 ft. This approach resulted in extensive fractures with a propagation radius of around 90 ft., leading to an 800% enhancement in airflow during subsequent well tests.

Propellant deflagration harnesses high energy to create a rapid gas-expanding cloud, propagating from the wellbore into the formation. It serves to generate new fractures or enhance existing fractures in the formation. Sandia National Laboratory has conducted several propellant fracturing tests (Cuderman, 1986, 1982) using gun propellants in a tunnel complex. Multiple fractures occurred for pressure rise times ranging from 0.05 to 5 ms. Borehole diameters varied from 0.048, 0.1, to 0.15 m, drilled to a depth of ~1 m. The rock material, composed of tuff, exhibited permeability within the range of 10^{-2} to 10^{-1} mD. The tests addressed considerations regarding propellant and rock responses necessary to achieve the desired pressure rise times for reliable multiple fracturing. The critical selection of rise time (the time to peak pressure during a given propellant burn) for optimal multiple fracturing, considering borehole size and expected peak pressure, was emphasized.

Explosive fracturing involves the use of high-energy explosives to rapidly generate fractures. This method is effective for creating extensive fracture networks in dense rock formations, thereby enhancing fluid flow. However, it presents several challenges, including a lack of precision, potential environmental and safety concerns such as groundwater contamination and air pollution, and difficulties in obtaining regulatory approval due to stringent safety measures. Conversely, propellant fracturing provides controlled and predictable fractures with reduced risks of unintended consequences. This method is perceived as having safety advantages and potentially a lower environmental impact. However, propellant fracturing tends to be slower and requires more complex equipment, which can lead to increased operational costs.

3.4 Electrical shock: Electro-hydraulic fracturing for EGS

Utilizing pulsed current to induce the electrical explosion of a metal wire is a commonly employed technique for producing (i) a plasma characterized by relatively high temperature and density, (ii) pulsed electromagnetic radiation, and (iii) shock waves (SWs). The electrical reservoir stimulation technology harnesses the capabilities of high-voltage electricity, affording precise control over the orientation of the fractures created. In contrast, conventional hydraulic fracturing often results in short-circuits and uncontrollable fractures, exerting dominance over fluid flow and hindering re-stimulation in areas requiring further fracturing.

This innovative approach enhances fluid flow, ensuring a more focused and efficient process for geologic reservoir stimulation and facilitating improved recovery of subsurface resources. The methodology is detailed in the patent by Mehrvand et al. (2021), outlining the application of electricity for fracture generation in EGS, as depicted in Fig. 6.

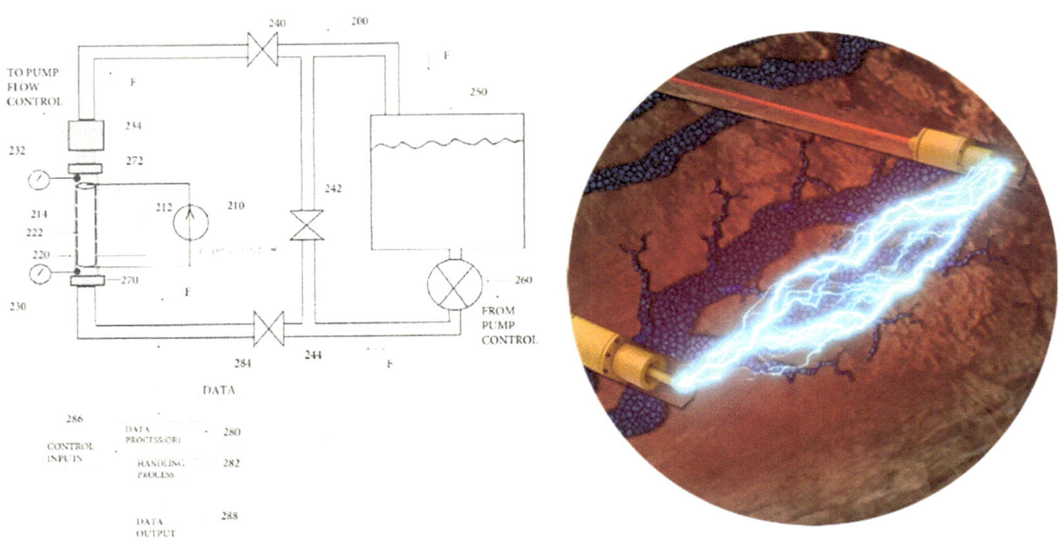

FIG. 6 System and method for pulsed electrical reservoir stimulation (Mehrvand et al., 2021) (https://www.zerogeo.energy/).

Zhou et al. (2015) conducted experiments to investigate the characteristics of underwater electrical wire explosions for reservoir stimulation. Fig. 7 illustrates the fracturing effects resulting from three different types of shock wave sources are applied to applied to coal cube specimens with dimensions equal to 30 cm, where water (water gap load), an electrical wire (underwater electrical wire, UEWE load), or an electrical wire immersed in an energetic material (EM load) are located in-between a fracture. The results demonstrate the formation of fractures induced by shock waves. However, the type of shock wave source significantly impacts the efficiency of fracture formation. A higher number of shock waves shots (200 shots) are needed to create some visible fractures via the water gap load approach (Fig. 7A), while a smaller number of shock wave shots (50 shots) are needed when using an electrical wire load (Fig. 7B), and only eight shock wave shots are needed when combining the electrical wire approach with an energetic material (Fig. 7C).

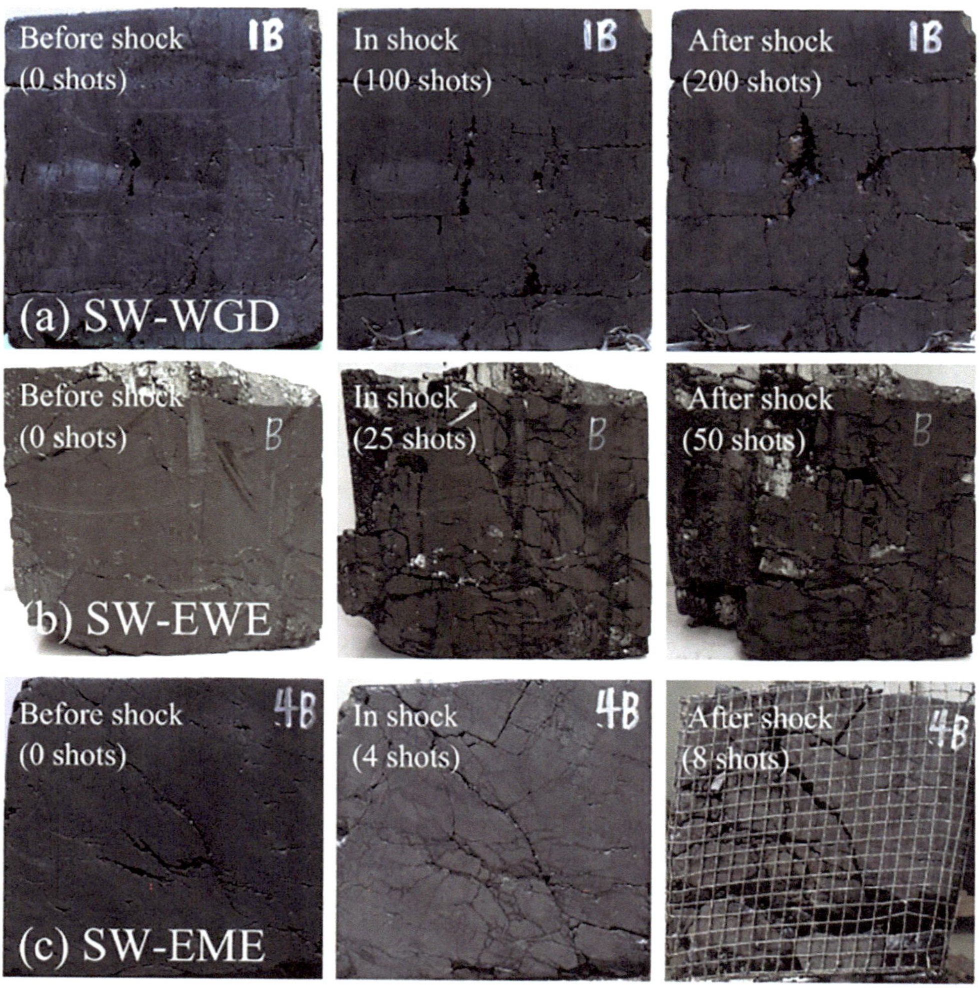

FIG. 7 Fracturing effects of three types of shock wave sources on coal cube specimens: (A) Shock wave by water gap discharge load. (B) Shock wave by underwater electrical wire load (UEWE), and (C) Shock wave by electrical wire immersed in an energetic material (EM) load (Zhou et al., 2015).

Electrical fracturing, a technique employed in electro-hydraulic fracturing for enhanced geothermal systems, also presents potential advantages and disadvantages if ever applied in a geothermal reservoir. One of its primary strengths lies in the precise control it affords over fracture initiation and propagation within the geothermal reservoir, allowing for targeted stimulation and improved reservoir performance. The method minimizes fluid flow disruption, contrasting with traditional hydraulic fracturing, which often leads to uncontrollable fractures and short-circuits. This results in more efficient geologic reservoir stimulation, enhancing resource recovery and increasing fluid flow for effective heat extraction and energy generation. Moreover, the potential for a reduced environmental impact, particularly in terms of water usage and groundwater contamination, adds to its appeal. However, the implementation of electrical fracturing requires advanced technology and expertise, possibly translating into higher upfront costs and operational complexities. The energy-intensive nature of the process, coupled with regulatory challenges and site-specific limitations, further underscores the need for careful consideration when adopting this innovative technique for enhanced geothermal systems.

In conclusion, the exploration of waterless fracturing techniques, encompassing methods such as N_2/CO_2 fracturing, explosive and propellant fracturing, CO_2 foam, and CO_2 responsive fluid fracturing, reflects a significant stride towards more sustainable and environmentally friendly hydraulic fracturing practices. These approaches address critical concerns related to water scarcity, environmental regulations, and the risk of formation damage near the wellbore.

The waterless fracturing methods, especially N_2/CO_2 fracturing, showcase advantages in terms of minimizing formation damage, enhancing well productivity, and avoiding substances that compromise reservoir permeability. However, challenges such as temperature and pressure considerations, limitations in proppant transport, and fluid phase transitions persist. CO_2 foam and CO_2-responsive fluid fracturing present promising alternatives, but their effectiveness relies on a comprehensive understanding of fluid properties and stability. Furthermore, it was recently reported that water saturation in granitic rocks plays an important role in permeability enhancement, with higher water saturation reducing fracturing fluid efficiency (Jian et al., 2023).

Explosive and propellant fracturing offer rapid fracture generation and controlled fractures, but concerns regarding precision, environmental and safety issues, and regulatory hurdles need to be carefully addressed. Meanwhile, electrical fracturing introduces precise control over fracture initiation, minimizing fluid flow disruption, and enhancing overall reservoir performance. However, the energy-intensive nature of this method and potential regulatory challenges emphasize the need for careful consideration.

To optimize these waterless fracturing techniques, in-depth studies are essential, including assessments of fracture initiation and propagation pressure, permeability enhancement under varying conditions, and the influence of different factors on fracture behavior. Field studies can provide valuable insights into real-world applications.

4 Proppants in geothermal

4.1 Background of high-temperature proppant and EGS

An EGS is a geothermal reservoir where fractures are created to allow fluids to flow and transfer heat, which is an enhanced recovery technology for rocks that have sufficient temperature but little permeability and/or fluid saturation. In an EGS, water is injected into the reservoir, circulates throughout the fractured rock, and transfers heat to the surface for heating and electricity production. EGS technologies will contribute clean and renewable energy to the United States, which is critical in promoting the transition to economy-wide net-zero carbon emissions. EGS represents a significant advancement in harnessing geothermal energy, one of the most sustainable and environmentally friendly energy sources. Unlike conventional geothermal systems that rely on natural hydrothermal resources, EGSs are engineered to extract heat from hot dry rocks deep beneath the Earth's surface. This is achieved by creating a network of artificial fractures in the rock, through which water is circulated to absorb the Earth's heat before being brought back to the surface to generate electricity. Superhot EGS takes this concept further by targeting extremely high-temperature rock formations, typically >450°C, which can potentially yield much higher energy outputs.

The primary motivation for developing EGS and superhot EGS is to expand the availability and efficiency of geothermal energy. Traditional geothermal resources are limited to specific geographic locations where natural heat, water, and permeability converge. In contrast, EGS technology can be utilized in a broader range of locations, significantly expanding the potential for geothermal energy development. Superhot EGS promises to tap into vast amounts of untapped thermal energy at greater depths, offering a higher energy return. The benefits of EGS and superhot EGS are substantial. They provide a reliable and constant energy source, unlike intermittent renewables like solar and wind. Geothermal energy is also low in greenhouse gas emissions, making it a cleaner alternative to fossil fuels. However, EGS superhot EGS have several drawbacks. The development of EGS, especially superhot EGS, involves high upfront costs and technological challenges, including drilling into hard rock at extreme depths and temperatures. There are also concerns about induced seismicity (earthquakes) due to the fracturing of rocks, and the long-term sustainability of the reservoirs. Despite these challenges, the pursuit of EGS and superhot EGS represents a promising frontier in the quest for sustainable and clean energy solutions.

Proppant, a crucial component in hydraulic fracturing, is used to keep fractures open in rock formations to facilitate the flow of oil or gas to the wellbore, as shown in Fig. 8. Especially in high-temperature environments, such as deep oil wells or geothermal energy extraction, the choice of proppant is critical due to the extreme conditions they must endure. High-temperature proppants are typically designed to withstand not just the intense heat, often exceeding 100°C, but also the substantial pressure and corrosive conditions found deep underground. Materials commonly used for high-temperature proppants include ceramics, resin-coated sand, and sintered bauxite, each selected for their ability to maintain structural integrity and permeability under heat stress. Ceramic proppants are favored for their strength and thermal stability. The effectiveness of these proppants directly impacts the efficiency of the extraction process, making their performance crucial in the energy sector. Additionally, understanding the behavior of proppants under such conditions is vital for minimizing environmental risks and ensuring the safety and longevity of the wells.

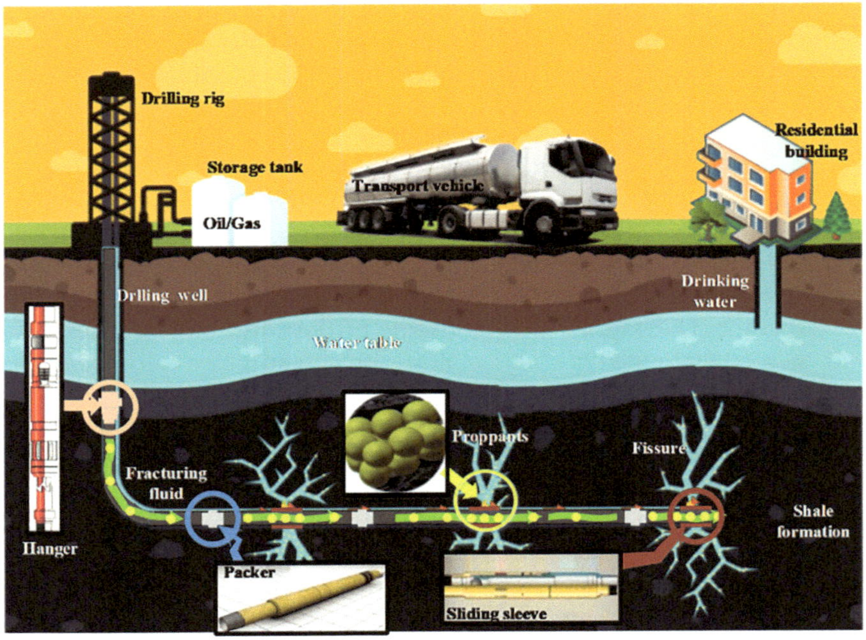

FIG. 8 Schematic of the hydraulic fracturing process with proppant application. *From Guo, Y., Zhang, M., Yang, H., Wang, D., Ramos, M.A., Hu, T.S., Xu, Q., 2013. Friction challenge in hydraulic fracturing. Lubricants 10 (2), 14.*

In EGS and superhot EGS, proppants are used to keep artificially created or enhanced fractures open, thereby facilitating a more effective heat exchange between the rock and circulating fluids. This is particularly vital in superhot EGS, where extreme temperatures (often exceeding 400°C) and pressures pose significant challenges to the integrity and performance of proppants. Fig. 9 illustrates the hierarchy of proppant based on conductivity. The ideal proppant in such environments would need to possess exceptional thermal stability, chemical inertness, and mechanical strength to withstand prolonged exposure to harsh conditions without significant degradation or crushing. To achieve this goal, Raysoni and Weaver (2013) investigated the durability of proppants used in hydraulic fracturing after 0, 15, 45, 90, and 180 days of hydrothermal exposure at 300 and 450°F (148 and 232°C). It focuses on the effects of high-temperature and geochemical interactions on proppant performance. The research reveals that diagenetic processes, such as zeolite formation, occur at realistic reservoir temperatures, leading to significant reductions in proppant permeability and strength within a year. The study emphasizes the importance of proppant material selection, noting that aluminum-based proppants show varying degrees of diagenesis. It suggests that surface coating with hydrophobic materials can mitigate geochemical degradation. The findings underscore the need for careful proppant selection and quality control to ensure long-term fracture conductivity and well productivity. In addition, Childers et al. (2017) explored the development of novel core–shell proppants for geothermal applications. These proppants, designed for enhanced dispersibility and thermal stability, are evaluated for their performance in geothermal reservoir conditions. This study focuses on the stability and behavior of bauxite-based proppants modified with polymer coatings, comparing them with unmodified proppants. They illustrated the improved stability of these proppants in various fluid environments, their ability to remain suspended in solution, and their resilience under high-temperature geothermal conditions. The research suggests that these advanced proppants could be more effective in geothermal energy extraction, offering potential advantages over traditional proppants in terms of environmental impact and efficiency.

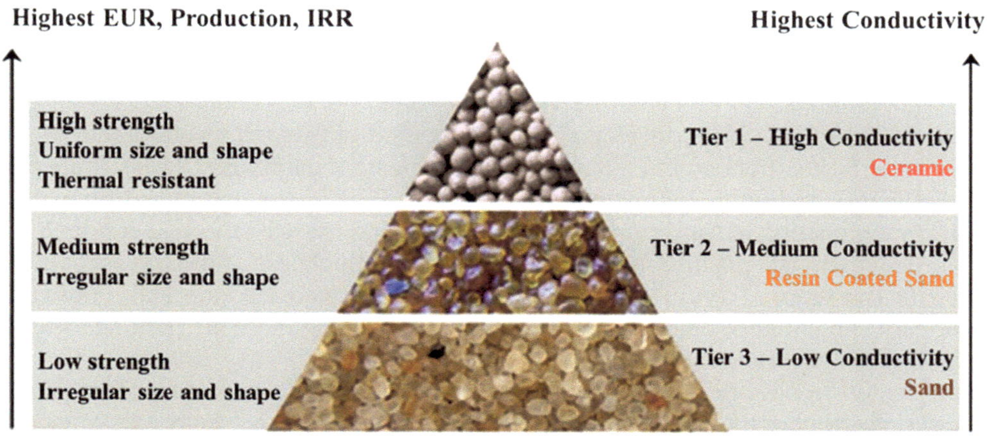

FIG. 9 The hierarchy of proppant based on conductivity (Gallagher, 2011).

Aside from proppant itself, another effect, such as temperature, is also critical for the proppant's application. Lima et al. (2019) investigated temperature effects on rock fracture apertures, key to geothermal energy systems. Using a unique experimental setup, it simulates thermal cycles on granite samples, measuring fracture aperture changes. Findings reveal that thermal cycling significantly alters fracture apertures: a maximum reduction of 90% at 60°C, with 70% recovery upon cooling. This behavior is attributed to thermal expansion and contraction of fracture walls. These insights are crucial for geothermal energy extraction, demonstrating the dynamic response of rock fractures to temperature variations and their implications for energy efficiency and sustainability. The difference between long term and short term should also be emphasized. In addition, Ko et al. (2023) focused on evaluating proppant performance in EGS, specifically at temperatures up to 320°C. It investigates the crush resistance and packing strength of various proppants, including non-coated, resin-coated, and hydrophobic polymer-coated ceramics, under dry and heated conditions. The findings reveal that high temperatures for prolonged periods (up to 14 days) significantly reduce the packing strength and increase the fines percentage of proppants, indicating decreased conductivity. Polymer coatings were found to degrade under these conditions, leading to reduced effectiveness in trapping fines, thus compromising the proppants' structural integrity and performance in EGS environments. Porlles et al. (2023) examined the long-term behavior of various proppants under high-pressure (up to 15,000 psi) and high-temperature (260°C) conditions in geothermal systems. It focuses on how proppants like ceramics, kryptospheres, and sand-based materials degrade over time when exposed to extreme geothermal conditions. The research includes crushing tests at various pressures and a numerical model for hydraulic fracturing design. The findings in this paper reveal that newer proppants, such as kryptospheres, offer better durability and effectiveness in maintaining fracture conductivity in harsh environments.

Complex chemical environment is an important feature of EGS. Guo et al. (2021) studied the coated proppants in unconventional reservoirs highlights their performance under high temperatures. Key findings include the impact of temperature on proppant solubility, particularly in alkaline conditions. As temperature increases, the solubility of the coated proppant also rises, weakening the coating layer's strength and reducing the stability of the proppant filling layer. Additionally, the permeability of fractures filled with coated proppants decreases with rising temperature. For instance, at 150°C, fracture permeability drops by over 30% compared with 20°C. This indicates that high temperatures significantly affect the physical and chemical properties of coated proppants, influencing their effectiveness in hydraulic fracturing operations. The efficiency of heat extraction in the EGS should be also considered. Balushi and Taleghani (2023) developed a novel method of thermally conductive proppants for improving heat extraction in EGS using discrete element methods (DEMs), finite element methods (FEMs), and finite volume methods (FVMs). They found that the addition of a thin layer of copper coating to sand proppant significantly increases its thermal conductivity, by >100%. Also, there is a positive correlation between the thickness of the conductive coating and the effective thermal conductivity of the proppant. The increased thermal conductivity of the proppant can enhance the rate of heat exchange, allowing more efficient heat extraction from geothermal reservoirs.

Therefore, proppants are used in EGS to keep the induced fractures open, thereby enhancing the permeability of the rock and improving the efficiency of heat exchange. The choice of proppant is crucial, especially in superhot EGS, due to the extreme temperatures and pressures involved. The proppant must withstand these conditions without degrading or reacting with the rock or fluid. While the use of proppants can significantly improve the efficiency of EGS, it adds to the complexity and cost of the system. Additionally, the environmental impact of proppant materials and the potential for groundwater contamination are concerns that need to be addressed.

4.2 Problem statement: Addressing challenges in EGS

Low fracture conductivity and thermal short-circuiting are two primary challenges in EGS developments. By the early 1990s, the EGS community had widely accepted the concept of shear stimulation (i.e., hydro-shearing), by which the aim of hydraulic fracturing stimulation was stimulating preexisting fractures by inducing slip, rather than creating new tensile fractures. In granite, rock surface roughness provides hydraulic conductivity after hydro-shearing, which is referred to as "self-propping." The challenge of low fracture hydraulic conductivity in an EGS can be solved by placing proppants in the fractures, and thermal short-circuiting can be minimized by multi-stage fracturing. Proppant is a granular material, typically sand, coated sand, or man-made ceramic materials, which has been successfully used in the recovery of unconventional oil and gas. Field injections of proppants in EGS projects have been conducted in both granitic and sedimentary formations, and improved well productivity was observed.

Rock creep is one of the major mechanisms that reduce a fracture's hydraulic conductivity in an EGS. By definition, creep is the continuously increasing deformation of a solid under constant loads. This slow deformation can lead to the loss of fracture area and thus the reduction of fracture width. The fracture closure is also influenced by extreme reservoir environments, such as varying stress and temperature conditions. Sone and Zoback (2014) performed triaxial creep experiments on shales and found that the rock creep magnitude is linearly proportional to the applied differential stress while is insensitive to the confining stress. Khosravi and Ghassemi (2017) conducted multistage triaxial creep tests on Mancos shale samples and discovered that the elevated temperature can lead to a significant increase in the shale creep rate. Therefore, failure to characterize the effect of time-dependent creep behaviors of an EGS rock on the fracture's hydraulic conductivity can result in significant mistakes in assessing the EGS reservoir quality. Fan et al. (2021) developed a numerical workflow that combined a 3D continuum-discrete mechanical coupling approach with the lattice Boltzmann (LB) method to simulate the coupled thermal–mechanical process in a rock/proppant system and to investigate the role of the time-dependent deformation (i.e., creep) of the rock on proppant embedment and fracture conductivity loss under varying temperature and stress conditions (Fig. 10). The numerical workflow was first compared with an experiment under varying temperature and stress conditions to calibrate the elastic, plastic, viscoelastic, and thermal properties of the rock, as well as the proppant properties. Then, the effect of fracture axial and confining stress, numbers of proppant layers, proppant size, proppant spatial distribution, and proppant crushing was systematically investigated. The simulation results indicate that when the rock creep is significant, large-size and a multilayer of proppant structure are suggested to maintain the fracture's hydraulic conductivity. The small percentage of particle breakage in a proppant assembly plays a less important role in the long-term maintenance of the fracture's hydraulic conductivity.

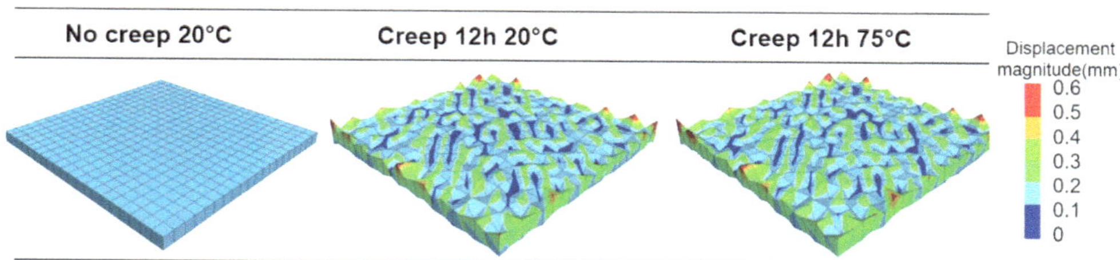

FIG. 10 Displacement magnitude on rock surfaces indented by proppant particles under different temperatures at the axial stress of 13.8MPa and confining stress of 20MPa (Fan et al., 2021).

The softening of rocks under EGS conditions may cause significant proppant embedment on rock surfaces, which may consequently lead to fracture closure and loss of the fracture's hydraulic conductivity and geothermal productivity. Fig. 11 illustrates the phenomena affecting damage to the fracture face surface during hydraulic fracturing of hydrocarbon deposits. To mitigate proppant embedment on rock surfaces, it is tempting to place more proppant particles within the fracture space, which increases the fracture width but in the meanwhile reduces the fracture porosity when the proppant structure is a partial monolayer. This competing process (i.e., increased fracture width vs. decreased fracture porosity) can lead to a non-monotonic evolution of fracture hydraulic conductivity with an increasing proppant areal concentration. Fan et al. (2019) developed a hybrid, experiment/simulation-integrated workflow, which combined the DEM and the LB method with laboratory-measured load-embedment correlations, to demonstrate the non-monotonic development of a fracture's hydraulic conductivity with increasing proppant concentration. This simulation/experiment-integrated method clearly identified and separated the fundamental mechanisms that regulated fracture conductivity evolution when the proppant assembly in the fracture space developed from a partial monolayer to a multilayer structure. Later, Li et al. (2022) conducted well-controlled experiments in the laboratory to validate the findings of Fan et al. (2019).

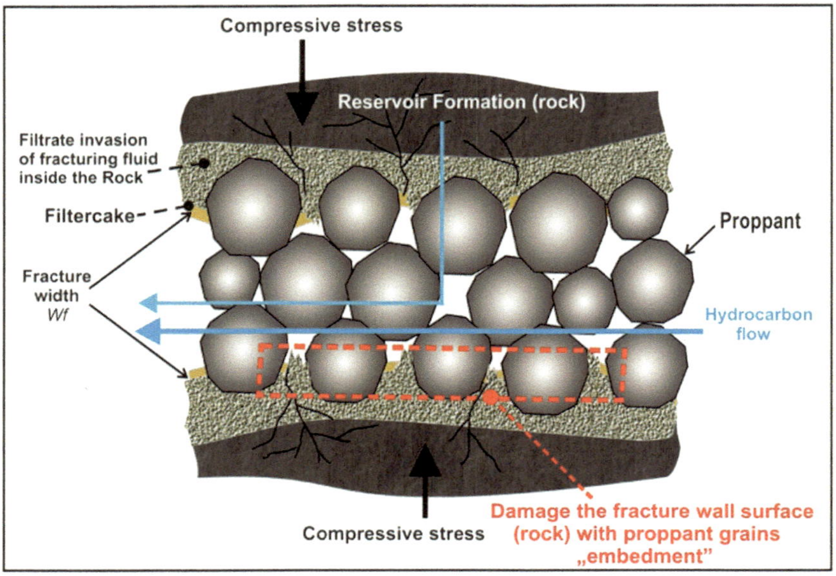

FIG. 11 The phenomena creating damage to the fracture wall surface during hydraulic fracturing of hydrocarbon deposits (Masłowski and Labus, 2021).

The application of proppants in EGS and superhot EGS represents a critical innovation in the pursuit of more efficient and sustainable geothermal energy extraction. Despite advancements, there remain significant knowledge gaps in the study of proppants for EGS applications. One primary area of concern is the long-term stability and performance of proppants under superhot conditions. Current research often focuses on short-term tests, leaving uncertainties about the durability and effectiveness of proppants over the lifespan of a geothermal well, which can span decades. Additionally, there is a lack of comprehensive understanding regarding the interaction between proppants and geothermal fluids, especially in terms of chemical reactions that might affect the proppant's integrity or the fluid's properties. Another area that requires further exploration is the optimization of proppant size, shape, and material composition to maximize conductivity and minimize environmental impact. Finally, the economic feasibility of deploying advanced proppants in EGS, particularly in superhot conditions, remains a critical factor that needs more in-depth analysis, considering the high costs associated with materials capable of withstanding such extreme environments. The following list is a detailed breakdown of the identified knowledge gaps:

- **Long-term durability and performance:** One of the most significant unknowns is the long-term behavior of proppants under the extreme conditions of superhot geothermal environments. Current studies often focus on short-term laboratory tests, which do not adequately simulate the decades-long operational lifespan of geothermal wells. Understanding how proppants withstand prolonged exposure to high temperatures, pressures, and corrosive fluids is crucial for ensuring the longevity and efficiency of EGS.

- **Chemical interactions and stability:** The interaction between proppants and geothermal fluids, particularly at superhot temperatures, is not well understood. There is a need for comprehensive research on the potential chemical reactions between proppants and these fluids, which could impact the mechanical integrity of the proppants and the chemical composition of the fluid. This includes studying the possible leaching of materials from the proppants and the resultant environmental impact.
- **Optimization of proppant properties:** The ideal size, shape, and material composition of proppants for EGS applications are still subjects of research. The challenge lies in balancing the need for high strength and thermal stability with the desire for minimal environmental impact and cost-effectiveness. Research into new materials or coatings that can enhance the performance of proppants under superhot conditions is needed.
- **Proppant transport and placement:** Effective transport and placement of proppants within the fractures of geothermal reservoirs are critical for optimal performance. The dynamics of proppant transport in high-temperature and high-pressure environments, and how these factors influence the settling and distribution of proppants within the fractures, are not fully understood.
- **Economic feasibility:** The cost-effectiveness of using advanced proppants in EGS, especially in superhot conditions, is a significant concern. High-performance proppants capable of withstanding extreme environments are often expensive, and their economic viability in large-scale geothermal operations remains a question. Studies focusing on the cost–benefit analysis of different proppant types in EGS are necessary to determine their practicality for widespread use.
- **Environmental impact:** There is a need for more comprehensive studies on the environmental impact of using different types of proppants, considering factors like the potential for groundwater contamination and the ecological footprint of proppant manufacturing and disposal.

4.3 Safety considerations and standards of EGS development

The development of EGS involves several safety considerations and must adhere to various international and national safety standards and regulations. For example, in the United States, these standards are designed to ensure environmental protection, public safety, and worker health. Here are some key regulations and standards:

- **National Environmental Policy Act (NEPA):** NEPA requires federal agencies to assess the environmental effects of their proposed actions prior to making decisions. This includes considering the environmental impacts of EGS projects, especially those on federal lands.
- **Clean Water Act (CWA):** This act regulates discharges of pollutants into the waters of the United States and quality standards for surface waters. For EGS, this is relevant for managing the discharge of fluids used in drilling and reservoir stimulation.

- **Safe Drinking Water Act (SDWA):** Particularly relevant is the Underground Injection Control (UIC) program under the SDWA, which regulates the injection of fluids underground to prevent contamination of drinking water sources.
- **Resource Conservation and Recovery Act (RCRA):** This act governs the disposal of solid and hazardous waste. EGS projects must ensure that any waste generated, including drilling muds and cuttings, is handled and disposed of in compliance with RCRA regulations.
- **Occupational Safety and Health Administration (OSHA) Regulations:** OSHA sets and enforces standards to ensure safe and healthful working conditions. This includes regulations on personal protective equipment, hazard communication, and general safety practices in industrial settings.
- **Seismic Monitoring and Reporting:** While there is no federal law specifically regulating seismicity induced by geothermal activities, the US Department of Energy (DOE) has guidelines for monitoring and managing induced seismicity. Projects may also be subject to state-specific seismic monitoring requirements.
- **Bureau of Land Management (BLM) Regulations:** For EGS projects on federal lands, BLM regulations regarding land use, drilling permits, and environmental protection are applicable.
- **Environmental Protection Agency (EPA) Regulations:** EPA regulations on air quality, water quality, and chemical use are relevant to EGS projects. This includes the management of greenhouse gas emissions and the use of chemicals in drilling and stimulation fluids.
- **Energy Policy Act:** This act includes provisions that are relevant to geothermal energy development, such as tax incentives and research funding.
- **State-Specific Regulations:** Individual states may have additional regulations. For example, California, a state with significant geothermal activity, has its own set of regulations governing geothermal development, including water use, air quality, and land use.

4.4 Investigation methods in high-temperature proppant and EGS

The study focuses on evaluating proppants under high-temperature conditions to understand their stability, effectiveness, and potential alterations in physical properties. The chosen methods involve a combination of laboratory testing and analytical procedures for simulating the extreme conditions that proppants would face in geothermal or deep oil well environments. The main laboratory testing include:

(1) Proppant aging tests: Proppants were subjected to hydrothermal aging in deionized water at high temperatures to simulate reservoir conditions in the laboratory.
(2) Scanning electron microscopy (SEM): This technique was used to analyze the surface morphology of proppants at reservoir conditions, observing mineral overgrowth and other changes. Fig. 12 shows the SEM images of ceramic proppant under subsurface conditions.

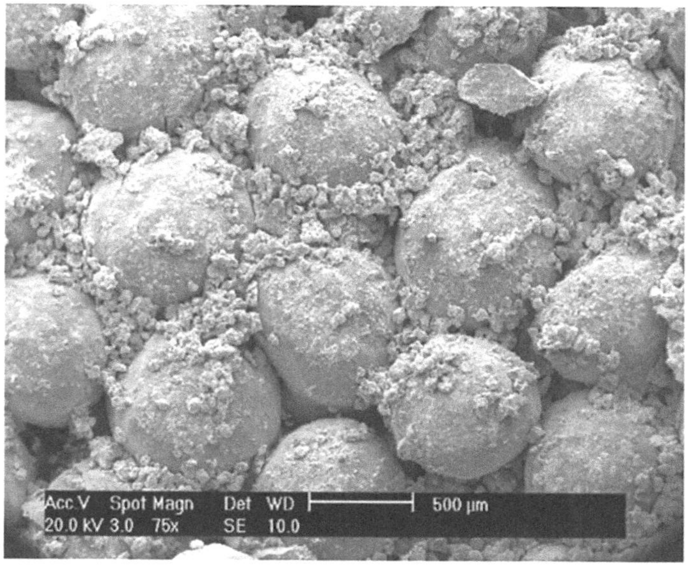

FIG. 12 Ceramic proppant (20/40 mesh, 2 lb/ft^2) after exposure to10,000 psi closure stress at 250°F for 140 h in 2% KCl solution under static flow condition (Weaver et al., 2007).

(1) X-ray diffraction (XRD): XRD analysis was conducted to identify the mineral composition of proppants and formation fines after aging.
(2) Energy-dispersive X-ray spectroscopy (EDX): This method was used for surface elemental analysis of proppants and formation fines at different stages of hydrothermal exposure.
(3) Computer tomography (CT): CT offers a nondestructive way to analyze and detect the microstructure, pore, and density measurement, and monitor changes over time of high temperature proppants. Fig. 13 shows the X-ray CT images of proppant particle crushing and embedment in the fracture of a shale rock under different effective stresses.
(4) Acid solubility tests: An important evaluation for high-temperature proppant for ensuring their chemical stability, durability, and overall performance in potentially harsh reservoir environments, particularly where acidic conditions are present.
(5) Crush strength tests: This test measures the crush strength of individual proppant grains at different intervals post-exposure to high temperatures to assess strength degradation.
(6) Fracture conductivity measurements by fracture conductivity cell: The conductivity and permeability of the proppant pack can be measured in the fracture conductivity cell at elevated temperatures to evaluate the impact on fluid flow. The rock slabs (granite, sandstone, shale, etc.) and loaded fluid (DI water, CO_2, nitrogen, etc.) can be adjustable based on the requirement.

The fracture conductivity by proppant is crucial in the study of the proppant in EGS applications. For investigating the various effects for the fracture conductivity of proppant, fracture conductivity cell plays an important role in such series of research. Li et al. (2022)

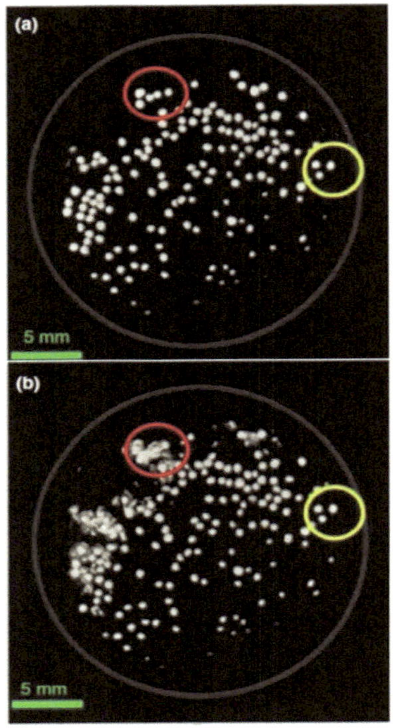

FIG. 13 X-ray CT images of proppant particle crushing and embedment in the fracture of a shale rock under effective stresses of (A) 0 and (B) 3000 psi (Chen et al., 2015).

displayed the schematic picture of the fracture conductivity cell used in the experiments (Fig. 14), which can measure fracture conductivity under different closure pressures and temperatures. In the test, two rock slabs and proppant particles sandwiched between them are placed in the testing cell, which is subjected to a closure pressure up to 17,000 psi with a precision of 0.2%. An electrical heater provides a testing temperature up to 250°C with a precision of 0.4%. The proppant concentration in the fracture was measured using the proppant mass and the surface area of the rock slabs. The fracture conductivity, C, according to the API RP-19D standard (API RP-19D, 2008), is the flow rate contributed by a unit length of fracture and calculated as:

$$C = kw_f = \frac{\mu \cdot Q \cdot L \cdot w_f}{\Delta p \cdot A} = \frac{\mu \cdot Q \cdot L}{\Delta p \cdot h} \tag{1}$$

where C is fracture conductivity (m^3); k is fracture permeability (m^2); w_f is fracture width (m); h is the size of the longer dimension of the fracture cross section (m); A is the area of fracture cross section and equal to $w_f h$ (m^2); μ is fluid viscosity (Pa·s); Q is flow rate (m^3/s); L is the length over which the pressure difference is measured (m); Δp is the pressure difference (Pa). Note that μ, L, and h in Eq. (1) are known. The measured flow rate, Q, and pressure difference, Δp, are then imported into Eq. (1) to calculate the fracture conductivity.

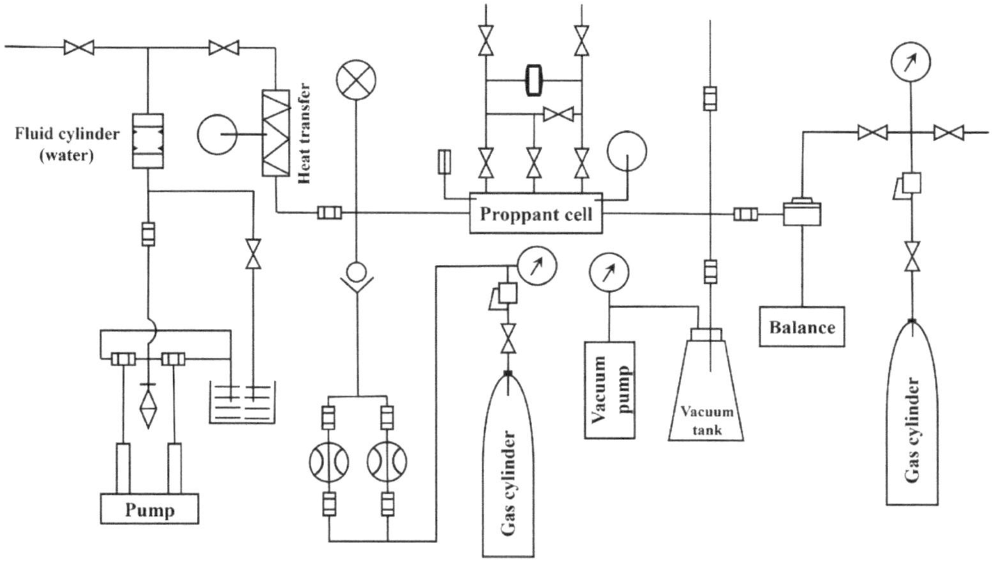

FIG. 14 Schematic picture of the fracture conductivity cell (Li et al., 2022).

In addition, the numerical approaches also play important roles in the proppant in EGS. These numerical and analytical methods are integral to the design, selection, and optimization of proppants for use in high-temperature reservoirs. They help in predicting performance, ensuring reliability, and optimizing the efficiency of hydraulic fracturing operations. The main numerical approaches include:

(1) FEM: FEM is used to simulate the mechanical behavior of proppants under stress. It helps in understanding how proppants will deform or fail under the high pressure and temperature conditions of deep reservoirs.
(2) DEM: DEM is used to simulate the behavior of individual proppant particles under mechanical stress. This helps in understanding how proppant packs will behave under reservoir conditions.
(3) Lattice Boltzmann (LB) method: LB method can be used for resolving pore-scale phenomena, which is essential for understanding how proppants interact with the porous rock at a microscopic level, including the effects of high temperature on these interactions.
(4) Thermal stress analysis: This method assesses how temperature changes impact the stress on proppants. It is crucial for understanding thermal expansion or contraction and its effects on proppant integrity and fracture conductivity.
(5) Material degradation modeling: This involves simulating the chemical interactions between proppants and reservoir fluids at high temperatures, which can lead to material degradation. It helps in predicting the long-term stability of proppants in harsh environments.
(6) Monte Carlo simulations: These are used for probabilistic analysis and risk assessment of proppant performance under varying conditions of temperature, pressure, and chemical exposure.

(7) Molecular dynamics (MD) simulations: MD simulation provides insights at the atomic or molecular level, particularly useful for understanding the chemical stability of proppants under high-temperature conditions.

5 Conclusions

This chapter has provided a comprehensive overview of the various stimulation methods and proppant technologies for EGS. The criticality of achieving economically viable flow rates in EGS through effective hydraulic stimulation has been underscored, highlighting the necessity for innovative approaches beyond conventional water fracturing. The exploration of alternative methods, such as supercritical CO_2, CO_2 foams, propellants, and liquid nitrogen, reflects the ongoing quest for more efficient and environmentally sustainable fracturing techniques. These alternatives, while promising, come with their unique challenges, such as foam stability under high-temperature conditions and the high costs associated with certain methods. The role of proppants in maintaining long-term reservoir permeability has been emphasized, especially considering the extreme thermo-mechanical and chemical stresses in EGS environments. The chapter has discussed the suitability of various materials like ceramics, resin-coated sand, and sintered bauxite for high-temperature proppants, stressing the importance of their thermal stability and structural integrity. Furthermore, the chapter has also explored advanced simulation and modeling techniques to better understand the behavior of proppants under EGS conditions. These methodologies are crucial for predicting the long-term stability and performance of proppants in harsh environments. In conclusion, while significant advancements have been made in the field of EGS, there remain substantial technical barriers that need to be addressed. The development of more efficient and cost-effective stimulation methods, along with the identification of durable and environmentally friendly proppants, are essential steps toward the successful implementation of EGS. The future of geothermal energy extraction lies in the continuous innovation and improvement of these technologies, ensuring a sustainable and reliable energy source for the future.

References

Action, E.C, 2008. Best Practice Handbook for the Development of Unconventional Geothermal Resources with a Focus on Enhanced Geothermal System. Collection Actes. BRGM Editions, Orléans.

Aksoy, N., Serpen, N.A., 2010. Acidizing in geothermal wells and HCl corrosion. In: Proceedings World Geothermal Congress 2010.

Allis, R., Moore, J., Davatzes, N., Gwynn, M., Hardwick, C., Kirby, S., McLennan, J., Pankow, K., Potter, S., Simmons, S., 2016. EGS concept testing and development at the Milford, Utah FORGE site. In: Proceedings, 41st Workshop on Geothermal Reservoir Engineering.

API RP-19D, 2008. Measuring the Long-Term Conductivity of Proppants. API, Washington, DC, USA.

Asanuma, H., Kumano, Y., Izumi, T., Soma, N., Kaieda, H., Tezuka, K., Wyborn, D., Niitsuma, H., 2004. Passive seismic monitoring of a stimulation of HDR geothermal reservoir at Cooper Basin, Australia. In: Technical Program Expanded Abstracts. 23. Society of Exploration Geophysists, pp. 556–559.

Axelsson, G., Thórhallsson, S., Björnsson, G., 2006. Stimulation of geothermal wells in basaltic rock in Iceland. In: Enhanced Geothermal Innovative Network for Europe Workshop. Vol. 3.

Baisch, S., et al., 2015. A review of induced seismicity in geothermal systems. Geothermics 52, 128–139.

Balushi, F.A., Taleghani, A.D., October **2023**. Thermally conductive proppants to improve heat extraction in geothermal systems. In: SPE Annual Technical Conference and Exhibition. Society of Petroleum Engineer. D021S016R002.

Barelli, A.C.G., Manetti, G., Peano, A., 1985. Well stimulation in latera field. Geotherm. Resour. Council Trans. 9 (Pt II), 213–219.

Barrios, L.A., Quijano, J.E., Romero, R.E., Mayorga, H., Castro, M., Caldera, J., 2002. Enhanced permeability by chemical stimulation at the Berlin Geothermal Field, El Salvador. Geotherm. Resour. Council Trans. 26, 73–78.

Batra, R., Albright, J.N., Bradley, C., 1984. Downhole Seismic Monitoring of an Acid Treatment in the Beowawe Geothermal Field. Los Alamos National Lab, NM (USA).

Baujard, C., Genter, A., Dalmais, E., Maurer, V., Hehn, R., Rosillette, R., Vidal, J., Schmittbuhl, J., 2017. Hydrothermal characterization of wells GRT-1 and GRT-2 in Rittershoffen, France: implications on the understanding of natural flow systems in the Rhine Graben. Geothermics 65, 255–268.

Bendall, B., Hogarth, R., Holl, H., Mcmahon, A., Larking, A., Reid, P., 2014. Australian experiences in EGS permeability enhancement—a review of 3 case studies. In: Proceedings of the 39th Workshop on Geothermal Reservoir Engineering. Stanford University, Stanford, CA.

Bertini, G., Casini, M., Gianelli, G., Pandeli, E., 2006. Geological structure of a long-living geothermal system, Larderello, Italy. Terra Nova 18 (3), 163–169.

Blöcher, G., Peters, E., Kluge, C., Ilangovan, N., Bruhn, D., Nick, H., 2016. The Horizon 2020 SURE Project: D3.1 Report on Stimulation Technologies for Geothermal Reservoirs. Helmholtz Centre Potsdam.

Bradford, J., McLennan, J., Moore, J., Podgorney, R., Plummer, M., Nash, G., 2017. Analysis of the thermal and hydraulic stimulation program at Raft River, Idaho. Rock Mech. Rock. Eng. 50, 1279–1287.

Breede, K., Dzebisashvili, K., Liu, X., Falcone, G., 2013. A systematic review of enhanced (or engineered) geothermal systems: past, present and future. Geotherm. Energy 1, 1–27.

Brown, D.W., February **2009**. Hot dry rock geothermal energy: important lessons from Fenton Hill. In: Proceedings of the Thirty-Fourth Workshop on Geothermal Reservoir Engineering. Stanford University, Stanford.

Burns, K., Weber, C., Perry, J., Harrington, H., 2000. Status of the geothermal industry in Australia. In: World Geothermal Congress. International Geothermal Association, pp. 99–108.

Callahan, T.J., February 1996. Reservoir investigation on the hot dry rock geothermal system, Fenton Hill, New, Mexico; tracer test results. In: Proceedings of the 21st Workshop on Geothermal Reservoir Engineering. Stanford University, Stanford, CA.

Cao, W., Huang, W., Jiang, F., 2016. Numerical study on variable thermophysical properties of heat transfer fluid affecting EGS heat extraction. Int. J. Heat Mass Transf. 92, 1205–1217.

Carpenter, A.T., 2013. Water and hydraulic fracturing. J. Am. Water Works Assoc. 105 (3), 56–59.

Chabora, E., Zemach, E., Spielman, P., Drakos, P., Hickman, S., Lutz, S., Boyle, K., Falconer, A., Robertson-Tait, A., Davatzes, N.C., 2012. Hydraulic stimulation of well 27-15, Desert Peak geothermal field, Nevada, USA. In: Proceedings of thirty-seventh workshop on geothermal reservoir engineering, Stanford University, Stanford. 30.

Chen, C., Martysevich, V., O'Connell, P., Hu, D., Matzar, L., 2015. Temporal evolution of the geometrical and transport properties of a fracture/proppant system under increasing effective stress. SPE J. 20 (03), 527–535.

Childers, I.M., Endres, M., Burns, C., Garcia, B.J., Liu, J., Wietsma, T.W., Fernandez, C.A., 2017. Novel highly dispersible, thermally stable core/shell proppants for geothermal applications. Geothermics 70, 98–109.

Cladouhos, T.T., Osborn, W.L., Petty, S., Bour, D., Iovenitti, J., Callahan, O., Nordin, Y., Perry, D., Stern, P.L., 2012. Newberry volcano EGS demonstration—phase I results. In: Proceedings of 37th Workshop on Geothermal Reservoir Engineering, Stanford, CA. Vol. 30.

Clauser, C., 2006. The Earth's thermal regime: group VIII, geothermal energy. Renew. Energy, 493–548.

Concha, D., Fehler, M., Zhang, H., Wang, P., February 2010. Imaging of the Soultz enhanced geothermal reservoir using microseismic data. In: Proceedings of the 35th Workshop on Geothermal Reservoir Engineering. Stanford University, Stanford, CA. SGP-TR-188.

Cornet, F., 1987. Results from Le Mayet de Montagne project. Geothermics 16 (4), 355–374.

Cuderman, J.F., 1982. Multiple fracturing experiments-propellant and borehole considerations. In: SPE Unconventional Resources Conference/Gas Technology Symposium. Society of Petroleum Engineer. SPE-10845.

Cuderman, J.F., 1986. Effects of wellbore liquids in propellant-based fracturing. In: ARMA US Rock Mechanics/Geomechanics Symposium. American Rock Mechanics Association. ARMA-86.

Dezayes, C., Genter, A., Valley, B., 2010. Structure of the low permeable naturally fractured geothermal reservoir at Soultz. Compt. Rendus Geosci. 342 (7–8), 517–530.

Dong, J., Chen, M., Li, Y., Wang, S., Zeng, C., Zaman, M., 2019. Experimental and theoretical study on dynamic hydraulic fracture. Energies 12 (3), 397.

Duan, K., et al., 2016. Numerical simulation of enhanced geothermal systems: a review. Renew. Sust. Energ. Rev. 56, 1–17.

Duchane, D.V., Winchester, W.W., 1993. Hot Dry Rock Energy Annual Report Fiscal Year 1992. No. LA-UR-93-1678, Los Alamos National Lab. (LANL), Los Alamos, NM (United States).

Economides, M., Martin, T., 2007. Modern Fracturing Enhancing Natural Gas Production. ET Publishing, Houston, TX, USA.

Eden, 2023. Eden Is Breaking Ground on Reservoir Stimulation Technology to Revolutionize the Geothermal Industry. Eden (accessed).

Egert, R., Korzani, M.G., Held, S., Kohl, T., 2020. Implications on large-scale flow of the fractured EGS reservoir Soultz inferred from hydraulic data and tracer experiments. Geothermics 84, 101749.

Ellsworth, W.L., 2013. Injection-induced earthquakes. Science 341 (6142), 1225942.

Elsworth, D., Spiers, C.J., Niemeijer, A.R., 2016. Understanding induced seismicity. Science 354 (6318), 1380–1381.

Evans, K.F., Zappone, A., Kraft, T., Deichmann, N., Moia, F., 2012. A survey of the induced seismic responses to fluid injection in geothermal and CO_2 reservoirs in Europe. Geothermics 41, 30–54.

Fan, M., McClure, J., Han, Y., Ripepi, N., Westman, E., Gu, M., Chen, C., 2019. Using an experiment/simulation-integrated approach to investigate fracture-conductivity evolution and non-Darcy flow in a proppant-supported hydraulic fracture. SPE J. 24 (04), 1912–1928.

Fan, M., Han, Y., Chen, C., 2021. Thermal–mechanical modeling of a rock/proppant system to investigate the role of shale creep on proppant embedment and fracture conductivity. Rock Mech. Rock. Eng. 54, 6495–6510.

Fernandez, C.A., Gupta, V., Dai, G.L., Kuprat, A.P., Bonneville, A., Appriou, D., Burghardt, J.A., 2019. Insights into a greener stimuli-responsive fracturing fluid for geothermal energy recovery. ACS Sustain. Chem. Eng. 7 (24), 19660–19668.

Flores, M., Davies, D., Couples, G., Palsson, B., 2005. Stimulation of geothermal wells, can we afford it (Proceedings World Geothermal Congress). Vol. 1.

Gallagher, D.G., 2011. The hierarchy of oily conductivity. J. Pet. Technol. 63 (04), 18–19.

Gaucher, E., Schoenball, M., Heidbach, O., Zang, A., Fokker, P.A., van Wees, J.D., Kohl, T., 2015. Induced seismicity in geothermal reservoirs: a review of forecasting approaches. Renew. Sust. Energ. Rev. 52, 1473–1490.

Ghassemi, A., 2012. A review of some rock mechanics issues in geothermal reservoir development. Geotech. Geol. Eng. 30 (3), 647–664.

Gischig, V.S., Giardini, D., Amann, F., Hertrich, M., Krietsch, H., Loew, S., Maurer, H., Villiger, L., Wiemer, S., Bethmann, F., 2020. Hydraulic stimulation and fluid circulation experiments in underground laboratories: stepping up the scale towards engineered geothermal systems. Geomech. Energy Environ. 24, 100175.

Gómez-Díaz, E., Scott, S., Ratouis, T., Newson, J., 2022. Numerical modeling of reinjection and tracer transport in a shallow aquifer, Nesjavellir Geothermal System, Iceland. Geotherm. Energy 10 (1), 1–28.

Grant, M.A., 2011. Geothermal reservoir engineering. In: Geothermal Energy: An Alternative Resource for the 21st Century. Elsevier, pp. 147–168.

Grant, M.A., Clearwater, J., Quinão, J., Bixley, P.F., Le Brun, M., 2013. Thermal stimulation of geothermal wells: a review of field data. In: Proceedings, Thirty-Eighth Workshop on Geothermal Reservoir Engineering.

Guo, T., Wang, Y., Du, Z., Chen, M., Liu, D., Liu, X., Rui, Z., 2021. Evaluation of coated proppant unconventional performance. Energy Fuel 35 (11), 9268–9277.

Hahne, B., Thomas, R., 2014. Deep geothermal energy for Lower Saxony (North Germany)—combined investigations of geothermal reservoir characteristics. Energy Procedia 59, 198–204.

Hawkins, A.J., Becker, M.W., Tester, J.W., 2018. Inert and adsorptive tracer tests for field measurement of flow-wetted surface area. Water Resour. Res. 54 (8), 5341–5358.

Hébert, R.L., Ledésert, B., Bartier, D., Dezayes, C., Genter, A., Grall, C., 2010. The enhanced geothermal system of Soultz-sous-Forêts: a study of the relationships between fracture zones and calcite content. J. Volcanol. Geotherm. Res. 196 (1–2), 126–133.

Hébert, R.L., Ledésert, B., Genter, A., Bartier, D., Dezayes, C., February 2011. Mineral precipitation in geothermal reservoir: the study case of calcite in the Soultz-sous-For.ts enhanced geothermal system. In: Proceedings of the 36th Workshop on Geothermal Reservoir Engineering. Stanford University, Stanford, CA. SGP-TR-191.

Huang, Z., Zhang, S., Yang, R., Wu, X., Li, R., Zhang, H., Hung, P., 2020. A review of liquid nitrogen fracturing technology. Fuel 266, 117040.

Jahan Bakhsh, K., 2017. Effects of Thermally-Induced Fractures on EGS Performance. 2017-Mines Theses & Dissertations, Colorado School of Mines.

Jia, Y., Tsang, C.-F., Hammar, A., Niemi, A., 2022. Hydraulic stimulation strategies in enhanced geothermal systems (EGS): a review. Geomech. Geophys. Geo-Energy Geo-Resour. 8 (6), 211.

Jian, G., Sarathi, R.S., Burghardt, J., Bonneville, A., Gupta, V., Fernandez, C.A., Garrison, G., 2021. Evaluation of a greener fracturing fluid for geothermal energy recovery: an experimental and simulation study. Geothermics 97, 102266.

Jian, G., Sarathi, R.S., Burghardt, J., Bonneville, A., Fernandez, C.A., 2023. Effect of initial water saturation on the performance of fracturing fluids with and without polyallylamine under simulated EGS conditions. Geothermics 111, 102715.

Jung, H.B., Carroll, K.C., Kabilan, S., Heldebrant, D.J., Hoyt, D., Zhong, L., Fernandez, C.A., 2015. Stimuli-responsive/rheoreversible hydraulic fracturing fluids as a greener alternative to support geothermal and fossil energy production. Green Chem. 17 (5), 2799–2812.

Jupe, A., Green, A., Wallroth, T., 1992. Induced microseismicity and reservoir growth at the Fjällbacka hot dry rocks project, Sweden. Int. J. Rock Mech. Min. Sci. Geomech. Abstr. 29, 343–354. Elsevier.

Kaieda, H., Sasaki, S., Wyborn, D., 2010. Comparison of characteristics of micro-earthquakes observed during hydraulic stimulation operations in Ogachi, Hijiori and Cooper Basin HDR projects. In: Proceedings of the World Geothermal Congress.

Karvounis, D., Jenny, P., February 2011. Modeling of flow and transport in enhanced geothermal systems. In: Proceedings of the 36th Workshop on Geothermal Reservoir Engineering. Stanford University, Stanford, CA. SGP-TR-191.

Khan, A.M., BinZiad, A., Alshammari, F.O., Glaznev, I., 2023. Fiber assisted enhanced CO_2 foam fracturing and proppant placement. In: SPE Annual Technical Conference and Exhibition, Society of Petroleum Engineer, October. D011S006R005.

Khosravi, E., Ghassemi, A., June **2017**. Influence of temperature and anisotropy on creep behavior of Mancos shale. In: In 51st US Rock Mechanics/Geomechanics Symposium. American Rock Mechanics Association.

Kim, K.I., Min, K.B., Kim, K.Y., Choi, J.W., Yoon, K.S., Yoon, W.S., Song, Y., 2018. Protocol for induced microseismicity in the first enhanced geothermal systems project in Pohang, Korea. Renew. Sust. Energ. Rev. 91, 1182–1191.

Knoblauch, T.A., Trutnevyte, E., 2018. Siting enhanced geothermal systems (EGS): heat benefits versus induced seismicity risks from an investor and societal perspective. Energy 164, 1311–1325.

Ko, S., Ghassemi, A., Uddenberg, M., 2023. Selection and Testing of Proppants for EGS. In Stanford University.

Koroncz, P., Vizhányó, Z., Farkas, M.P., Kuncz, M., Ács, P., Kocsis, G., Mucsi, P., Fedorné Szász, A., Fedor, F., Kovács, J., 2022. Experimental rock characterisation of upper Pannonian sandstones from Szentes Geothermal Field, Hungary. Energies 15 (23), 9136.

Ladner, F., Häring, M.O., 2009. Hydraulic characteristics of the basel 1 enhanced geothermal system. GRC Trans. 33, 199–203.

Law, R., Cotton, L., Ledingham, P., 2019. The united downs deep geothermal power project. In: Proceedings of the European Geothermal Congress, Den Haag, The Netherlands, pp. 11–14.

Ledésert, B., Hébert, R.L., Grall, C., Genter, A., Dezayes, C., Bartier, D., Gérard, A., 2009. Calcimetry as a useful tool for a better knowledge of flow pathways in the Soultz-sous-Forêts Enhanced Geothermal System. J. Volcanol. Geotherm. Res. 181 (1–2), 106–114.

Li, S., Zhang, S., Ma, X., Zou, Y., Li, N., Chen, M., Bo, Z., 2019. Hydraulic fractures induced by water-/carbon dioxide-based fluids in tight sandstones. Rock Mech. Rock. Eng. 52, 3323–3340.

Li, Z., Zhao, Q., Teng, Y., Fan, M., Ripepi, N., Yin, X., Chen, C., 2022. Experimental investigation of non-monotonic fracture conductivity evolution in energy georeservoirs. J. Pet. Sci. Eng. 211, 110103.

Lima, M.G., Vogler, D., Querci, L., Madonna, C., Hattendorf, B., Saar, M.O., Kong, X.Z., 2019. Thermally driven fracture aperture variation in naturally fractured granites. Geotherm. Energy 7 (1), 1–28.

Lummer, N.R., Rauf, O., Gerdes, S., 2015. Next generation acid system–first field trial of a new biodegradable fluid in a sandstone/granite formation. In: SPE European Formation Damage Conference and Exhibition. SPE. SPE-174242-MS.

Lutz, S.J.Z., Ashwani, Robertson-Tait, A., Drakos, P., Zemach, E., 2011. Lithologies, hydrothermal alteration, and rock mechanical properties in wells 15-12 and BCH-3, Bradys Hot Springs Geothermal Field, Nevada. Geotherm. Resour. Council Trans. 35, 469–576.

Malate, R., Austria, J., Sarmiento, Z., Di Lullo, G., Sookprasong, P., Francia, E., 1998. Matrix stimulation treatment of geothermal wells using sandstone acid. In: Proc. 23rd Workshop on Geothermal Reservoir Engineering.

Masłowski, M., Labus, M., 2021. Preliminary studies on the proppant embedment in Baltic basin shale rock. Rock Mech. Rock. Eng. 2021 (54), 2233–2248.

Matsunaga, I., Sugita, H., Tao, H., February 2001. Tracer monitoring by a fiber-optic fluorometer during a long-term circulation test at the Hijiori HDR site. In: Proceedings of the 26th Workshop on Geothermal Reservoir Engineering. Stanford University, Stanford, CA.

Matsunaga, I., Yanagisawa, N., Sugita, H., Tao, H., February 2002. Reservoir monitoring by tracer test during a long-term circulation test at the Hijiori HDR site. In: Proceedings of the 27th Workshop on Geothermal Reservoir Engineering. Stanford University, Stanford, CA.

Mattson, E., Zhang, Y., Hawkins, A., Johnson, T., Ajo-Franklin, J., Neupane, G., et al., February 2019. Preliminary Collab fracture characterization results from flow and tracer testing efforts. In: Proceedings of the 44th Workshop on Geothermal Reservoir Engineering. Stanford University, Stanford, California.

Maury, J., Hamm, V., Loschetter, A., et al., 2022. Development of a risk assessment tool for deep geothermal projects: example of application in the Paris Basin and Upper Rhine graben. Geotherm. Energy 10, 26.

McClure, M.W., Horne, R.N., 2013. An investigation of stimulation mechanisms in enhanced geothermal systems. Int. J. Rock Mech. Min. Sci. 61, 1–12.

McClure, M.W., Horne, R.N., 2014. An investigation of stimulation mechanisms in enhanced geothermal systems. Int. J. Rock Mech. Min. Sci. 72, 242–260.

Mehrvand, M., Akhmadullin, I., Alali, A., Smalls, P., 2021. System and Method for Pulsed Electrical Reservoir Stimulation. US11091991B1.

Michelet, S., Toksoz, M.N., 2006. Fracture Mapping in the Soultz-sous-Forêts Geothermal Field from Microearthquake Relocation. Massachusetts Institute of Technology, Earth Resources Laboratory.

Miller, J.S., Johansen, R.T., 1976. Fracturing oil shale with explosives for in situ recovery. In: Shale Oil, Tar Sands, and Related Fuel Sources. Advances in Chemistry, American Chemical Society, pp. 98–111.

Moeck, I., Bloch, T., Graf, R., Heuberger, S., Kuhn, P., Naef, H., Sonderegger, M., Uhlig, S., Wolfgramm, M., 2015. The St. Gallen project: development of fault controlled geothermal systems in urban areas. In: Proceedings World Geothermal Congress, p. 5.

Nami, P., Schindler, M., Tischner, T., Teza, D., 2007. Hydraulic and chemical stimulation experiments in deep crystalline formation for geothermal power production. In: First European Geothermal Review—Geothermal Energy for Power Production, pp. 80–92.

Nami, P., Schellschmidt, R., Schindler, M., Tischner, T., 2008a. Chemical stimulation operations for reservoir development of the deep crystalline HDR/EGS at Soultzsous-For.ts (France). In: Proceedings of the Thirty-Third Workshop on Geothermal Reservoir Engineering, Stanford University, Stanford, California, January. SGP-TR-185.

Nami, P., Schindler, M., Tischner, T., Teza, D., 2008b. Chemical stimulation results at Soultz and overview of the flow performance of the wells. In: Proceedings, EHDRA Scientific Conference, pp. 24–25.

Park, S., Xie, L., Kim, K.-I., Kwon, S., Min, K.-B., Choi, J., Yoon, W.-S., Song, Y., 2017. First hydraulic stimulation in fractured geothermal reservoir in Pohang PX-2 well. Procedia Eng. 191, 829–837.

Parker, R., 1999. The rosemanowes HDR project 1983–1991. Geothermics 28 (4–5), 603–615.

Pemecker, G., 1999. Altheim Geothermal Plant for Electricity Production by ORC-Turbogenerator. Centre d'hrydrogiologie Universitd de Neuchatel, Editions Peter Lang, Altheim, Austria.

Plummer, M., Bradford, J., Moore, J., Podgorney, R., 2016. Reservoir Response to Thermal and High-Pressure Well Stimulation Efforts at Raft River, Idaho. No. INL/CON-16-39543, Idaho National Lab. (INL), Idaho Falls, ID (United States).

Porlles, J.W., Tomomewo, O.S., Afari, S.A., Gyimah, E., Laalam, A., Bakelli, O., 2023. An experimental study of the effect of long-term time-dependent proppant behavior under HP-HT reservoir conditions. In: SPE Annual Technical Conference and Exhibition. Society of Petroleum Engineer. D021S016R003.

Portier, S., Vuataz, F.D., 2010. Developing the ability to model acid-rock interactions and mineral dissolution during the RMA stimulation test performed at the Soultz-sous-Forêts EGS site, France. Compt. Rendus Geosci. 342 (7–8), 668–675.

Portier, S., André, L., Vuataz, F.-D., 2006. Review of chemical stimulation techniques and results of acid injection experiments at Soultz-sous-Forêts. In: Proceedings of EHDRA Scientific Conference.

Portier, S., André, L., Vuataz, F.-D., 2007. Review on chemical stimulation techniques in oil industry and applications to geothermal systems. In: ENGINE, Work Package. 4. Deep Heat Mining Association (DHMA), Switzerland, p. 32.

Portier, S., Vuataz, F.D., Nami, P., Sanjuan, B., Gérard, A., 2009. Chemical stimulation techniques for geothermal wells: experiments on the three-well EGS system at Soultz-sous-Forêts, France. Geothermics 38 (4), 349–359.

Radilla, G., Sausse, J., Sanjuan, B., Fourar, M., February 2010. Tracer tests in the enhanced geothermal system (EGS) of Soultz–Sous–Forêts. What does the stratified medium approach tell us about the fracture permeability in the reservoir. In: 35th Stanford Workshop on Geothermal Reservoir Engineering (WGRE-35). Stanford University, Stanford, CA.

Rafiee, M., Soliman, M., Pirayesh, E., Meybodi, H., 2012. Geomechanical considerations in hydraulic fracturing designs. In: Proceedings of the SPE Canadian Unconventional Resources Conference, Calgary, AB, Canada, November.

Raysoni, N., Weaver, J., 2013. Long-term hydrothermal proppant performance. SPE Prod. Oper. 28 (04), 414–426.

Riffault, J., Dempsey, D., Karra, S., Archer, R., 2018. Microseismicity cloud can be substantially larger than the associated stimulated fracture volume: the case of the Paralana Enhanced Geothermal System. J. Geophys. Res. Solid Earth 123 (8), 6845–6870.

Rodrigues, N., 1993. Tracer experiment results during the long-term flow test of the Fenton Hill reservoir. In: Proceedings of the 18th Workshop on Geothermal Reservoir Engineering. Stanford University, Stanford, CA.

Rose, P.E., 2013. Creation of an Enhanced Geothermal System Through Hydraulic and Thermal Stimulation. Energy and Geoscience Institute at the University of Utah.

Sanjuan, B., Rose, P., Foucher, J., Brach, M., 2004. Tracer testing at Soultz-Sous-Forêts (France) using Na-benzoate, 1,5 and 2,7-naphthalene disulfonate. In: Proceedings of the 29th Workshop on Geothermal Reservoir Engineering. Stanford University, Stanford, CA.

Sanjuan, B., Michel, B., Genter, A., Sanjuan, R., Scheiber, J., Touzelet, S., 2015. Tracer testing of the EGS site at Soultz-sous-Forêts (Alsace, France) between 2005 and 2013. In: World Geothermal Congress 2015, p. 12.

Satman, A., February 2011. Sustainability of geothermal doublets. In: Proceedings of the 36th Workshop on Geothermal Reservoir Engineering. Stanford University, Stanford, CA. SGP-TR-191.

Sausse, J., Dezayes, C., Dorbath, L., Genter, A., Place, J., 2010. 3D fracture zone network at Soultz based on geological data, Image logs, microseismic events and VSP results. CR Geosci. 342, 531–545.

Schindler, M., Baumgärtner, J., Gandy, T., Hauffe, P., Hettkamp, T., Menzel, H., Penzkofer, P., Teza, D., Tischner, T., Wahl, G., 2010. Successful hydraulic stimulation techniques for electric power production in the Upper Rhine Graben, Central Europe. In: Proceedings World Geothermal Congress. Vol. 2010.

Schrage, C., Bems, C., Kreuter, H., Hild, S., Volland, S., 2012. Overview of the Enhanced Geothermal Energy Project in Mauerstetten, Germany. T&A (Technology & Advice) Survey, Amsterdam, Netherlands.

Shao, H., Kabilan, S., Stephens, S., Suresh, N., Beck, A.N., Varga, T., Martin, P.F., Kuprat, A., Jung, H.B., Um, W., Bonneville, A., Heldebrant, D.J., Carroll, K.C., Moore, J., Fernandez, C.A., 2015. Environmentally friendly, rheoreversible, hydraulic-fracturing fluids for enhanced geothermal systems. Geothermics 58, 22–31.

Sharma, M.M., Gadde, P.B., Sullivan, R., Sigal, R., Fielder, R., Copeland, D., Griffin, L., Weijers, L., 2004. Slick water and hybrid fracs in the Bossier: some lessons learnt. In: SPE Annual Technical Conference and Exhibition. OnePetro.

Sone, H., Zoback, M.D., 2014. Time-dependent deformation of shale gas reservoir rocks and its long-term effect on the in situ state of stress. Int. J. Rock Mech. Min. Sci. 69, 120–132.

Song, X., Guo, Y., Zhang, J., Sun, N., Shen, G., Chang, X., Xue, Z., 2019. Fracturing with carbon dioxide: from microscopic mechanism to reservoir application. Joule 3 (8), 1913–1926.

Strawn, J., 1980. Results of Acid Treatments in Hydrothermal Direct Heat Experiment Wells. EG and G Idaho, Inc., Idaho Falls (USA).

Tenzer, H., Schanz, U., Stang, H., Baisch, S., Weidler, R., 2004. Potential for geothermal power production from deep hot crystalline rock at Bad Urach (Southern Germany). Z. Angew. Geol. 50 (2), 43–49.

Tester, J.W., et al., 2006a. The Future of Geothermal Energy: Impact of Enhanced Geothermal Systems (EGS) on the United States in the 21st Century. Massachusetts Institute of Technology.

Tester, J.W., Anderson, B.J., Batchelor, A.S., Blackwell, D.D., DiPippo, R., Drake, E.M., Garnish, J., Livesay, B., Moore, M., Nichols, K., 2006b. The future of geothermal energy. Massachusetts Inst. Technol. 358, 1–3.

Thakore, V., Wang, H., Wang, J.A., Polsky, Y., Ren, F., 2024. Stability study of aqueous foams under high-temperature and high-pressure conditions relevant to enhanced geothermal systems (EGS). Geothermics 116, 102862.

Toews, M., Riddell, D., Vany, J., Schwarz, B., 2020. Case study of a multilateral closed-loop geothermal system. In: Proceedings World Geothermal Congress, Reykjavik, Iceland, April.

Tulinius, H., Correia, H., Sigurdsson, O., 2000. Stimulating a high enthalpy well by thermal cracking. In: World Geothermal Congress.

Walters, M., Dobson, P., 2013. Demonstration of an Enhanced Geothermal System at the Northwest Geysers Geothermal Field, CA. Geothermal Technologies Office.

Wang, C., Lo, W., Song, S.-R., 2018. Geothermal play fairway analysis at a populated rifting basin area of Taiwan. Geothermics 75, 146–153.

Wasantha, P.L.P., Xu, T., Xu, B., October 2021. Effect of fluid injection rate on the hydraulic fracture propagation characteristics. In: IOP Conference Series: Earth And Environmental Science. Vol. 861. IOP Publishing, p. 032095. No. 3.

Weaver, J., Parker, M., van Batenburg, D., Nguyen, P., 2007. Fracture-related diagenesis may impact conductivity. SPE J. 12 (03), 272–281.

Wu, H., Fu, P., Hawkins, A.J., Tang, H., Morris, J.P., 2021a. Predicting thermal performance of an enhanced geothermal system from tracer tests in a data assimilation framework. Water Resour. Res. 57 (12). e2021WR030987.

Wu, Y., Cao, M., Zhao, Q., Wu, X., Guo, F., Tang, L., Dai, C., 2021b. Novel high-hydrophilic carbon dots from petroleum coke for boosting injection pressure reduction and enhancing oil recovery. Carbon 184, 186–194.

Xiaofei, Q., Shuantong, S., Guobin, Z., Miaomiao, P., Ye, S., Lanlan, T., Xiang, L., Yongchao, Q., Jianyong, Z., 2020. Site selection and developmental prospect of a hot dry rock resource project in the Matouying Uplift, Hebei Province. Earth Sci. Front. 27 (1), 94.

Xing, P., Damjanac, B., Radakovic-Guzina, Z., Finnila, A., Podgorney, R., Moore, J., McLennan, J., June 2021. Numerical simulation of hydraulic fracturing stimulation enhanced geothermal system well at Utah forge site. In: Paper presented at the 55th U.S. Rock Mechanics/Geomechanics Symposium, Virtual.

Xu, J., Feng, B., Cui, Z., Liu, X., Ke, Z., Feng, G., 2021. Comparative study of acid and alkaline stimulants with granite in an enhanced geothermal system. Acta Geol. Sin. Engl. Ed. 95 (6), 1926–1939.

Yanagisawa, N., 2009. First tracer test at Cooper-basin, Australia HDR reservoir. Geotherm. Resour. Council Trans. 33, 281–284.

Yanagisawa, N., Matsunaga, I., Tao, H., Sugita, H., 2001. Evaluation by tracer experiments in the Hijiori HDR Reservoir. J. Geotherm. Res. Soc. Jpn. 23 (4), 293–301.

Yanagisawa, N., Ngothai, Y., Rose, P., Wyborn, D., 2011. Geochemical behavior of EGS reservoir during first circulation test at Habanero Site, Cooper Basin, Australia. J. Geotherm. Res. Soc. Jpn. 33 (3), 125–130.

Yang, F., Nasr-El-Din, H.A., Al-Harbi, B., 2012. Acidizing sandstone reservoirs using HF and organic acids. In: SPE International Production and Operations Conference and Exhibition. SPE: pp SPE-157250-MS.

Yang, Z., Zhang, B., Bi, J., 2023. Laboratory investigation of cryogenic fracturing of HDR wellbores under triaxial-confining stresses. J. Phys. Conf. Ser. 2520 (1), 012016. IOP Publishing.

Zeng, C., Wu, C., Deng, W., 2019. Thermal mechanical analysis of porous granite asperities in contact using probabilistic damage model. Int. J. Rock Mech. Min. Sci. 124, 104130.

Zeng, C., Deng, W., Wu, C., Insall, M., 2020. Thermal effect of cold fluid injection on fracture integrity of host rock: a microscale analysis of single asperity. Geothermics 87, 101872.

Zhong, C., Xu, T., Yuan, Y., Feng, B., Yu, H., 2022. The feasibility of clean power generation from a novel dual-vertical-well enhanced geothermal system (EGS): A case study in the Gonghe Basin, China. J. Clean. Prod. 344, 131109.

Zhou, H., Zhang, Y., Li, H., Han, R., Jing, Y., Liu, Q., Qiu, A., 2015. Generation of electrohydraulic shock waves by plasma-ignited energetic materials: III. Shock wave characteristics with three discharge loads. IEEE Trans. Plasma Sci. 43 (12), 4017–4023.

Zimmermann, G., Hurter, S., Saadat, A., Köhler, S., Trautwein, U., Holl, H., Wolfgramm, M., Winter, H., Legarth, B., Huenges, E., 2003. The in-situ geothermal laboratory Groß Schönebeck-stimulation experiments of sandstones in 4200 m depth. In: Proc. Twenty-Eight Workshop on Geothermal Reservoir Engineering. Stanford University, Stanford, California, pp. 27–29.

Zimmermann, G., Tischner, T., Legarth, B., Huenges, E., 2009. Pressure-dependent production efficiency of an enhanced geothermal system (EGS): stimulation results and implications for hydraulic fracture treatments. In: Rock Physics and Natural Hazards. Birkhäuser, Basel, pp. 1089–1106.

Zimmermann, G., Moeck, I., Blöcher, G., 2010. Cyclic waterfrac stimulation to develop an enhanced geothermal system (EGS)—conceptual design and experimental results. Geothermics 39 (1), 59–69.

Zimmermann, G., Blöcher, G., Reinicke, A., Brandt, W., 2011. Rock specific hydraulic fracturing and matrix acidizing to enhance a geothermal system—concepts and field results. Tectonophysics 503 (1–2), 146–154.

CHAPTER

9

Production and operation of geothermal systems

Pejman Shoeibi Omrani[a,b] *and Govert de With*[c]

[a]Heat Transfer and Fluid Dynamics Department, TNO, The Hague, The Netherlands
[b]Environmental Technology, Wageningen University and Research, Wageningen, The Netherlands [c]NRG, Arnhem, The Netherlands

Key terms and definitions

Individual explanation of each key term needed to understand this topic.

Abbreviations

ADC	analog-to-digital converter
CRA	corrosion-resistant alloys
DIS	direct ion storage
ESP	electrical submersible pump
MCA	multichannel analyzer
MIC	microbiologically influenced corrosion
NCG	noncondensable gases
NORM	naturally occurring radioactive materials
OLS	optically stimulated luminescence
TLD	thermoluminescent dosimeter

1 Introduction

An efficient production of geothermal energy requires a deep understanding of the operational issues of these assets and defining accurate and robust operational strategies. A schematic of a typical geothermal system for heating applications is shown in Fig. 1. The figure shows a specific doublet configuration, consisting of a production well to produce the geothermal brine from the geothermal reservoir and an injection well to inject the brine back into the reservoir. The process of production and injection can be enhanced and

https://doi.org/10.1016/B978-0-443-21662-6.00002-5

Copyright © 2025 Elsevier Inc. All rights are reserved, including those for text and data mining, AI training, and similar technologies.

controlled using production pumps and injection pumps, respectively. On the surface, there are a number of equipment, including degassers to separate the by-produced gas from the brine, filters to separate particles from the brine system, and a heat exchanger to harvest the thermal energy from the geothermal source. It is optional to include all this equipment, and some of the plants have other surface facilities, including a heat pump, CHP unit (combined heat and power), corrosion and scaling inhibitor line, and other monitoring systems.

As can be seen from the plant schematic, in the journey from production to injection within a geothermal reservoir, the brine undergoes significant changes in both pressure and temperature. During the production phase, as the brine produces from the subsurface reservoir to the surface, it experiences a decrease in pressure. This reduction in pressure occurs as the fluid ascends through the wellbore, transitioning from the high-pressure environment of the reservoir to the lower-pressure conditions at the surface. Simultaneously, the temperature of the geothermal brine tends to decrease, reflecting the heat exchange with the surrounding rocks during its production and transport through the surface heat exchanger. In contrast, during the injection phase, as the brine is returned to the reservoir to sustain reservoir pressure and optimize heat recovery, it encounters an increase in pressure. Furthermore, the temperature of the reinjected brine may rise as it interfaces with the high-temperature reservoir rocks. The interaction of pressure and temperature variations with the chemistry of the geothermal brine during the production and injection cycles is a key consideration in effectively managing geothermal reservoirs.

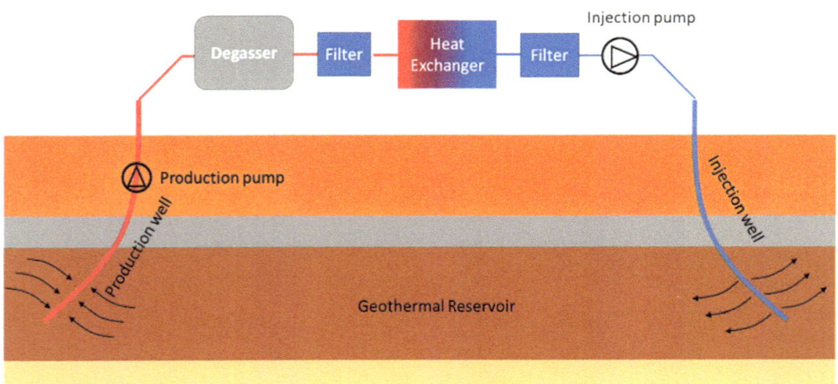

FIG. 1 Schematic of a geothermal plant for direct use heating application.

Geothermal energy production systems are susceptible to an array of operational and production arises from the changes in the pressure and temperature that the fluid experiences and the change in the local chemical properties of the brine. These challenges can intricately influence their efficiency, cost, and safety. Issues encompass a spectrum from declines in productivity and injectivity to flow assurance challenges such as scaling, corrosion, ALS failures, and the emergence of naturally occurring radioactive materials (NORMs). Operational details evident from equipment failures, maintenance concerns, and power plant performance anomalies are capable of disrupting the seamless functioning of geothermal systems.

Operational disruptions, such as ALS failures, can lead to production downtimes, highlighting the need for proactive intervention. Maintenance issues, characterized by pipe obstructions or leaks, introduce inefficiencies, undermining the overall effectiveness of these systems. The remediation of such challenges necessitates a complex approach involving continuous monitoring, targeted maintenance, and strategic upgrades within geothermal production systems.

This chapter provides a brief technical highlight on several challenges in geothermal production and operations of geothermal assets and methods to monitor and mitigate these challenges. The main focus of this chapter will be on low-enthalpy geothermal systems for heating applications. However, wherever possible, links to the challenges in high-enthalpy systems are made. The operational challenges during geothermal energy production which are covered in this chapter are scaling, corrosion, NORMs, challenges in ALSs, sand production, and noncondensable gases (NCGs).

2 Scaling

The composition and mineral content of the geothermal fluids produced are dependent on geological conditions, formation properties, depth (pressure), and temperature. Because the topside conditions differ from the conditions in the reservoir, this can impact the solubility of dissolved minerals and gases in the geothermal brine. Solids can be formed from the minerals and precipitate when the concentration of the dissolved minerals exceeds their solubility in the brine. Each of the minerals (scales) has a particular solubility which is dependent on pressure and temperature and the ion composition of the fluid.

There are several parameters which impact the precipitation including saturation ratio, crystallization kinetics, pipe or substrate surface properties, and flow regimes (Raheem et al., 2021). The main thermodynamic mechanism for the crystallization which is the main cause of scaling is either in the fluid bulk or in interaction with a pipe surface. In order to better explain the mechanisms of scaling, the concept of supersaturation needs to be defined as this is important for both crystal nucleation and growth. Supersaturation is a nonequilibrium, physical state in which a solution contains more solute than the equilibrium solubility allows at a given condition such as temperature, pressure, and composition of the fluid, defined in a simple manner as follows:

$$\Delta C_{\text{super saturation}} = C_{\text{actual concentration}} - C^{*}{}_{\text{solubility}}$$

The Gibbs free energy change from the supersaturated state to equilibrium is a driving force for the precipitation. For the scaling to occur, the supersaturation ratio of crystalline should exceed the unity that is described by the saturation index. Saturation index (Ω) is a parameter which describes the "tendency" of the brine to precipitate the dissolved minerals. The saturation index is given by the ion activity product, divided by the solubility product:

$$\Omega = \frac{I_{\text{AP}}}{K_{\text{SP}}}$$

where K_{sp} is the solubility product and I_{AP} is ion activity. For $\Omega > 1$, the brine is supersaturated and the respective ions will react and precipitate.

Scaling can have several significant impacts on the performance and efficiency of geothermal systems. One of the most common observed issues is the clogging of the wells, installations, equipment such as pumps and filters, and reservoir pores which will directly reduce the productivity and injectivity (Hartog, 2016). Scale deposits act as an insulating layer which can reduce the heat transfer efficiency of geothermal assets if they precipitate in the heat exchanger (Scheiber et al., 2014). Scale deposits can accumulate on the surfaces of pumps, valves, and other equipment, leading to mechanical issues such as increased wear and tear, reduced equipment lifespan, and potential operational failures. In addition, scaling can create localized corrosion which will be further discussed in the corrosion section. This necessitates increased maintenance and can lead to production downtime.

2.1 Type of scaling in geothermal sites

2.1.1 *Carbonates*

Carbonate scaling especially calcium carbonate is one of the most common types of scaling in geothermal assets. Next to the mineral content, also the dissolved gases in the brine can influence the precipitation of the minerals. The most common gases in geothermal assets are methane, ethane, carbon dioxide, and hydrogen sulfide. Especially CO_2 can play an important role in the precipitation of calcite (or one of its variants such as aragonite). In an equilibrium state, calcite equilibrates with the brine following this reaction:

$$CaCO_3 = Ca^{2+} + CO_3{}^{2-}$$

And the solubility product of the calcite will be determined by the product of ion activities (determined by ion concentrations):

$$K_{sp} = \left(Ca^{2+}\right)\left(CO_3{}^{2-}\right)$$

Calcite solubility is retrograde, i.e., it decreases with increasing temperature. However, reducing the pressure causes the dissolved CO_2 to be degassed from the geothermal brine (increasing $CO_3{}^{2-}$ concentration) and initiate the calcite deposition which occurs according to (increasing the calcium carbonate saturation).

$$Ca^{2+} + 2HCO_3{}^{-} \leftrightarrow CaCO_3 + CO_{2(g)} + H_2O$$

It is important to note that the concentration of calcium ions depends on temperature, amount of CO_2 dissolved (partial pressure of CO_2), and ionic strength.

2.1.2 *Sulfates*

Barite, celestine, and Sr-sulfate are among the common sulfates scaling in geothermal systems. Barite is a precipitated mineral from the class "barium sulfates" with the chemical formula $BaSO_4$. Barium sulfate precipitation was also observed in oil fields mainly caused by seawater injection (Hardy and Simm, 1996). The precipitation of barite in geothermal fields is dependent on temperature, pressure, and salinity. The temperature drop, which is one of the main mechanisms of barite precipitation in geothermal assets, reduces the solubility of barite (Tranter et al., 2020) and leads to the precipitations after the reaction stated in the following:

$$Ba^{2+}{}_{(aq)} + SO_4{}^{-}{}_{(aq)} \leftrightarrow BaSO_{4(s)}$$

In addition, reducing the pressure will also reduce the solubility of barite. At higher salinity, the solubility of barite will increase. In geothermal fluids, the impact of pH on solubility of barite is negligible (Zhen-Wu, 2016). As mentioned, barite precipitation in geothermal site is strongly correlated to the temperature drop, and thus for this reason, the barite has been mainly observed in heat exchangers, injection wells, and downstream filters. In some cases, the barite precipitation was observed in the production well and tubing even though the calculations suggested that the chance of barite precipitation should be low. This can be explained either by the chance of barite precipitation during the well shut-in which the fluid cools down (Regenspurg et al., 2015) or high ionic strength of the fluid at the specific site that impacts the calculation of ion supersaturation (Kristensen et al., 2020).

2.1.3 *Heavy metal*

Heavy metal scaling in geothermal assets refers to the undesirable deposition of metal compounds, particularly sulfides and oxides, within the geothermal infrastructure. In geothermal systems, heavy metals such as lead, copper, iron, manganese, and mercury can form scales leading to several operational challenges. These precipitates have a high density, are poorly soluble, and can pose safety risks in geothermal plants due to their toxic (arsenic) or radioactive (^{210}Pb) components. Various heavy metal scales can be observed in geothermal assets including lead carbonate, lead hydroxychloride, elemental lead, pyrite (FeS_2), cerussite ($PbCO_3$), or magnetite (Fe_3O_4). As mentioned, these heavy metal ions are hardly soluble, but they can form soluble complexes with chloride or other organic compounds in brine. The formation of scales, such as lead, copper, or mercury, is attributed to ion oversaturation or electrochemical reactions with the piping. The precipitation can be impacted by changes in pH or redox value of the brine.

The heavy metal scaling was mainly observed in the fields producing from Rotliegend sedimentary, such as in Altmark gas field in which scales with the highly toxic lead mercury was observed (Read et al., 2004) or lead scales which were found in Dutch geothermal wells (producing from Rotliegend sandstone) or German geothermal sites (located in Upper Rhine valley) (Scheiber et al., 2012).

Galena composed of lead sulfide (PbS) is a mineral which forms when lead and sulfide react with each other at high temperatures:

$$Pb^{2+}{}_{(aq)} + S^{2-}{}_{(aq)} \leftrightarrow PbS_{(s)}$$

Pyrite (FeS_2) and magnetite (Fe_3O_4) are both iron scales which can be formed in geothermal assets. Pyrite usually undergoes oxidation in geothermal assets which can form iron oxide and an acidic environment and consequently development of sulfide scales. In geothermal assets, magnetite was mainly observed by temperature reduction. When the dissolved heavy metal component reacts with iron of the casing, an electrochemical reaction takes place. By taking copper as an example, Cu^{2+} will be reduced to a solid native copper (Cu_0) and iron can oxidize to magnetite. Both magnetite and pyrite can be initiated by dissolved H_2S

corrosion in carbon steel as observed in some sites in Molasse Basin. These types of scaling were mainly observed in the well casing and filter residues.

The corrosion of steel worsens heavy metal scaling as it leads to the deposition of elemental lead through the reduction of lead ions, utilizing electrons released during the oxidation of iron from the steel components in wells or surface installations. Consequently, equipment such as heat exchangers, injection wells, and areas near the well are more prone to cerussite scaling than other carbonate scales.

2.1.4 *Silica scaling*

Silica scaling is primarily composed of amorphous silica (SiO_2), and it forms when the geothermal fluid is saturated with silica and undergoes temperature changes. Dissolved silica is one of the main components in geothermal brine from volcanic rock reservoirs with a typical concentration between 300 and 600 ppm. The solubility curve of silica in water is shown in Fig. 2. Silica scaling can occur above the amorphous solubility curve. This scaling exhibits almost in all the high-enthalpy geothermal fluids; however, it was also observed in some medium enthalpy geothermal wells (Andritsos et al., 2002), and also the deposition was observed in any part of geothermal installation. The silica deposition is developed through the polymerization of silicic acid (Ellis and Majon, 1977), and due to their slow kinetics, the deposits can form in the order several minutes to hours after the supersaturation condition is met.

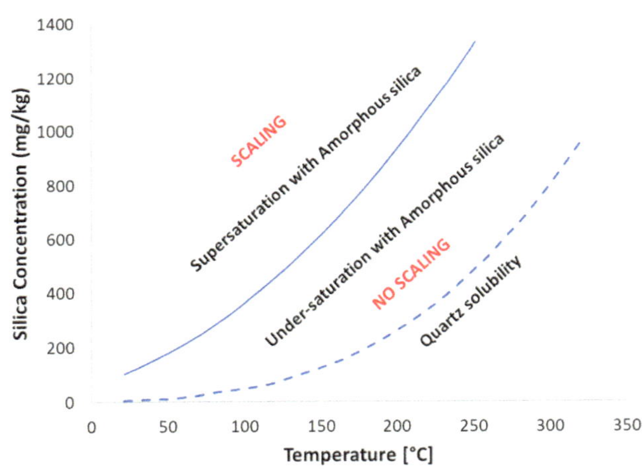

FIG. 2 Solubility of silica in water; due to the kinetic limitation of quartz precipitation, amorphous silica can precipitate.

Metal silicate scaling involves the deposition of silicate compounds containing metal ions such as aluminum and iron. The conditions leading to metal silicate scaling are similar to those of silica scaling. Changes in temperature, pressure, or fluid composition can cause supersaturation of the fluid with respect to silicate compounds containing metals. The presence of aluminum and iron ions accelerates the rate of silica deposition. Usually, iron silicate scale does not form at higher pressures (typically above 16 bars).

2.2 Monitoring

Scaling monitoring is often done in an indirect manner in geothermal assets. Scaling can cause clogging and blockage in the system and hence reduces the production rate or pressure in the system. It can also show higher pressure drop across the components with clogging at a given production rate. Monitoring production rate and pressure drop across the components are the common methods to monitor scaling. However, it is important to note that a detailed analysis of the production rate decline or pressure drop increase needs to be done to identify the true cause of the clogging. For better identifying and monitoring of scaling risk in geothermal assets, brine sampling and analysis is recommended. Brine chemistry analysis over time and at different location of the geothermal asset can provide insights on the potential type and location of mineral precipitation. This analysis is often done periodically but not continuous. At this stage, the online monitoring of brine chemistry is not commonly used in geothermal assets, except for pH monitoring which is observed in few geothermal side. pH monitoring can assist operators to better estimate the calcite scaling potential in their systems. Analysis of the precipitated materials in the clogged components such as heat exchangers and filters can also provide valuable information about the composition of the precipitates.

2.3 Mitigation measures

Several methods are reported in the literature and from operational experiences for mitigating and cleaning the scaling. There are several remedies depending on the type of scaling; thus, as the first point, an accurate prediction of the type, location, and amount of scaling is crucial to determine a suitable mitigation measure. In addition, the operational and plant constraints, such as availability of the chemical injection and access to the location of scaling especially if downhole, need to be considered for selecting a suitable mitigation strategy. An overview of the mitigation measures for each type of scaling and a potential location observed in geothermal assets is provided in Table 1. In this section, some details about the possible control or mitigation measures for each type of scaling are discussed.

Some of the common practices to prevent or control the carbonate scales are as follows:

- Pressure management: As explained in the scaling type, calcite precipitation is directly related to the degassing of the brine solution and partial pressure of CO_2. By keeping the solution and system under pressure especially on the hot side of the geothermal system and prevent CO_2 degassing, the calcite precipitation can be controlled. From the operational experiences, it was found that the system pressure can be lower than the bubble point since partial degassing did not show much calcite precipitation problem.
- For the cases that calcite precipitation can occur without degassing, adding acid or CO_2 to control brine pH can be done. By lowering the pH of solution, the carbonate solubility will increase and hence the calcite precipitation can be mitigated.
- Chemical inhibitor injection with the aim of keeping the calcite dissolved in the brine solution.
- Periodic workover, including the acid job, high-pressure washing, or even reaming with a rig.
- For high-enthalpy wells, gradually decreasing the wellhead pressure could shift the scaling point.
- Cation filters to remove major cations can be an effective way to minimize calcite precipitation.

For mitigating sulfate scales, there are less options available compared with carbonate scaling. The most common mitigation measure is the inhibitor injection. There are new filter developments to adsorb cations, e.g., barium prior to the locations which sulfate precipitation is expected. Similar to carbonate scaling, cation filters can also be an effective way to minimize barite. The common types of materials for the filter bed are chitosan and zeolite (Kristensen et al., 2022). Novel techniques based on decomposition of the barite scale layer with laser are under developments and demonstration.

Regarding mitigation of heavy metal scales in geothermal systems, as described, one of the main mechanisms of the heavy metal scale is the electrochemical reaction between the casing materials and the dissolved metal ions. This reaction can be prevented by using noble materials in the casing and piping such as stainless steel or Ni-based alloys or deposition of coatings such as polymers or using nonmetallic materials such as glass-fiber reinforced plastics. For existing systems in which the change in the casing and piping materials will be costly and often not possible, other methods can be used. Corrosion inhibitor injection was found to be an effective method to minimize heavy metal scaling. However, practically there is not much option to mitigate heavy metal scaling. Novel methods for cation filters are currently being developed and tested (Kristensen et al., 2022).

For preventing or controlling silica scales in the field, there are limited mitigation measures possible. A comprehensive overview of silica removal methods is provided by Milne et al. (2014). Inhibitor injection is found to be ineffective due to silica scale structures. One of the most common is pH control or acidizing of the water (Sigfusson and Gunnarsson, 2011). By lowering the pH of the geothermal fluid, the polymerization and aggregation of silica will be controlled which in turn increases the corrosion rate of the geothermal plant. Another observed method is diluting the separated water with condensate which can also increase the corrosion rate. Other precipitation methods including seed-induced, lime precipitation, and metal salt addition were also developed and tested for silica scaling control by which lime precipitation is the most common method so far.

TABLE 1 Overview of scaling types, their potential precipitation location in a geothermal doublet, and mitigation measures.

Type	Potential precipitation location	Mitigation measure/solution
Carbonates (calcite)	Production wells, ESP, filters, heat exchangers, pipes, degasser	Controlling degassing, inhibitor injection, injecting CO2 or acids, cleaning with acids
Sulfates (Ba, Sr)	Production well, heat exchangers, surface pipes and equipment	Inhibitors for sulfates
Amorphous silica, silicates[a]	Observed in all components	Pressure and temperature management (e.g., rapid cooling and pH control), chemical treatment
Sulfides (Zn, Pb)	Surface pipes, ESP and equipment such as heat exchangers, injection wells	Different pipe materials, corrosion inhibitors, methods under development such as selective ion extraction filters
Oxides (Fe, Mn)	Wells and wellhead (mainly iron oxides)	Different pipe materials, corrosion inhibitors, regular cleaning

[a] *Common in high-enthalpy geothermal fluids.*

3 Corrosion

Corrosion involves the gradual breakdown of metallic materials due to chemical, electrochemical, or other reactions with elements such as oxygen, moisture, or various corrosive substances. This phenomenon can weaken structures and equipment, affecting their functionality and lifespan. Corrosion in geothermal installations is a dominant challenge that can significantly impact the efficiency and longevity of the wells and installations. The concentration of salt can be extremely high in the geothermal brine (~10%–30% of Na-Ca-Cl), compared with seawater which is around 3% salinity. Combination of high salinity of geothermal brine, with other elements such as dissolved metal ions (Pb^{2+}), dissolved gases (CO_2), gas ingress (O_2), and presence of different metals in the geothermal installation, can lead to the presence of different types of corrosion in geothermal assets. There are several parameters which can impact the corrosion rate:

- Metallic material, such as alloys and surface treatments.
- Chemical compositions, such as salinity, pH, chloride content, sulfide, CO_2 content, and concentrations of dissolved metals.
- Physical and flow parameters, such as temperature, pressure, and fluid velocities.
- Bacterial species, mainly sulfidogenic bacteria.

In a geothermal system, corrosion can occur at several locations including well, wellhead, electrical submersible pump (ESP), pipelines, and top-side equipment. The corrosion in the well and wellhead could be uniform or localized due to acidity and salinity of the geothermal fluid. In case of steam geothermal system, the HCl content is also very crucial to the corrosion rate. Additionally, external corrosion of the casing resulting from the cooling during the shut-in periods was observed in geothermal wells. In the surface facilities, the flow topology changes due to elbow, tees, etc., and degassing of the dissolved gas will affect the corrosion rate. Thus, a good understanding of corrosion mechanisms and risks and methods to monitor and mitigate them is of crucial importance.

In the next section, first, an overview of different type of corrosion in geothermal installations will be described. Second, methods to monitor the corrosion in geothermal assets are provided together with method to mitigate these issues. Most of the current technologies to monitor and mitigate corrosion is derived from petroleum and water sector which emphasizes the link between these sectors and geothermal sector.

3.1 Corrosion in geothermal sites

3.1.1 Corrosion mechanisms

Before deep diving into different corrosion mechanisms in geothermal assets, one needs to identify the main parameters and conditions which will enhance the corrosion. Geothermal brine has a high electrical conductivity (~150 mS/cm) and also contains high salinity with the main salt ion being Na^+, Ca^{2+}, and Cl^-. The combination of these conditions could increase the chance of certain electrochemical reaction to occur over extended distances. In this case, since most of the geothermal pipes, wells, and facilities are made out of metals, the presence of salts and high electrical conductivity will enhance the conditions for the development of electrochemical processes involving metals (e.g., iron or steel). As an example, the electrochemical reactions of aqueous corrosion of iron which involves release of electrons are as follows (anodic reaction):

$$Fe \rightarrow Fe^{++} + 2e^{-}$$

As soon as this reaction is balanced with a cathodic reaction (to consume electron), then the corrosion can occur. Examples of cathodic reaction for oxygen reduction and hydrogen reduction is as follows:

$$\frac{1}{2}O_2 + H_2O + 2e^{-} \rightarrow 4OH^{-}$$

$$2H^{+} + 2e^{-} \rightarrow H_2(g)$$

The balancing cathodic reaction in geothermal installation needs either the presence of oxygen, e.g., due to ingress of O_2 to the geothermal brine or driven by the presence of acid that can be due to dissolved gases in the brine (e.g., CO_2 or H_2S). In very rare cases, the anoxic corrosion derived from the natural dissociation of water and leading to H^+ can happen, but this mechanism in pH-neutral brine is not considerable.

In some of the brines, corrosion can be initiated by the deposition of dissolved metal ions, such as lead and copper, which further enhances the potential for the corrosion of metal components within geothermal systems. This occurrence is particularly problematic in geothermal environments, as the elevated temperatures and pressures can accelerate the dissolution and precipitation of these metal ions, creating an environment conducive to corrosion. The interaction of these metal ions with the geothermal infrastructure, particularly metallic components, poses a significant challenge in ensuring the long-term integrity and reliability of geothermal assets. The reaction for occurrence of the corrosion is shown in the following:

$$Pb^{++} + 2e^{-} \rightarrow Pb(s)$$

$$Cu^{++} + 2e^{-} \rightarrow Cu(s)$$

3.1.2 Corrosion types

As explained in the previous section, several mechanisms can initiate and enhance the corrosion. Depending on the presence of conditions for each mechanism, pipe wall materials, and flow conditions, different types of corrosion can occur in geothermal systems. An overview of main corrosion types in geothermal assets is shown in Fig. 1. As it can be seen, different corrosion types can occur in geothermal systems initiated by either dissolved ions in the brine (salt or metal ions), dissolved gas, type of wall materials, or presence of different types of metals in the geothermal installations. Another notable point from the figure is that the corrosion in corrosion-resistant alloys (CRA), which are often used for subsurface and surface piping of geothermal plants, is mainly triggered by the presence of oxygen, while carbon steel can lead to all types of corrosion mechanisms.

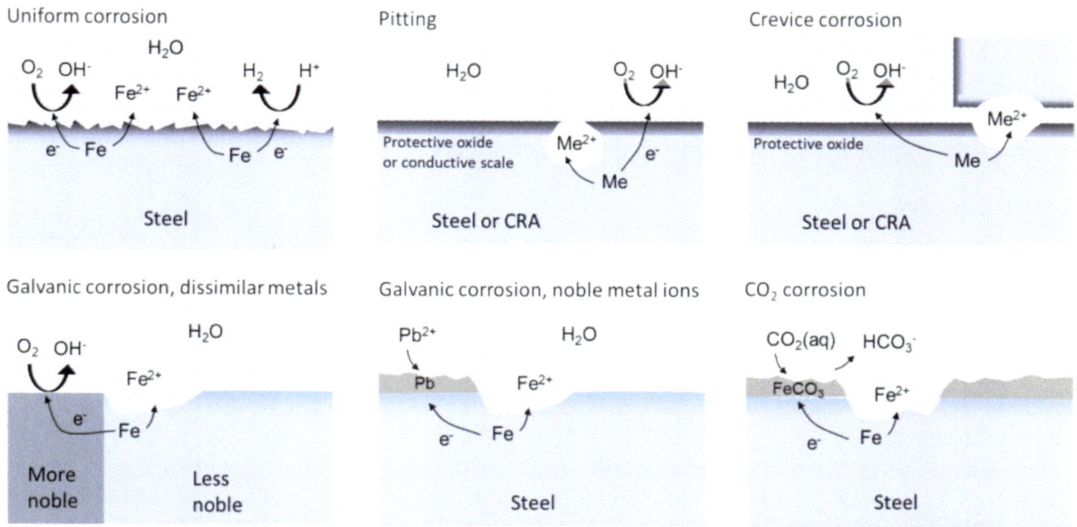

FIG. 3 Schematic of different corrosion mechanisms in geothermal assets (Mathiesen et al., 2021). (A) Uniform corrosion, (B) pitting corrosion, (C) crevice corrosion, galvanic corrosion due to (D) dissimilar metals and (E) noble metal ions, (F) CO_2 corrosion.

Uniform corrosion

Uniform corrosion is a type of corrosion that occurs evenly across the surface of a material, causing it to gradually degrade and thin out. In uniform corrosion, the entire surface of the material is exposed to the corrosive environment, and the metal or material undergoes a consistent and predictable loss of material over time. Uniform corrosion can be a relatively predictable form of corrosion, making it easier to estimate the rate of material loss and plan for maintenance or replacement of affected components. The uniform corrosion is mainly possible in carbon steel in contact with the brine solution. Oxygen and hydrogen reduction can enhance the uniform corrosion. It is often prevented or mitigated through the use of protective coatings and corrosion-resistant materials. In some cases, that part of the surface is covered by solids such as iron oxide—the uniform corrosion rate can be reduced.

Pitting corrosion

Pitting or localized corrosion is a common form of corrosion that occurs on the surface of both steel and CRA. It is characterized by the formation of small pits, craters, or cavities in the material. These pits can be quite small but have the potential to penetrate deep into the metal, compromising its structural integrity. Pitting corrosion can occur even in environments where the overall corrosion rate is relatively low. Pitting corrosion typically starts when a small area of the metal's protective passive oxide layer is compromised, as shown in Fig. 3. Since, in the location of the pit, a metal-rich and low-pH solution is formed, this can accelerate the rate of corrosion. Pitting corrosion can happen due to factors such as exposure to chloride ions, localized differences in oxygen concentration, or the presence of impurities. Once the protective oxide layer is detached, the corrosion process accelerates at the pit site, leading to the formation of the characteristic pits. Even a small concentration of oxygen can lead to pitting which gets even worse in low alloy grades such as 17Cr. The stainless steel can be protected by pitting corrosion if in contact with carbon steel, since the carbon steel acts as an anode protection for the stainless steel. However, the risk of severe pitting becomes relevant when larger sections of stainless steel, such as pipework, heat exchangers, or pumps, are installed. As a mitigation measure, often highly corrosion-resistant materials such as titanium are employed for surface facilities like heat exchangers.

Crevice corrosion

Crevice corrosion is a localized form of corrosion that occurs within confined spaces or crevices between two surfaces. It is often associated with stagnant or restricted environments where the access of oxygen or the removal of corrosive by-products is limited. Crevice corrosion can affect various materials, including metals such as stainless steel and aluminum. Common examples of crevices include gaps between fasteners, flanges, under deposits or deposits in narrow spaces, and areas where protective coatings break down, leaving small, enclosed regions.

The corrosion process in crevices is initiated by differences in the concentration of ions or dissolved oxygen (DO) between the crevice and the surrounding environment. This concentration cell corrosion can lead to accelerated degradation of the material within the crevice. Depletion of oxygen in crevice or buildup of aggressive ion species, such as chloride ions, leading to acidic conditions can initiate the crevice corrosion. In geothermal assets, the surface facilities and injection wells are at a greater risk of crevice corrosion due to oxygen ingress.

Prevention measures for crevice corrosion often involve design considerations to minimize crevice formation, the use of corrosion-resistant materials, and appropriate coatings or treatments to protect vulnerable areas. Regular inspection and maintenance are also essential to identify and address potential crevice corrosion issues. One common mitigation measure for crevice corrosion is the use of oxygen scavengers to remove the cathodic reagent.

Galvanic corrosion

Galvanic corrosion, also known as bimetallic corrosion or dissimilar metal corrosion, is a type of electrochemical corrosion that occurs when two different metals are in electrical contact with each other in the presence of an electrolyte. The electrolyte in geothermal fields is the geothermal brine. This type of corrosion can lead to the deterioration of one of the metals involved. The process of galvanic corrosion starts with the anodic and cathodic reactions between the two metals. The anodic metal will sacrifice itself to protect the cathodic metal, due to the electrons flowing from the anodic metal to cathodic metal. The galvanic corrosion can be enhanced in the presence of oxygen. One important factor impacting the galvanic corrosion is the type of metals. If the electrochemical potentials of two metals are not different, then the chance of galvanic corrosion is low. As an example, contact of different grades of stainless steels is less likely to cause galvanic corrosion.

Another type of galvanic corrosion is caused by metallic deposits and noble metal ions. This type is also very common in geothermal systems. In some of the geothermal reservoirs, the brine contains dissolved metal ions, including lead, copper, arsenic, and antimony. The deposition of the metal ions can be enhanced at certain temperature and flow rate conditions which consequently leads to galvanic corrosion. The conditions of noble metal ion deposition are further explained in the scaling section.

CO_2 corrosion

Dissolution of CO_2 in the brine can lead to a weak acid formation and consequently corrode the steel. The corrosion rate of CO_2 corrosion is relatively easy to estimate as it is mainly related to the patrial pressure of CO_2. This type of corrosion is electrochemical in nature and initiates by the formation of carbonic acid (H_2CO_3) due to the dissolution of CO_2 in the brine. Afterward, the carbonic acid dissociates into hydrogen (H^+) and bicarbonate ions (HCO_3^-). The hydrogen ions will promote the corrosion of metal surfaces due to electrochemical reactions of metal ions and electrons. The metal ions produced contribute to the overall corrosion process.

CO_2 corrosion is particularly concerning in geothermal systems where high temperatures and pressures are present. Higher pressure will increase the partial pressure of CO_2, due to higher concentrations of dissolved CO_2. The carbonate film also becomes more stable at higher temperatures. The corrosion rates can be influenced by factors such as fluid velocity, turbulence, and the composition of the geothermal fluid. The areas most susceptible to CO_2 corrosion in geothermal systems include pumps, tubing, valves, and other components where fluid flow and CO_2 exposure are prevalent.

Mitigation strategies for CO_2 corrosion in geothermal systems often involve the use of corrosion-resistant materials, inhibitors, and modifications to the system design to minimize turbulence and promote protective surface films. Regular monitoring and maintenance are crucial to detect and address potential corrosion issues before they lead to significant damage.

Oxygen (O_2) corrosion

Oxygen corrosion is a type of corrosion that occurs in the presence of oxygen dissolved in the brine. It is a common form of corrosion for metals that are primarily composed of iron, such as carbon steel. The main parameters impacting the oxygen corrosion are dissolved content of oxygen, temperature, and flow velocity. In case of DO in the brine, the metal surface acts as an anode and oxygen reduction occurs at cathodic sites. This sets up an electrochemical cell on the metal surface. The metal at the anodic sites undergoes oxidation, resulting in the release of electrons and metal ions into the surrounding electrolyte. The process is followed by oxygen reduction at the cathodic sites by gaining electrons and forming hydroxide ions. The metal ions and hydroxide ions react to form corrosion products.

Flow-related corrosion

Corrosion in geothermal systems can be enhanced by flow conditions. Flow-related corrosion in geothermal fields is a critical consideration due to its significant impact on corrosion rates, particularly in the presence of CO_2. The velocity and shear forces associated with fluid flow play an essential role in transporting reactants and corrosion products to and from surfaces, influencing the overall corrosion process. The removal of protective corrosion products and corrosion inhibitor films is affected by local flow conditions, with inhibitors becoming less effective under high shear conditions. Localized corrosion with elevated rates is often observed in areas experiencing high velocity, shear, or turbulent flow, commonly found in environments containing CO_2. These locations include ESP, pumps, pump inlet/outlet piping, tubing perforations, sand-screens, valves and bends. In some cases, the change in the flow areas can lead to the exsolution of the dissolved gases which can impact the surface integrity.

Mitigating flow-related corrosion in geothermal fields involves an optimum hydraulic design of the system and continuous monitoring and measurements. This includes minimizing locations in the pipeline configurations with sharp bends and changes in the flow area and uneven surface profiles, as these factors contribute to increased corrosion risk. Additionally, the use of corrosion-resistant materials, such as stainless steel or nonmetallic options, is crucial in high-risk locations, such as sand–screens, pumps, valves, and heat exchangers.

Microbiologically influenced corrosion

Microbiologically influenced corrosion (MIC), also known as biotic induced corrosion, poses a significant challenge in geothermal systems, where the interaction between microorganisms and solid materials leads to the gradual loss of structural integrity. A wide range of organisms can withstand geothermal conditions such as high temperature, salinity, and even lack of oxygen. As an example for sulfate-reducing bacteria (SRB), the range of activity of different parameters is shown in Table 2.

TABLE 2 Physicochemical ranges of parameters to expect SRB activity (Maxwell and Campbell, 2006).

Parameter	Range	Unit
pH	5.0–9.5	–
Temperature	0–80	°C
Salinity	1–11	w%
Sulfate	>10	mg/L
Organic carbon	>1	mg/L

Every microorganism necessitates a food source for its growth and reproduction. Various organisms flourish on distinct nutrients, encompassing both inorganic substances such as sulfur, ammonia, and sulfate ions and organic substances like hydrocarbons and organic acids. Furthermore, for their growth, all organisms demand a supply of carbon, nitrogen, and phosphorus.

While a range of materials used in well construction, including carbon and low alloy steels, stainless steels, and certain nickel-based alloys, are susceptible to MIC, practical observations indicate that stainless steels with alloy content higher than 316/316L are rarely affected. MIC is most commonly encountered in shut-in wells, particularly water injectors, or in wells where packer fluids receive incorrect chemical treatment. In the latter scenario, MIC may manifest in nonprocess locations such as the outer annulus spaces of the well.

In surface equipment within geothermal systems, MIC tends to proliferate in stagnant or low-flow areas, exemplified by the bottoms of separator vessels and piping dead-legs. The corrosion is not limited to internal surfaces, as MIC can also occur externally when materials are in contact with specific soils or are subjected to damaged coatings. The presence of MIC in geothermal systems underscores the importance of appropriate chemical treatments and maintenance protocols to mitigate the risk of corrosion-induced structural damage and ensure the longevity of the infrastructure.

MIC is often very challenging to be predicted due to the complex interaction between microbial, chemical, and physical processes which can impact the corrosion rate of localized and general corrosion. In cases and reservoirs that MIC is expected, use of biocides is suggested to control microorganism activity and MIC. As explained, stagnant areas are more prone to MIC, thus ensuring a minimum flow or recirculation in the stagnant areas which can greatly reduce the chance of MIC.

3.2 Monitoring

Several methods are employed for corrosion monitoring, including the use of corrosion coupons, electrical resistance (ER) probes, electrochemical probes, DO, pH, and oxidation-reduction potential (ORP). Corrosion coupons are one of the most common methods for monitoring corrosion. They are metal samples placed in the system to be monitored. These samples are made of the same material as the system components. Periodically, the coupons are removed and analyzed for signs of corrosion, such as metal loss or pitting. Comparing the initial and final conditions of the coupons allows for the calculation of corrosion rates over a specific period.

ER probes measure changes in the ER of a metal, which is related to its corrosion rate. In principle, any loss in the metal material can increase the ER. It is important to note that the changes in the ER is relatively small and can be impacted by temperature or any conductive layer formed on the metal probe (e.g., iron sulfides).

Electrochemical corrosion probes are devices that measure the electrochemical behavior of metals in a corrosive environment. They typically consist of a metal electrode that is inserted into the system being monitored. The corrosion rate is determined by measuring parameters such as corrosion current and polarization resistance. One of the common electrochemical techniques is linear polarization resistance (LPR) method. This method works by applying a small potential to the metal surface to measure the resulting current. These probes provide real-time data on corrosion activity and rate without a need for a reference electrode.

DO levels in a corrosive environment can impact the corrosion rate. Low oxygen levels can reduce corrosion, while high levels may accelerate it. Monitoring DO provides insights into the environmental conditions affecting corrosion. Sensors or probes are used to measure the concentration of DO in liquids. For geothermal systems, due to a low level of oxygen present, high-sensitive and accurate DO sensors need to be used.

pH is a measure of the acidity or alkalinity of the brine. Corrosion rates can be influenced by the pH of the environment. Acidic conditions often accelerate corrosion, while alkaline conditions may inhibit it. pH sensors or probes are used to monitor the acidity or alkalinity of a solution. In geothermal systems, degassing of or gas dissolution in the geothermal brine can change the pH level and impact the chance of corrosion in the system.

ORP measures the tendency of a solution to gain or lose electrons. In corrosive environments, ORP can provide information about the likelihood of corrosion by monitoring the redox potential of the brine. It is particularly relevant in systems where the electrochemical environment plays a significant role in corrosion processes. ORP sensors or probes measure the potential difference between a reference electrode and a working electrode.

3.3 Mitigation measures

Controlling and mitigating corrosion in geothermal installations is crucial for the sustainable operation of these systems, as it can lead to costly maintenance and repair, reduced energy production, and potential environmental hazards for the systems which are designed to last for at least 30 years. In case of any corrosion issue in the wells and downhole equipment, the maintenance will be complex and costly which will interrupt the production. Corrosion-related concerns extend to the decreased efficiency of surface equipment such as heat exchangers, particularly if corrosion induces scaling and reduces the heat transfer. Equally critical is the risk of corrosion in well tubing, which could potentially result in the leaking and contamination of shallower reservoirs, such as groundwater reservoirs containing brine or chemicals (if applicable).

In the design phase, the selection of suitable piping, well materials, and pipe layout can have a significant impact on controlling and mitigating corrosion issues. Using high quality materials, such as stainless steel alloy (Cr, Ni) and titanium, could control the corrosion rate. It is recommended to avoid local turbulences to reduce the corrosion rate.

Regarding the mitigation or control of the corrosion rate, corrosion inhibitor injection is one of the common practices. Inhibitors are chemical additives, mainly amine-based, which are injected in the system and can form a protection layer (an organic film) on the pipe surfaces to prevent or minimize the contact of corrosive fluid with the surfaces. Currently, these inhibitors are injected at the production well head or at surface pipeline; however, the recommended practice is to inject the inhibitors deeper in the production well to also protect the production well and downhole equipment such as downhole pumps. For this purpose, it is recommended to test the stability and compatibility of the inhibitors at high temperatures. In addition, different types of corrosion inhibitors might be required for different types of corrosion, e.g., metal ions or CO_2 corrosion.

As explained above, one of the parameters promoting corrosion is the presence of oxygen. Deaeration of system is necessary in all phases of operation including start-up, production, shutdown, and maintenance periods. In some cases, during the start-up phase, inert gases such as nitrogen or argon is used for pressure buildup and in minimizing oxygen ingress. During the operation, since the system is pressurized, the probability of oxygen ingress is very low. At some of the surface facilities, such as injection pump, the chance of oxygen ingress can be higher which can be mitigated by applying pressurized liquid and nitrogen outside of the seal. During the shutdown of the system, when the chance of oxygen ingress is high, blanketing or pressurizing with inert gas is a common practice to displace air. Finally, injecting oxygen scavengers is another method to control the oxygen corrosion throughout the geothermal asset.

As explained in the previous section, bacteria could cause MIC in geothermal systems. This is specially an issue during the standstill periods or at locations where stagnant fluid is present. Biocides are often recommended to control the rate of MIC. Biocides are substances or chemical agents that are used to control or inhibit microorganisms. A suitable selection of biocides based on the process conditions and brine composition is of key importance. For geothermal applications, organic biocides are recommended to control anaerobic bacteria and not oxidizing biocides (e.g., hypochlorite) as they will accelerate corrosion.

4 Naturally occurring radioactive materials

The Earth's crust naturally contains small amounts of radioactive uranium (^{235}U, ^{238}U), thorium (^{232}Th), and potassium (^{40}K). These so-called primordial radionuclides were present during the formation of planet Earth 4.55 billion years ago and have a half-life time comparable with the age of the Earth. For this reason, measurable quantities of these nuclides are still present in the subsurface. Such radionuclide will undergo a process by which the nuclide decays by the emission of radiation. This radiation consists of particles (α- and β-radiation) that in most cases are accompanied by electromagnetic waves (γ-radiation and X-rays). In the decay process, the so-called parent nuclide transforms to a nucleus of a different type, named the daughter nuclide. Thus, a change in the chemical element occurs or an isotope is created according to the decay type. The SI derived unit of radioactivity is Becquerel (symbol Bq), which is defined as the activity of a quantity of radioactive material, in which one nucleus decays per second. The old unit is Curie (symbol Ci), which is still used in some countries (e.g., the USA). The conversion is $1\,Ci = 3.7 \times 10^{10}$ Bq. The energy of the emitted radiation is sufficient to ionize atoms by stripping electrons along the path of the radiation. This process occurs either direct (α- and β-radiation) or indirect (γ- and X-rays).

^{40}K is a radionuclide that decays to stable calcium and argon. Uranium and thorium are heading a decay series; each member of this series is itself unstable and continues to decay until a stable isotope of the element lead is eventually formed, creating chains of radionuclides (see Fig. 4). These chains include elements such as radium, radon, polonium, bismuth, and lead, which can manifest itself in the geothermal installation. Within these chains of radioactive nuclides, α-, β-, and γ-radiation can be observed. Due to their origin, these radionuclides are also called naturally occurring radionuclides (NORs).

NORM is defined in the IAEA Safety Glossary as a "radioactive material containing no significant amounts of radionuclides other than naturally occurring radionuclides" where the exact definition of "significant amounts of naturally occurring radionuclides" would be a regulatory decision.

During geothermal operation, dissolved radionuclides such as radium (^{228}Ra, ^{226}Ra, ^{224}Ra), radon (^{222}Rn, ^{220}Rn), and lead (^{210}Pb) together with carbonates and sulfates as well as resuspended NOR-containing mineral particles will be carried along with the formation water. The uranium and thorium isotopes, including ^{238}U and ^{232}Th, are predominantly present as insoluble species and will therefore remain in the rock formation. From the formation water, the (mineral) particles and/or dissolved radionuclides will deposit via various chemical and physical mechanisms on the walls of the casing or the aboveground installation parts. Thus, the ^{210}Pb and ^{210}Po formed from the released ^{222}Rn will precipitate, e.g., as PbS or via electrochemical reaction with the steel casing. However, the radium isotopes settle on the wall via a coprecipitation with, e.g., Ba^{2+}, Sr^{2+}, and Ca^{2+}. This precipitation leads to scaling and radioactive contamination. In addition, residual particles together with the radionuclides present will be trapped in the filters and lead to radioactive accumulation. Finally, as a result of natural decay, the daughter nuclides that did not come from the underground formation can manifest themselves in the installation via ingrowth.

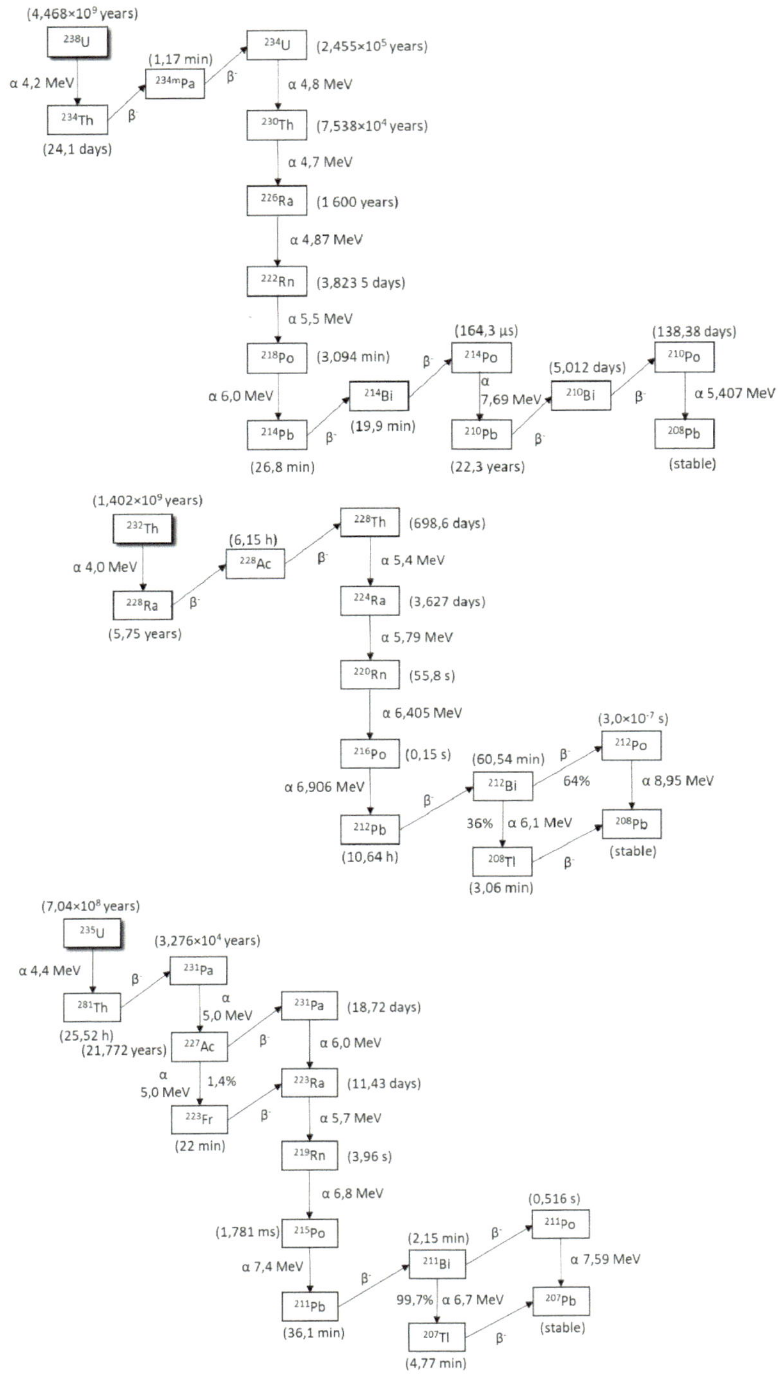

FIG. 4 Natural radioactive decay series, starting from ^{238}U, ^{232}Th, and ^{235}U.

4.1 NOR in the geothermal installation

Uranium, thorium, and potassium are present everywhere either as main constituents of some rock materials or as trace elements. This means NOR exists nearly everywhere and depending on its amount and use can cause radiological problems. Because of their geochemical behavior, some NOR can concentrate in certain minerals.

The concentrations of the primordial radionuclides in the subsurface have been reported by UNSCEAR. Generally, the activity concentration of ^{40}K in the subsurface is an order of magnitude higher than that of ^{238}U or ^{232}Th. In UNSCEAR's first publication (UNSCEAR, 1982), it reported representative concentrations of these radionuclides in soil and suggested the values of 370, 25, and 25 Bq/kg for ^{40}K, ^{238}U, and ^{232}Th, respectively. On the basis of the higher levels reported for China and the USA, UNSCEAR (1993) revised the values for both ^{238}U and ^{232}Th to 40 Bq/kg. A more recently completed countrywide survey in China indicated somewhat lower values. These and the results for many more countries resulted in median values of 400, 35, and 30 Bq/kg for ^{40}K, ^{238}U, and ^{232}Th, respectively. A more detailed overview of the concentrations of the NORs K, Th, and U in sedimentary rock is shown in Table 3.

TABLE 3 Concentration of the naturally occurring radionuclides K, Th, and U in sedimentary rock.

Sedimentary rock	**Potassium (K)**		**Thorium (Th)**		**Uranium (U)**	
Class	%w	**Bq/g**	%w	**Bq/g**	%w	**Bq/g**
Detrital	2.1	0.6	12.4	0.05	4.8	0.06
Sandstone	1.2	0.4	9.7	0.04	4.1	0.05
Orthoquartzites	2.0	0.6	1.5	0.01	0.5	0.01
Arkoses	2.5	0.8	5.0	0.02	1.5	0.02
Shale						
Gray/green	3.5	1.1	16.3	0.07	5.9	0.07
Black	3.0	0.9	13.0	0.05	3.0	0.04
Clay	0.6	0.2	8.6	0.03	4.0	0.05
Chemical	0.6	0.2	14.9	0.06	3.6	0.04
Carbonates	0.3	0.1	1.8	0.01	2.0	0.02
Limestone	0.5	0.2	3.0	0.01	13.0	0.16
Evaporites					<0.1	<0.01
Soil (UNSCEAR)	1.3	0.4	7.4	0.03	2.8	0.04

4.1.1 Chemistry and transport

As previously mentioned, ^{232}Th and ^{238}U are primordial radionuclides that are heading a decay series containing a total of around 25 radionuclides belonging to different chemical elements. These decay series tend to be in equilibrium when they remain undisturbed.

Nevertheless, the extraction of water from the formation and the change in pressure and temperature will disturb the decay series with certain radionuclides and sections of the decay series appearing in different phases (liquid, solid, gas) and parts of the installation. This phenomenon emanates from the different chemical and physical properties associated with the radionuclides. Depending on its solubility and its ability to precipitate as a mineral phase, a radionuclide can (1) stay trapped in the rock matrix within the reservoir, (2) be present in the geothermal brine (gas, water), (3) precipitate along the surface of the installation, or (4) accumulate in the (filter) residues. Table 4 provides a brief summary of the relevant chemical features and its diagenesis in the geothermal installation.

TABLE 4 Chemical features of the relevant radionuclides.

Element	^{232}Th	^{238}U	Chemical behavior	Diagenesis/accumulation
Uranium (U)		^{238}U, ^{234}U	Reducing environment: U^{4+} insoluble, no UO_2^{2+}	Remains inside or at surface of reservoir grains
Thorium (Th)	^{232}Th, ^{228}Th	^{234}Th, ^{230}Th	Reducing environment: Th^{4+} very insoluble	
Radium (Ra)	^{228}Ra, ^{224}Ra	^{226}Ra	Alkaline earth (Mg, Ca, Sr, Ba)$^{2+}{}_{aq}$ carrier for Ra^{2+} ions	Preference for and geochemically transported by aqueous phase
Radon (Rn)	^{220}Rn	^{222}Rn	Noble gas, very polarizable atoms	Water/gas phase partitioning
Lead (Pb)		^{210}Pb	Stable Pb_{aq}^{2+} carrier for $^{210}Pb^{2+}$ ions	Preference for and geochemically transported by aqueous phase

As shown in Table 4, U and Th are both highly insoluble, and therefore no significant concentrations of these radionuclides are found in the geothermal brines nor in any of the deposits in the installation. In contrast, radium isotopes (^{224,226,228}Ra) are found in the geothermal water due to leaching from the rock formation and subsequent transport in the installation where the radium can coprecipitate with other scaling minerals on the inner surfaces of the installation. Among those, the most prominent scaling mineral is radiobarite ((Ba,Ra)SO_4). From the various radium isotopes, ^{224}Ra has a half-life of only 4 days, and no significant amounts of this radionuclide that originates from the water formation will be found once operations have stopped for more than a month. In contrast, ^{228}Ra has a much longer half-life and is part of the same ^{232}Th series—its presence in the installation will be responsible for ingrowth of ^{228}Th, which subsequently will provide a source for ^{224}Ra and its progeny. Any ^{226}Ra with a half-life of 1600 yr will remain present for a prolonged period and lead to the formation of progeny nuclides. The first progeny is ^{222}Rn (radon) with a half-life of 3.8 days. Radon is a noble gas, which can emanate from its solid structure and is then transported together with the geothermal water or gas. Subject to the composition of the rock or scale (e.g., thickness, porosity), a fraction of the ^{222}Rn will emanate and subsequently deposit to form a ^{210}Pb film. Similarly, ^{224}Ra decays to ^{220}Rn (thoron) to progenies; however, due to the much shorter half-life of ^{220}Rn, this isotope is generally of less significance. Finally, there is ^{210}Pb which is leached from the formation rock where it is then transported upstream and found as a deposit due to electrochemical or mineral deposition. Due to nuclear decay of the ^{210}Pb, also ^{210}Bi and ^{210}Po will be present. A sketch on the migration pathways of the radionuclides from the ^{238}U decay series is shown in Fig. 5.

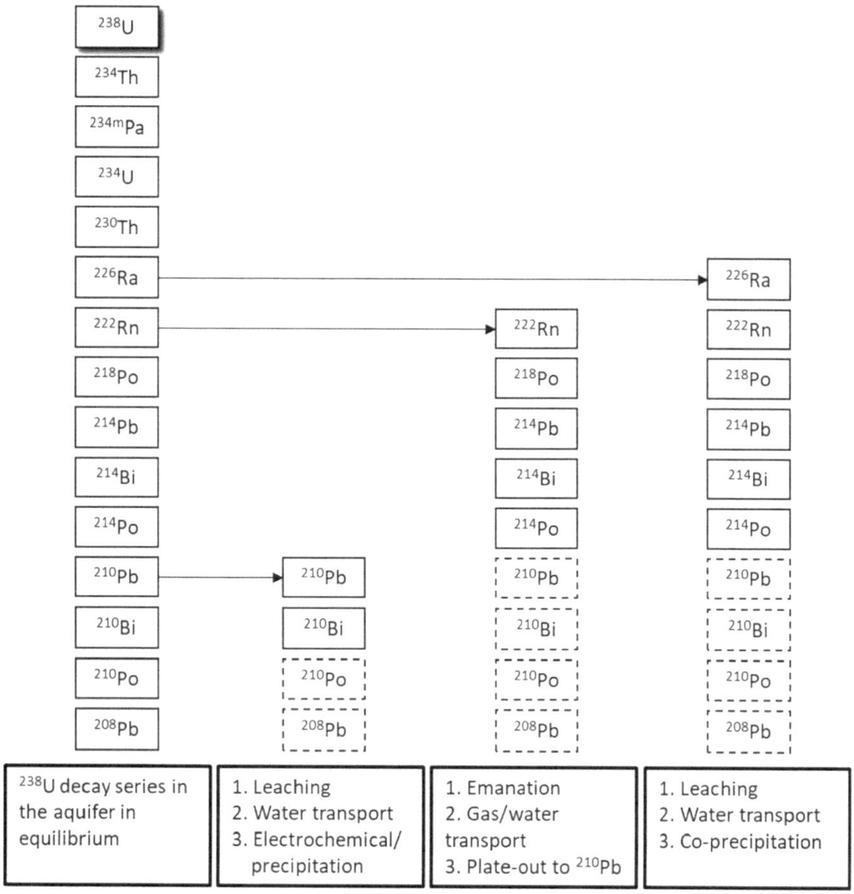

FIG. 5 Migration pathways and phases of the main radionuclides in the ^{238}U decay series that can induce occurrences of NOR in the geothermal installation. 1. Migration pathway. 2. Transport mode. 3. Deposition mechanism. Isotopes with dashed lines indicate ingrowth after a prolonged period of time.

4.1.2 Brines

Similar to oil and gas production, uranium and thorium and their radioactive decay products may be present in formations from which geothermal fluids are extracted. However, available information indicates that no significant quantities of uranium and thorium are dissolved or entrained in geothermal fluids. The primary radionuclides present in geothermal fluids are ^{224}Ra, ^{226}Ra, ^{228}Ra, ^{210}Pb, ^{212}Pb, and ^{222}Rn from the ^{238}U and ^{232}Th decay chains (CDC, 1996). More recent studies from Regenspurg et al. (2014) and Kölbel et al. (2020)

indicated concentrations of ^{226}Ra and ^{228}Ra in the order of some tens of Bqs per gram with maximum values of up to 100 Bq/g (Regenspurg et al., 2014). The samples were taken at geothermal sites located in Germany in the North German Basin and the Upper Rhine Graben. Please note that concentrations can vary considerably between different regions and doublets.

4.1.3 Residues

Limited information is available on the concentrations of NOR in geothermal solid waste, such as filter cakes, scale in piping and production equipment, and gases. Concentrations of NOR in geothermal residues will vary with the geology and mineralogy of the geothermal resource area, along with the physical and chemical changes that occur during energy extraction. Table 5 summarizes typical radionuclide concentrations found in residues from geothermal energy production. However, it is important to stress that activity concentrations of all radionuclides can vary by more than three orders of magnitude.

TABLE 5 Radionuclide concentrations in geothermal energy production waste (EPA, 1993).

Nuclide	Residue concentrations (Bq/kg)
^{228}Th	930
^{228}Ra	3440
^{226}Ra	4880
^{210}Po	3550
^{210}Pb	3550

A more recent inventory carried out in the Netherlands provided a detailed overview of the NOR concentrations in filter waste and indicated that the activity concentration of ^{210}Pb is by far the highest of all radionuclides present. These concentrations were in most cases as much as 100–1000 Bq/g. For other radionuclides from the ^{238}U and ^{232}Th series, namely, ^{226}Ra, ^{228}Ra, and ^{228}Th concentrations were mostly below the 1 Bq/g. No significant levels of ^{40}K were found in the filter waste. A number of geothermal sites were identified with higher concentrations of natural radioactivity. However, no correlation was found with regard to the levels of radioactivity and the used water formation.

Radium

An important aspect with $^{226,228}Ra$ is its natural presence in barium sulfate, known as radiobarite in the oil exploration jargon. The formation of $(Ba,Ra)SO_4$ solid solutions in natural environments has long been known, especially in geothermal areas, where barite was early recognized to be radioactive (e.g., Knett, 1904). Radium, a highly soluble component found naturally in low concentrations in the environment, is a radioactive alkaline earth metal exhibiting similar environmental behavior to calcium, barium, strontium, and magnesium (Ca, Ba, Sr, and Mg).

The mobility, bioavailability, and toxicity of radium in the aqueous environment depend on its phase as an aqueous species Ra^{2+}, $RaOH^{+}$, $RaCl^{+}$, $RaCO_3$, or $RaSO_4$; sorbed to clays, organic matter, and ferric and manganese oxides; or coprecipitated into sulfate and carbonate minerals. Radium preferentially coprecipitates with sulfate and barium or strontium when supersaturated to form radiobarite $(Ba,Ra)SO_4$ and radiocelestite $(Sr,Ra)SO_4$ compared with coprecipitation with supersaturated carbonate minerals (Poonoosamy et al., 2021; Kulik et al., 2017).

The affinity of the radium ion with the barite crystal lattice comes from the similarity of barium and radium ionic radius (1.36 and 1.43 Å, respectively), so that it is expected to be well accommodated in this structure. In addition, the fact that the polarizability of both ions is almost identical (see Table 6) indicates a clear preference for Ra^{2+} to substitute for Ba^{2+} in the ionic crystal lattice of Ba(II) sulfate (Goldschmidt, 1938).

TABLE 6 Polarizability of selected divalent cations (Å3) (Goldschmidt, 1938).

Be^{2+}	Mg^{2+}	Ca^{2+}	Sr^{2+}	Ba^{2+}	Ra^{2+}
5.6	10.6	22.8	27.6	39.7	38.3

This similarity in polarizability lies behind the fact that $(Ba,Ra)SO_4$ solid solutions are thermodynamically more stable than the simple mechanical mixture between solid solution end-members ($BaSO_4$ and $RaSO_4(s)$) even at trace concentrations of one of them. The reason for this higher stability is found in the reduction of the Gibbs free energy when a solid solution forms. Consequently, radium will preferentially be incorporated into barite, lowering the aqueous radium concentrations in equilibrium.

On this basis, it is assumed that in the event of a simultaneous release of Ra with Ba, the former will be readily incorporated into the precipitating $BaSO_4$ to build a Ra-Ba sulfate solid solution. All the existing evidence indicates, that in this case, the behavior of the system can be described by the established aqueous–solid solution thermodynamic formalism and assuming that the system behaves ideally (Grandia et al., 2008).

The barium and radium content in waters from different origins were compared (Grandia et al., 2008) and presented in Fig. 6. In many cases, Ra-Ba positive correlation is evidenced. However, the extremely low measured Ra(II) concentrations together with the small range of variability of the Ra-Ba ratios would hamper any further analysis. The figure shows that radium concentrations are approximately six orders of magnitude less and therefore the contribution of Ra sulfate to the total scale volume will be minimal.

Ra might also manifest itself in $RaCO_3$. However, according to the findings by Bruno et al. (1997), the solubility limiting solid phase is $RaSO_4(s)$ in the pH range studied (pH, 5–10). Therefore, it is expected that radium sulfate would be the dominant portion of the complexed Ra(II).

4.1.4 Lead

Lead deposits in the geothermal installation are the most common radioactive contaminants due to the presence of the NOR ^{210}Pb. The radioactive isotope ^{210}Pb normally appears together with isotopes of stable lead such as $^{206,207,208}Pb$. Due to presence of ^{226}Ra in the rock

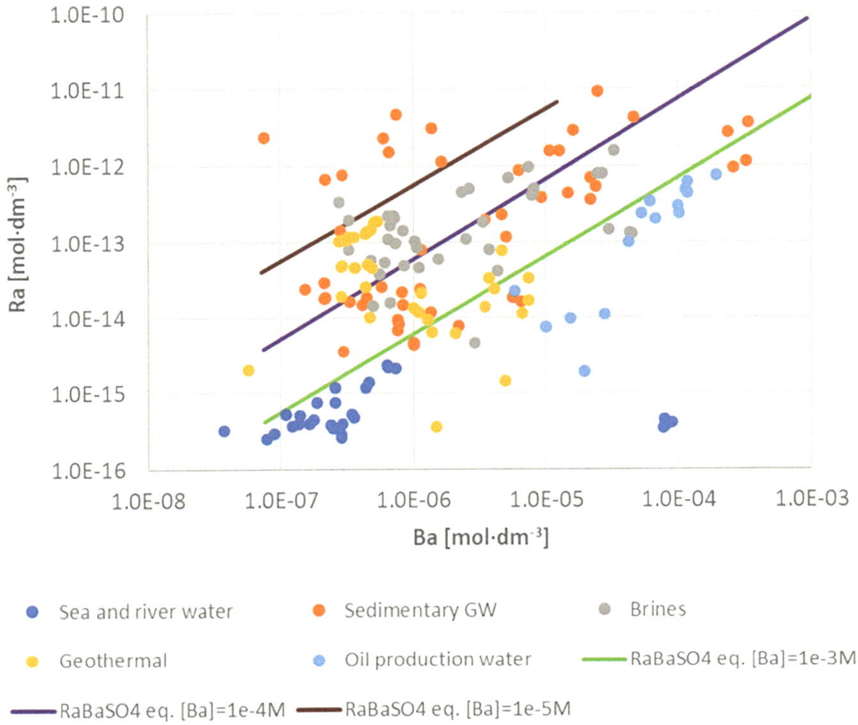

FIG. 6 Radium versus barium concentrations in waters from natural and anthropogenic environments. Solid lines show the calculated1 Ra and Ba concentrations in a mixture of two granitic groundwaters with variable barium and sulfate concentrations ($[Ba_{tot}] = 10^{-3}$ to 10^{-5} M; $[SO_4^{2-}] = 0.1$–10^{-4} M and fixed initial $[Ra_{tot}] = 10^{-10}$). These lines show similar slope in major natural Ra-Ba trends. *Redrawn from the data of Grandia et al., 2008.*

formation, there is a continuous production of ^{210}Pb, which is found as a trace element in different lead deposits. These include minerals such as $PbCO_3$, Galena (PbS), and Laurionite ($Pb(OH)Cl$) or elementary lead deposits. The latter deposits are a result of electrochemical ion exchange with the steel of the geothermal installation that takes the following chemical reaction $Fe(s) + Pb^{2+} \rightarrow Pb(s) + Fe^{2+}$. These deposits typically occur in brines that are very much reduced in terms of its redox potential. To mitigate such type of deposits and thereby also reducing radioactive contamination, stainless steel or glass-fiber reinforced epoxy (GRS) should be applied.

4.2 Risk and exposure

Exposure to ionization radiation emitted from NORM has a harmful impact on the health of humans and the environment, hence the need for minimizing its exposure and adequate regulatory control. By and large most of the harmful impact from radiation exposure is of a stochastic nature. The general hypothesis for stochastic health effects is the principle of linear no-threshold, where the health impact or detriment is zero at zero exposure followed by a linear relation between exposure and detriment. The specific concerns related to health are dependent on the nature of the radiation emission (α-, β-, and γ-radiation).

- Alpha (α-) radiation consists of α-particles containing two protons and two neutrons, and it is identical to an ionized ^{4}He nuclei. Its path in biological tissues is no longer than a few tens of micrometers, where its exposure is primarily of concern in case of direct contact. For this reason, alpha emitters are primarily responsible for internal exposure, which includes inhalation, ingestion, and skin contact.
- Beta (β-) radiation is made up of β-particles that are electrons, with a negative elementary charge, or, not so often, positrons both with positive elementary electric charge. Exposure to beta particles will cause more external irradiation hazard and less internal radiation hazard than alpha particles. The external irradiation brought by beta particles is mostly confined to the epidermis and outer skin tissue, but it can also have a harmful impact on the eye lens.
- Gamma (γ-) radiation is composed of high-energy photons (electromagnetic waves) coming from the nucleus. This type of radiation is only weakly ionizing but has a high penetrating power and can travel through hundreds of meters of air. Usually, thick concrete or lead shielding is used to shield γ-radiation. It is primarily responsible for external exposure.

4.2.1 *External radiation exposure*

Exposure to external radiation might occur from an accumulation of NORs in the above-ground installation. The radionuclides that are responsible for external exposure are, e.g., ^{214}Pb, ^{214}Bi from the ^{238}U decay series and ^{228}Ac, and ^{208}Tl from the ^{232}Th decay series. These radionuclides emit multiple high-energy photons with energies in the range of 100–1000 keV. The photons will penetrate through the material of the installation parts, and therefore exposure may occur during regular operation of the installation. However, adequate material thickness of the installation parts where accumulation of NORs may occur, such as the filter holders, as well as ensuring sufficient distance of workers—mostly a couple of meters—is more than sufficient to minimize its exposure.

Increased levels of exposure may occur when the installation is opened, for example, during a well intervention and maintenance or when replacing the filters. During such activities, monitoring of the radiation dose for the workers involved is recommended. Generally, exposure levels are very limited, particularly when standard mitigation measures against external exposure are applied. These are minimizing the duration of the activities, maintaining sufficient distance from the source, and application of shielding material. These rules equally apply during transport of any waste and any subsequent cleaning of the contaminated equipment or handling during waste disposal.

4.2.2 *Internal radiation exposure*

Internal exposure is of no concern during regular operation when the installation is fully closed and there is no emission of radionuclides. Only when the installation is opened exposure can be expected. Radioactive elements such as Ra, Po, Pb, and Bi, e.g., present as minerals may become airborne in the form of dust particles leading to inhalation or direct contact on the skin. An alternative exposure pathway is through direct contact of the contaminated water or solids, e.g., on the hand and subsequent ingestion. Mostly this kind of exposure can be mitigated by adequate working protocols, such as ventilation of the workspace and clean working. These mitigation measures apply during handling of any equipment or waste streams on-site but also during transport and subsequent handling at a cleaning facility or waste storage facility.

Another radiation dose comes from the exposure to radon. Radon is a noble gas and can be released during any of the abovementioned activities. For NORs, there are two isotopes of radon that are of concern. These are ^{222}Rn with a half-life of 3.8 days and ^{220}Rn, also called thoron, with a half-life of 56 s. ^{222}Rn is mostly widely known and originates from the decay of ^{238}U decay series, while ^{220}Rn (thoron) originates from ^{232}Th. In most circumstances, ^{222}Rn is of primary concern, and in most countries there are regulatory provisions on the exposure to ^{222}Rn at workplaces. Due to the short half-life of ^{220}Rn, this radionuclide normally does not pose a health risk. However, monitoring can be needed in case of high concentrations of the ^{228}Ra and ^{228}Th progenies from ^{232}Th.

Even though radon is mostly named as an exposure risk, exposure to radon itself is relatively harmless. The risk emanates from its decay products (progeny) polonium (Po), lead (Pb), and bismuth (Bi). These are primarily alpha-emitters, which can be deposited and accumulated in the bronchi and lungs, causing an internal radiation exposure. Radon primarily acts as a generator and transporter of these progenies. Furthermore, the presence of radon (progeny) can be determined best by the measurement of radon gas. It is for these reasons that it is common practice to refer to radon exposure.

4.2.3 *Environment*

It is important to avoid waste disposals from where radioactive material can migrate uncontrolled in the environment. Typically, these risks are associated with any waste containing NORs. Its transport, handling, processing, and storage may result in radionuclide emissions to the environment, such as releases to atmosphere and ground- and surface water. Due to the long half-life of some of the radionuclides such as ^{226}Ra (1600 yr) and ^{210}Pb (22 yr) these radionuclides together with their short-lived progenies may transfer through the environment manifesting itself in locally produced vegetables, meat, fish, and drinking water. For this reason, any waste that is classified as NORM must be stored and/or handled by a classified facility for managing NORM.

4.3 Regulation and compliance

The protection of workers and public against the exposure to ionizing radiation is laid down in national regulation. These mostly find their roots in the international safety requirements of the IAEA or the European EURATOM directives. The IAEA has established requirements on radiation protection for NORM industries under the International Basic Safety Standards described in GSR Part 3 (IAEA, 2014). For member states of the European Union, the Basic Safety Standards described in the EURATOM council directive 2013/59 (EU, 2014) apply. The safety requirement of the IAEA and the EURATOM directive provides minimum requirements for the member states. The transposition into national legislation may therefore vary; nevertheless, the regulatory requirements for NORM industries will be broadly similar in many countries.

4.3.1 *Regulatory requirements*

Regulatory control for handling materials containing NORs is generally applicable when the material exceeds the activity concentrations of general exemption and clearance. These are specified for the NORs as 1 Bq/g for radionuclides in the decay series of ^{238}U, ^{235}U, and ^{232}Th and 10 Bq/g for ^{40}K. In many geothermal installations, materials such as scales or sludges exceed these exemption and clearance levels. As a result regulatory requirements regarding the protection against the dangers from ionizing radiation apply.

The three general principles of radiation protection are justification, optimization of protection, and application of dose limits. The justification deals with the question if the practice produces sufficient benefit to the exposed individuals or to society to offset the harm that it might cause. Optimization is to limit the radiation exposure as low as reasonably achievable. Finally radiation exposure should not exceed the dose limits for workers and public. For a member of the public and an unexposed worker, the limit for the effective dose E is set to 1 mSv per year. For an exposed worker, the limit for effective dose is 20 mSv per year. In the regular operation of a geothermal installation, workers should not be classified as an exposed worker, meaning that their effective dose should not exceed the 1 mSv effective dose.

If the material of concern is classified as NORM, it will fall under regular control, and a.o. a detailed radiation protection program should be established (IAEA, 2018). The general objective of the program is to fulfill the management's responsibility for protection and safety through the adoption of management structures, policies, procedures, and organizational arrangements that are commensurate with the nature and extent of the risks. The radiation protection program should cover all the main elements contributing to protection and safety. As part of a radiation protection program, the following aspects should be considered:

- Prior radiological evaluation and safety assessment
- Qualified experts
- Accountability for radiation generators and radioactive sources
- Classification of areas
- Local rules and supervision
- Personal protective equipment
- Work planning and work permits
- Monitoring and assessment of exposures
- Information, instruction, and training
- Workers' qualification and certification
- Audits and reviews

4.3.2 *International guidance*

IAEA General Safety Guide No. GSG-7

The Safety Guide GSG-7 (IAEA, 2018) is prepared to provide guidance on fulfilling the requirements of GSR Part 3 with respect to occupational exposure. It provides general guidance on the exposure conditions for which radiation protection programs are required to be established, including the setting up of monitoring programs to assess radiation doses to workers arising from exposure due to external sources of radiation and from exposure due to intakes of radionuclides. It also provides specific guidance on the assessment of doses from exposure due to external sources of radiation and from exposure due to intakes of radionuclides.

IAEA Safety Reports Series No. 34

The IAEA SRS No. 34 (IAEA, 2003a) provides practical information, based on good working practices, on radiation protection and radioactive waste management in the oil and gas industry. It is intended to assist in meeting the relevant radiation safety requirements and promoting a common understanding between the industry and regulatory bodies. The Safety Report will be of interest to regulatory bodies; oil and gas field operators and support companies; workers and their representatives; health, safety, and environmental professionals; and health and safety training officers.

IAEA Technical Report Series No. 419

The IAEA TRS No. 419 (IAEA, 2003b) is drafted to help member states gain perspectives regarding the industrial processes that may lead to NORM generation and where, within those processes, NORM can accumulate and eventually may lead to increased potential for exposure of the general public to radioactivity. The report covers all relevant NORM industries and includes a dedicated section (5.1) on geothermal energy production.

EU radiation protection 122 part II

The RP 122 part II of the European Commission (EC, 2002) provides guidance on establishing regulatory control for materials arising from industries which mine or process ores or other materials for which the presence of NORs are of concern. Such NORM materials are dealt with in EU BSS (work activities for which enhanced exposure to natural radiation sources is of concern). The document is of relevance as multiple countries have adopted the recommendations of the RP 122 part II for material stream including those for geothermal operation.

4.4 Measurement of NORM

To comply with regulatory requirements related to the handling of NORM, multiple types of radiation measurement and monitoring is required. This includes measurement to determine the radioactive content in the material of concern, radiation dose in the direct vicinity and beyond, and presence of unexpected contaminations. For this purpose, there is an extensive choice of measurement equipment available using various detection techniques and tailored for the particular needs of the end user. This overview is by no means complete; however, it provides information on the type of equipment that is of primary interest to the geothermal operator.

4.4.1 Measurements by gamma-ray spectrometry

Many of the regulatory limits for materials containing radioactivity are expressed in terms of an activity concentration or massic activity (Bq/g). Typical examples in geothermal operation are the sludges, scales, and filter waste found in the installation. The requirements on regulatory control at the geothermal site or if applicable at a waste storage facility will depend on the activity concentration of the material of concern. For this reason, gamma-ray spectrometry measurements are regularly performed. The main advantage of this method is the possibility to measure many radionuclides simultaneously and the limited need concerning sample preparation. Moreover, well-developed software analyzing gamma-ray spectra are commercially available. Up to date gamma spectrometers, depending on configuration, can be widely applied for precise laboratory quantitative measurement as well as for qualitative screening or monitoring.

Gamma-ray spectrometry relies on the generation of a measurable pulse, either electrical or optical, by a photon (a gamma-ray) in a radiation detector. Independent of the detector type employed, the detector output signal must be converted into a current or voltage pulse that is proportional in magnitude to the gamma-ray emission energy produced by the decay of the radioactive material being measured. The amplitude of a registered pulse must be measured, usually by means of an analog-to-digital converter (ADC), and the measured pulses are then sorted by their amplitude into the so-called channels of a multichannel analyzer (MCA). The number of channels necessary for the MCA should be as much as of the ADC resolution; common values include consecutive nonnegative powers of 2 (i.e., 512, 1024, 2048, 4096, 8192, or 16,384 channels). All pulses registered in channels create a sample spectrum with characteristic peaks (photopeaks) that reflect energies of gamma-rays emitted by radionuclides enclosed in a sample. Each gamma-emitting radionuclide has its own characteristic gamma-ray energy emissions, and these can be used both to establish the presence of the radionuclide in a sample and to quantitatively determine its activity concentration in the sample.

Semiconductor spectrometry

At present, the most widely used type of semiconductor for gamma-ray spectrometry is a high purity germanium (Ge) crystal, which is typically cooled to around −160°C using liquid nitrogen. An electric voltage is applied to the detector, and any interaction of the detector with gamma radiation will result in an electric pulse. Due to the requirements on infrastructure (e.g., N_2 cooling, low radiation background, climate control) and expertise to operate such detector, these types of measurements are mostly performed by an accredited laboratory. In most countries, there are at least one or more organizations that run such laboratory, and geothermal operators can send their samples for measurement. Recently, semiconductor spectrometry is now also available using electric cooling and reduced temperature sensitivity. As a result, in situ spectrometry is now also available.

Scintillation spectrometry

A scintillation detector consists of a scintillator which emits light pulses when exposed to radiation and a device, usually a photomultiplier tube (PM), which transforms these light pulses into electrical signals. Still the most widely used type of scintillation detector for γ-spectrometry is the NaI(Tl) detector. Such detector is relatively low in cost, small, and mobile. However, its major drawback in the measurement of NORM samples is its low energy resolution, making it challenging to distinguish the different photons from the individual radionuclides. Nevertheless, in the last decade, there have been developments in the use of scintillation spectrometry for screening and on-site verification measurements.

It is important to remember that gamma-ray spectrometry is suitable for measuring radionuclides with an emission of γ-radiation. Radionuclides which are pure alpha or beta emitters or do not have any γ-ray emission can only be measured indirectly using its progenies. For this reason, measurement of, e.g., ^{228,226}Ra, ^{232,228}Th, is mostly performed using its progeny radionuclides. However, such indirect measurement is only feasible if the progeny is in equilibrium with its parent nuclide.

A particular challenge in gamma-ray spectrometry is the measurement of ^{210}Pb. Its measurement is highly relevant as many of the deposits and waste streams contain traces of ^{210}Pb. Nevertheless, the radionuclide has only one gamma emission with a low energy of 46 keV. This means that the emitted photons can be absorbed by the sample itself, depending on the sample geometry and chemical composition. Therefore, if a ^{210}Pb contamination is expected, particular attention should be given to the applied measurement protocol to ensure the measured activity is not underestimated. Typically, protocols include a correction for low energy photons in the applied calibration or a reduced sample size. More details on the challenges in performing gamma-ray spectrometry for NORs are described by Friedman et al. (2017). Alternative methods to determine the activity concentration of pure alpha or beta emitters using alpha spectrometry is challenging and requires advanced sample preparation.

4.4.2 *Dose rate measurement*

Radiation level measurements are carried out to monitor the radiation exposure of the workers and to ensure that the applicable limit values for the effective dose as laid down in legislation are not exceeded. However, the effective dose E cannot be measured directly—most dose rate measurements instead report the ambient dose equivalent rate H*(10). Although there are some differences between the H*(10) and the effective dose E, which is the limiting quantity, the result is directly interpreted as the effective dose. Constant ambient dose rates are usually measured by active dose rate meters, based either on ionization chambers, Geiger counters, or scintillation counters. For legal purposes, these devices must be calibrated, and in some countries, it is necessary to have them stamped by a national metrology institute. In such a certification, the conditions for the use of the measurement device are specified, and the uncertainty of the reading is given when used within the limits of the conditions of use. Often also correction factors are given for a use outside the limits of the conditions of use. Typical conditions concern ambient temperature, humidity, power supply voltage, air pressure, linearity, etc. which usually are not problematic. Much more difficulties come with energy dependence and angle dependence.

4.4.3 *Personal dosimetry*

For those workers that are classified as radiological worker, individual monitoring must be considered. The most common detectors are the thermoluminescent dosimeter (TLD); however, other types of detectors such as optically stimulated luminescence (OLS) and direct ion storage (DIS) technology are rapidly becoming popular. These dosimeters are calibrated for measurement of the Hp(10), which is a direct measure of the effective dose E. Personal dosimeters can be obtained from a nationally approved dosimetry service, and increasingly member states require an accreditation under ISO 17005 as part of the approval.

4.4.4 *Contamination monitors*

Radioactive contamination occurs during work with open activity, for example, through splashes, spills, or leakage. Usually, the activity involved is so low that an increase in the external radiation field in relation to the background is not or hardly measurable. Contaminations are detected with contamination monitors that are moved closely over the surface to be examined, and the measurement result is displayed in counts per second (cps). Maximum permitted contaminations in regulation are generally described in terms of Bq/m^2. Unfortunately, a conversion from cps to Bq/m^2 is generally not easy. The instruments need to be more sensitive than the dose rate monitors, and proper use requires some knowledge of the expected contamination and the appropriate monitor. A GM tube or NaI scintillator is often used for the contamination monitors.

4.4.5 *Radon measurement*

For the monitoring of radon, multiple types of detectors are available. For practical purpose, the available methods are distinguished based on the duration of the sampling: (i) integrated measurement methods, (ii) continuous measurement methods including measurements with registration periods from 1 to 6h, and (iii) spot measurement methods.

Integrated measurement method

This method provides an average radon concentration in the air. For an adequate results the duration of the measurement must be at least a couple of weeks and ideally for as long as a full year. The most widely used device is a so-called radon beaker with passive air sampling containing a solid-state nuclear track detector (SSNTD). The detector is a material such as a CR-39 plastic. When exposed to radon progenies, the material is etched, and the tracks in the material yield information about the presence of radon in the indoor environment.

Continuous measurement method

In continuous monitoring, the radon concentration is reported at set time intervals providing information on the temporal changes. Multiple measurement techniques are available; however, alpha spectrometry technology is widely used in many of the commercially available radon monitors, and other techniques include scintillation and gamma spectrometry. Most of the monitors are equipped with an air pump for active sampling. However, there also exist monitors without active sampling, making the equipment much cheaper and less prone to necessary repairs.

Spot measurement method

Spot measurements are performed using the monitors described above. However, to ensure sufficient detector signal, an air pump is part of the setup to enable a high air sampling rate. Results from a spot measurement must always be interpreted carefully, as concentrations are expected to vary during the day and during the seasons.

5 Challenges of artificial lift systems

5.1 Introduction to artificial lift systems

Artificial lift refers to a set of equipment or technologies to increase the production rates and flow in the wellbores. These technologies can enhance the production either by decreasing the bottomhole pressure (BHP) or decreasing the surface pressure or reducing the fluid density. This can enable the production from subhydrostatic wells or increase the flow rate of flowing wells.

In order to explain the basic working principle of ALSs, an example of downhole pumps is used. Firstly, the definitions of inflow performance relation (IPR) and vertical lift performance (VLP) are provided:

- IPR is a relationship that describes the rate at which fluids can flow into a wellbore as a function of the flowing BHP. It represents the well's natural ability to produce the geothermal brine under specific reservoir conditions. The IPR curve typically shows the relationship between the production rate (Q) and the flowing BHP at the sandface of the reservoir. As the BHP decreases, the inflow performance of the reservoir increases, allowing more fluids to flow into the well.
- VLP is a relationship that describes how the production rate from a well varies with the tubing head pressure (THP) or surface pressure. The VLP curve includes the effects of the wellbore and surface equipment, such as tubing, wellhead, and surface flowlines. It represents the ALS's ability to lift fluids to the surface. The VLP curve typically shows the relationship between the production rate (Q) and the THP.

Both of these relations are solved together to estimate the production rates and/or flowing BHP. The working principle of downhole pumps as an ALS is shown schematically in Fig. 7. It can be seen that the interception of IPR and VLP curves will provide the production rate and/or flowing BHP. The use of pumps can provide a pressure boost in the VLP curve which eventually leads to a lower BHP and a higher production rate, shown by ΔQ in Fig. 7. It is important to note that the shape of IPR and VLP can be different depending on reservoir and wellbore parameters, such as reservoir permeability, multiphase flow, wellbore trajectory, etc.

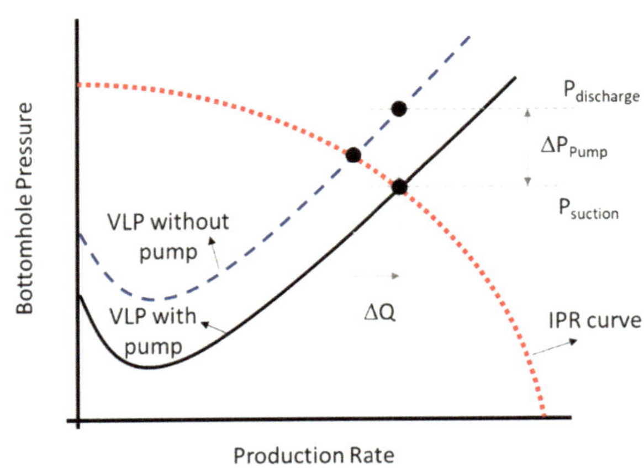

FIG. 7 Schematic of the working principle of downhole pumps.

5.2 Different types of ALS

The petroleum industry uses a portfolio of lift methods including the ESP but also makes use of, e.g., gas lift, chemical lift (foam lift), and a range of pump types (jet pumps, beam pumps, etc.) (Lea and Nickens, 1999). There are several types of ALSs, and the choice of which to use depends on factors such as well characteristics, production rates, and the type of fluids being produced. Some common types of ALSs include downhole pumps and surface pump gas lift.

Downhole pumps, also known as subsurface pumps or submersible pumps, are pumps which are installed in the wellbore. These pumps are often multistage centrifugal pumps, consisting of an impeller and a diffuser at each stage, which are operating in the wellbores and at a vertical configuration in which the production fluid is subjected to a centrifugal force caused by the rotational speed of the impellers. The kinetic energy is converted to pressure energy in the diffusers which is required for the fluid to be lifted upward in the wellbore. There are several types of downhole pumps which are categorized based on the location of the motor and the type of system used for driving the pumps. ESPs refer to downhole pumps which the electrical motor is also in the wellbore and powered by an electrical cable. Line shaft pumps (LSPs) are downhole pump with an electrical motor on the surface powered by the connecting shaft from the surface to the installed pump downhole.

There is a broad spectrum of downhole pumps which are categorized based on the different mechanisms to drive them. Hydraulic surface pumps (HSPs) are driven by a turbine, progressive cavity pumps (PCPs) are driven by a surface motor and a rod, and hydraulic piston pumps (HPPs) are using a piston and pressurized fluid to drive the pump. In case reservoir pressure is high enough to drive the fluid, surface pumps can be utilized to control the production rates.

Gas lift lowers the BHP by lowering the density of the mixture and with that the hydrostatic head in the production tubing by injecting a gas into the production stream. The operating valve should be placed as deep as possible to increase the efficiency. A conduit for the gas, usually the production annulus in O&G wells, is required (Fig. 8).

5.3 Comparison of ALS in geothermal systems

LSP and ESP are the most common ALSs in geothermal wellbores. With the current technological developments, LSPs are mostly used for high-enthalpy geothermal systems and ESPs for low and medium enthalpy geothermal systems. LSPs are mature technology in the geothermal industry, and since there is no downhole motor, they can handle higher temperatures and their maintenance is simpler. Thus, for high-temperature geothermal systems in which a significantly high production rate is not required, LSP is a suitable option. However, these pumps are limited to shallower installation depth and nearly vertical wellbores. Because of a long shaft connecting the surface motor with the downhole pump, there is a high chance of shaft mechanical failure and breakage, and a continuous shaft lubrication is required which can pose fouling challenges.

Unlike LSPs, ESPs can be installed and operated at much deeper depths even in deviated and inclined wellbore. ESPs can also provide much higher flow rates. The surface equipment and footprint of ESPs are lower compared with LSP due to the installation of the submersible

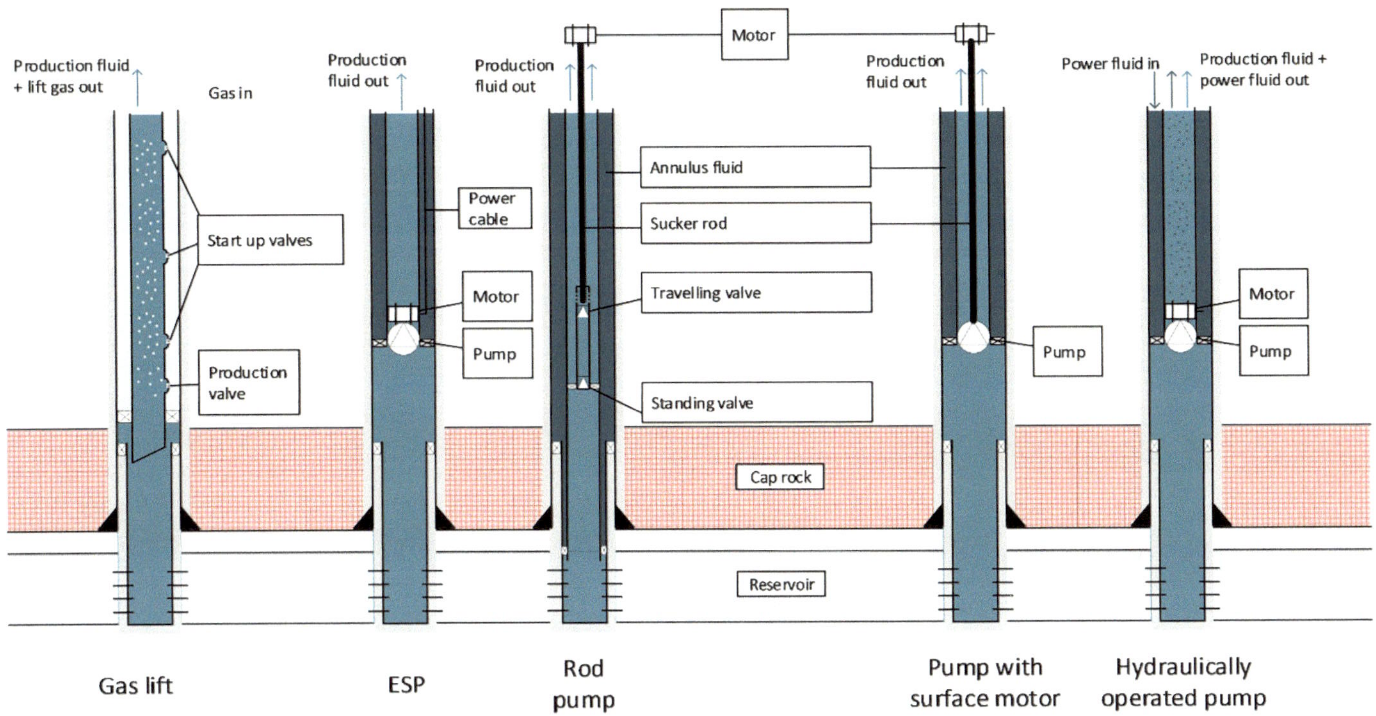

FIG. 8 Impression of different types of artificial lift systems applicable to liquid-dominated wells.

motor in ESP. However, ESP selection, installation, and operation require a significant attention and knowledge to improve its reliability (Shoeibi Omrani et al., 2021). One of the major challenges in geothermal ESP is electrical failures due to several reasons including motor overheating, water infiltration, and insulation failures. Even though ESPs can be installed in the deeper depth and in inclined wells, their installation at locations with a severe dogleg can impact ESP and wellbore integrity. Since for each maintenance and inspection the entire system (including pump, motors, and cables) needs to be pulled, the servicing procedure can be time-consuming and costly. In addition, the efficiency of the LSP is higher than ESP due to the surface motor.

Both ESP and LSP operating in geothermal conditions need to be carefully designed, installed, and operated due to geothermal brine composition (high salinity brine) and conditions (high temperature). Downhole pumps are susceptible to scaling and corrosion issues, and if not treated correctly, their lifetime can be significantly reduced. The presence of gas and abrasive solids can impact LSP and ESP operation. Comparing downhole pumps in geothermal and oil wells, one outstanding difference is that most of geothermal ESPs are installed at shallower depth <1000 m true vertical depth (TVD). Shallower installed depths can provide high thermal stresses during the production and shut-in periods which is not a case in oil wells.

In geothermal wells, higher production rate is required, which means larger pumps with a higher horse power compared with oil wells are required. Combination of higher production rates, with high temperature and high salinity of geothermal brine, will pose challenges on the performance of the downhole pumps. Several flow assurance issues were observed in geothermal ESP and LSP, such as scaling and corrosion, caused by the geothermal brine composition and operational conditions. Specifically for low and medium enthalpy geothermal wells for the direct use and heating applications, the production rate of geothermal systems is driven by the heat demand which can vary significantly during winter and summer time. An example of the ESP production rate in a geothermal well throughout the year is shown in Fig. 5. If such a variability is not considered during the design of the pumps, it can lead to a significant decline in the lifetime of the pumps. Thus, despite similarities of installing and operating ESP in geothermal and oil wells, there are some distinct differences that require building and sharing knowledge in the sector. One of the similarities identified between ESP used in petroleum and geothermal sector is the use of ESP for water producing wells employed in early stages of oilfield production to maintain reservoir pressure or supply water for injection. These ESPs often supply water based on the demand which is similar to geothermal ESP.

Continuous monitoring of downhole pumps and understanding their performance degradation, failures, and parameters impacting the lifetime of ALS are a crucial step in improving the reliability and performance of these systems. In addition, training operators to operate these pumps in an optimum manner is important. As an example, sudden start-up and shut-ins of the pump can severely harm the pumps and shorten their lifetime. Finally, adaption of the pumps from oil and water applications to geothermal applications will require a detailed study to modify their designs based on the impact of higher production rate, higher power, higher salinity, and temperatures which are observed in geothermal fields (Fig. 9).

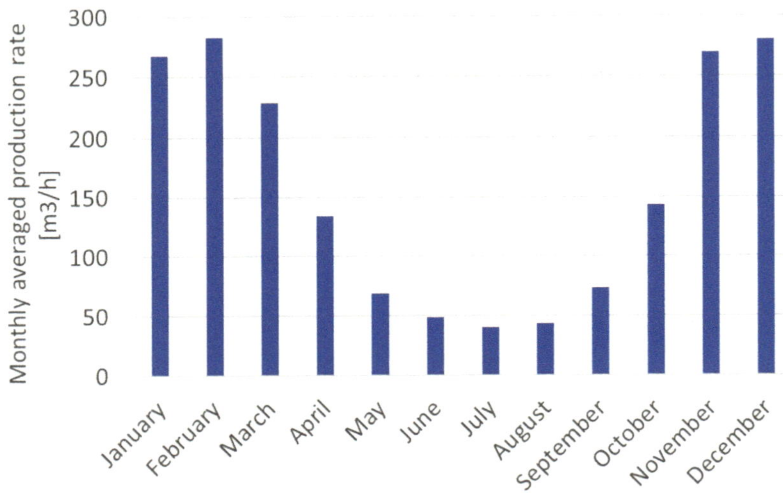

FIG. 9 An example of a low-enthalpy geothermal well production rate in 1 year which is impacted by the seasonality of the heat demand.

6 Other production challenges

6.1 Sand production

Sand production is a common issue both in hydrocarbon and geothermal wells. The issues with sand production have been observed both in sandstone and carbonate reservoirs (Wulan et al., 2007). In the case of consolidated rock, the initial step is the mechanical failure of the rock, followed by the transport of the failed sand to the wellbore. The transport of sand is directly linked to flow-induced strain, with a correlation to applied drawdown which is defined as the difference between the far-field reservoir pressure and the BHP at the wellbore during production. In unconsolidated or soft formations in which there is marginal or no natural cementation, the sand production is mainly explained by fine migration and transport. There are several parameters impacting the sand production both from the reservoir and well sides such as formation type and characteristics which will determine the strength of the rock, reservoir depth, production conditions such as flow rate, pressure and temperature, wellbore trajectory, and completion (Fjaer et al., 2008; Van den Hoek et al., 2000). In geothermal wells, the sand production was found to be more critical in shallow formations caused by unconsolidated formation (Peters et al., 2022).

One of the main impacts of sand production is erosion. Erosion is the process of surface material removal due to the impact or relative movement between the surface and solid particles, liquid droplets (in gaseous phase), or bubble (in liquid phase). The solid particles contributing to erosion may originate from the migration of particles from the reservoirs to the wells and surface facilities, or they may result from abrasive particles generated during material corrosion. Particle erosion can be a potential challenge in geothermal wells and topside facilities. The particle and sand erosion are mainly observed in (shallow) geothermal systems or systems producing from unconsolidated formation. This issue has been observed near geometrical features where flow area reduction or increase exists, such as screens, ESP internals, bends, restrictions/valves, and pipe joints. The locations near the ESP's and at the well head are subjected to higher erosion risks, due to higher flow velocities and increased levels of turbulence due to a rapid change in the flow velocities (Zwinkels, 2015). In simple terms, erosion is proportional to particle production rate, fluid velocity, hardness of the surface and particles, and angle of particle impingement. Erosion-corrosion is when particle erosion enhances corrosion by exposing "fresh metal" to the corrosive medium or when the corroded material is more easily being eroded away. As an example, the protective oxide layer or inhibitor layers can be eroded by abrasive particles which enhances corrosion of the surface, and this is an issue encountered in geothermal wells more frequently.

In addition, the produced sand can accumulate downhole and in the surface facilities which can reduce or completely cease the production. If sand production continues, it can eventually lead to the formation collapse and can severely impact the well productivity and integrity. Surface subsidence was also observed in petroleum fields as a result of large volume of sand production which leads to a reduced pore pressure and compaction of the reservoir rock.

In order to control or mitigate sand production, the possibility and impact of sand production should be assessed ideally at the early stage of the design. The detrimental impact of erosion on the system integrity can be minimized in the early stages by proper sizing of the equipment and pipelines. If no sand production is expected, often API RP 14E (1991) is used as a method to check the sizing of the components and determine the maximum velocities in the system. In case sand and solids are expected, then other methods and guidelines such as DNV RP 0501 (2011) are recommended for the design assessment. In components more susceptible to particle erosion, such as in the pumps or valves, the use of high hardness or non-metallic materials is recommended.

There are several methods to control the sand production during the completion design with the aim of producing from the reservoir sections that are sufficiently strong to the desired production rates and drawdown. Due to the correlation between the sections with higher compressive strength (capable of handling higher drawdown) and lower permeability, this method can have an impact on the production rate and economics of the asset. Thus, methods to produce from the full reservoir section and being capable of controlling the sand production are often employed in the industry. Stand-alone screens or slotted liners are one of the most adapted methods to control sand in hydrocarbon and geothermal wells. They act as a downhole filter to control formation sand production ensuring optimal well performance by preventing sand or other particles from entering the production stream. Historically they showed poor reliability in heterogenous reservoirs (Bellarby, 2009).

Another sand control method is gravel packing. Gravel packing is a well completion method by placing a layer of specifically sized gravel or sand, referred to as the gravel pack, around the external surface of a screen or slotted liner within the wellbore. By serving as an effective filter, the gravel pack facilitates the passage of hydrocarbons and geothermal brine while obstructing the migration of formation particles. There are two type of gravel packs used, open hole gravel pack and cased hole gravel pack each with their pros and cons. Cased hole gravel pack can provide excellent zonal isolation and can be broadly used for different formation strengths, while they will lead to a higher skin factor.

One of the common methods to handle sand production during the operation is by reducing drawdown. This can be done by controlling the wellhead production choke which in turn impacts the production rates. Accurate estimation of critical drawdown to initiate sand production is required together with the sand handling capability of the asset to optimize the production rate. In case the produced sand cannot be completely avoided or controlled, particle filters with a wide range of particle size are being used in the surface upstream of the critical components such as heat exchanger and injection well. This requires a frequent exchange of the particle filters as soon as the pressure drop over the filter unit reaches a certain threshold. The accumulated sand in the wells and surface facilities needs to be cleaned up. Downhole cleaning for sand loaded wells is often done using wireline (with a bailer installed). If the sand loading is severe, then several wirelines will be required to perform a thorough cleaning. Other methods such as coiled tubing or running a smaller tubing string are often used for downhole cleaning of wells with sand loading.

6.2 Noncondensable gases

Another issue in geothermal production is unwanted by-production of gases. The produced NCGs in the geothermal systems could contain mainly carbon dioxide, methane, hydrogen sulfide, hydrogen, nitrogen, and even ammonia. The concentration and composition of the dissolved and produced gas from geothermal reservoirs need to be assessed carefully for operation and risk management. A summary of the CO_2/CH_4 ratio in different European geothermal brine is shown in Fig. 10. The concentration of the formation gas can be different in different assets. In the Dutch geothermal systems, the gas/liquid ratio of 0.3–1 nm^3/m^3 is reported (Hartog, 2015) which is mainly composed of CH_4. In other geothermal assets in Europe, CO_2 is the dominant component. In the presence of NCGs and their concentration, several safety and integrity measures need to be considered. The associated gas in the geothermal systems could greatly impact the scaling and corrosion rates (Vorum and Fritzler, 2000). In addition, NCGs in power production systems can reduce the heat transfer efficiency in the condensers which will require a higher condenser operating pressure and ultimately reduces the turbine power output (Gokcen and Yildirim, 2008). One of the first studies on this topic reported that 10% of NCGs in the geothermal steam can reduce the power output by approximately 25% (Khalifa and Michaelides, 1978). Another issue could be the multiphase flow-induced production fluctuations due to flow instabilities or gas expansion which can happen for higher gas to liquid ratios. Finally, the by-produced gases can impact the environmental footprint of the geothermal operation if the gas content is not handled properly. It is important to note that emissions from geothermal plants are much lower compared with other fossil sources for heating and power. The emission factor from oil and gas sources is estimated around 700 g/kWh and 400 g/kWh, respectively, while for low-enthalpy geothermal systems, it is estimated at around 15–40 g/kWh (Dinkelman et al., 2021). However, for power production systems, the emission factor can be higher if the NCGs are not handled properly.

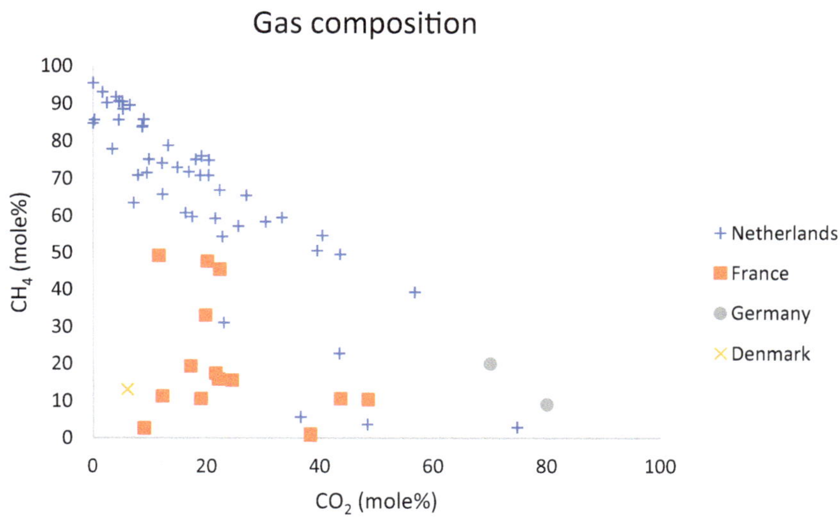

FIG. 10 CH_4 and CO_2 concentration of geothermal brine in four different European countries.

Produced CH_4 can cause safety issues in geothermal assets. On the other hand, utilization of the produced methane, for instance, by using CHP units and gas motors, can provide additional power source to run the geothermal system. As an example, in Túrkeve plant in Hungary with a production temperature of 75 °C, the separate gas was used to drive gas motors to supply 413 MWh/yr electricity (OPERA, 2019). Thus, handling the gas content properly can create additional value for the geothermal assets.

Several studies are performed to propose various methods to handle by-produced gases in geothermal systems and create an emission-free geothermal plant (Nimwegen et al., 2022). For the low-enthalpy geothermal systems with CH_4 and CO_2 as major NCGs, the proposed methods are pressurization of the system (to keep the gases dissolved), CO_2 capture, and formation of gas reinjection. Other alternative applications of NCG utilization especially for CO_2 are for greenhouse use, food and beverage, and sequestration (McIntush et al., 2022). For each of the application, there is a requirement specification and concentration of the gas mixture. As an example, the threshold limit value (TLV) for CO_2 for the greenhouse application is 5000 ppm (OSHA, 2019), while for the beverage application, a very high purity level of CO_2 is required with a minimum value of 99.9 v% (McIntush et al., 2022).

Supplementary online elements

Nontextual elements which are published alongside or as components of the work in online formats only.

Multimedia

Nontextual online content which can only be provided non-textually, principally to help visualize key points, drive pedagogy or to otherwise support the work. Examples include videos, audio presentations, slideshows, and animations.

Reference data

Raw data gathered, used, or produced in relation to work discussed in the text. Often too large to reproduce directly. Used to show results for experimental validation purposes.

References

Andritsos, N., Karabelas, A.J., Koutsoukos, P.G., 2002. Scale formation in geothermal plants. In: International Summer School on Direct Application of Geothermal Energy.

API RP 14E, 1991. Recommended Practice for Design and Installation of Offshore Production Platform Piping Systems, fifth ed. American Petroleum Institute.

Bellarby, J., 2009. Well Completion Design, first ed. vol. 56, ISBN: 9780444532107.

Bruno, J., Cera, E., de Pablo, J., Duro, L., Jordana, S., Savage, D., 1997. Determination of Radionuclide Solubility Limits to be used in SR 97 – Uncertainties Associated to Calculated Solubilities. SKB Technical Report 97-33.

CDC, 1996. A Study of NORM Associated with Oil and Gas Production Operations in California, Division of Oil, Gas, and Geothermal Resources. California Department of ConservatioN, Sacramento, CA. 62 pp.

Dinkelman, D., Dijkstra, H., Shoeibi Omrani, P., 2021. Sustainability of geothermal energy. In: GeoTherm Congress, June 2021, Offenburg, Germany.

DNV RP 0501, 2011. Erosive Wear in Piping Systems.

EC, 2002. Radiation Protection 122 Part II, Application of the Concepts of Exemption and Clearance to Natural Radiation Source. European Commission, Luxembourg, ISBN: 92-894-3315-9.

Ellis, A.J., Majon, W.A.J., 1977. Chemistry and Geothermal Systems. Academic Press, London.

EPA, 1993. Diffuse NORM Wastes — Waste Characterization and Preliminary Risk Assessment, Draft RAE-9232/1-2, SC&A. US Environmental Protection Agency, Office of Radiation and Indoor Air, Inc. and Rogers & Associates Engineering Corporation, Salt Lake City, UT.

EU, 2014. Council Directive 2013/59/Euratom, 2014 laying down basic safety standards for protection against the dangers arising from exposure to ionizing radiation and repealing Directives 89/618, 90/641, 96/29, 97/43 and 2003/122/Euratom. Off. J. Eur. Union L13.

Fjaer, E., Holt, R.M., Horsrud, P., Raaen, A.M., Risnes, R., 2008. Petroleum Related Rock Mechanics, second ed. Elsevier.

Friedman, H., Nuccetelli, C., Michalik, B., Anagnostakis, M., Xhixha, G., Kovler, K., De With, G., Gascó, C., Schroeyers, W., Trevisi, R., Antropov, S., Tsapalov, A., Kunze, C., Petropoulos, N.P., 2017. Measurement of NORM. In: Schroeyers, W. (Ed.), Naturally Occurring Radioactive Materials in Construction. Elsevier, Woodhead Publishing Series in Civil and Structural Engineering, United Kingdom (Chapter 5).

Gokcen, G., Yildirim, N., 2008. Effect of non-condensable gases on geothermal power plant performance. Case study: Kizildere Geothermal Power Plant-Turkey. Int. J. Exergy 5 (5/6).

Goldschmidt, B., 1938. Sur la précipitation mixte des sulfates de baryum et strontium. Hebd. Séances Acad. Sci. 206, 1110.

Grandia, F., Merino, J., Amphos, J.B., 2008. Assessment of the radium-barium co-precipitation and its potential influence on the solubility of Ra in the near-field. SKB Technical Report TR-08-07.

Hardy, J.A., Simm, I., 1996. Low sulphate seawater mitigates barite scale. Oil Gas J. 94, 64–67 (PennWell Publications).

Hartog, N., 2015. Geochemical Assessment of Injectivity Problems in Geothermal Wells—A Case Study for Several Greenhouse Geothermal Systems in the Netherlands. KWR 2015.012, KWR Watercycle Research Institute.

Hartog, 2016. Carbonate scalings in deep geothermal systems. In: Schreiber, S., Lapanje, A., Ramsak, P., Breemboek, G. (Eds.), Operational issues in Geothermal Energy in Europe. Coordination Office Geothermal ERA NET (Chapter: 2.2).

IAEA, 2003a. Protection and the Management of Radioactive Waste in the Oil and Gas Industry. Safety Reports Series No. 34, International Atomic Energy Agency, Vienna, Austria.

IAEA, 2003b. Extent of Environmental Contamination by Naturally Occurring Radioactive Material (NORM) and Technological Options for Mitigation. Technical Report Series No. 419, International Atomic Energy Agency, Vienna, Austria.

IAEA, 2014. Radiation Protection and Safety of Radiation Sources: International Basic Safety Standards, General Safety Requirements Part 3 No. GSR Part 3. International Atomic Energy Agency, Vienna, Austria.

IAEA, 2018. Occupational Radiation Protection, General Safety Guide No. GSG-7. International Atomic Energy Agency, Vienna, Austria.

Khalifa, H.E., Michaelides, E., 1978. The Effect of Noncondensable Gases on the Performance of Geothermal Steam Power Systems. US Department of Energy, Rhode Island, USA. Report No. CATMEC/28.

Knett, J., 1904. Indirekter Nachweis von Radium in den Karlsbader Thermen–Sitzungsber. Akad. Wis. Wien 113, Abt. II.a, 753–762.

Kölbel, L., Kölbel, T., Maier, U., Sauter, M., Schäfer, T., Wiegand, B., 2020. Water-rock interactions in the Bruchsal geothermal system by U-Th series radionuclides. Geotherm. Energy 8, 24.

Kristensen, L., Dideriksen, K., Holmslykke, H.D., Kjøller, C., Larsen, U., Mathiesen, T., Dijkstra, H., Poort, J., Wasch, L., Shoeibi Omrani, P., Regenspurg, S., 2020. PERFORM WP1: learn and understand knowledge database for understanding the current problems. PERFORM Work Package Report.

Kristensen, L., Dideriksen, K., Kjøller, C., Zotzmann, J., Regenspurg, S., Wasch, L., Shoeibi Omrani, P., van Pul, V., 2022. Improving geothermal system performance through collective knowledge building and technology development. PERFORM Work Package 5 Report.

Kulik, D., Brandt, F., Klinkenberg, M., Weber, J., Winkler, B., Bosbach, D., 2017. Thermodynamics of the solid solution—aqueous solution system (Ba,Sr,Ra)SO_4 + H_2O. I. The effect of strontium content on radium uptake by barite. Appl. Geochem. 89.

Lea, J.F., Nickens, H.V., 1999. Selection of artificial lift. In: SPE Mid-continent operations symposium, Oklahoma City, SPE 52157.

Mathiesen, T., Nielsen, J.B., Shoeibi Omrani, P., Wasch, L., Dideriksen, K., Kristensen, L., Holmslykke, H., Kjøller, C., Regenspurg, S., 2021. PERFORM best practice for geothermal plants to minimize scaling and corrosion. In: Deliverable GEOTHERMICA PERFORM Project, Link.

Maxwell, S., Campbell, S., 2006. Monitoring the mitigation of MIC risk in pipelines, corrosion NACExpo 2006. In: 61st Annual Conference and Exposition, Paper No. 06662.

McIntush, K.E., Mamrosh, D.L., Piggot, B.D., Beitler, C.A.M., 2022. Considerations for alternative use of noncondensable gas at geothermal power plants. In: Proceedings of Geothermal Rising Conference 2022, Reno, NV.

Milne, N.A., O'Reilly, T., Sanciolo, P., Ostarcevic, E., Beighton, M., Taylor, K., Mullett, M., Tarquin, A.J., Gray, S.R., 2014. Chemistry of silica scale mitigation for RO desalination with particular reference to remote operations. Water Res. 65, 107–133.

Nimwegen, D., van't Westende, J., Shoeibi Omrani, P., Wasch, L., Peters, E., 2022. Handling Greenhouse Gases From Geothermal Wells. WarmingUP program Report.

OPERA, 2019. Operational Issues in Geothermal Energy in Europe, Status and Overview.

OSHA, 2019. United States Department of Labor.

Peters, E., Dinkelman, D., van Nimwegen, D., Shoeibi Omrani, P., 2022. Sand Production in Shallow Geothermal Applications. WarmingUP Program Report.

Poonoosamy, J., Mahrous, M., Curti, E., et al., 2021. A lab-on-a-chip approach integrating in-situ characterization and reactive transport modelling diagnostics to unravel (Ba,Sr)SO_4 oscillatory zoning. Sci. Rep. 11, 23678.

Raheem, K., Charpentier, T., Sanni, O., Neville, A., 2021. Surface Precipitation and Growth Kinetics of Calcium Carbonate ($CaCO_3$) Scale Using a Noven Capillary Flow Rig. Association for Materials Protection and Performance (AMPP), Leeds.

Read, D., Black, S., Ceccarello, S., Weiss, H., Schubert, M., Kunze, C., Grossmann, J., 2004. Radioactive scales from a natural gas production facility in the Altmark region, Germany. In: ICAM 2004, Int. Congress on Applied Mineralogy, Águas de Lindóla, Brazil, September 19–22.

Regenspurg, S., Dilling, J., Mielcarek, J., Korte, F., Schkade, U., 2014. Naturally occurring radionuclides and their geochemical interactions at a geothermal site in the North German Basin. Environ. Earth Sci. 72, 4131–4140.

Regenspurg, S., Feldbusch, E., Byrne, J., Deon, F., Legesse Driba, D., Henninges, J., Kappler, A., Naumann, R., Reinsch, T., Schubert, C., 2015. Mineral precipitation during production of geothermal fluid from a Permian Rotliegend reservoir. Geothermics 54, 122–135.

Scheiber, J., Nitschke, F., Seibt, A., Genter, A., 2012. Geochemical and mineralogical monitoring of the geothermal power plant in Soultz-sous-Forêts (France). In: Proceedings, Thirty-Seventh Workshop on Geothermal Reservoir Engineering, Stanford University, Stanford, California.

Scheiber, J., Seibt, A., Birner, J., Cuenot, N., Genter, A., Moeckes, W., 2014. Barite scale control at the Soultz-Sous-Forêts (France) EGS Site. In: PROCEEDINGS, Thirty-Eighth Workshop on Geothermal Reservoir engineering, Stanford university, February, 2014.

Shoeibi Omrani, P., Van der Valk, K., Bos, W., Nizamutdinov, E., Van der Sluijs, L., Eilers, J., Pereboom, H., Castelein, K., Van Bergen, F., 2021. Overview of opportunities and challenges of electrical submersible pumps ESP in the geothermal energy production systems. In: Paper presented at the SPE Gulf Coast Section Electric Submersible Pumps Symposium, Virtual and The Woodlands, Texas, USA, October., https://doi.org/10.2118/204524-MS.

Sigfusson, B., Gunnarsson, I., 2011. Scaling prevention experiments in the Hellisheiði power plant. In: Proceedings Thirty-Sixth Workshop on Geothermal Reservoir Engineering, Iceland.

Tranter, M., de Lucia, M., Wolfgramm, M., Kuhn, M., 2020. Barite scale formation and injectivity loss models for geothermal systems. Water 12 (11), 3078.

UNSCEAR, 1982. Ionizing Radiation: Sources and Biological Effects—Annex B Exposures to Natural Radiation Sources. United Nations Scientific Committee on the effects of Atomic Radiation, United Nations, New York.

UNSCEAR, 1993. Sources and Effects of Ionizing Radiation—Annex a Exposures from Natural Sources of Radiation. United Nations Scientific Committee on the effects of Atomic Radiation, United Nations, New York.

Van den Hoek, P.J., Kooijman, A.P., De Bree, P., Kenter, C.J., Zheng, Z., Khodaverdian, M., 2000. Horizontal-wellbore stability and sand production in weakly consolidated sandstones. SPE Drill. Complet. 15 (04), 274–283.

Vorum, M., Fritzler, E.A., 2000. Comparative Analysis of Alternative Means for Removing Non-Condensable Gases from Flashed-Steam Geothermal Power Plants, NREL/SR-550-28329. National Renewable Energy Laboratory (NREL), Colorado, USA.

Wulan, R.S., Susilo, R.Y., Hendra, Y.S., et al., 2007. Development Strategy of Soft Friable Carbonate Gas Reservoir through Horizontal Open Hole Gravel Packed Completion: APN Field Offshore West Java. SPE 104532.

Zhen-Wu, B.Y., 2016. A Combined Thermodynamic and Kinetic Model for Prediction of Barite Precipitation at Oil Reservoir Conditions. PhD Dissertatio), Faculty of Science, University of Copenhagen, Denmark. 223 pp.

Zwinkels, T.M.A.J., 2015. Verstoppingen in Geothermische Productieputten, Ministerie van EZ - Stichting Programmafonds Glastuinbouw. End Report E15014.

CHAPTER

10

Simulation of geothermal resources management

Denis Voskov

Department of Geoscience & Engineering, Delft University of Technology, Delft, The Netherlands; Department of Energy Science and Engineering, Stanford University, Stanford, CA, United States

1 Introduction

In this chapter, different types of geothermal systems are presented and the challenges they introduce to reservoir modeling are explained. Different physical phenomena involved in the successful forecasting, optimization, and risk analysis of geothermal applications are discussed as well. This chapter reflects more than 20 years of experience of the author and his group in developing modeling techniques for various subsurface applications including geothermal; the majority of the results are based on the PhD research by Mark Khait, Yang Wang, and Aleks Novikov. While this overview is not complete and may miss some scenarios relevant to geothermal, all described principles and techniques are applicable to a wide range of geothermal problems.

1.1 Types of geothermal systems

Geothermal energy is a renewable energy source that utilizes the natural heat from the Earth's crust to produce electricity and heat. Several types of modern geothermal energy systems exist, each with unique characteristics and applications. Depending on these parameters, the modeling approach can be altered for different geothermal systems. In general, there are two major types of geothermal systems based on the utilization of produced heat: direct-use and electricity-generating geothermal systems.

The direct use of geothermal resources is the utilization of underground hot water to heat buildings, growing plants in greenhouses, dehydration, heat water for fish farming, industrial applications, etc. It uses the constant heat from the Earth's crust, often based on the

Geothermal Energy Engineering
https://doi.org/10.1016/B978-0-443-21662-6.00008-6

Copyright © 2025 Elsevier Inc. All rights are reserved, including those for text and data mining, AI training, and similar technologies.

natural gradient of temperature distribution with depth. Direct-use systems can be categorized into heat pumps, heat piles, and deep geothermal direct-use systems. Heat pumps and piles are mostly implemented as closed-loop systems and based on heat adsorbed from the ground in winter and heat released to the ground in summer. These systems operate at relatively shallow depths, with a temperature range between 15°C to 30°C.

Deep geothermal direct-use systems involve accessing hot water or steam from deep underground for direct use in heating, cooling, or industrial processes. These systems typically operate at temperatures ranging from 40°C to 150°C and require drilling deep wells to access the geothermal resources. Deep geothermal direct-use systems offer several advantages over heat pumps and piles. They offer a constant and reliable heat source, with low operating costs and a long lifespan. They are particularly suitable for large-scale heating and cooling applications, where the high initial capital costs of drilling wells can be justified by long-term energy savings. However, deep geothermal direct-use systems are limited by the availability and quality of geothermal resources. They require access to high-temperature geothermal reservoirs, which may be limited in certain regions and at required depths especially when no thermal anomaly is present. Besides, such systems are operated in convection-dominated mode forcing the cold front to slowly propagate toward the production well. Once the cold front breaks through, the system needs to be stopped and abandoned, as the thermal recharge, mostly driven by conductive reheating, will take a significant amount of time to restore the original temperature in the reservoir.

There is an entire variety of electricity-generating geothermal systems, also known as geothermal power plants. Conventional geothermal power plants are most commonly used in areas with high heat flow and significant geothermal resources usually supported by thermal anomalies (hot spot). They work by drilling deep wells into hot geothermal reservoirs and pumping hot water or steam to the surface to drive turbines connected to generators. They can operate in dry steam, flash steam, and binary cycle regimes depending on available geothermal resources, subsurface conditions and surface facilities. All these technologies range in power production and economic efficiency often constrained by available geothermal resources.

Enhanced geothermal systems (EGS), also known as hot dry rock systems, are a type of geothermal energy technology that aims to create geothermal reservoirs in areas where high-temperature conditions are present but the permeability of the in situ rock is significantly low. EGS are usually created by drilling wells into hot rock formations and injecting fluid (often cold water) under high pressure to create a fractured network in the rock. EGS has the potential to significantly expand the availability of geothermal energy as a renewable energy source, as it can be applied in areas that are not traditionally associated with geothermal energy production. EGS also offers greater flexibility in terms of the location and size of the geothermal resource, as well as the ability to scale up the system to meet increasing energy demand. The main challenge associated with EGS is the difficulty in creating and maintaining the fractured system in the rock, and the high potential for induced seismicity (earthquakes) due to hydraulic and thermal stimulation.

There are several promising technologies for geothermal energy production and electricity generation using novel experimental methodologies. These methods involve closed-loop systems of different types, such as U-shaped well loops with multiple deviations, single vertical or deviated wells where fluids circulate in coaxial wellbores, and a combination of the coaxial wellbore and hydraulic fractures with independent convective circulations in both regions.

Closed-loop systems are particularly attractive due to the significantly reduced risks of induced seismicity, and the potential use of alternative fluids in place of water to improve conductivity. However, most of these technologies face challenges related to insufficient heat exchange between the fluid and the reservoir rock due to a slow rate of conductive heating.

1.2 Modeling of geothermal systems

In this overview, we will mostly focus on deep geothermal systems often developed by a pair of wells (injection and production) called doublet. However, a lot of discussed modeling techniques are also applicable to shallower geothermal resources. While deep geothermal energy is an attractive energy resource, the technical and economic feasibility of geothermal projects can be directly impacted by high heterogeneity and uncertainty in the characterization of subsurface geothermal reservoirs (Vogt et al., 2013). The heterogeneity depends on the type of rock used for geothermal development which, in turn, depends on the sedimentary environment.

Geological characterization of fluvial reservoirs is intricate as it comprises heterogeneity in various sediment bodies. Fluvial reservoirs primarily comprise impermeable floodplains and permeable sandstone bodies with very uncertain structures. Shale and other low-permeable facies are typically excluded from the reservoir characterization in the case of hydrocarbon extraction modeling to reduce the size of the simulation model. However, this simplified approach can lead to inaccuracies in the simulation results for geothermal applications. This is because heat transfer in geothermal reservoirs relies on both thermal convection and conduction. The impermeable facies act as a buffer for heat, influencing the recharge of fluid flowing convectively. Using process-based models, Crooijmans et al. (2016) discovered that shale facies are critical in determining the doublet performance when the net-to-gross ratio is less than 50%.

In addition to their ability to seal and trap fluid, burden layers within geothermal reservoirs can also provide thermal recharge to the payzone, as noted by Shukla et al. (2010) and Poulsen et al. (2015). However, early studies have not consistently recognized or treated burden layers. In some cases, they are considered nonconductive thermal insulation (Gong et al., 2020). In contrast, in other studies, they are found to have a weak impact on doublet performance, for example, in the case of synthetic process-based heterogeneous models (Shetty et al., 2018). Willems et al. (2017) found that burden layers have a positive impact on heat production by charging the target aquifer based on a simplified homogeneous model. It is important to note that subsurface systems are inherently heterogeneous (Krevor et al., 2011; Miller et al., 1998; Pini et al., 2012), and investigating the influence of burden layers in realistic contexts is more meaningful and representative. Furthermore, a sufficient characteristic resolution is crucial for this investigation, as oversimplification or coarser representation of burden layers can weaken or ignore their influence on geothermal doublet performance.

The success of a geothermal project, especially in the presence of subsurface heterogeneity and uncertainties, relies on reasonable operational strategies. For instance, the discharge rate directly impacts thermal propagation and energy output, as reported by Asai et al. (2019) and Mottaghy et al. (2011). Saeid et al. (2015) found that a higher discharge rate results in a shorter doublet lifetime, sharper thermal front, and lower energy extraction. In a multiwell system, well interference can alter the development of the cold water plume and affect doublet lifetime. To maximize doublet lifetime, optimized doublet configuration and well spacing are generally required, as noted by Willems et al. (2016, 2017). Daniilidis et al. (2021) performed systematic sensitivity analyses of doublet performance to operational parameters and doublet configuration in a faulted reservoir and highlighted the importance of well placement strategy for the profitability of geothermal projects. Additionally, Shetty et al. (2018) observed significant changes (up to 50%) in doublet lifetime with a small variation (less than 50 m) in doublet location based on systematic studies using conceptual synthetic models.

Fractured media usually poses major complexity to uncertainty quantification (UQ). Normally, the model parameters for fractured reservoirs are estimated from available outcrops and extended to the reservoir scale. There is limited data available for quantitative characterization of fractured reservoirs in depth. One of the biggest challenges is related to the distribution and morphology of the fractured network. From a dynamic point of view, a deterministic solution to model flow in fractured porous media is needed while the data limitations force us to go for a stochastic solution. In response, the ensemble of fractured reservoir models is regularly upscaled to effective media models (e.g., dual-porosity (Barenblatt et al., 1960; Warren and Root, 1963) or embedded discrete fracture models (Li and Lee, 2008)) to reduce the computational time and still attempt to make meaningful predictions. Another strategy is to apply a multiscale approach to improve the performance (Hajibeygi et al., 2011). However, in the case of realistically complicated fracture orientation and connectivity, all of these models might not accurately reflect reality. Therefore, the use of a discrete fracture matrix (DFM) model, which explicitly represents the fracture geometry, is preferred (Moinfar et al., 2011).

Energy application management of subsurface resources should take into account uncertain parameters adequately (Bickel and Bratvold, 2008). Various industrial applications have demonstrated that developing strategies for subsurface systems can be adjusted by considering uncertainties (Scheidt and Caers, 2009; Ballio and Guadagnini, 2004; Dai et al., 2014; James and Oldenburg, 1997). Uncertainties can also significantly affect energy production from geothermal reservoirs (Shetty et al., 2018; Saeid et al., 2020; Cui et al., 2019). Those uncertainties can arise from multiple sources. First, due to geological measurement and observation limitations, knowledge about a target reservoir is usually based on data interpretation and empirical correlations. For instance, the relationship between permeability and porosity is typically derived from empirical petrophysical correlations (Willems et al., 2020), while porosity values are usually interpreted from core analysis and log curves (Chang et al., 1994).

To aid reservoir management and improve developing schemes, quantifying the influence of key simulation parameters beforehand is crucial (O'Sullivan and O'Sullivan, 2016). A numerical simulation is an effective tool that can be utilized to evaluate the reservoir's response to various uncertainties during development (Jansen et al., 2009; Xu et al., 2007). Large numerical models can be constructed to fully capture reservoir heterogeneity in subsurface geology, enabling easy characterization of uncertainties in geological information within different numerical models. Furthermore, Monte Carlo (MC) simulation can be employed for uncertainty analysis in geothermal production. MC simulation is widely used to quantify uncertainties in subsurface applications with ensembles of forward models (Ballio and Guadagnini, 2004; Cremon et al., 2020b). This method performs simulations with input data sampled stochastically within a given distribution in the parameter space.

The representativeness of MC results depends on the dimension of the uncertain parameter space and the nonlinearity of the underlying model. However, to obtain converging results, MC typically requires a large number of forward simulations, necessitating a high-performance numerical framework and significant computational resources, particularly for large models with detailed geological characterization (Athens and Caers, 2019).

2 Governing relationships for geothermal modeling

In this section, we will review the most general governing equations and auxiliary relations which can describe a full variety of geothermal applications present in the modern geo-energy market. The equations are formulated in generic form using discrete representation. We assume that the investigated domain is split by control volumes Ω, bounded by surface Γ. The conservation equations can be expressed for this domain in a uniformly integral way, as

$$\frac{\partial}{\partial t}\int_{\Omega} M^c d\Omega + \int_{\Gamma} \boldsymbol{F}^c \cdot \boldsymbol{n} d\Gamma = \int_{\Omega} Q^c d\Omega. \tag{1}$$

Here, M^c denotes the accumulation term for the cth component ($c = 1, \ldots, n_c$, indexing for the mass components [e.g., water, CO_2] and $c = n_c + 1$ for the energy quantity); $\boldsymbol{F}_c$ refers to the flux term of the cth component; $\boldsymbol{n}$ refers to the unit normal pointing outward to the domain boundary; and Q_c denotes the source/sink term of the cth component.

2.1 Conservation of mass and energy

First, we adapt Eq. (1) to represent the conservation of mass and energy in the reservoir. The mass accumulation term collects each component distribution over n_p fluid phases in a summation form,

$$M^c = \phi \sum_{j=1}^{n_p} x_{cj}\rho_j s_j, \quad c = 1, \ldots, n_c, \tag{2}$$

where ϕ is porosity, s_j is phase saturation, ρ_j is phase density (kmol m^{-3}), and x_{cj} is molar fraction of c component in j phase.

The energy accumulation term contains the internal energy of fluid and rock,

$$M^{n_c+1} = \phi \sum_{j=1}^{n_p} \rho_j s_j U_j + (1-\phi)\rho_r U_r, \tag{3}$$

where U indicates specific internal energy (kJ kmol^{-1}). The rock is assumed compressible and represented by the change of porosity through:

$$\phi = \phi_0\left(1 + c_r(p - p_{\text{ref}})\right), \tag{4}$$

where ϕ_0 is the initial porosity, c_r is the rock compressibility (1 bar^{-1}), and p_{ref} is the reference pressure (bars).

The mass flux of each component is represented by the summation over n_p fluid phases,

$$F^c = \sum_{j=1}^{n_p} \left(x_{cj}\rho_j \boldsymbol{v}_j + s_j \mathbf{J}_{cj} \right), \quad c = 1, \ldots, n_c. \tag{5}$$

Here, the velocity $\boldsymbol{v}_j$ follows the extension of Darcy's law to multiphase flow,

$$\boldsymbol{v}_j = -\mathbf{K}\frac{k_{rj}}{\mu_j}(\nabla p_j - \gamma_j \nabla z), \tag{6}$$

where $\mathbf{K}$ is the permeability tensor (mD), k_{rj} is the relative permeability of phase j, μ_j is the viscosity of phase j (mPa s), p_j is the pressure of phase j (bars), $\gamma_j = \rho_j g$ is the specific weight (Nm^{-3}), and z is the depth vector (m). The $\mathbf{J}_{cj}$ is the diffusion flux of component c in phase j, which is described by Fick's law as

$$\mathbf{J}_{cj} = -\phi \mathbf{D}_{cj} \nabla (\rho_j x_{cj}), \tag{7}$$

where $\mathbf{D}_{cj}$ is the diffusion coefficient (m^2 day^{-1}).

The energy flux includes the thermal convection and conduction terms,

$$F^{n_c+1} = \sum_{j=1}^{n_p} \left(h_j \rho_j \boldsymbol{v}_j + \phi \kappa_j s_j \nabla T \right) + (1-\phi)\kappa_r \nabla T, \tag{8}$$

where h_j is phase enthalpy (kJ $kmol^{-1}$) and κ_j is thermal conductivity for phase j ($kJm^{-1}day^{-1}K^{-1}$) (including solid).

These governing equations describe the most general reservoir model for modern geothermal applications at the continuous scale. Depending on the problem, the number of components can vary from one (e.g., when the impact of heterogeneity is studied) to multiple (e.g., when sensitivity to chemical interactions is investigated). Several closing relations are also required to fully define the problem. They are usually formulated based on thermodynamic assumptions and will be described later.

2.2 Momentum conservation for geomechanics

Here, we adapt Eq. (1) to represent the momentum equation that governs mechanic effects and compliments conservation equations of mass and energy introduced earlier. The equation will represent a conservation of stationary linear momentum in an elastic body that omits the accumulation term. The system of equations following (Coussy, 2003) reads

$$-\nabla \cdot \boldsymbol{\sigma} = \gamma_t \nabla z, \tag{9}$$

where $\boldsymbol{\sigma}$ is the stress tensor and $\gamma_t = \left((1-\phi)\rho_s + \phi\sum_j^{n_p} s_j\rho_j\right)g$ is the total specific weight of saturated medium and ρ_s is the density of unsaturated porous rock. The system is subject to the relations defined in Coussy (2003) and Zhao and Borja (2020), that represent

- the stress changes within an anisotropic poroelastic medium

$$\boldsymbol{\sigma} - \boldsymbol{\sigma}_0 = \mathbb{C} : \nabla^s \boldsymbol{u} - (p - p_0)\zeta \mathbf{I} + (T - T_0)\xi \mathbf{I}, \tag{10}$$

- the porosity associated with anisotropic thermoporoelasticity

$$\phi - \phi_0 = \frac{(\zeta - \phi)(1 - \zeta)}{K_s}(p - p_0) + \zeta \nabla \cdot (\boldsymbol{u} - \boldsymbol{u}_0) + \xi_\phi (T - T_0). \tag{11}$$

In the previous equations, $\nabla\cdot$ is the divergence operator and ∇^s is the symmetric gradient operator. Moreover, $\mathbb{C}$ is the rank-four drained stiffness tensor, ζ is the Biot's coefficient, ξ is the thermal expansion coefficient, $\mathbf{I}$ is the identity matrix, K_s is the bulk modulus of drained porous rock, and ξ_ϕ is the volumetric coefficient of thermal expansion of porous rock.

The subscript "0" denotes the reference state of a variable, that is,

$$\boldsymbol{\sigma}_0 = \boldsymbol{\sigma}(\boldsymbol{u}_0, p_0, T_0), \quad \phi_0 = \phi(\boldsymbol{u}_0, p_0, T_0). \tag{12}$$

In the applications, the effective stress tensor is frequently used. It is defined by the following relation:

$$\boldsymbol{\sigma} = \boldsymbol{\sigma}' - p\mathbf{I} = \boldsymbol{\sigma}'' - p\zeta, \tag{13}$$

where $\boldsymbol{\sigma}'$ is the Terzaghi effective stress tensor and $\boldsymbol{\sigma}''$ is the Biot effective stress tensor.

We use the following definition for the total traction vector $\boldsymbol{f}$ over an interface δ with unit normal $\boldsymbol{n}$:

$$\boldsymbol{f}|_\delta = -\boldsymbol{\sigma}|_\delta \cdot \boldsymbol{n}_\delta = -(\mathbb{C} : \nabla^s \boldsymbol{u} - \zeta p \mathbf{I} + \xi T \mathbf{I})|_\delta \cdot \boldsymbol{n}_\delta. \tag{14}$$

2.3 Momentum conservation for well

Here, we formulate a momentum conservation equation for well modeling. In reservoir simulation, the well can be modeled in two ways: using a single segment or multiple segments to represent the wellbore. In the single-segment approach, we assume that all mass and energy uniformly mix in the wellbore. In these assumptions, we do not need additional equations for the wellbore and all mass and energy conservation can be preserved explicitly (Peaceman, 1990). However, in this approach, we cannot account for complex thermal multiphase flow, heat losses, and chemical alterations in the wellbore.

To overcome the previous drawback, the multisegmented well (ms-well) model was proposed (Holmes et al., 1998; Jiang, 2008). In the ms-well model, the borehole is discretized into several segments with the fluid velocity, node pressure, and other properties simulated along with the wellbore geometry. The benefits of the general ms-well model are its flexible approximation of the actual geometry of each wellbore and its handling of the complex well topology and controls in the pipeline network. Moreover, introducing velocity as an additional degree of freedom allows us to account for various pressure losses by solving the momentum equation.

Multiple solution strategies for coupled reservoir and well models were proposed in the literature. Livescu et al. (2008) introduced a semianalytical thermal multiphase wellbore-flow model for general-purpose simulation. Pan and Oldenburg (2014) proposed the hybrid implicit (semiexplicit) solution for the transient momentum equation in the geothermal wellbore. In addition, a few fully coupled numerical models with different strategies for thermal wellbore flow have been suggested (Holmes et al., 1998; Jiang, 2008; Livescu et al., 2010). All these models operate at various simplifying assumptions for wellbore physics and wellbore coupling with the reservoir. There are different formulations for geothermal wellbore simulation exist. Multiple solution strategies for stand-alone geothermal wellbore simulation that neglect wellbore coupling have been recommended. Tonkin et al. (2021) make an in-depth review of the geothermal wellbore simulation.

Here, we formulate a one-dimensional (1D) momentum balance equation to describe fluid flow in the wellbore. Assuming the coordinate z points along the well, this equation is given by Hibiki and Ishii (2003) and Issa and Kempf (2003).

$$\frac{\partial p^w}{\partial z} = -\rho_m g - \frac{\partial p^w}{\partial t}(\rho_m V_m) - \frac{\partial(\rho_m V_m^2)}{\partial z} - \frac{2f_{tp}\rho_m V_m^2}{d_{in}}. \tag{15}$$

Here, ρ_m is the total mixture density, g is the gravitational acceleration in the z-direction, d_{in} is the internal diameter of the well, where f_{tp} is the friction factor, which is a function of the dimensionless Reynolds number Re (ratio of inertial forces to viscous forces), d is the diameter of the segment, and V_m is the velocity of the fluid mixture in the segment.

The friction factor is usually constrained by the following relations:

$$f_{tp} = \begin{cases} 16/Re, & Re \leq 2000 \text{ (laminar flow)}, \\ 1\Big/\left(-3.6\log\left(\frac{6.9}{Re} + \left(\frac{\epsilon}{3.7d}\right)^{10/9}\right)\right)^2, & Re \geq 4000 \text{ (turbulent flow)}, \\ 16/2000 + k_f(R_e - 2000), & 2000 < Re < 4000. \end{cases} \tag{16}$$

Here, ϵ is the roughness of the well surface. The general momentum equation (15) can be simplified to the steady-state momentum equation as follows:

$$\frac{\partial p^w}{\partial z} = \left(\frac{\partial p}{\partial z}\right)_h + \left(\frac{\partial p}{\partial z}\right)_f + \left(\frac{\partial p}{\partial z}\right)_a. \tag{17}$$

The earlier equation simply states the total pressure loss over any control volume (segment) of the well as the sum of three components' forces (hydrostatic, friction, and acceleration) which can be defined as

$$\left(\frac{\partial p}{\partial z}\right)_h = \rho_t g, \quad \left(\frac{\partial p}{\partial z}\right)_f = \frac{2f_{tp}\rho_t V_m^2}{d}, \quad \left(\frac{\partial p}{\partial z}\right)_a = \frac{2m_{in} V_m}{A}. \tag{18}$$

Here, $m_{in} = \sum_p \rho_p(\omega) Q_p$ is the mass flow rate of the mixture entering the segment, ρ_t is the total mass density, and $A = \pi d^2/4$ is the cross-section of well.

2.4 Constitutive relationships and properties

For main geothermal applications, the simulation can be formulated as a single-component system present at different states (phases). Several equations of state exist for such representation. The most commonly used is the IAPWS-97 equation of state (Wagner and Kretzschmar, 2008). It was developed by the International Association for the Properties of Water and Steam, which allows for the accurate calculation of the thermodynamic properties of water. These equations take into account the complex behavior of water under a wide range of conditions for temperature and pressure, including the critical state and the region of superheated steam. The IAPWS-97 formulation is based on extensive experimental data and is widely used in engineering and scientific applications, such as the design of power plants and reservoir simulation. The validity range of the IAPWS-97 equations is divided into distinct regions, each represented by a different fundamental equation with a smooth transition at the boundaries between regions. Using appropriate combinations of derivatives from these equations, it is possible to determine any desired thermodynamic property, such as density or specific volume, heat capacity, enthalpy, and entropy.

Some geothermal systems may require the involvement of more than one component for fluid representation. This will impose evaluating component partitioning x_{cj} among different phases, the so-called multiphase flash calculation. This evaluation is usually performed using a thermodynamic equilibrium assumption. At thermodynamic equilibrium, pressure and temperature are uniform throughout the system. Furthermore, a necessary condition for equilibrium is that the chemical potential for each component is the same throughout all phases, or, equivalently, that the fugacities of the individual components are equal:

$$\hat{f}_i^j(T, P, \boldsymbol{n}_j) = \hat{f}_i^k(T, P, \boldsymbol{n}_k), \quad i = 1, \dots, n_c, \tag{19}$$

between all phases j and k. Finally, the system is at equilibrium if and only if the total Gibbs energy is at its global minimum. This requires finding the conditions at which the Gibbs energy is at a minimum for all possible combinations of phases. The Gibbs free energy is given by

$$G^* = G/RT = \sum_{j=1}^{n_p} \sum_{i=1}^{n_c} n_{ij} \ln \hat{f}_{ij} \tag{20}$$

with n_{ij} the number of moles of species i in phase j.

A material balance for each component yields

$$x_{i0} + \sum_{j=1}^{n_p - 1} \nu_j (x_{ij} - x_{i0}) = z_i, \quad i = 1, \dots, n_c \tag{21}$$

with phase 0 the reference phase, z_i is the overall molar fraction, and nu_j is the phase fraction. Finally, mole fractions in the different phases must sum to unity, yielding

$$\sum_{i=1}^{n_c} (x_{ij} - x_{i0}) = 0, \quad j = 1, \dots, n_p - 1. \tag{22}$$

For two phases, the system to be solved amounts to n_c material balances, n_c equilibrium equations, and one summation of mole fractions constraint, yielding $2n_c + 1$ relations between the $2n_c + 3$ unknowns, $T, p, \boldsymbol{x}, \boldsymbol{y}, \nu_1$. In pT-flash, the phase equilibrium is defined by fixed temperature and pressure. The ph-flash usually requires an additional equation constraining the total enthalpy of the system to the enthalpy computed for a given p and T since most of the constitutive properties involved in multiphase flash are explicit functions of p and T. Each new phase would add another n_c mole fraction and one-phase fraction to be determined, n_c equal-fugacity constraints, and a summation of mole fractions constraint.

The quantities required for multiphase flash (e.g., fugacity) can be derived from different equations of states required for an accurate representation of a particular geothermal system. Besides, many properties (e.g., density and phase enthalpies) can be also derived from the same equations of state. It is important to keep consistency between constitutive relations used in simulation and multiphase flash since it guarantees the conservation of the numerical scheme and the uniqueness of the fully implicit solution.

3 Discretization of mass and heat conservation equations

Two methods for discretization of governing partial differential equations (PDE) usually dominate in engineering which are finite element and finite volume discretizations. The key difference between these two methods lies in how they partition and discretize the domain of the PDE solution. The finite element method (FEM) involves dividing the computational domain into smaller geometric elements, which can vary in shape and size to accommodate irregular boundaries of different domains of the solution. Within each element, an approximation of the solution is constructed using polynomial basis functions, allowing for a flexible approximation of the solution behavior. These basis functions are usually defined by nodal points, where continuity conditions are enforced between neighboring elements. A system of algebraic equations is formed representing the contribution of all elements. This system is then solved to obtain the numerical solution. FEM excels in handling problems with varying material properties, irregular geometries, and complex boundary conditions, making it a versatile tool for simulating and analyzing a wide range of engineering problems.

In the finite volume method (FVM), the computational domain is divided into a set of control volumes or grid cells similar to FEM. However, unlike the FEM, FVM focuses on the conservation of quantities in the integral form across the boundaries of each control volume. The solution variable is approximated at the center of each cell, and the fluxes of the variable across the cell faces are computed. This approach ensures that the net change of the conserved quantity within a control volume is balanced by the fluxes across its boundaries. Similarly to FEM, a system of algebraic equations is generated, which is then solved to obtain the numerical solution. FVM excels in capturing conservation properties, making it particularly suitable for problems involving flow and transport where maintaining the overall balance of components is crucial. FVM is very similar to the finite difference method (FDM) with the only difference being that FVM is conservative by construction while FDM may be conservative or not.

The most successful geothermal simulators usually utilize FVM. It happening for several reasons:

- The conservation of FVM guarantees the uniqueness and consistency of the nonlinear solution.
- FVM usually has very limited nonzero entries in the resulting linear matrix which helps with performance.
- Low accuracy of FVM is often compensated by large uncertainties in the subsurface characterization.

In the following description, we will be focusing on FVM for the discretization of all governing equations in geothermal problems.

3.1 Discretization of mass and energy conservation

The thermodynamic state of a geothermal system is fully defined by the nonlinear variables chosen for the solution. They depend on nonlinear formulation and closing relations and will be discussed later. Here, we assume that the unified state vector ω in a given control volume fully defines the thermodynamic state of the geothermal system. The upstream weighting of the physical state is used to determine the flux-related fluid properties determined at the interface l. The discretized mass conservation equation in compact form for a given gridblock (here we omit i) reads:

$$V\phi_0[\alpha_c(\omega) - \alpha_c(\omega_n)] - \Delta t \sum_{l \in L(i)} \sum_{j=1}^{n_p} \left[\Gamma_a^l \beta_{cj}^l(\omega^u) \Delta\psi_j^l + \Gamma_d^l \gamma_j^l(\omega) \Delta\chi_{cj} \right] + \Delta t V \delta_c(\omega) = 0, \tag{23}$$

where V is the control volume, ω_n is the physical state of block i at the previous timestep, ω is the physical state of block i at the new timestep, ω^u is the physical state of the upstream block, Γ_a^l and Γ_d^l are the advective and diffusive transmissibilities, respectively, assuming two-point flux approximation, and $L(i)$ is a set of interfaces for gridblock i.

Here, we defined the following state-dependent aggregate terms, representing the thermodynamics property of the system

$$\alpha_{cf}(\omega) = \left(1 + c_r(p - p_{ref})\right) \sum_{j=1}^{n_p} x_{cj}\rho_j s_j, \quad c = 1, \ldots, n_c; \tag{24}$$

$$\beta_{cj}(\omega) = x_{cj}\rho_j\lambda_j/\Lambda, \quad c = 1, \ldots, n_c, \; j = 1, \ldots, n_p; \tag{25}$$

$$\gamma_j(\omega) = \left(1 + c_r(p - p_{ref})\right) s_j, \quad j = 1, \ldots, n_p; \tag{26}$$

$$\chi_{cj}(\omega) = D_{cj}\rho_j x_{cj}, \quad c = 1, \ldots, n_c, \; j = 1, \ldots, n_p; \tag{27}$$

$$\delta_c(\omega) = \sum_{j=1}^{n_p} v_{cj} r_j(\omega), \quad c = 1, \ldots, n_c. \tag{28}$$

Here, $\lambda_j = k_{rj}/\mu_j$ is mobility of the phase j and $\Lambda = \sum_j \lambda_j$ is the total mobility. As the default technique, the phase-potential-upwinding (PPU) strategy is often applied to model the convective flux, gravity, and capillary effect (Khait and Voskov, 2018; Lyu et al., 2021). The potential difference of phase j on the interface l between blocks 1 and 2 can be written as

$$\Delta\psi_j^l = \Lambda \left[p_1 - p_j^c(\omega_1) - (p_2 - p_j^c(\omega_2)) - \frac{\rho_j(\omega_1) + \rho_j(\omega_2)}{2} g(z_2 - z_1) \right], \tag{29}$$

where p_j^c is the capillary pressure.

The discretized energy conservation equation in compact form can be written as

$$\begin{aligned} V\phi_0[\alpha_{ef}(\omega)-\alpha_{ef}(\omega_n)] - \Delta t \sum_{l\in L(i)} \sum_{j=1}^{n_p} \left[\Gamma^l_a \beta^l_{ej}(\omega^u)\Delta\psi^l_j + \Gamma^l_c \gamma_j(\omega)\Delta\chi_{ej}\right] + \Delta t V \delta_e(\omega) \\ + (1-\phi_0)V[\alpha_{er}(\omega) - \alpha_{er}(\omega_n)] - \Delta t \sum_l (1-\phi_0)\Gamma^l_c \kappa_r \Delta\chi_{er} = 0, \end{aligned} \quad (30)$$

where

$$\alpha_{ef}(\omega) = \left(1 + c_r(p - p_{ref})\right) \sum_{j=1}^{n_p} \rho_j s_j U_j; \quad (31)$$

$$\beta_{ej}(\omega) = h_j \rho_j \lambda_j / \Lambda, \quad j = 1, \ldots, n_p; \quad (32)$$

$$\chi_{ej}(\omega) = \kappa_j T_j, \quad j = 1, \ldots, n_p; \quad (33)$$

$$\delta_e(\omega) = \sum_{j=1}^{n_j} v_{ej} r_{ej}(\omega). \quad (34)$$

In addition, to account for the energy of rock, three additional aggregate terms should be defined:

$$\alpha_{er}(\omega) = \rho_r U_r, \quad \chi_{er}(\omega) = T_r. \quad (35)$$

Here, α_{er} represents the multiplier for rock internal energy and conduction and U_r is a state-dependent parameter.

As mentioned earlier, for discretization we apply here a two-point flux approximation which is reflected in the following expression for advective, diffusive, and conductive transmissibilities:

$$\Gamma^l_a = \frac{(a_i K_i/d_i)(a_j K_j/d_j)}{a_i K_i/d_i + a_j K_j/d_j}, \quad \Gamma^l_d = \frac{(a_i D_i/d_i)(a_j D_j/d_j)}{a_i D_i/d_i + a_j D_j/d_j}, \quad \Gamma^l_c = \frac{(a_i \kappa_i/d_i)(a_j \kappa_j/d_j)}{a_i \kappa_i/d_i + a_j \kappa_j/d_j}. \quad (36)$$

Here, i and j are the adjusting cells of interface l, a_i is the block interface area, d_i is the distance from block to interface centers, and K_i, D_i, and κ_i are permeability, diffusivity, and conductivity of the block i, respectively. Note that this approximation is accurate only for k-orthogonal grids (Settari and Aziz, 1972).

3.2 Discretization of geomechanics

We use the FVM to discretize the momentum balance equations. The FVM scheme for Eq. (9) in cell i can be written in the following vector form:

$$-V_i \gamma_{t,i}^{n+1} \nabla z + \sum_{j \in Ł_i} \delta_{ij} \boldsymbol{f}_{ij}^{n+1} = \boldsymbol{0}, \quad (37)$$

where subscript j denotes the neighbors of the cell i, Δt is the time step size, V_i is the volume of cell i, and δ_{ij} denotes the area of the interface between cells i and j. Here, superscripts $n+1$ denote the variables taken from the next time step, respectively.

The approximation of traction vector f_{ij} at the interface is derived from the local problem. It postulates the continuity of unknown pressures p, temperatures T, and displacements $\boldsymbol{u}$ along with the continuity of Darcy's fluxes, heat conduction fluxes and tractions over the interface that can be written as

$$p_{\delta i} = p_i + (\mathbf{x}_{ij} - \mathbf{x}_i)^T \nabla p_i = p_j + (\mathbf{x}_{ij} - \mathbf{x}_j)^T \nabla p_j = p_{\delta j}, \tag{38}$$

$$T_{\delta i} = T_i + (\mathbf{x}_{ij} - \mathbf{x}_i)^T \nabla T_i = T_j + (\mathbf{x}_{ij} - \mathbf{x}_j)^T \nabla T_j = T_{\delta j}, \tag{39}$$

$$\boldsymbol{u}_{\delta i} = \boldsymbol{u}_i + \mathbf{I} \otimes (\mathbf{x}_{ij} - \mathbf{x}_i)^T [\nabla \otimes \boldsymbol{u}]_i = \boldsymbol{u}_j + \mathbf{I} \otimes (\mathbf{x}_{ij} - \mathbf{x}_j)^T [\nabla \otimes \boldsymbol{u}]_j = \boldsymbol{u}_{\delta j}, \tag{40}$$

$$-\mathbf{K}_i \boldsymbol{n}_{ij} \cdot (\nabla p_i - \gamma_{\mathrm{f},ij} \nabla z) = -\mathbf{K}_j \boldsymbol{n}_{ij} \cdot (\nabla p_j - \gamma_{\mathrm{f},ij} \nabla z), \tag{41}$$

$$-\boldsymbol{\Lambda}_i \boldsymbol{n}_{ij} \cdot \nabla T_i = -\boldsymbol{\Lambda}_j \boldsymbol{n}_{ij} \cdot \nabla T_j, \tag{42}$$

$$-\left(\mathbb{C}_i : \nabla^s \boldsymbol{u}_i - \zeta_i p_{\delta i} \mathbf{I} + \xi_i T_{\delta i} \mathbf{I}\right) \cdot \boldsymbol{n}_{ij} = -\left(\mathbb{C}_j : \nabla^s \boldsymbol{u}_j - \zeta_j p_{\delta j} \mathbf{I} + \xi_j T_{\delta j} \mathbf{I}\right) \cdot \boldsymbol{n}_{ij}, \tag{43}$$

where $p_{\delta i}$ and $p_{\delta j}$, $T_{\delta i}$ and $T_{\delta j}$, and $\boldsymbol{u}_{\delta i}$ and $\boldsymbol{u}_{\delta j}$ are the single-side approximations of pressure, temperature, and displacements, respectively, at the center of interface $\mathbf{x}_{ij}$. Moreover, $\mathbf{x}_i$ and $\mathbf{x}_j$ are the centers of cells i and j, $\otimes$ denotes the Kronecker product, $\mathbf{I} \otimes (\mathbf{x}_{ij} - \mathbf{x}_i)^T$, $\mathbf{I} \otimes (\mathbf{x}_{ij} - \mathbf{x}_j)^T$ are 3×9 matrices and $\nabla \otimes \boldsymbol{u}_i$, $\nabla \otimes \boldsymbol{u}_j$ are 9×1 vectors constructed as

$$\mathbf{I} \otimes (\mathbf{x}_{ij} - \mathbf{x}_i)^T = \begin{pmatrix} (\mathbf{x}_{ij} - \mathbf{x}_i)^T & & \\ & (\mathbf{x}_{ij} - \mathbf{x}_i)^T & \\ & & (\mathbf{x}_{ij} - \mathbf{x}_i)^T \end{pmatrix}, \quad [\nabla \otimes \boldsymbol{u}]_i = \begin{pmatrix} \nabla u_{ix} \\ \nabla u_{iy} \\ \nabla u_{iz} \end{pmatrix}. \tag{44}$$

Furthermore, $\boldsymbol{n}_{ij}$ is unit normal vector to the interface between cells i and j, $\mathbf{K}_i$ and $\mathbf{K}_j$ are permeability tensors in cells i and j,$\gamma_{\mathrm{f},ij}$ is the approximation of specific fluid weight $\gamma_{\mathrm{f}} = g\sum_{p=1}^{n_p} s_p \rho_p$ at the center of interface $\gamma_{\mathrm{f},ij} = (d_i + d_j)^{-1}(d_i \gamma_{\mathrm{f},j} + d_j \gamma_{\mathrm{f},i})$ with $d_i = \left|(\mathbf{x}_{ij} - \mathbf{x}_i) \cdot \boldsymbol{n}_{ij}\right|$ where $d_j = \left|(\mathbf{x}_{ij} - \mathbf{x}_j) \cdot \boldsymbol{n}_{ij}\right|$ are distances from the centers of cells i, j and the center of interface $\mathbf{x}_{ij}$, $\boldsymbol{\Lambda}_i$, $\boldsymbol{\Lambda}_j$ are the heat conduction tensors in cells i and j, $\mathbb{C}_i$, $\mathbb{C}_j$ are the stiffness tensors in cells i and j, ζ_i, ζ_j are Biot's coefficients in cells i and j, ξ_i, ξ_j are thermal expansion coefficients in cells i and j.

Local problem defined in Eqs. (38)–(43) can be utilized for the reconstruction of gradients of unknowns and, subsequently, leading to the gradient-based variant of coupled multipoint flux and multipoint stress approximation (MPFA-MPSA) of Darcy's, heat conduction fluxes and traction vector. For more details on how this discretization has been derived, see Terekhov (2020). For the extension to fractured media, see Novikov et al. (2022).

Geomechanics affects fluid flow and energy through porosity defined in Eq. (11). The following approximation of porosity is utilized:

$$\phi_i = \frac{(\zeta_i - \phi_{0,i})(1 - \zeta_i)}{K_{s,i}} p_i + \xi_\phi T_i + \frac{1}{V_i} \sum_{j \in \partial V_i} \delta_{ij} \zeta_i (\boldsymbol{u}_{ij} - \boldsymbol{u}_i) \cdot \boldsymbol{n}_{ij}, \tag{45}$$

where $K_{s,i}$ is the bulk modulus of drained porous rock in cell i and ξ_ϕ is the volumetric coefficient of thermal expansion of porous rock in cell i.

3.3 Discretization of well momentum equation

The discretization of the wellbore is relatively straightforward assuming that the connectivity in the wellbore model is most often 1D. The momentum equation (17) can be rewritten in discretized form:

$$\Delta P_w + \Delta P_h + \Delta P_f + \Delta P_h = 0, \tag{46}$$

where $\Delta P_w = p_i - p_j$ is the pressure drop between two nodes i and j, and ΔP_h^w, ΔP_f^w, and ΔP_a^w are the hydrostatic, frictional, and acceleration components of the pressure drop, respectively.

The frictional pressure difference between two nodes is defined as

$$\Delta P_f = \left(\frac{2 f_{tp} \rho^{seg} V_m^2}{d} \right) \Delta z, \tag{47}$$

where f_{tp} is the friction factor which is a function of the dimensionless Reynolds number Re and defined in Eq. (16), d is the diameter of the segment, and V_m is the velocity of the fluid mixture in the segment connection. The hydrostatic pressure difference between the two nodes is

$$\Delta P_h = \rho_t g \Delta z. \tag{48}$$

Acceleration pressure losses are given by the following formula:

$$\Delta P_a = \frac{2 m_{in} V_m}{A} \Delta z, \tag{49}$$

where m_{in} is the mass flow rate of the mixture entering the segment from the reservoir, which is defined as

$$m_{in} = \sum_{i}^{Nperf} \left(\Gamma_i^w (p_u^w - p_i) \Lambda(\omega) \rho_t(\omega) \right)_i. \tag{50}$$

Here, Γ_i^w is the well index (Peaceman, 1990) for the perforation i, ρ_t is the total mass density, p_u^w is the pressure of the upwind node for connection between two well segments, and p_i is the pressure in the perforated reservoir block.

4 Numerical approaches for modeling of geothermal applications

This section will describe different numerical approaches common for most geothermal simulators. In the following description, we assume that only one component (water) is present in two possible phases (liquid and vapor). However, all described methodologies can be extended for more components and phases in more complex geothermal systems.

4.1 Types of nonlinear formulations

The nonlinearity of the reservoir simulation solution mostly depends on the time approximation of conservation equations (Aziz and Settari, 1979; Coats, 1980a). The most common is a fully implicit method (FIM) of approximation since it is unconditionally stable (does not depend on timestep size), which is especially important in highly resolved heterogeneous reservoirs. However, when the number of components in the simulation study is large (e.g., when chemical interactions are taken into account), the application of FIM is less attractive due to the size of the Jacobian matrix. The performance can be improved by using the adaptive implicit method (AIM), which is a combination of FIM and partially explicit schemes (e.g., implicit pressure explicit composition [IMPEC]) where some of the degrees of freedom, in addition to pressure and temperature, are treated implicitly (Cao, 2002). For the same timestep profile, the nonlinear behavior of AIM systems is expected to be quite similar to the FIM method, if one uses rigorous adaptive stability criteria to decide on the implicit level (Cao, 2002; Moncorgé and Tchelepi, 2011).

The implicit approximation of governing equations in reservoir simulation often entails solving a highly nonlinear system of algebraic equations. This process involves formulating and solving a set of equations that describe fluid flow, mass, and energy conservation within the reservoir. The system of equations is typically nonlinear due to the presence of terms involving fluid properties, saturation-dependent parameters, and closing assumptions on equilibrium or nonequilibrium thermodynamics. To obtain a solution, the nonlinear variables within the system must be specified, which can include pressure, temperature, phase saturations, and/or mass variables. Reservoir simulation offers two primary options for specifying these nonlinear variables: natural and molar variables.

The natural-variables formulation, first introduced by Coats (1980a) utilizes pressure, saturations, and phase mole fractions as the nonlinear unknowns in reservoir simulation. This approach involves constructing a complete nonlinear algebraic system of equations that encompass the conservation of mass and energy, equilibrium relations, and constraints for grid blocks or control volumes in the two-phase region (such as vapor-liquid). If, during a time step or nonlinear iteration, one of the phases disappears, the corresponding saturation variable is removed, leading to a reduction in the equation set. This crucial process, known as "variable substitution," is an integral part of the natural-variables formulation. Alternatively, in the mass (mole) variable formulation, first introduced by Fussell and Fussell (1979), the nonlinear unknowns consist of pressure and enthalpy for single-component problems, with additional molar fractions considered when multiple components are present. While several formulations were originally proposed for mixed-implicit treatment (e.g., AIM), all the formulations discussed here are solved using the FIM.

The nonlinear formulations discussed in the context of petroleum-related simulation software have also been examined for their application to geothermal systems, as explored in Faust and Mercer (1979a, b). The primary distinction between these formulations lies in how the variables behave during phase transitions. In the pressure-enthalpy formulation, the variables remain independent and constant throughout both single- and two-phase regions. Conversely, the natural variable approach, utilizing pressure and temperature, necessitates a switch in the primary variables. This distinction results in different Jacobian matrices and nonlinear convergence characteristics for the two formulations. The pressure-enthalpy approach has been previously implemented in multiphase simulators (Faust and Mercer, 1979a; Pruess et al., 1979). However, one drawback of these approaches is that thermodynamic properties are often expressed in terms of natural variables (temperature and saturation), requiring iterative methods for calculations involving persistent variables (such as enthalpy), as noted later by Pruess et al. (1999). On the other hand, the natural formulation requires switching primary variables between different single- and two-phase conditions which bring additional complexity to the numerical implementation. Next, we are going to discuss some implementation details for both types of variables in geothermal simulation.

4.1.1 Natural variables

The precursive of natural variables formulation has been originally proposed by Coats et al. (1974) for the three-phase steam-flood problem. It was one of the first truly fully implicit formulations allowed to solve such complex problems. In addition, it was shown that the Cartesian grid has very strong grid orientation effect when a two-point flux approximation is used. Later, this formulation was adapted for the geothermal problem in Coats (1977). In addition, the highly implicit formulation allowed including fractures and wellbore model in several multiphase test problems. Finally, the term "natural formulation" was introduced in Coats (1980a) for a thermal multicomponent multiphase problem.

In a single-component geothermal problem that allows liquid water to transfer to the vapor phase, the pressure, temperature, and steam saturation play the role of unknowns. However, the choice depends on the status of a block. A status is defined to be the composition and phase state of the block. For a system containing only pure water under two-phase assumptions, there are three possible statuses: pure liquid water, pure superheated steam, and a two-phase system. In single-phase conditions (compressed water or superheated steam), pressure (p) and temperature (T) are the primary variables. However, in two-phase conditions, pressure and temperature are not independent anymore and $p = p_{sat}(T)$. In this case, pressure p and steam saturation S_g become the primary variables in the two-phase region.

Based on the status of a grid block from the previous timestep, the status of a grid block for a new timestep is defined after each nonlinear update. In the nonlinear iteration, we assume that the pressure update defines the phase behavior of the system rather than the temperature update. This happens since in most flow problems, the temperature changes often happen slower than the pressure changes. The nonlinear update preserves the following logic:

- If the status is 0 (liquid water) and $p < p_{sat}(T)$, then the status becomes 2 (two-phase), $p = p_{sat}(T)$ and $S_g = \varepsilon$.
- If the status is 1 (steam) and $p > p_{sat}(T)$, then the status becomes 2 (two-phase), $p = p_{sat}(T)$ and $S_g = 1 - \varepsilon$.
- If the status is 2 (two-phase) and $S_g > 1$, then the status becomes 1 (steam phase).
- If the status is 2 (two-phase) and $S_g < 0$, then the status becomes 0 (liquid water).

Here, ε refers to a small number of orders 10^{-4}, which plays the role of initial guess for the nonlinear solver.

When more than one component is present in the system, the variable substitution follows a similar logic. To complement more conservation equations and equilibrium constraints, molar phase fractions are used as nonlinear variables, see examples in Voskov (2011) and Zaydullin et al. (2014).

4.1.2 Molar variables

An alternative to natural variables is molar variables which are often called pressure-enthalpy formulation in the case of the geothermal system with a single component only

(Faust and Mercer, 1979a). The advantage of molar variables is that it does not require variable substitution as in the natural variables formulation. The main reason for this is the fact that the enthalpy of the system is continuously changed during the phase change when temperature is not and requires tracking of phase saturation for continuous transition.

For single-phase conditions, the enthalpy (h) represents the single-phase fluid enthalpy. For the two-phase region, h represents the enthalpy of the fluid mixture defined as

$$h = \frac{\rho_w S_w h_w + \rho_g S_g h_g}{\rho_w S_w + \rho_g S_g}, \tag{51}$$

where h_w and h_g correspond to water and steam enthalpy, respectively. The denominator of this relation is often called total density $\rho_T = \rho_w S_w + \rho_g S_g$. For a known enthalpy of a fluid mixture, the saturations can be evaluated using

$$S_w = \frac{(h_g - h)\rho_g}{h(\rho_w - \rho_g) - h_w \rho_w + h_g \rho_g}, \quad S_g = 1 - S_w. \tag{52}$$

To account for the phase transition, the total enthalpy (or enthalpy of the mixture) should be compared with the saturated enthalpies for water and steam, respectively. If enthalpy crosses the saturated enthalpies, phase presence should be adjusted. In molar formulation, all constructive relations, including those depending on temperature, should depend on pressure and enthalpy to get sensitivity to the nonlinear unknown changes. This often introduces additional complications for the implementation of molar variables, see Voskov and Tchelepi (2012) and Wong et al. (2015) for more details.

4.2 Linearization techniques

FIMs are conventionally used in reservoir simulation because of their unconditional stability (Aziz and Settari, 1979). On the other hand, after discretization is applied to governing PDEs of a problem, the resulting nonlinear system represents different tightly coupled physical processes, which are difficult to solve. Usually, a Newton-based iterative method is applied, which demands an assembly of the Jacobian and the residual for the combined system of equations (i.e., linearization) at every iteration forming a linear system (often ill-conditioned). Precisely the solution of such systems takes most of the simulation time in the majority of practical applications.

Several conventional linearization approaches exist, though neither of them is robust, flexible, and computationally efficient all at once. Numerical derivatives provide flexibility in the nonlinear formulation (see Xu et al., 2011, for example), but a simulation based on numerical derivatives may lack robustness (Vanden and Orkwis, 1996). Besides, the performance of numerical differentiation is usually bad since one needs to recompute the residual as many times as the number of nonlinear unknowns in the problem. Straightforward hand differentiation is the state-of-the-art strategy in the most modern simulators (Schlumberger, 2011; Cao et al., 2009). However, this approach requires the introduction of a complicated framework for storing and evaluating derivatives for each physical property, which in turn reduces the flexibility of a simulator to incorporate new physical models and increases the possibility of potential errors.

The development of the automatic differentiation (AD) technique allows for preserving both flexibility and robustness in derivative computations. In reservoir simulation, the Automatically Differentiable Expression Templates Library (ADETL) was introduced by Younis (2011). Using the capabilities of ADETL, an Automatic Differentiation General Purpose Research Simulator (AD-GPRS) was developed (Voskov and Tchelepi, 2012; Zhou et al., 2011). Later, the AD technique became more demanding in research frameworks for reservoir simulation (Krogstad et al., 2015). Being attractive from the perspective of flexibility, the AD technique by design inherits computational overhead, which often affects the performance of reservoir simulation (Khait and Voskov, 2017).

Recently, another linearization technique has been proposed. As it can be seen, the structure of governing equations for thermal-compositional flow and transport described by Eqs. (23), (30) can be factored into the product of the most nonlinear state-dependent terms present in equations by Greek letters and the rest of parameters. The complexity of the evaluation of state-dependent terms is also increased by closing assumptions such as thermodynamic equilibrium (multiphase flash). In the operator-based linearization (OBL) approach, the most nonlinear part of equations—state-dependent terms—are replaced by their multilinear approximation using a uniform grid in parameter space (Voskov, 2017).

The OBL approach has several benefits. The first of them is the simplicity of the Jacobian assembly for complex physical models since it can be represented following a chain rule as a sum of products of linear operators representing partial derivatives of state-dependent terms and the rest, less nonlinear, terms. The Jacobian and residual construction can also be adjusted to modern computational architectures which can improve the performance of linearization by almost two orders of magnitude (Khait and Voskov, 2021). Another benefit of the proposed method is related to a nonlinear solver for generic thermal-compositional problems which can be implemented based on a trust-region method (Pour et al., 2023). The truly unique opportunity is the fact that any nonlinear procedure used for the evaluation of governing equations can be integrated as a black box without any derivative information.

4.3 Linear solution

As mentioned before, in practical reservoir modeling, a majority of simulation time is taken up by linear solutions. The effectiveness of these solvers does not only depend on the algorithms themselves or how well they are implemented but also on selecting the foundational data storage and relies on the number of algebraic operations it experiences at different stages of the linear solution. A prevalent storage format for large sparse matrices is the compressed sparse row (CSR) (Saad, 2003). It aims to minimize data transfers between memory and the CPU by storing contiguous nonzero entries within a single row.

In situations where linear systems arise from problems with n degrees of freedom per control volume (typical for fully implicit schemes), matrix portraits often showcase dense $n \times n$ blocks of nonzero entries. The block compressed sparse row (BCSR) matrix format, introduced by Pinar and Heath (1999), capitalizes on this characteristic to further decrease memory access. This reduction is achieved by a factor of n in the count of row and column indices. More significantly, operations involving the matrix, such as sparse matrix-vector multiplications, are executed in blocks. These operations work on dense $n \times n$ submatrices, leading to enhanced efficiency on modern CPU architectures due to more intensive utilization of register and cache storage. Given that the computational mesh mostly remains constant throughout the simulation and adjustments in well opening or closing do not involve altering the number of variables, the BCSR structure is calculated once at the start and remains unchanged during the process. While matrix values are recalculated at each iteration, the matrix's sparse structure remains consistent. This assumption permits applying a uniform linear solution approach: internal setups are initialized just once, thereby enhancing the efficiency of the linear solution.

The commonly preferred linear solution strategy for large reservoir models has usually been based on Krylov subspace iterative solvers paired with an advanced preconditioning strategy (Saad, 2003). Preconditioning stands as a crucial element aimed at reducing the condition number of the linear system, significantly enhancing the convergence rate of iterative solvers. Typically, different preconditioners exhibit varying efficiencies when applied to systems with diverse characteristics. For instance, the algebraic multigrid method (AMG) proves effective for near-elliptic problems (Stüben, 1983), while the incomplete LU factorization with zero level of fill-in ILU(0) finds successful application for near-hyperbolic equations (Behie and Forsyth, 1983).

Unfortunately, equations in reservoir simulation using a fully implicit approximation create a linear system that contains both types of unknowns. This system in isothermal assumptions includes a near-elliptic pressure equation and near-hyperbolic composition (or saturation) equations. To effectively handle these systems, an approach called constrained pressure residual (CPR) was developed as detailed in Wallis (1983) and Wallis et al. (1985). The CPR method functions as a two-stage preconditioner. First, it decouples the pressure system from the overall system using IMPEC (or saturation) approximation and solves it separately using an AMG-based scheme. Often, a single V-cycle proves to be sufficient for effective preconditioning at this stage. Subsequently, the full system undergoes processing via an ILU(0) preconditioner, utilizing the pressure solution obtained in the first stage. This strategy has demonstrated remarkable robustness and efficiency, especially in highly heterogeneous reservoirs where a strong coupling exists between the elliptic and hyperbolic segments of the linear system. As a result, it ensures stable convergence within a limited number of linear solver iterations, even with significantly large simulation time steps (Cao et al., 2005).

For thermal flow, including geothermal, the temperature (or enthalpy) unknowns often demonstrate a mixed nature when used to solve the energy balance equation. The flow regime is often represented in terms of thermal Péclet number (Pe), which is defined to quantify the interplay between thermal convection and conduction. At a small Pe number, temperature or enthalpy behaves as an elliptic variable while at a high Pe, it behaves as hyperbolic. In the first case, the standard CPR preconditioning is inefficient and some adjustments to the standard CPR procedure can be applied (Cremon et al., 2020a).

When geomechanics effects are involved in the solution, the preconditioning strategy is complicated even more. The mechanical displacements often behave as elliptic variables and the AMG preconditioning proves its efficiency for the solution or preconditioning of linear elasticity problem (Griebel et al., 2003; Baker et al., 2010). One strategy to solve a coupled problem is to use a sequential coupling strategy when thermal-compositional flow and mechanics are solved separately using one of the iterative coupling schemes (Kim et al., 2011a, b). However, when the coupling between flow and geomechanics becomes stronger, this strategy usually is not efficient (Garipov et al., 2018). Following the ideas of sequential coupling, the efficient preconditioning strategy for a fully coupled thermal-compositional flow and mechanics problem was proposed by White et al. (2019). When the contact mechanic is involved, additional degrees of freedom introduced for contacts may require extra logic in constructing efficient preconditioning for a fully coupled problem (Franceschini et al., 2022).

5 Typical challenges in geothermal simulation

In this section, we look into some details of geothermal simulation which makes it different from other types of reservoir modeling.

5.1 Single-phase geothermal flow

We start with a simple geothermal model. Here, we look into the reinjection of cold brine into a reservoir with hot brine and dissolved CH_4 which is a typical situation in direct-use applications in the Netherlands. Here, CH_4 concentration can be used as a passive tracer to estimate the velocity of reinjected water. The details for simulation can be found in Table 1. Simulation results are shown in Fig. 1. The upper figure corresponds to pressure, second to CH_4 molar fraction, and last one to temperature in K. Due to the low compressibility of the brine, the pressure solution changes mostly due to the changes in the viscosity of the brine as a function of the temperature. Two jumps in molar fraction and temperature going from injection to production well. However, you can see that the velocity of temperature shock is much lower than degassing shock. This is happening due to the difference in characteristics of Eqs. (23), (30).

TABLE 1 Parameter settings for a single-phase 1D geothermal model.

Parameter	Value
Grid dimensions	$1000 \times 1 \times 11$
Grid size, m	$1 \times 10 \times 50$
Permeability kx, mD	100
Porosity	0.2
Rock heat capacity, kJ m^{-3} K^{-1}	2200
Rock heat conductivity, kJ m^{-1} day^{-1} K^{-1}	100
Initial temperature, K	350
Initial pressure, bars	200
Initial CH_4 molar fraction	10^{-4}
Injection temperature, K	300
Injection pressure, bars	220
Production pressure, bars	180

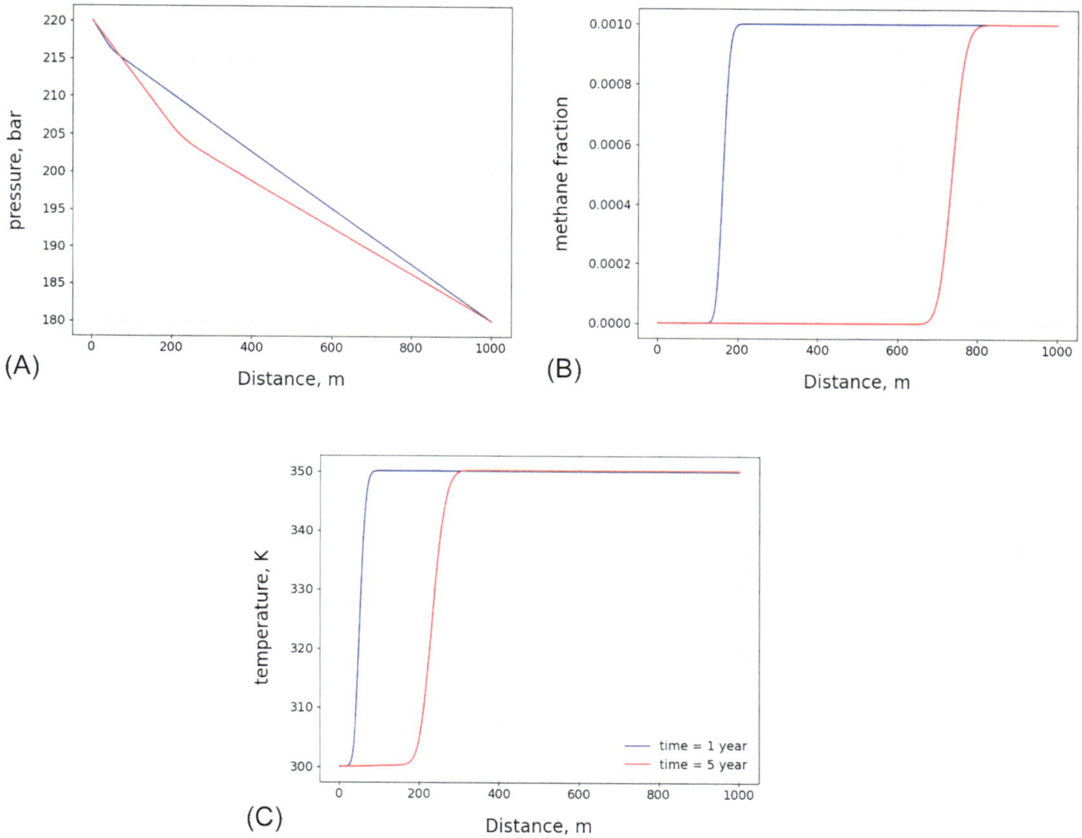

FIG. 1 One-dimensional simulation results with a distribution of (A) pressure, (B) methane molar fraction, and (C) temperature after 1 and 5 years of operation.

To illustrate this effect, we are going to make several simplifying assumptions. First, we assume that the convection dominates the single-phase displacement in the described model and the rest of the phenomena including compressibility and thermal conduction can be ignored. In the following derivations, we will skip the phase index assuming only an aqueous brine phase is present. For an incompressible fluid, the velocity $v = const$ which yields

$$v = -\sum_{l \in L(i)} \Lambda^l \Gamma_a^l \Delta \psi^l = const. \tag{53}$$

In the simplified assumptions, we can rewrite Eq. (23) for the first (CH_4) component and Eq. (30) in the following continuous 1D form:

$$\begin{aligned} \phi_0 \frac{\partial \alpha_{1f}}{\partial t} + v \frac{\partial \beta_1}{\partial x} &= 0, \\ \phi_0 \frac{\partial \alpha_{ef}}{\partial t} + (1 - \phi_0) \frac{\partial \alpha_{er}}{\partial t} + v \frac{\partial \beta_e}{\partial x} &= 0. \end{aligned} \tag{54}$$

In the earlier-mentioned assumptions, $\alpha_{1f} = \rho z$, $\alpha_{ef} = \rho U$, $\alpha_{er} = \rho_r U_r$, $\beta_1 = \rho z$, and $\beta_e = \rho h$ where z is an overall molar fraction of methane component. Since variation in composition is low, we can assume that $\rho = \rho(p)$ and independent from changes in molar fractions. The final simplifying assumption yields $h = U = C_f(T - T_0)$ and $U_r = C_r(T - T_0)$ where C_f and C_r are specific heat capacity of water and rock, respectively. Within these assumptions, Eq. (54) becomes

$$\begin{aligned} \phi_0 \rho \frac{\partial z}{\partial t} + v\rho \frac{\partial z}{\partial x} &= 0, \\ \left[\phi_0 \rho C_f + (1 - \phi_0)\rho_r C_r\right] \frac{\partial T}{\partial t} + v\rho C_f \frac{\partial T}{\partial x} &= 0. \end{aligned} \tag{55}$$

From these derivations, it is obvious that characteristic velocity for z and T are controlled by

$$v_z = \frac{v}{\phi_0}, \quad v_T = \frac{v}{\phi_0 + \widetilde{\phi}_r}, \tag{56}$$

where $\widetilde{\phi}_r = (1 - \phi_0)\rho_r C_r / (\rho C_f) > 0$. It is clear that $v_z > v_T$ due to the additional energy in the rock volume present by $\widetilde{\phi}_r$ term.

5.2 Influence of nonreservoir lithology

In most of the subsurface reservoirs, the dynamics of heat flow exhibit a fascinating dichotomy. Within the permeable layers, heat transfer is dominated by convective forces, where energy is transmitted through the fluid motion of the porous medium. Contrastingly, in the nonpermeable parts of the reservoir, heat is predominantly transferred by conduction and recharging the fluid passing by. The interplay between these two distinct mechanisms forms a complex transmission of heat in subsurface reservoirs. Geothermal simulations, attempting to capture these intricate interactions, must contend with the sensitivity of the system to small-scale heterogeneities including nonpermeable facies.

To illustrate this phenomenon, we consider a small three-layer homogeneous reservoir with injection and production wells operating for 20 years with resulting temperature distribution shown in Fig. 2A. All physical parameters for this model are present in Table 2. Here, the upper layer plays the role of low-permeable overburden and the other two layers represent a permeable reservoir. The corresponding thermal breakthrough time (a two-degree drop in production temperature) for this case is 16.1 years. The temperature front propagates faster in the reservoir since it is driven by convection while the propagation in the overburden is only to conductive depletion from the reservoir. It is also important to remember that the typical overburden thickness is much larger than the thickness of the reservoir and representing it with a single reservoir layer is not accurate. In real reservoirs, the temperature depleted in the overburden will be compensated from the upper overburden layers.

One simple solution can be just to increase the thickness of the overburden layer. The simulation results for the model with the thickness of the overburden increased by 10 times are shown in Fig. 2B. Here, the overburden temperature is depleted by less than 1°C (not visible in the figure) and the thermal breakthrough happens after 15.0 years which are close to the scenario when the overburden is not taken into account at all. That can be explained by the nature of finite volume approximation used for the simulation of heat conduction. In Eq. (36), the

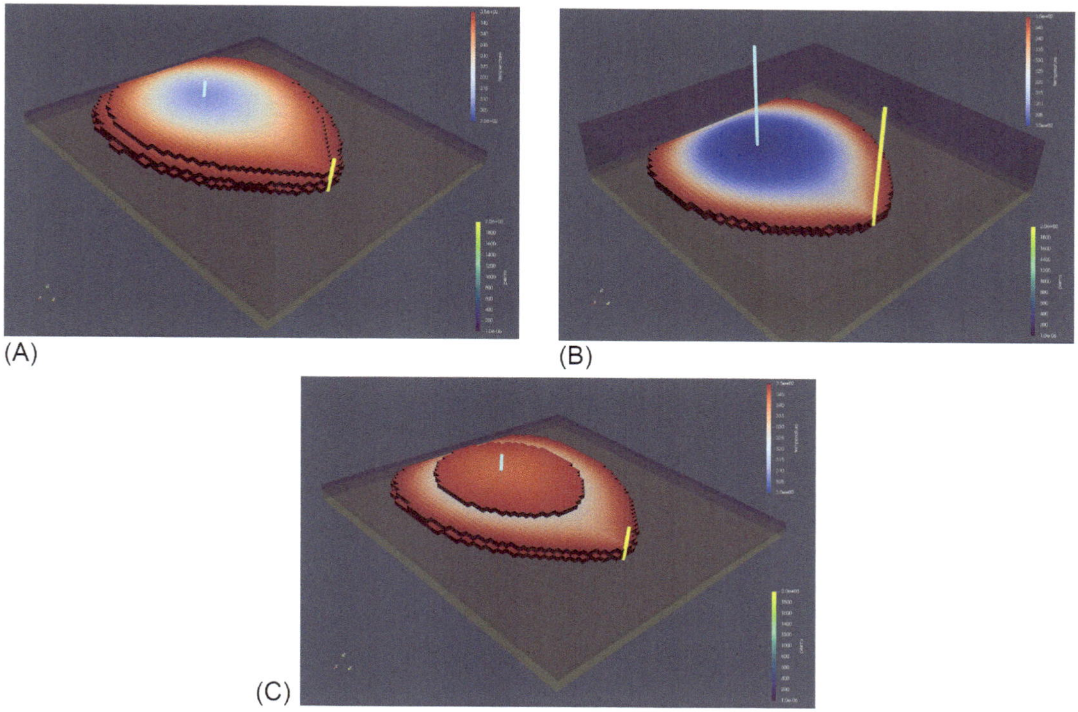

FIG. 2 Temperature distribution in a homogeneous three-layer model. *Light blue* and *yellow lines* correspond to injection and production wells, respectively. (A) Overburden with small layer. (B) Oveburden with large layer. (C) Overburden with large volume.

TABLE 2 Parameter settings for basic 3D geothermal model.

Parameter	Value
Grid dimension	$60 \times 60 \times 3$
Grid size, m	$30 \times 30 \times 30$
Permeability kz, mD	2000
Porosity	0.2
Rock heat capacity, kJ m^{-3} K^{-1}	2200
Rock heat conductivity, kJ m^{-1} day^{-1} K^{-1}	500
Initial temperature, K	350
Initial pressure, bars	200
Injection temperature, K	300

distance from the interface to the center is located in the denominator of each term. When this distance is 10 times larger, the transmissibility decreases significantly which almost neglects the thermal conduction. The easy solution is to increase the volume of the overburden layer keeping the thickness constant. The simulation results for this arrangement are shown in Fig. 2C with the corresponding breakthrough of the cold front at 16.4 years. Here, due to the larger volume (equivalent to the 10 times larger layer thickness), the temperature in the overburden is depleted insignificantly. Of course, this approach is a very rough approximation. For accurate overburden (and underburden) modeling, they should be discretized and approximated with multiple layers (Wang et al., 2021b).

Next, let us check how these results will be affected if heterogeneity is present in the reservoir. For this, we will be using one of the realizations of the EGG model proposed by Jansen et al. (2014) within the same modeling setup. The original model contains seven layers with several disactivated gridblocks forming egg shape. Here, we will be using only the first three of them with all gridblocks active. The simulation results are shown in Fig. 3A. Here, the thermal breakthrough time drops to 9.7 years (8.6 without an influence of overburden) due to the faster brine propagation through high-permeable channels oriented in the direction of flow. To improve the performance of such a system, wells can be drilled in the direction across the permeable channels (Willems and Nick, 2019). Fig. 3B shows the temperature distribution for this scenario. Here, much more heat is swept from the reservoir significantly delaying the thermal breakthrough: 19.2 years with overburden and 16.8 years without it. This example indicates the significance of acquiring a comprehensive grasp of geology when undertaking geothermal operations.

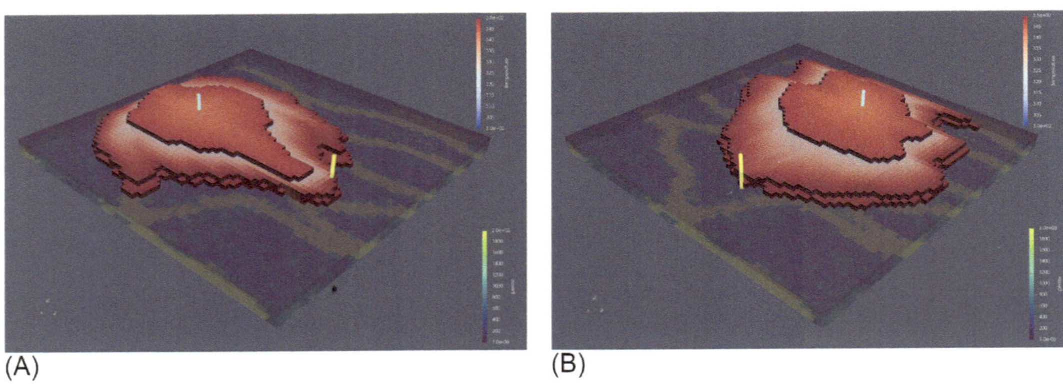

FIG. 3 Temperature distribution in a heterogeneous three-layer model with corresponding permeability pattern. *Light blue* and *yellow lines* correspond to injection and production wells, respectively. (A) Displacement along the channels. (B) Displacement across the channels.

5.3 Two-phase geothermal flow

To illustrate the complexity of a two-phase system flow, we will consider a 1D reservoir described in Section 5.1 with few parameters changed to emphasize two-phase behavior, see Table 3 for details. Here, the structure of the solution is much more complex than in the single-phase case. In Fig. 4, you can see pressure, temperature, and saturation distributions.

TABLE 3 Parameter settings for a two-phase 1D geothermal model.

Parameter	Value
Grid dimension	$1000 \times 1 \times 1$
Grid size, m	$1 \times 10 \times 50$
Permeability kz, mD	100
Porosity	0.2
Rock heat capacity, kJ m^{-3} K^{-1}	0
Rock heat conductivity, kJ m^{-1} day^{-1} K^{-1}	100
Initial temperature, K	520
Initial pressure, bars	30
Injection temperature, K	320
Injection pressure, bars	40

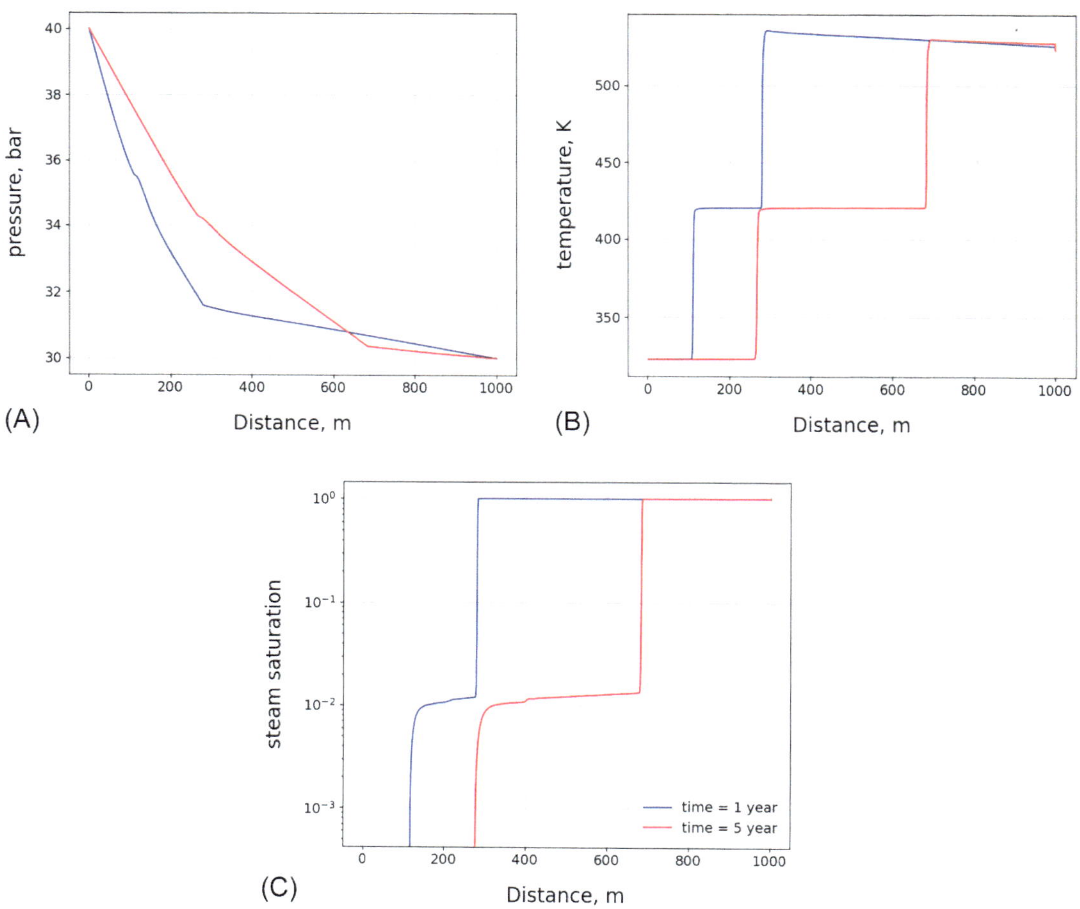

FIG. 4 One-dimensional simulation results with a distribution of (A) pressure, (B) temperature, and (C) steam saturation after 1 and 5 years of operation.

It is clear from the pressure distribution that the compressibility of water is changing quite significantly between multiphase regions. These regions can be recognized by temperature changes from injection (single-phase liquid region) to initial (single-phase gas region) and two-phase in between. Saturation distribution, shown in the log scale, contains two distinguished condensation (left) and vaporization (right) shocks which is similar to the structure of gas injection solution for binary system (Orr, 2007). The changes in saturation for the leading condensation shock are insignificant (less than 1%) when vaporization shock covers the entire change in saturation. Depending on the conditions and other parameters of the model, the structure of the solution can vary a lot (Wang et al., 2021a). However, the distinguished features, driven by the hyperbolic nature of the convective-dominated geothermal problem, are always present in the solution.

Modeling of high-enthalpy geothermal applications also introduces several challenges related to the complex thermodynamics of such systems. One of the most complex challenges is a phenomenon commonly referred to as apparent negative compressibility. This problem was originally spotted and explained by Coats (1980b). A single-cell configuration is used for the explanation where the cold water was injected into a cell containing saturated steam under fixed pressure. The introduction of cold water triggers the condensation of hot steam, leading to a decline in cell pressure following fluid volume shrinkage due to the phase transition. The cell pressure steadily diminishes until all the steam is condensed, causing the cell pressure to return to the injection pressure. To ensure the convergence of nonlinear iterations, it becomes imperative to reduce the timestep every time this phenomenon occurs which is indicated as "stalling behavior" in Pruess et al. (1999).

To illustrate the apparent negative compressibility problem, we will use the setup described in Wang and Voskov (2022). This phenomenon can be illustrated using a single-cell model setup with a cold-water injection at fixed pressure (Coats, 1980b; Wong et al., 2018). In addition, similar simplifying assumptions used to derive Eq. (54) are applied except that fluid now can form two phases, and rock energy is neglected. Therefore, the following mass and energy conservation equations for a single-cell problem are obtained:

$$
\begin{aligned}
V\frac{\partial \rho_t}{\partial t} - \Upsilon(p_{\text{in}} - p) &= 0, \\
V\frac{\partial \rho_t h}{\partial t} - H_{\text{in}}\Upsilon(p_{\text{in}} - p) &= 0, \\
\rho_t &= \rho_w s_w + \rho_s s_s.
\end{aligned}
\tag{57}
$$

Here, Υ is the flow transmissibility at the injection boundary, p_{in} is the fixed pressure of the injection boundary, ρ_t is the total fluid density, H_{in} is the injected enthalpy, ρ_w and ρ_s are the density of water and steam, and s_w and s_s are the saturation of water and steam, respectively. We will use the following parameters to illustrate this problem: $p = 50$ bar, $h = 2000$ kJ kg^{-1}, $s_s = 0.97$ and $p_{\text{in}} = 90$ bar, and $H_{\text{in}} = 345$ kJ kg^{-1}.

Fig. 5 shows the residual map and the nonlinear solution path with an initial guess chosen at the initial conditions. Here, the nonlinear solver follows the gradient provided by a standard Newton's method. Because of the large variations in thermodynamic properties between water and steam phases, the residual equation becomes highly nonlinear with two distinguished minima seen in Fig. 5. The second minimum does not satisfy the residual equation (the norm of residual is far from 0) and the nonlinear solution will not converge

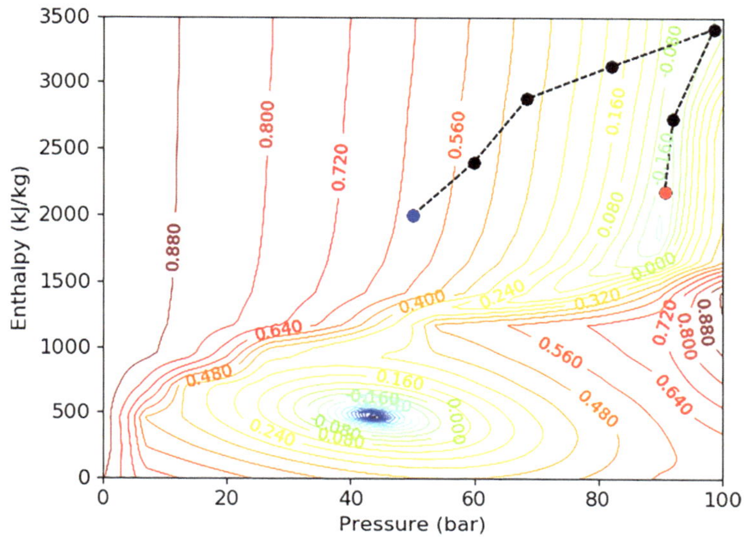

FIG. 5 Newton path (*dash line*) starts from the initial condition; the contours (in l_2-norm) show the residuals, *blue dot* represents the initial condition of the cell, *black dots* are the Newton updates, and *red dot* shows the solution for a current timestep. *Adapted from Wang, Y., Voskov, D., 2022. High-enthalpy geothermal simulation with continuous localization in physics. Mathematics 10 (22). https://doi.org/10.3390/math10224328.*

for this case. Once the timestep is cut significantly, the initial guess becomes closer to the solution, and the nonlinear solver will finally converge.

Pruess et al. (1987) examined this issue, linking the apparent negative compressibility problem to the treatment of thermodynamic equilibrium. Wang (2015) analyzed this problem within fully implicit formulations. The analysis resulted in the development of a stability criterion for timestep management, aiming to enhance convergence by wasted timesteps. However, the stability criterion still imposes strong limitations on allowable timestep sizes which can significantly increase simulation time. Later, Wong et al. (2018) introduced a nonlinear preconditioner to the fully coupled, fully implicit solution. This approach involved utilizing a sequential fully implicit (SFI) approach, with the solutions serving as the initial guess for the FIM (Wong et al., 2017). While this method reduced the severity of timestep restrictions in certain practical scenarios, a robust strategy for achieving convergence in high-enthalpy nonlinear solutions at a target timestep, especially in the presence of the "negative compressibility" phenomenon, still remains challenging.

An interesting solution was proposed in Wang and Voskov (2022). Using the operator-based parameterization (OBL) approach (Voskov, 2017), a continuous localization in physics is proposed to solve the governing equation. In the OBL approach, the governing terms of conservation equations are parameterized in the solution space of the problem (e.g., p and h for a geothermal problem). Next, the parameterization accuracy varies from a very coarse to fine representation. As the physics parameterization transitions from coarse to fine resolution, the behavior of state-dependent terms of conservation equations and resulting residuals shifts from nearly linear and monotonic to highly nonlinear and nonmonotonic form.

In the proposed nonlinear strategy, the solution obtained at a coarser physics parameterization serves as the initial guess for the solution at a finer physics resolution. This continuous localization approach enhances the robustness of the nonlinear convergence process. Fig. 6 shows the convergence paths for moderate and large timestep with corresponding residual maps. Besides, Wang and Voskov (2022) also demonstrate that continuous localization in physics allows for significant improvement in the nonlinear convergence of challenging realistic problems.

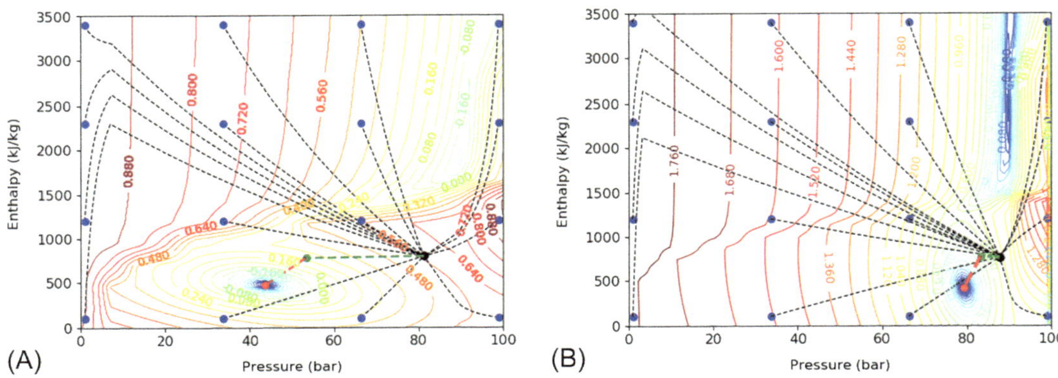

FIG. 6 Nonlinear solution path (*dashed line*) starting from various initial guesses in pressure-enthalpy space with (A) moderate and (B) large timestep; *blue dots* represent the initial guess, the *black dot* is the solution of the coarsest physics resolution, the *green dot* is the intermediate solution, and the *red dot* is the true solution; residual contours (in L_2-norm) are plotted for the reference physics. *Adapted from Wang, Y., Voskov, D., 2022. High-enthalpy geothermal simulation with continuous localization in physics. Mathematics 10 (22). https://doi.org/10.3390/math10224328.*

5.4 Geomechanics in geothermal reservoirs

The geomechanics of subsurface reservoirs has garnered significant attention in recent years, primarily owing to the induced seismicity associated with several geothermal projects. The intricate dynamics of the subsurface reservoir, resulting from geothermal operations, present a complex challenge that requires thorough analysis. To better understand the geomechanics response of geothermal reservoirs, let us delve into a detailed examination using a simplified 1D setup. In this setup, we mimic the injection of cold fluid into a warm reservoir which is mostly associated with the area near a geothermal injection well.

Thus, we consider the vertical column of 1000 m height subjected to instant vertical loading with $F = 10$ bar at the top boundary. The constant pressure $p = 100$ bar and temperature $T = 20°C$ are maintained at the top boundary while the initial pressure and temperature are equal to $p_0 = 100$ bar and $T_0 = 70°C$, respectively. All other sides of the domain are impermeable to fluid and heat and subjected to the roller boundary condition.

Furthermore, the stiffness tensor is defined by Young's modulus $E = 10$ GPa and Poisson's ratio $\nu = 0.4$. Isotropic permeability $\mathbf{K} = k\mathbf{I}$ and heat conductivity $\boldsymbol{\Lambda} = \lambda\mathbf{I}$ tensors are defined by its scalar values $k = 100$ mD and $\lambda = 1$ W m^{-1} k^{-1}, whereas Biot's coefficient and thermal dilation coefficient are defined as $\beta = 0.6$ and $\alpha = 10^{-6}$ K^{-1}, respectively. The total volumetric heat capacity is equal to $c = 2200\text{kJ m}^{-3}\text{ K}^{-1}$. Single-phase fluid is maintained incompressible with fluid viscosity $\mu = 1$ cP.

Fig. 7 illustrates the profiles of temperature and vertical displacement over the column's height evaluated at two moments in time. An instant applied extension continues with fluid influx inside the column. However, heat conduction transfers energy out of the column through the top boundary, which is maintained under low temperature $T = 20°\text{C}$, throughout the whole domain. Temperature decrease causes thermal contraction, competing with extension. Depending on reservoir thickness, this dynamic can be detected even at the surface by InSAR observations or other surface monitoring methods. This information in combination with other observations can be used for monitoring of the cold front propagation as well as utilized in data assimilation schemes.

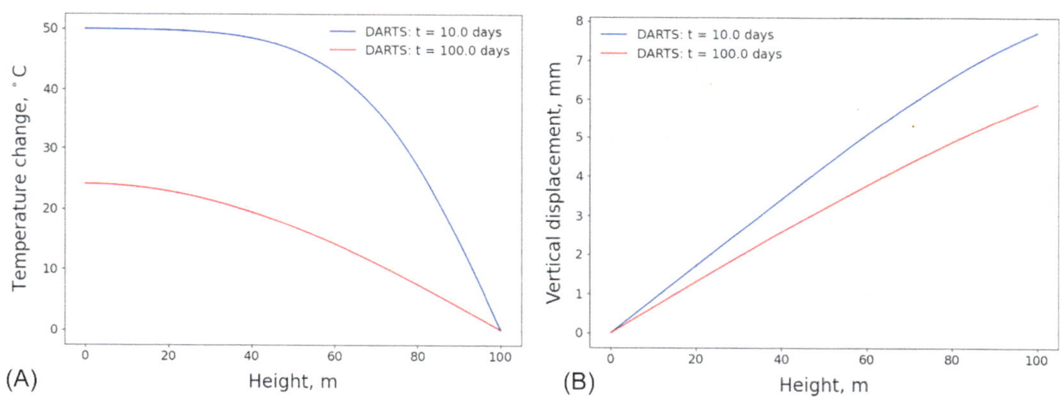

FIG. 7 The profiles of temperature (A) and vertical displacement (B) over the column's height after 10 and 100 days of operation.

This is a typical situation with two types of forces affecting the geothermal reservoir. Further from the injection well, the extension will be reduced while the contraction, caused by the temperature changes, will remain at a similar amplitude. This phenomenon is relatively safe until the cold front reaches existing faults. At critically stressed faults, even a small contraction may cause fault slip, which in turn will create a seismic event. In a fractured system, this risk is higher since fracture openings can redistribute the stresses and generate seismic events even before the cooling front reaches critically stressed faults.

6 Modeling of different geothermal resources

In this section, we will discuss the modeling of different types of sediments used for geothermal energy production. Starting from fluvial sediments mostly relevant to direct heat geothermal systems, moving to naturally fractured systems with different permeability of matrix rock.

6.1 Fluvial sediments

Large sandstone aquifers provide a significant potential for direct heating purposes worldwide. In most places, the structure of such reservoirs is based on fluvial sediments which often introduce significant heterogeneity with a highly uncertain structure. One example of such sediments is the West Netherlands Basin (WNB) which is a 60-km-wide basin in the southwest of the Netherlands (Donselaar et al., 2015). Currently, WNB is responsible for 70% of all deep geothermal doublets operational in the Netherlands. This sedimentary structure corresponds to a shallow marine depositional setting with good lateral continuity and sand average permeability up to 1 Darcy.

The main challenge in modeling such sediments is the uncertainty in their structure and properties of facies which can directly impact the technical and economic feasibility of geothermal projects (Vogt et al., 2013). The lifetime of geothermal doublets and the heat recovery rate of the reservoir often vary a lot with both reservoir parameters and operational management (Willems and Nick, 2019). Shale formations are commonly seen as barriers to flow due to their low permeability. In simulations at isothermal assumptions, these shale formations are frequently excluded from reservoir descriptions to reduce the size of the model. However, for geothermal reservoirs, this simplified approach can lead to inaccuracies in simulation outcomes. As heat transport in subsurface reservoirs relies on both thermal convection and conduction, the impermeable shale formations often act as heat sources recharging the fluid convectively transferred in permeable (sand) facies.

It is important to note that most conclusions on the influence of reservoir heterogeneity on geothermal production were drawn from conceptual synthetic models at resolution typical for reservoir simulation studies in the hydrocarbon industry. The first detailed sensitivity study of the performance of geothermal reservoirs at a high-resolution geologic scale has been done by Wang et al. (2021b). This work employed an extensive set of forward simulations based on a highly resolved geological model. The base-case porosity distribution is shown in Fig. 8 with a 3 times exaggeration along the z-axis. The physical model dimensions are [9000 m $\times$ 4200 m] with a variable vertical thickness of up to 130 m.

FIG. 8 Base-case porosity distribution of the target geothermal reservoir. In both doublets, the injection wells are shown in *blue* and the production wells in *orange*.

The base simulation model at the geological scale contains around 3.2 million grid cells with 0.8 million corresponding to sand and 2.4 million corresponding to shale facies. The grid block sizes in the model are variable with a horizontal resolution of around 50 m and a vertical resolution varying between 10 and 20 cm. The open boundary conditions are used on the sides of the model with confining over- and under-burden layers on top and bottom. The simulation time includes a 100-year period to achieve a breakthrough in both doublets at different sensitivity parameters. The resulting cold plums are shown in Fig. 9. It can be seen that plum distribution is strongly affected by reservoir heterogeneity and is quite far from the "cylindrical" form usually predicted by simulation in homogeneous assumptions.

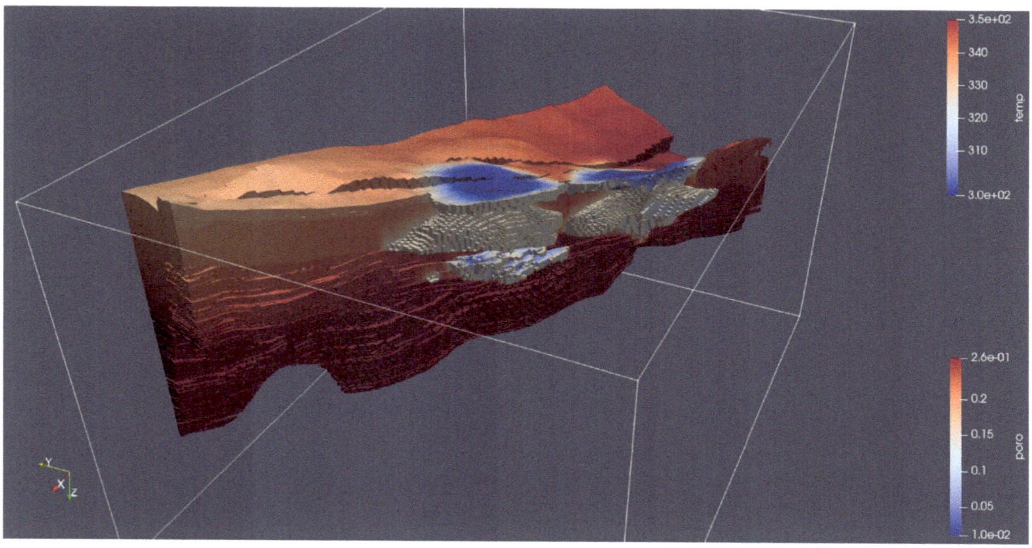

FIG. 9 Thermal plume propagation inside the reservoir for both geothermal doublets.

Several major conclusions have been derived from this study:

- Shale layers should not be excluded when modeling geothermal reservoirs. The energy contained in shales can conductively recharge cold water passing through permeable facies. Additionally, the distribution of shale and sand facies plays a pivotal role in the efficient exploitation of energy stored in a geothermal reservoir.
- Over- and under-burden confining layers can thermally recharge fluid flowing in the reservoir. However, the effect of confining layers to a doublet lifetime strongly depends on reservoir heterogeneity.
- Even small changes in the position of the doublet can significantly (up to 25%) change the lifetime of the system. In Fig. 10, you can see production temperature in doublet 1 and 2 with doublet position changes by 50 m around original position C.
- The total energy production increases with the production rate due to the better sweep efficiency. However, due to the larger pumping energy at higher production rates, these benefits may not be significant.

- The porosity-permeability correlation can affect the lifetime of the doublet but this effect strongly depends on the local heterogeneity.

Notice that the sensitivity of lifetime to doublet position in this study is lower than the sensitivity observed in Shetty et al. (2018). That can be explained by a higher NTG ratio of these heterogeneous model (close to 60%). Other differences between these two studies correspond to the vertical resolution of the models (895 vs 40 layers) and different techniques used for the generation of heterogeneous reservoirs (multipoint statistics vs process-based approach). To address this last aspect, Major et al. (2023) performed a comparison study between process-based and statistical approaches for another fluvial reservoir and concluded that both techniques produce comparable results in terms of stochastic response.

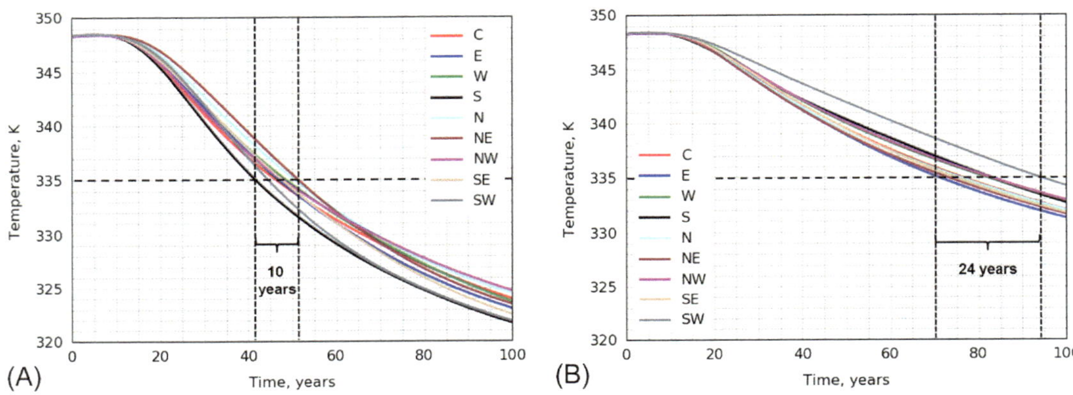

FIG. 10 Production temperature of each doublet for different doublet positions. (A) Production in doublet 1. (B) Production in doublet 2. *Adapted from Wang, Y., Voskov, D., Daniilidis, A., Khait, M., Saeid, S., Bruhn, D., 2023. Uncertainty quantification in a heterogeneous fluvial sandstone reservoir using GPU-based Monte Carlo simulation. Geothermics 114. https://doi.org/10.1016/j.geothermics.2023.102773.*

Using the same setup as Wang et al. (2021a), systematic UQ has been performed in Wang et al. (2023). The geological uncertainties in this study have been limited by a fixed reservoir structure and fluvial scenario with 100 statistical realizations. Besides the geological realizations, uncertainties in porosity-permeability correlation were also included. In addition, the sampling of the most sensitive physical parameters identified in Wang et al. (2021a) such as brine salinity and rock conductivity was included in uncertainty analysis. Operational uncertainties include injection temperature and well injectivity while economic uncertainties cover heat and electricity prices.

Using these uncertain parameters, full Monte Carlo (MC) simulations have been run on 16 GPU devices (Titan RTX) using a high-fidelity geological model. The run time of a single realization with 3.2 million grid blocks for 100 years took an average of 2.2 min, allowing for running MC simulation until convergence in less than a day. The convergence of MC has been evaluated based on the norm of the difference and statistical moments of cumulative energy and net present value (NPV). With 10,000 MC iterations, full convergence was achieved. The resulting probability density functions for total energy produced and NPV are shown in Fig. 11.

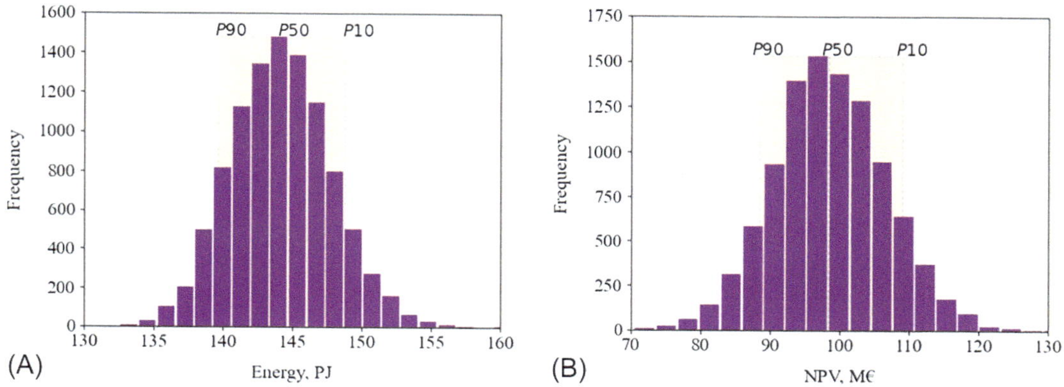

FIG. 11 The histograms of produced energy and NPV of the geothermal project with samples falling between P10 and P90 quantiles of both parameters are marked in *orange. Adapted from Wang, Y., Voskov, D., Daniilidis, A., Khait, M., Saeid, S., Bruhn, D., 2023. Uncertainty quantification in a heterogeneous fluvial sandstone reservoir using GPU-based Monte Carlo simulation. Geothermics 114. https://doi.org/10.1016/j.geothermics.2023.102773.*

Another interesting finding in this study corresponds to the effect of upscaling on the prediction of reservoir dynamics. Using upscaled models is quite common in reservoir simulation due to a limited performance when geological scale models are used for simulation. The common engineering approach is to use limited two-dimensional (2D) models for the geothermal simulations (Veldkamp et al., 2015). The study (Wang et al., 2023) investigates the impact of upscaling in the vertical direction using the most convenient averaging methodology. The vertical porosity and permeability have been upscaled from 895 layers to 101 layers, 9 layers, and 1 layer. The effect of this upscaling on the plume distribution is shown using a zoomed version of the cold plume in Fig. 12.

Strong heterogeneity of the original model is smoothed down with an upscaling ratio which reduces the spread of cold plume and changes the lifetime expectation in the upscaled models to more optimistic scenarios. It is shown in Wang et al. (2023) that UQ in the simplified 2D model overpredicts the mean of total energy recovery by more than 25% which is a significant difference for the project economy. This happens due to the lumping of shale and sand layers in the process of upscaling which completely changes the heat transport in the model and under-predicts the cold front propagation and breakthrough which is seen in Fig. 12.

As the conclusion to modeling of geothermal energy production in fluvial sediments, the accurate representation of reservoir heterogeneity plays a significant role in the evaluation of reservoir dynamics. A complex interplay between convective cold plume propagation in sand channels and its conductive recharge from nonpermeable shale facies requires detailed uncertainty analysis using high-fidelity models at geological resolution. The commonly used upscaling technique should be carefully used due to the introduced bias.

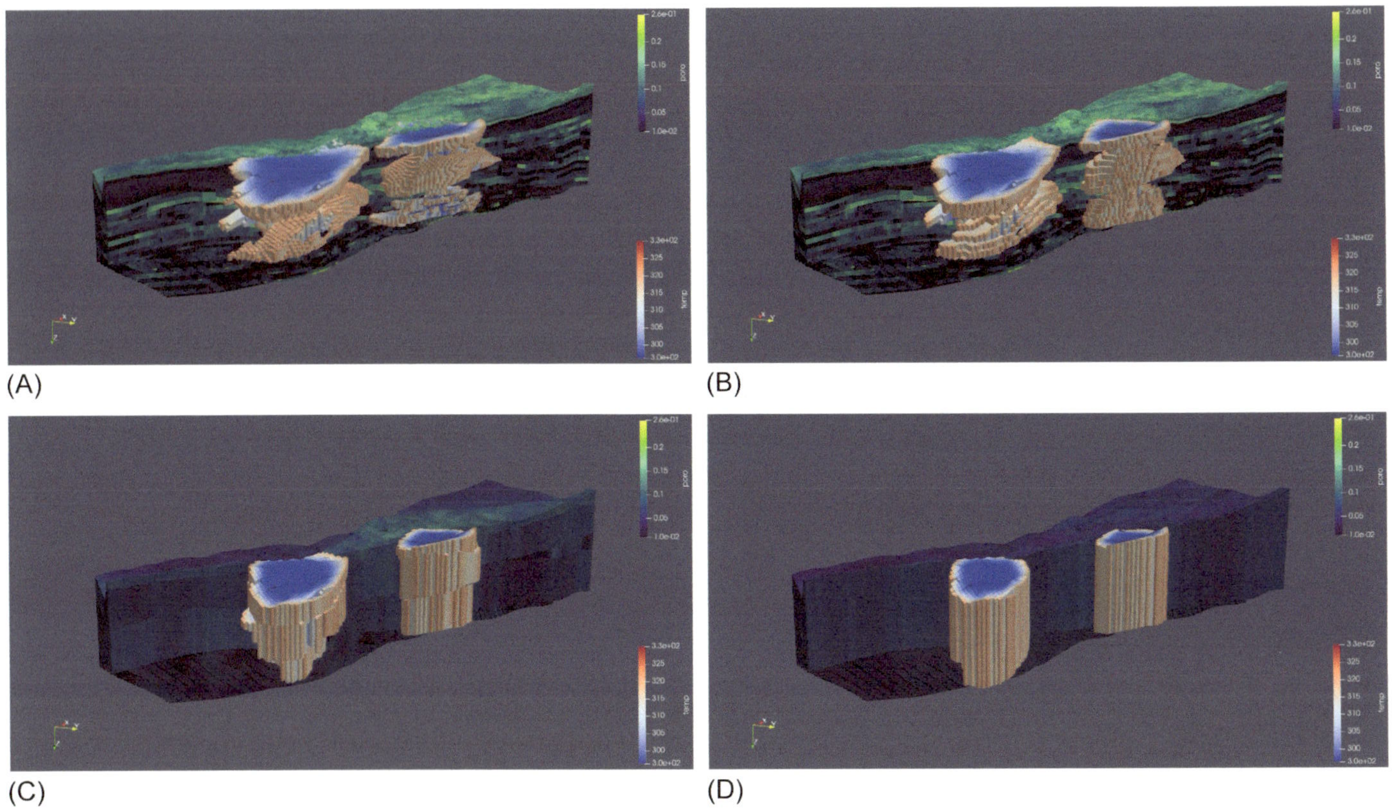

FIG. 12 Distribution of cold plum in vertically upscaled models exaggerated three times in vertical direction with (A) 895 layers, (B) 101 layers, (C) 9 layers, and (D) 1 layer.

6.2 Fractured media

Here, we will discuss a computational framework for UQ and data assimilation of fractured reservoirs. Based on experience, the DFM model usually produces the most accurate response and it will be used in this study. However, the most of discussed techniques can be also applied to multiple continua and embedded-discrete fracture matrix (EDFM) models.

The DFM method employs numerical approximation, representing fracture segments as lower-dimensional objects within the mesh. Early concepts of DFM emerged in Gureghian (1975), where FEMs were utilized, and in Helmig et al. (1997), which employed FVM. In modern reservoir simulation, characterized by an implicit time approximation and finite-volume discretization on unstructured grids, Karimi-Fard et al. (2004) formulated the DFM methodology. This approach is favored in detailed geological studies for its precision (Moinfar et al., 2011; Flemisch et al., 2018; Berre et al., 2019; Wong et al., 2020). However, DFM models necessitate extensive meshing accuracy to capture the intricate geometry of fracture networks. This significantly escalates computational complexity, making them impractical for UQ purposes (Jung et al., 2013; Nejadi et al., 2017; Spooner et al., 2021). Recent advancements have enhanced the DFM approach in reservoir simulation by integrating geomechanics (Garipov et al., 2016; Garipov and Hui, 2019) and fracture propagation (Gallyamov et al., 2018) for practical applications. Yet, these intricate physical processes demand fine modeling resolution, accentuating the challenges of incorporating UQ.

These constraints significantly hamper the pursuit of low-risk and sustainable subsurface operations. The substantial computational complexity in DFM models primarily stems from meshing issues, such as skinny triangles, small control volumes, and a high degree of freedom, arising from conformal meshes and associated nonlinear convergence challenges (Geiger and Matthäi, 2014; Li and Li, 2019; Koohbor et al., 2020). Fracture network input data typically originate from outcrop analysis or statistical models, where raw output—whether manually or automatically interpreted—poses challenges for meshing software. These meshing challenges are widely recognized in the existing literature (Reichenberger et al., 2006; Mustapha and Mustapha, 2007; Mallison et al., 2010; Karimi-Fard and Durlofsky, 2016; Berre et al., 2019).

A strategy to simplify complex fracture networks in terms of flow response based on a robust preprocessing approach using graph theory has been proposed in de Hoop et al. (2022). This scale-dependent strategy has undergone testing on both synthetic and actual (outcrop-based) fractured networks. The technique streamlines the fracture network's topology by consolidating fracture nodes or vertices within a predefined radius, essentially combining the rows of the incidence matrix for each vertex. This process maintains the network's connectivity while explicitly addressing problematic fracture intersections occurring below a specified angle threshold. Additionally, this framework expands on existing preprocessing methods, such as those detailed in Mustapha and Mustapha (2007) and Karimi-Fard and Durlofsky (2016), by integrating an aperture correction when merging vertices. This correction aims to better preserve the original connectivity and manage heterogeneous aperture distributions.

The proposed framework enables the generation of a fully conformal grid uniformly distributed across a realistic fracture network with varying apertures, achieving the necessary precision. This method's impact is assessed across several parameters: the fracture network's geometry, including its angle distribution; the meshing outcomes, considering element volume and quality; and the dynamic response of the reservoir under high-enthalpy geothermal production conditions. By reducing the computational intricacies of DFM models, this approach paves the way for employing efficient DFM models with a comparative computational cost to EDFM or even dual-porosity/dual-permeability models. Importantly, it accurately captures the discrete nature of fracture networks, thereby facilitating UQ and data assimilation endeavors. An example of the dynamic responses from three different fracture networks at two different scales (top and bottom rows) is shown in Fig. 13. One can see that even though the changes in the scale introduce bias to the dynamic response, all correct dynamic trends are still captured by a coarser representation.

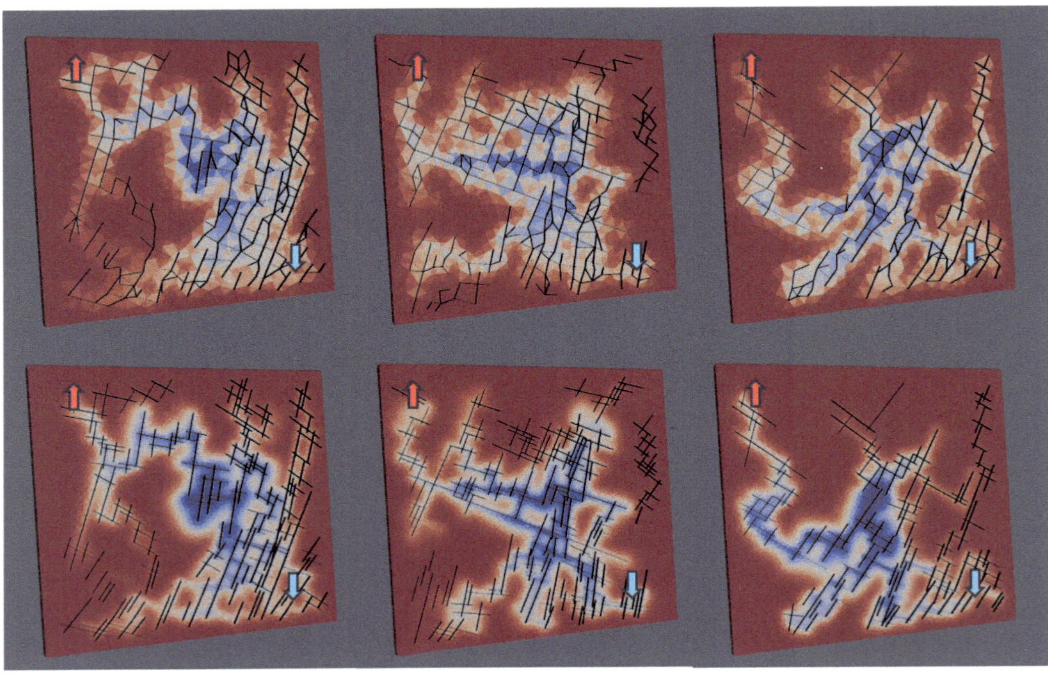

FIG. 13 Temperature distribution for three different fractured networks (*columns*) at two different scales with the coarser resolution on *top*. The *red* and *blue arrows* correspond to production and injection wells, respectively.

This multiscale representation for fractured networks can be used for efficient UQ. Rather than resorting to simplified effective media models like dual-porosity, there is an opportunity to employ high-fidelity DFM models that precisely depict the intricate geometry of fracture networks. Achieving this performance involves leveraging coarse-scale data and opting for a subset of high-fidelity models capable of representing the same statistical features as the entire ensemble of models. This selection process can be facilitated through the use of the multidimensional scaling (MDS) algorithm (Borg and Groenen, 2005). A more detailed description of the approach can be found in de Hoop and Voskov (2021).

The successful application of this technique is illustrated in Fig. 14. Here, we show an ensemble response fractured network (in gray) and stochastic moments for fine and coarse scales (blue and green, respectively). For the coarser scale ensemble, the distances between dynamic responses of the ensemble members are projected into lower-dimensional space using MDS and clustering is performed using the k-medoids algorithm. The centroids of each cluster are chosen as a representative and comprise the subset of models. These models are then simulated on the fine-scale and for the computation of the cumulative distribution function the responses are weighted by the size of the cluster which they represent. As a result, we can reconstruct statistic moments for high fidelity ensemble (shown in red in Fig. 14C) in a fraction of the time needed for full fine-scale evaluation.

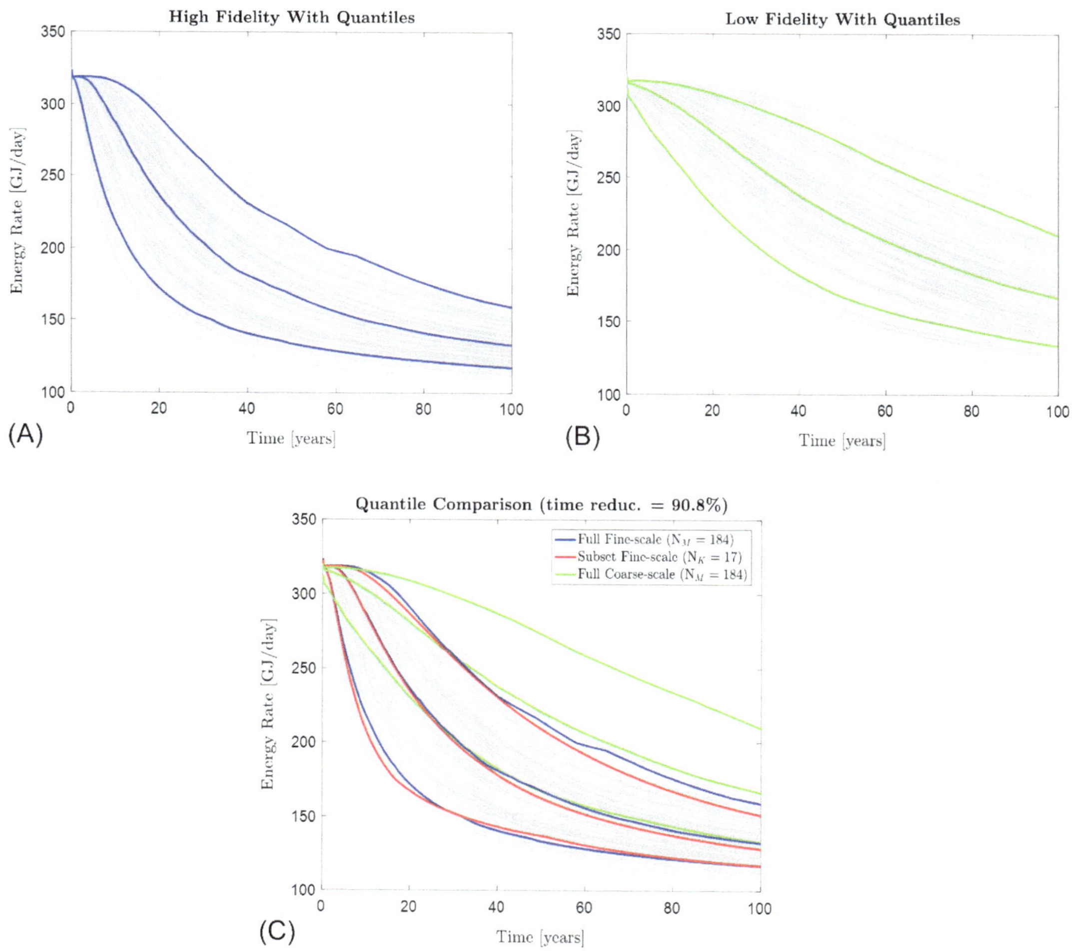

FIG. 14 Energy rate distribution for an ensemble of models (184 members) with corresponding stochastic moments for high-fidelity representation (A), coarse representation (B), and reconstructed (C).

Another application of the multiscale DFM framework is the evaluation of geothermal energy production from fractured reservoirs with different rock types. The discretized fractured model is shown in Fig. 15, where the fracture aperture is assumed to be constant at 7500 Darcy. The temperature distribution for different matrix permeability corresponding to sandstone (0.1 Darcy), carbonate (1 milli-Darcy), and granite (1 micro-Darcy) are shown in Fig. 15 after 20 years of production. The impact of fractures on temperature distribution has almost no effect with the ratio between matrix and fractured rock less than three orders of magnitude in the convective-dominated regime. When this ratio reaches six orders, more heat is driven by fluid in fractures with still significant energy depletion to the matrix. With nine orders, heat flow is predominantly going through the fractures with a very limited energy depletion from the matrix mostly driven by conduction. A more detailed analysis of flow in fractured systems can be found in Wang et al. (2021a).

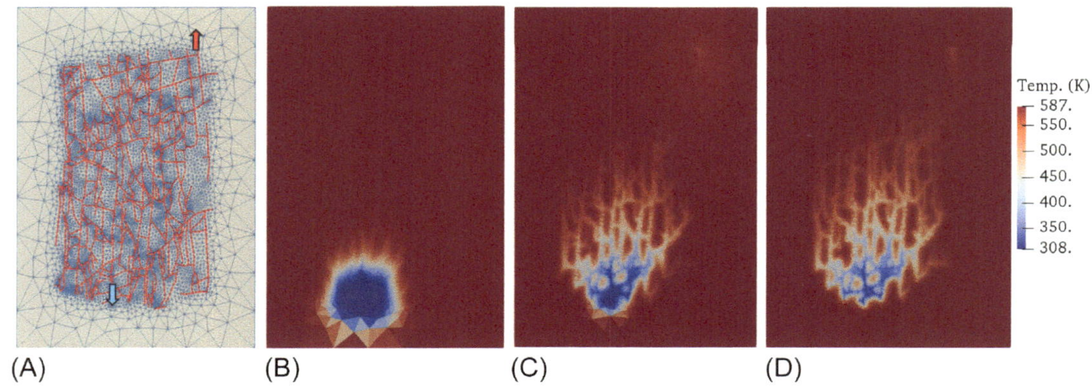

FIG. 15 Discretized DFM model (A) and temperature distributions for sandstone (B), carbonate (C), and granite (D) matrix rock. *Red* and *blue arrows* correspond to production and injection wells, respectively. *Adapted from Wang, Y., de Hoop, S., Voskov, D., Bruhn, D., Bertotti, G., 2021. Modeling of multiphase mass and heat transfer in fractured high-enthalpy geothermal systems with advanced discrete fracture methodology. Adv. Water Resour. 154. https://doi.org/10.1016/j.advwatres.2021.103985.*

In most examples earlier, the fracture aperture and corresponding permeability are assumed to be constant. However, in real fractured networks, the aperture distribution can have significant variation. The next example illustrates the impact of heterogeneity in fracture aperture distribution on the dynamic response of a geothermal system. Here, the same outcrop is assumed to be a carbonate reservoir at a depth of 4000 m with a natural gradient of pressure and temperature. A normal distribution of fracture aperture is used with a mean of 0.6 mm and a standard deviation of 0.3 mm to generate 100 realizations to address uncertainties in aperture distribution. Fig. 16 illustrates the sensitivity of pressure and temperature distribution to P10, P50, and P90 with respect to NPV which includes a full evaluation of energy produced and spent. This study concludes that depending on connectivity provided by the stochastic distribution of fracture apertures, an optimal rate of production can vary by 15%, see details in Van Der Kooij et al. (2020).

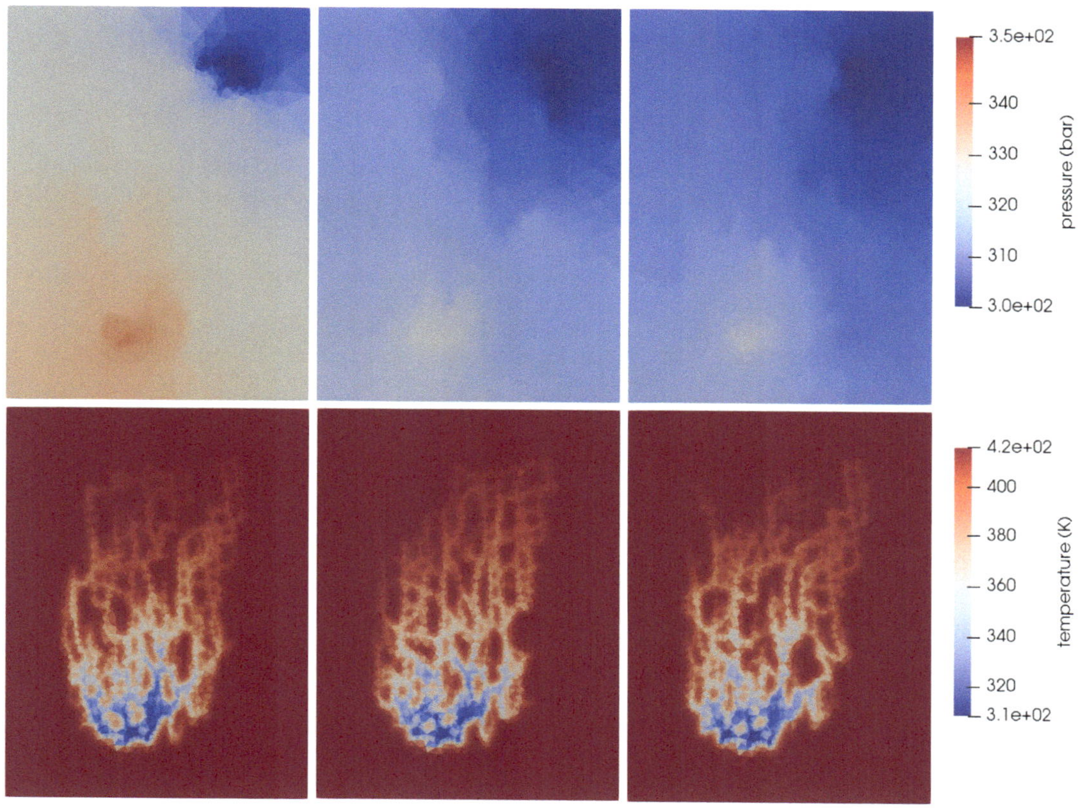

FIG. 16 Dynamic pressure (*upper row*) and temperature (*lower row*) solutions after 10 years of simulation for P10, P50, and P90 aperture distributions with respect to NPV. *Adapted from Van Der Kooij, R., Wang, Y., Daniilidis, A., Voskov, D., 2020. Performance influences of a deep high temperature fractured geothermal system. In: 1st Geoscience and Engineering in Energy Transition Conference, GET 2020.*

References

Asai, P., Panja, P., McLennan, J., Deo, M., 2019. Effect of different flow schemes on heat recovery from enhanced geothermal systems (EGS). Energy 175, 667–676. https://doi.org/10.1016/j.energy.2019.03.124.

Athens, N.D., Caers, J.K., 2019. A Monte Carlo-based framework for assessing the value of information and development risk in geothermal exploration. Appl. Energy 256, 113932. https://doi.org/10.1016/j.apenergy.2019.113932.

Aziz, K., Settari, T., 1979. Petroleum Reservoir Simulation. Applied Science Publishers.

Baker, A.H., Kolev, T.V., Yang, U.M., 2010. Improving algebraic multigrid interpolation operators for linear elasticity problems. Numer. Linear Algebra Appl. 17 (2–3), 495–517. https://doi.org/10.1002/nla.688.

Ballio, F., Guadagnini, A., 2004. Convergence assessment of numerical Monte Carlo simulations in groundwater hydrology. Water Resour. Res. 40 (4), W046031–W046035. https://doi.org/10.1029/2003WR002876.

Barenblatt, G.I., Zheltov, I.P., Kochina, I.N., 1960. Basic concepts in the theory of seepage of homogeneous liquids in fissured rocks [strata]. J. Appl. Math. Mech. 24 (5), 1286–1303. https://doi.org/10.1016/0021-8928(60)90107-6.

Behie, A., Forsyth Jr., P., 1983. Comparison of fast iterative methods for symmetric systems. IMA J. Numer. Anal. 3 (1), 41–63.

Berre, I., Doster, F., Keilegavlen, E., 2019. Flow in fractured porous media: a review of conceptual models and discretization approaches. Transp. Porous Media 130 (1), 215–236.

Bickel, J.E., Bratvold, R.B., 2008. From uncertainty quantification to decision making in the oil and gas industry. Energy Explor. Exploit. 26, 311–325. https://doi.org/10.1260/014459808787945344.

Borg, I., Groenen, P.J.F., 2005. Modern Multidimensional Scaling: Theory and Applications. Springer Science & Business Media.

Cao, H., 2002. Development of Techniques for General Purpose Simulators (Ph.D. thesis). Stanford University.

Cao, H., Tchelepi, H.A., Wallis, J.R., Yardumian, H.E., et al., 2005. Parallel scalable unstructured CPR-type linear solver for reservoir simulation. In: SPE Annual Technical Conference and Exhibition, Society of Petroleum Engineers.

Cao, H., Crumpton, P.I., Schrader, M.L., 2009. Efficient general formulation approach for modeling complex physics. In: SPE Reservoir Simulation Symposium Proceedings, vol. 2, pp. 1075–1086.

Chang, D., Harold, J., Chris, M., Chris, S., 1994. Effective porosity, producible fluid and permeability in carbonates from NMR logging. In: SPWLA 35th Annual Logging Symposium. https://onepetro.org/SPWLAALS/proceedings-split/SPWLA-1994/SPWLA-1994-A/19156.

Coats, K.H., 1977. Geothermal reservoir modelling. In: Proceedings—SPE Annual Technical Conference and Exhibition, 1977-October. vol.

Coats, K.H., 1980a. Equation of state compositional model. SPE J. (5), 363–376. https://doi.org/10.2118/8284-pa.

Coats, K.H., 1980b. Reservoir simulation: a general model formulation and associated physical/numerical sources of instability. In: Boundary and Interior Layers-Computational and Asymptotic Methods.

Coats, K.H., George, W.D., Marcum, B.E., Chu, C., 1974. Three-dimensional simulation of steamflooding. Soc. Pet. Eng. AIME J. 14 (6), 573–592. https://doi.org/10.2118/4500-pa.

Coussy, O., 2003. Thermoporoelasticity. In: Poromechanics, John Wiley & Sons, Ltd, pp. 71–112.

Cremon, M.A., Castelletto, N., White, J.A., 2020a. Multi-stage preconditioners for thermal-compositional-reactive flow in porous media. J. Comput. Phys. 418. https://doi.org/10.1016/j.jcp.2020.109607.

Cremon, M.A., Christie, M.A., Gerritsen, M.G., 2020b. Monte Carlo simulation for uncertainty quantification in reservoir simulation: a convergence study. J. Pet. Sci. Eng. 190. https://doi.org/10.1016/j.petrol.2020.107094.

Crooijmans, R.A., Willems, C.J.L., Nick, H.M., Bruhn, D.F., 2016. The influence of facies heterogeneity on the doublet performance in low-enthalpy geothermal sedimentary reservoirs. Geothermics 64, 209–219. https://doi.org/10.1016/j.geothermics.2016.06.004.

Cui, T., Fox, C., O'Sullivan, M.J., 2019. A posteriori stochastic correction of reduced models in delayed-acceptance MCMC, with application to multiphase subsurface inverse problems. Int. J. Numer. Methods Eng. 118 (10), 578–605. https://doi.org/10.1002/nme.6028.

Dai, Z., Viswanathan, H., Fessenden-Rahn, J., Middleton, R., Pan, F., Jia, W., Lee, S.-Y., Mcpherson, B., Ampomah, W., Grigg, R., 2014. Uncertainty quantification for CO_2 sequestration and enhanced oil recovery. Energy Procedia 63, 7685–7693. https://doi.org/10.1016/j.egypro.2014.11.802.

Daniilidis, A., Saeid, S., Doonechaly, N., 2021. The fault plane as the main fluid pathway: geothermal field development options under subsurface and operational uncertainty. Renew. Energy 171, 927–946. https://doi.org/10.1016/j.renene.2021.02.148.

de Hoop, S., Voskov, D., 2021. Fast and robust scheme for uncertainty quantification in naturally fractured reservoirs. In: SPE Reservoir Simulation Symposium Proceedings.

de Hoop, S., Voskov, D.V., Bertotti, G., Barnhoorn, A., 2022. An advanced discrete fracture methodology for fast, robust, and accurate simulation of energy production from complex fracture networks. Water Resour. Res. 58 (5). https://doi.org/10.1029/2021WR030743.

Donselaar, M.E., Groenenberg, R.M., Gilding, D.T., 2015. Reservoir geology and geothermal potential of the Delft Sandstone Member in the West Netherlands Basin. In: World Geothermal Congress. https://pangea.stanford.edu/ERE/db/WGC/papers/WGC/2015/12054.pdf.

Faust, C.R., Mercer, J.W., 1979a. Geothermal reservoir simulation: 1. Mathematical models for liquid and vapor-dominated hydrothermal systems. Water Resour. Res. 15, 23–30. https://doi.org/10.1029/WR015i001p00023.

Faust, C.R., Mercer, J.W., 1979b. Geothermal reservoir simulation: 2. Numerical solution techniques for liquid and vapordominated hydrothermal systems. Water Resour. Res. 15, 31–46. https://doi.org/10.1029/WR015i001p00031.

Flemisch, B., Berre, I., Boon, W., Fumagalli, A., Schwenck, N., Scotti, A., Stefansson, I., Tatomir, A., 2018. Benchmarks for single-phase flow in fractured porous media. Adv. Water Resour. 111, 239–258.

Franceschini, A., Gazzola, L., Ferronato, M., 2022. A scalable preconditioning framework for stabilized contact mechanics with hydraulically active fractures. J. Comput. Phys. 463. https://doi.org/10.1016/j.jcp.2022.111276.

Fussell, L.T., Fussell, D.D., 1979. Iterative technique for compositional reservoir models. Soc. Pet. Eng. AIME J. 19 (4), 211–220.

Gallyamov, E., Garipov, T., Voskov, D., Van den Hoek, P., 2018. Discrete fracture model for simulating waterflooding processes under fracturing conditions. Int. J. Numer. Anal. Methods Geomech. 42 (13), 1445–1470. https://doi.org/10.1002/nag.2797.

Garipov, T.T., Hui, M.H., 2019. Discrete fracture modeling approach for simulating coupled thermo-hydro-mechanical effects in fractured reservoirs. Int. J. Rock Mech. Min. Sci. 122, 104075. https://doi.org/10.1016/j.ijrmms.2019.104075.

Garipov, T.T., Karimi-Fard, M., Tchelepi, H.A., 2016. Discrete fracture model for coupled flow and geomechanics. Comput. Geosci. 20 (1), 149–160. https://doi.org/10.1007/s10596-015-9554-z.

Garipov, T.T., Tomin, P., Rin, R., Voskov, D.V., Tchelepi, H.A., 2018. Unified thermo-compositional-mechanical framework for reservoir simulation. Comput. Geosci. 22, 1039–1057. https://doi.org/10.1007/s10596-018-9737-5.

Geiger, S., Matthäl, S., 2014. What can we learn from high-resolution numerical simulations of single- and multi-phase fluid flow in fractured outcrop analogues? Geol. Soc. Spec. Publ. 374 (1), 125–144. https://doi.org/10.1144/SP374.8.

Gong, F., Guo, T., Sun, W., Li, Z., Yang, B., Chen, Y., Qu, Z., 2020. Evaluation of geothermal energy extraction in enhanced geothermal system (EGS) with multiple fracturing horizontal wells (MFHW). Renew. Energy 151, 1339–1351. https://doi.org/10.1016/j.renene.2019.11.134.

Griebel, M., Oeltz, D., Schweitzer, M.A., 2003. An algebraic multigrid method for linear elasticity. SIAM J. Sci. Comput. 25 (2), 385–407. https://doi.org/10.1137/S1064827502407810.

Gureghian, A.B., 1975. A study by the finite-element method of the influence of fractures in confined aquifers. SPE J. 15 (2), 181–191.

Hajibeygi, H., Karvounis, D., Jenny, P., 2011. A hierarchical fracture model for the iterative multiscale finite volume method. J. Comput. Phys. 230 (24), 8729–8743.

Helmig, R., et al., 1997. Multiphase Flow and Transport Processes in the Subsurface: A Contribution to the Modeling of Hydrosystems. Springer-Verlag.

Hibiki, T., Ishii, M., 2003. One-dimensional drift-flux model for two-phase flow in a large diameter pipe. Int. J. Heat Mass Transf. 46 (10), 1773–1790. https://doi.org/10.1016/S0017-9310(02)00473-8.

Holmes, J.A., Barkve, T., Lund, O., 1998. Application of a multisegment well model to simulate flow in advanced wells. In: Proceedings of the European Petroleum Conference, vol. 2, pp. 171–181.

Issa, R.I., Kempf, M.H.W., 2003. Simulation of slug flow in horizontal and nearly horizontal pipes with the two-fluid model. Int. J. Multiph. Flow 29 (1), 69–95. https://doi.org/10.1016/S0301-9322(02)00127-1.

James, A.L., Oldenburg, C.M., 1997. Linear and Monte Carlo uncertainty analysis for subsurface contaminant transport simulation. Water Resour. Res. 33 (11), 2495–2508. https://doi.org/10.1029/97WR01925.

Jansen, J.D., Douma, S.D., Brouwer, D.R., Van den Hof, P.M.J., Bosgra, O.H., Heemink, A.W., 2009. Closed-loop reservoir management. In: SPE Reservoir Simulation Symposium.

Jansen, J.D., Fonseca, R.M., Kahrobaei, S., Siraj, M.M., Van Essen, G.M., Van den Hof, P.M.J., 2014. The EGG model—a geological ensemble for reservoir simulation. Geosci. Data J. 1 (2), 192–195. https://doi.org/10.1002/gdj3.21.

Jiang, Y., 2008. Techniques for Modeling Complex Reservoirs and Advanced Wells (Ph.D. thesis). Stanford University.

Jung, A., Fenwick, D.H., Caers, J., 2013. Training image-based scenario modeling of fractured reservoirs for flow uncertainty quantification. Comput. Geosci. 17 (6), 1015–1031.

Karimi-Fard, M., Durlofsky, L.J., 2016. A general gridding, discretization, and coarsening methodology for modeling flow in porous formations with discrete geological features. Adv. Water Resour. 96, 354–372.

Karimi-Fard, M., Durlofsky, L.J., Aziz, K., 2004. An efficient discrete-fracture model applicable for general-purpose reservoir simulators. SPE J. 9 (2), 227–236. https://doi.org/10.2118/88812-PA.

Khait, M., Voskov, D., 2017. GPU-offloaded general purpose simulator for multiphase flow in porous media. In: SPE Reservoir Simulation Conference.

Khait, M., Voskov, D., 2018. Adaptive parameterization for solving of thermal/compositional nonlinear flow and transport with buoyancy. SPE J. 23, 522–534. https://doi.org/10.2118/182685-PA.

Khait, M., Voskov, D., 2021. A GPU-based integrated simulation framework for modelling of complex subsurface applications. In: SPE Reservoir Simulation Symposium Proceedings.

Kim, J., Tchelepi, H.A., Juanes, R., 2011a. Stability, accuracy, and efficiency of sequential methods for coupled flow and geomechanics. SPE J. 16 (2), 249–262. https://doi.org/10.2118/119084-PA.

Kim, J., Tchelepi, H.A., Juanes, R., 2011b. Stability and convergence of sequential methods for coupled flow and geomechanics: fixed-stress and fixed-strain splits. Comput. Methods Appl. Mech. Eng. 200 (13–16), 1591–1606. https://doi.org/10.1016/j.cma.2010.12.022.

Koohbor, B., Fahs, M., Hoteit, H., Doummar, J., Younes, A., Belfort, B., 2020. An advanced discrete fracture model for variably saturated flow in fractured porous media. Adv. Water Resour. 140, 103602.

Krevor, S.C.M., Pini, R., Li, B., Benson, S.M., 2011. Capillary heterogeneity trapping of CO_2 in a sandstone rock at reservoir conditions. Geophys. Res. Lett. 38. https://doi.org/10.1029/2011GL048239.

Krogstad, S., Lie, K.-A., Møyner, O., Nilsen, H.M., Raynaud, X., Skaflestad, B., et al., 2015. MRST-AD—an open-source framework for rapid prototyping and evaluation of reservoir simulation problems. In: SPE Reservoir Simulation Symposium.

Li, L., Lee, S.H., 2008. Efficient field-scale simulation of black oil in a naturally fractured reservoir through discrete fracture networks and homogenized media. SPE Reserv. Eval. Eng. 11 (4), 750–758.

Li, X., Li, D., 2019. A numerical procedure for unsaturated seepage analysis in rock mass containing fracture networks and drainage holes. J. Hydrol. 574, 23–34.

Livescu, S., Durlofsky, L.J., Aziz, K., 2008. A semianalytical thermal multiphase wellbore flow model for use in reservoir simulation. In: Proceedings—SPE Annual Technical Conference and Exhibition, vol. 4, pp. 2345–2360.

Livescu, S., Durlofsky, L.J., Aziz, K., Ginestra, J.C., 2010. A fully-coupled thermal multiphase wellbore flow model for use in reservoir simulation. J. Pet. Sci. Eng. 71 (3–4), 138–146. https://doi.org/10.1016/j.petrol.2009.11.022.

Lyu, X., Khait, M., Voskov, D., 2021. Operator-based linearization approach for modeling of multiphase flow with buoyancy and capillarity. SPE J., 1–18. https://doi.org/10.2118/205378-PA.

Major, M., Daniilidis, A., Hansen, T.M., Khait, M., Voskov, D., 2023. Influence of process-based, stochastic and deterministic methods for representing heterogeneity in fluvial geothermal systems. Geothermics 109. https://doi.org/10.1016/j.geothermics.2023.102651.

Mallison, B.T., Hui, M.-H., Narr, W., 2010. Practical gridding algorithms for discrete fracture modeling workflows. In: ECMOR XII—12th European Conference on the Mathematics of Oil Recovery, European Association of Geoscientists & Engineers, p. cp-163.

Miller, C.T., Christakos, G., Imhoff, P.T., McBride, J.F., Pedit, J.A., Trangenstein, J.A., 1998. Multiphase flow and transport modeling in heterogeneous porous media: challenges and approaches. Adv. Water Resour. 21, 77–120. https://doi.org/10.1016/S0309-1708(96)00036-X.

Moinfar, A., Narr, W., Hui, M.-H., Mallison, B.T., Lee, S.H., 2011. Comparison of discrete-fracture and dual-permeability models for multiphase flow in naturally fractured reservoirs. In: SPE Reservoir Simulation Symposium.

Moncorgé, A., Tchelepi, H.A., 2009. Stability criteria for thermal adaptive implicit compositional flows. SPE J. 14 (2), 311–322. https://doi.org/10.2118/111610-PA.

Mottaghy, D., Pechnig, R., Vogt, C., 2011. The geothermal project Den Haag: 3D numerical models for temperature prediction and reservoir simulation. Geothermics 40 (3), 199–210. https://doi.org/10.1016/j.geothermics.2011.07.001.

Mustapha, H., Mustapha, K., 2007. A new approach to simulating flow in discrete fracture networks with an optimized mesh. SIAM J. Sci. Comput. 29 (4), 1439–1459.

Nejadi, S., Trivedi, J.J., Leung, J., 2017. History matching and uncertainty quantification of discrete fracture network models in fractured reservoirs. J. Pet. Sci. Eng. 152, 21–32.

Novikov, A., Voskov, D., Khait, M., Hajibeygi, H., Jansen, J.D., 2022. A scalable collocated finite volume scheme for simulation of induced fault slip. J. Comput. Phys. 469. https://doi.org/10.1016/j.jcp.2022.111598.

Orr, F., 2007. Theory of Gas Injection Processes. Tie-Line Publications, Holte.

O'Sullivan, M.J., O'Sullivan, J.P., 2016. Reservoir modeling and simulation for geothermal resource characterization and evaluation. In: Geothermal Power Generation: Developments and Innovation, pp. 165–199.

Pan, L., Oldenburg, C.M., 2014. T2Well—an integrated wellbore-reservoir simulator. Comput. Geosci. 65, 46–55. https://doi.org/10.1016/j.cageo.2013.06.005.

Peaceman, D.W., 1990. Interpretation of wellblock pressures in numerical reservoir simulation. Part 3. Off-center and multiple wells within a wellblock. SPE Reserv. Eng. (Soc. Pet. Eng.) 5 (2), 16976. https://doi.org/10.2118/16976-pa.

Pinar, A., Heath, M.T., 1999. Improving performance of sparse matrix-vector multiplication. In: SC'99: Proceedings of the 1999 ACM/IEEE Conference on Supercomputing, p. 30.

Pini, R., Krevor, S.C.M., Benson, S.M., 2012. Capillary pressure and heterogeneity for the CO_2/water system in sandstone rocks at reservoir conditions. Adv. Water Resour. 38, 48–59. https://doi.org/10.1016/j.advwatres.2011.12.007.

Poulsen, S.E., Balling, N., Nielsen, S.B., 2015. A parametric study of the thermal recharge of low enthalpy geothermal reservoirs. Geothermics 53, 464–478. https://doi.org/10.1016/j.geothermics.2014.08.003.

Pour, K.M., Voskov, D., Bruhn, D., 2023. Nonlinear solver based on trust region approximation for CO_2 utilization and storage in subsurface reservoir. Geoenergy Sci. Eng. 225. https://doi.org/10.1016/j.geoen.2023.211698.

Pruess, K., Zerzan, J.M., Schroeder, R.C., Witherspoon, P.A., 1979. Description of the three-dimensional two-phase simulator shaft78 for use in geothermal reservoir studies. In: SPE Reservoir Simulation Symposium Proceedings, 1979-January, pp. 253–261. vol.

Pruess, K., Calore, C., Celati, R., Wu, Y.S., 1987. An analytical solution for heat transfer at a boiling front moving through a porous medium. Int. J. Heat Mass Transf. 30 (12), 2595–2602. https://doi.org/10.1016/0017-9310(87)90140-2.

Pruess, K., Oldenburg, C.M., Moridis, G.J., 1999. TOUGH2 User's Guide. http://tough.lbl.gov/assets/docs/TOUGH2_V2_Users_Guide.pdf.

Reichenberger, V., Jakobs, H., Bastian, P., Helmig, R., 2006. A mixed-dimensional finite volume method for two-phase flow in fractured porous media. Adv. Water Resour. 29 (7), 1020–1036.

Saad, Y., 2003. Iterative Methods for Sparse Linear Systems, second ed. Society for Industrial and Applied Mathematics, Philadelphia, PA.

Saeid, S., Al-Khoury, R., Nick, H.M., Hicks, M.A., 2015. A prototype design model for deep low-enthalpy hydrothermal systems. Renew. Energy 77, 408–422. https://doi.org/10.1016/j.renene.2014.12.018.

Saeid, S., Wang, Y., Daniilidis, A., Khait, M., Voskov, D., Bruhn, D., 2020. Lifetime and energy prediction of geothermal systems: uncertainty analysis in highly heterogeneous geothermal reservoirs (Netherlands). In: World Geothermal Congress. https://pangea.stanford.edu/ERE/db/WGC/papers/WGC/2020/22157.pdf.

Scheidt, C., Caers, J., 2009. Uncertainty quantification in reservoir performance using distances and kernel methods—application to a West Africa deepwater turbidite reservoir. SPE J. 14 (4), 680–692. https://doi.org/10.2118/118740-PA.

Schlumberger, 2011. ECLIPSE Reference Manual 2011.2. Schlumberger, Houston.

Settari, A., Aziz, K., 1972. Use of irregular grid in reservoir simulation. Soc. Pet. Eng. J. 12 (2), 103–114. https://doi.org/10.2118/3174-pa.

Shetty, S., Voskov, D., Bruhn, D.F., 2018. Numerical strategy for uncertainty quantification in low enthalpy geothermal projects. In: Workshop on Geothermal Reservoir Engineering. https://pangea.stanford.edu/ERE/pdf/IGAstandard/SGW/2018/Shetty.pdf.

Shukla, R., Ranjith, P., Haque, A., Choi, X., 2010. A review of studies on CO_2 sequestration and caprock integrity. Fuel 89, 2651–2664. https://doi.org/10.1016/j.fuel.2010.05.012.

Spooner, V.E., Geiger, S., Arnold, D., 2021. Dual-porosity flow diagnostics for spontaneous imbibition in naturally fractured reservoirs. Water Resour. Res., e2020WR027775.

Stüben, K., 1983. Algebraic multigrid (AMG): experiences and comparisons. Appl. Math. Comput. 13 (3–4), 419–451.

Terekhov, K.M., 2020. Cell-centered finite-volume method for heterogeneous anisotropic poromechanics problem. J. Comput. Appl. Math. 365. https://doi.org/10.1016/j.cam.2019.112357.

Tonkin, R.A., O'Sullivan, M.J., O'Sullivan, J.P., 2021. A review of mathematical models for geothermal wellbore simulation. Geothermics 97. https://doi.org/10.1016/j.geothermics.2021.102255.

Van Der Kooij, R., Wang, Y., Daniilidis, A., Voskov, D., 2020. Performance influences of a deep high temperature fractured geothermal system. In: 1st Geoscience and Engineering in Energy Transition Conference, GET 2020.

Vanden, K.J., Orkwis, P.D., 1996. Comparison of numerical and analytical Jacobians. AIAA J. 34 (6), 1125–1129.

Veldkamp, J.G., Pluymaekers, M.P.D., van Wees, J.D.A.M., 2015. DoubletCalc 2D 1.0 User Manual.

Vogt, C., Iwanowski-Strahser, K., Marquart, G., Arnold, J., Mottaghy, D., Pechnig, R., Gnjezda, D., Clauser, C., 2013. Modeling contribution to risk assessment of thermal production power for geothermal reservoirs. Renew. Energy 53, 230–241. https://doi.org/10.1016/j.renene.2012.11.026.

Voskov, D.V., 2011. Modeling of multi-component flow in porous media with arbitrary phase changes. In: Proceedings—SPE Annual Technical Conference and Exhibition, vol. 4, pp. 2896–2910.

Voskov, D.V., 2017. Operator-based linearization approach for modeling of multiphase multi-component flow in porous media. J. Comput. Phys. 337, 275–288. https://doi.org/10.1016/j.jcp.2017.02.041.

Voskov, D.V., Tchelepi, H.A., 2012. Comparison of nonlinear formulations for two-phase multi-component EoS based simulation. J. Pet. Sci. Eng. 82–83, 101–111. https://doi.org/10.1016/j.petrol.2011.10.012.

Wagner, W., Kretzschmar, H.-J., 2008. IAPWS Industrial Formulation 1997 for the thermodynamic properties of water and steam. In: International Steam Tables. Springer, Berlin, Heidelberg. https://doi.org/10.1007/978-3-540-74234-0_3.

Wallis, J., 1983. Incomplete Gaussian elimination as a preconditioning for generalized conjugate gradient acceleration. In: Proceedings of 7th SPE Symposium on Reservoir Simulation.

Wallis, J., Kendall, R., Little, T., Nolen, J., 1985. Constrained residual acceleration of conjugate residual methods. In: Proceedings of 8th SPE Symposium on Reservoir Simulation.

Wang, Y., 2015. A Stability Criterion for the Negative Compressibility Problem in Geothermal Simulation and Discrete Modeling of Failure in Oil Shale Pyrolysis Process (Master's thesis). Stanford University. https://pangea.stanford.edu.

Wang, Y., Voskov, D., 2022. High-enthalpy geothermal simulation with continuous localization in physics. Mathematics 10 (22). https://doi.org/10.3390/math10224328.

Wang, Y., de Hoop, S., Voskov, D., Bruhn, D., Bertotti, G., 2021a. Modeling of multiphase mass and heat transfer in fractured high-enthalpy geothermal systems with advanced discrete fracture methodology. Adv. Water Resour. 154. https://doi.org/10.1016/j.advwatres.2021.103985.

Wang, Y., Voskov, D., Khait, M., Saeid, S., Bruhn, D., 2021b. Influential factors on the development of a low-enthalpy geothermal reservoir: a sensitivity study of a realistic field. Renew. Energy 179, 641–651. https://doi.org/10.1016/j.renene.2021.07.017.

Wang, Y., Voskov, D., Daniilidis, A., Khait, M., Saeid, S., Bruhn, D., 2023. Uncertainty quantification in a heterogeneous fluvial sandstone reservoir using GPU-based Monte Carlo simulation. Geothermics 114. https://doi.org/10.1016/j.geothermics.2023.102773.

Warren, J.E., Root, P.J., 1963. The behavior of naturally fractured reservoirs. SPE J. 3 (3), 245–255.

White, J.A., Castelletto, N., Klevtsov, S., Bui, Q.M., Osei-Kuffuor, D., Tchelepi, H.A., 2019. A two-stage preconditioner for multiphase poromechanics in reservoir simulation. Comput. Methods Appl. Mech. Eng. 357. https://doi.org/10.1016/j.cma.2019.112575.

Willems, C.J.L., Nick, H.M., 2019. Towards optimisation of geothermal heat recovery: an example from the West Netherlands Basin. Appl. Energy 247, 582–593. https://doi.org/10.1016/j.apenergy.2019.04.083.

Willems, C.J.L., Goense, T., Nick, H.M., Bruhn, D.F., 2016. The relation between well spacing and net present value in fluvial hot sedimentary aquifer geothermal doublets; a West Netherlands Basin case study. In: Workshop on Geothermal Reservoir Engineering. https://pangea.stanford.edu/ERE/pdf/IGAstandard/SGW/2016/Willems.pdf.

Willems, C.J.L., Nick, H.M., Weltje, G.J., Bruhn, D.F., 2017. An evaluation of interferences in heat production from low enthalpy geothermal doublets systems. Energy 135, 500–512. https://doi.org/10.1016/j.energy.2017.06.129.

Willems, C.J.L., Vondrak, A., Mijnlieff, H.F., Donselaar, M.E., van Kempen, B.M.M., 2020. Geology of the upper jurassic to lower cretaceous geothermal aquifers in the West Netherlands Basin—an overview. Neth. J. Geosci. 99, 1–13. https://doi.org/10.1017/njg.2020.1.

Wong, Z.Y., Horne, R., Voskov, D., 2015. A geothermal reservoir simulator in AD-GPRS. In: World Geothermal Congress. https://pangea.stanford.edu/ERE/db/WGC/papers/WGC/2015/22043.pdf.

Wong, Z.Y., Rin, R., Tchelepi, H., Horne, R., 2017. Comparison of fully implicit and sequential implicit formulation for geothermal reservoir simulations. In: Workshop on Geothermal Reservoir Engineering. https://pdfs.semanticscholar.org/b31c/873caeee5d759f4e1ea498c5f05ec897b1ab.pdf.

Wong, Z.Y., Horne, R.N., Tchelepi, H.A., 2018. Sequential implicit nonlinear solver for geothermal simulation. J. Comput. Phys. 368, 236–253. https://doi.org/10.1016/j.jcp.2018.04.043.

Wong, D.L.Y., Doster, F., Geiger, S., Francot, E., Gouth, F., 2020. Fluid flow characterization framework for naturally fractured reservoirs using small-scale fully explicit models. Transp. Porous Media 134 (2), 399–434.

Xu, T., Apps, J.A., Pruess, K., Yamamoto, H., 2007. Numerical modeling of injection and mineral trapping of CO_2 with H_2S and SO_2 in a sandstone formation. Chem. Geol. 242, 319–346. https://doi.org/10.1016/j.chemgeo.2007.03.022.

Xu, T., Spycher, N., Sonnenthal, E., Zhang, G., Zheng, L., Pruess, K., 2011. Toughreact version 2.0: a simulator for subsurface reactive transport under non-isothermal multiphase flow conditions. Comput. Geosci. 37 (6), 763–774. https://doi.org/10.1016/j.cageo.2010.10.007.

Younis, R., 2011. Modern Advances in Software and Solution Algorithms for Reservoir Simulation (Ph.D. thesis). Stanford University.

Zaydullin, R., Voskov, D.V., James, S.C., Henley, H., Lucia, A., 2014. Fully compositional and thermal reservoir simulation. Comput. Chem. Eng. 63, 51–65. https://doi.org/10.1016/j.compchemeng.2013.12.008.

Zhao, Y., Borja, R.I., 2020. A continuum framework for coupled solid deformation-fluid flow through anisotropic elastoplastic porous media. Comput. Methods Appl. Mech. Eng. 369, 113225. https://doi.org/10.1016/j.cma.2020.113225.

Zhou, Y., Tchelepi, H.A., Mallison, B.T., 2011. Automatic differentiation framework for compositional simulation on unstructured grids with multi-point discretization schemes. In: SPE Reservoir Simulation Symposium. SPE 141592.

CHAPTER

11

Artificial intelligence in the geothermal energy systems

Pejman Shoeibi Omrani[a,b], *Jonah Poort*[a], *and Sadegh Shahmohammadi*[c]

[a]Heat Transfer and Fluid Dynamics Department, TNO, The Hague, The Netherlands
[b]Environmental Technology, Wageningen University and Research, Wageningen, The Netherlands [c]Healthy Living & Work, Netherlands Organisation for Applied Scientific Research (TNO), The Hague, The Netherlands

1 Introduction

Harnessing geothermal energy involves intricate processes such as exploration, resource characterization, drilling, production, operation, and maintenance. Optimizing numerous parameters and factors within this complex landscape is essential to enhance the efficiency, safety, and cost-effectiveness of production of geothermal energy. Despite the wealth of data generated in every stage of geothermal energy production, its full potential still needs to be utilized in process optimization and design innovation.

In addressing this untapped potential, integrating artificial intelligence and data analytics emerges as a pivotal force in exploring, developing, and operating geothermal production systems. The abundant data generated throughout the geothermal energy's life cycle can be harnessed to unlock valuable insights. Through comprehensive procedures to unlock the value from the data, different stakeholders in the development and exploitation of geothermal energy can pinpoint new resources, fine-tune production processes, and anticipate potential equipment failures. Machine learning models serve as vital tools in this endeavor, empowering organizations to make informed decisions regarding resource development, production strategies, and maintenance protocols. These models leverage the data to create predictive algorithms that optimize performance and prevent downtime, thereby enhancing the overall reliability of geothermal production systems.

Furthermore, the expanding realm of natural language processing (NLP) and utilizing large language models present significant opportunities within the geothermal industry.

Geothermal Energy Engineering
https://doi.org/10.1016/B978-0-443-21662-6.00004-9

Copyright © 2025 Elsevier Inc. All rights are reserved, including those for text and data mining, AI training, and similar technologies.

These technologies facilitate the extraction of relevant information and the automation of document processing, including logs, publications, and textual data. The synergy of data analytics, machine learning, and artificial intelligence technologies can accelerate the geothermal industry toward more outstanding advancements, ensuring a more robust and resilient energy production ecosystem.

This chapter introduces machine learning, laying the groundwork for a further deep dive into various machine learning models. We then explore practical examples of machine learning and deep learning applications across different sectors. A focused examination of AI's role in geothermal systems follows, highlighting specific applications. This chapter concludes by examining the application of large language models, illustrating their transformative potential in handling textual data in the geothermal sector. This structured progression offers readers a concise yet comprehensive insight into the dynamic interplay between AI and machine learning in geothermal production applications.

2 Introduction to artificial intelligence, data analytics, and machine learning

The world of artificial intelligence (AI) is enormous and, at first sight, seems hard to get a grip on. In this section, an overview of some of the most used and well-established machine learning approaches and model types will be provided. Initially, we will first define the different terms the reader will typically encounter when discussing AI, machine learning, deep learning, and data analytics.

2.1 Definitions

There are a lot of terms associated with artificial intelligence, such as machine learning, deep learning, natural language processing, and data analytics. In addition, many of these terms show considerable connection to each other. Therefore, the aim of this section is to provide the definitions of these terms and explain their relationships. To help with the discussion, Fig. 1 provides a high-level schematic overview of the relations between the terms covered. Since most of these terms do not have a single unified definition used by the entire industry, definitions are given from multiple sources.

The first term to define is artificial intelligence, as it encompasses and/or is closely related to all other definitions to be discussed. As defined by Google, AI is "[...] a field of science concerned with building computers and machines that can reason, learn, and act in such a way that would normally require human intelligence or that involves data whose scale exceeds what humans can analyze" (Google, n.d.-a). Similarly, the Encyclopedia Britannica defines it as: "the ability of a digital computer or computer-controlled robot to perform tasks commonly associated with intelligent beings" (Copeland, n.d.). Both of these definitions emphasize the main aim of AI, which is to provide ways through which computers can solve problems that classically required human intervention. A good example of this is image recognition. Differentiating between an image of a dog or a cat is a trivial task for a human; something even a kid can do. However, writing a computer algorithm that can do the same has historically been a challenging task, and only with the advent of AI and machine learning computer systems have been able to do so.

In addition, a fundamental aspect of AI covered in the definition by Google is the scale. Not only does AI aim to mimic human problem-solving skills, but it also aims to surpass it, allowing for processing of amounts of information no human could reasonably expect or want to handle. It is possible to hire a full-time employee to transcribe long audio files, but it is more efficient to develop an AI system that could do so automatically on much larger scales. Examples of AI subfields, as defined by the European Commission's High-Level Expert Group on Artificial Intelligence, are machine learning (learning relationships from data), machine reasoning (optimization, planning, scheduling, search, and reasoning), and robotics (control, perception, sensing, actuating, and the integration of the digital and physical world; The European Commission's High-Level Expert Group On Artificial Intelligence, 2018).

The part of AI that is primarily focused on learning from data is that of machine learning (ML). As Fig. 1 shows, ML is a subset of AI and includes the bulk of the techniques and models discussed in this chapter, such as supervised, unsupervised, semisupervised, and reinforcement learning (Bishop, 2006). As IBM defines it: "machine learning is a branch of artificial intelligence (AI) and computer science which focuses on the use of data and algorithms to imitate the way that humans learn [...]" (IBM, n.d.-a; [1]). ML thus aims to perform a given task (prediction, clustering, control, and generation) by learning and generalizing from the example data of that task. This sets it apart from other fields within AI (machine reasoning and robotics) that are not inherently related to data-driven learning. The power of ML is in the fact that the complexities of the processes behind a given task are implicitly learned from the example data and do not need to be encoded explicitly. On the other hand, when example data are scarce, it can be difficult to create ML models that generalize well to new instances of the same task.

Deep learning (DL) is a further subset of the ML field and refers to machine learning that uses artificial neural network (ANN) models and architectures. The term "deep" specifically refers to the number of hidden layers in the architecture of the ANN model. IBM classifies any neural network with three or more layers as deep (IBM, n.d.-b). In a paper on deep learning, Jürgen Schmidhuber discusses the topic of the cutoff between shallow and deep networks with multiple experts and eventually concludes that no consensus on such a cutoff exists, yet that any network with more than 10 layers "requires very deep learning" (Schmidhuber, 2015). Currently, many modern machine learning models that are referred to as deep (e.g., ChatGPT) go far beyond the cutoff points mentioned previously, having more than a hundred layers and billions to trillions of parameters.

Natural language processing (NLP) is a field within AI and computer science dedicated to the (automated) processing of text and speech. Historically, NLP was largely separate from ML and focused on computational methods for processing text/speech such as regular expressions, n-grams, and Markov models (Jurafsky and Martin, 2023). However, as AI technologies got more powerful, they started to become one of the main methods used for NLP tasks. By now, they are almost inseparable from each other, with the most advanced method for NLP almost exclusively being deep learning models, typically referred to as large language models (LLM). This strong connection between ML and NLP is further underlined by the fact that both Google and Amazon include the term machine learning in their definitions of NLP (Google, n.d.-b; Amazon Web Services, n.d.-a). To provide a more in-depth discussion on the potential for NLP in the geothermal sector, the final section of this chapter is entirely dedicated to the different uses of LLMs.

Finally, a term oftentimes used together with artificial intelligence is that of data analytics (DA). As the name implies, data analytics includes all techniques that focus on analyzing datasets in order to gain further insights into the data. According to Amazon, data analytics "converts raw data into actionable insights. It includes a range of tools, technologies, and processes used to find trends and solve problems by using data" (Amazon Web Services, n.d.-b). The inherent multidisciplinary nature of data analytics is further underlined by Mathar in the book of *Fundamentals of Data Analytics* in which they describe data analytics as "a fast developing interdisciplinary branch of science combining methods from exploratory statistics, algorithm and information theory to reveal structures in large data sets" (Mathar et al., 2020). Thus, while AI is nowadays hard to separate from data analytics, data analytics includes many more techniques beyond AI, from simple statistics to visualization methods.

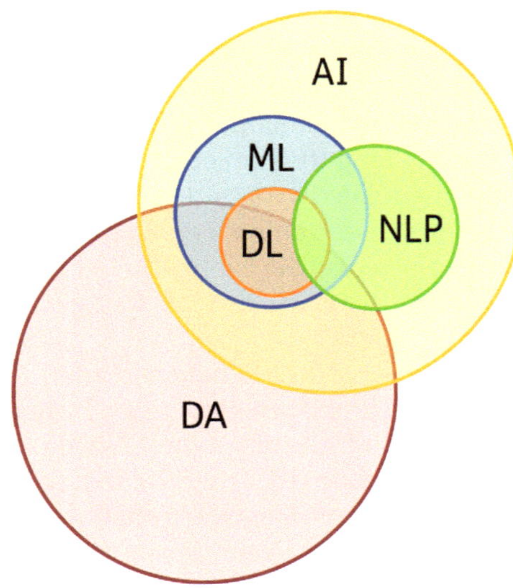

FIG. 1 Schematic depiction of the relationships between data analytics (DA), artificial intelligence (AI), machine learning (ML), deep learning (DL), and natural language processing (NLP).

3 Different types of machine learning techniques

In this section, we give a short introduction to the main techniques used in the field of machine learning, including supervised, unsupervised, semisupervised, active, reinforcement, and transfer learnings.

3.1 Supervised, unsupervised, semi-supervised, and active learning

The main aspect that separates the majority of machine learning techniques is on how the learning process is approached. Primarily, the distinction is made between "labeled" and "unlabeled" data (Bishop, 2006). Here, labeled data describes any dataset in which samples have been given a target (e.g., the label) that has to be predicted using the remaining "features" (descriptors) of the sample. When such targets/labels are not present, the dataset is thus unlabeled. The other subset is for the dataset which are partially labeled. Fig. 2 shows a schematic depiction of the differences between supervised, unsupervised, semisupervised, and active learnings which all will be further explained in this section.

Typically, labeled datasets consist of inputs (the features) and outputs (the labels), and the machine learning goal is to find the relationship between input and output. For instance, using the features of wellhead temperature, wellhead pressure, and valve opening, one might want to predict the label flowrate. Problems in which the dataset is labeled and the goal is to predict the labels using the features are called "supervised." The two main types of supervised learning methods are regression (predicting continuous output values) and classification (predicting classes). A simplified supervised classification problem is schematically depicted in Fig. 2A.

When a dataset is unlabeled, one is not able to learn direct input–output relations. Instead, "unsupervised" learning techniques are used to extract relationships between the features themselves in order to provide further insights into the clusters formed within the dataset. Examples of unsupervised learning include dimension reduction (reducing the size/complexity of a dataset to make it easier to process/interpret) and clustering (dividing the data into groups of high similarity; see Fig. 2B). For instance, based on unlabeled data coming from a large collection of electrical submersible pumps (ESPs), one could use unsupervised clustering to divide these assets into groups that show similar performance or behavior to help in diagnostics of potential issues. In many cases, unsupervised learning can be used as a preprocessing step before using a supervised learning method.

The other method is semisupervised learning, which combines the two methods to leverage information from datasets that are only partially labeled. For instance, in self-training (Amini et al., 2023), a model is trained on limited labeled data before being used to predict new labels for the unlabeled samples, which are then added to the labeled dataset and used to train a new model. Another approach to semisupervised learning is to use an unsupervised method to cluster the unlabeled data into groups, before using the limited labeled data to generalize the labels to the entire clusters. For instance, if the majority of the labeled samples in a particular cluster have one label, the remainder of the cluster can be given the same label (see Fig. 2C). This newly labeled data can then be used for supervised learning. Semisupervised learning does come with certain risks, particularly in generalizing the few labels to the unlabeled data and in making certain assumptions on the unlabeled data structure/distribution (Zhu, 2005). In addition, partially labeled datasets often contain natural biases, as certain samples might be easier to label than others, such that labels are primarily available for specific groups of samples. This can potentially lead to incorrect generalizations of the available labels to the unlabeled instances. Thus, care must be taken when applying semisupervised methods to ensure model results can be trusted.

Closely related to the topic of semisupervised learning is "active learning" (Settles, 2009). In active learning, statistical and/or optimization methods are used to define where to gather or label additional data points (see Fig. 2D). The underlying assumption for a validated model is the possibility to generate new data or labels during the training of the ML model, which might not be always possible. But in the case that it is possible, active learning can help in making the data generation process more efficient.

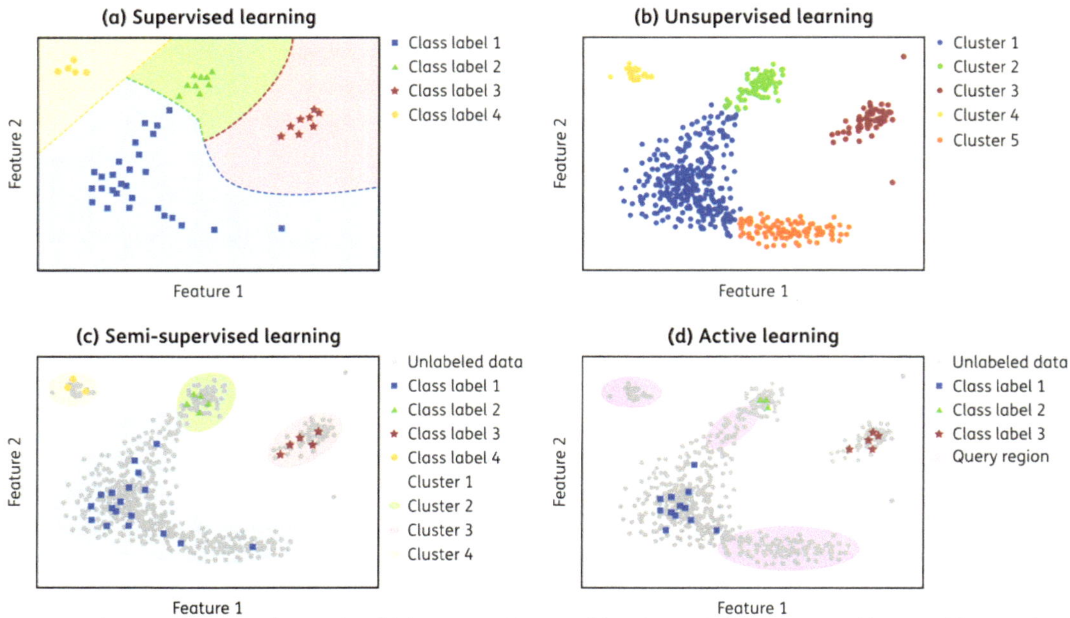

FIG. 2 Schematic depiction of examples of (A) supervised learning (classification), (B) unsupervised learning (clustering), (C) semisupervised learning (clustering with limited labels), and (D) active learning.

3.2 Reinforcement learning

A distinct approach from those mentioned earlier is "reinforcement learning" (RL; Sutton and Barto, 2018), which is employed to address issues related to the control or design of complex, dynamic, and non-linear systems. A reinforcement learning method consist of an "agent" which learns to achieve a certain goal within an "environment" (i.e., physical system) by taking "actions" (i.e., changing control settings) that can directly influence the "state" of this environment (Fig. 3). Each time the agent takes a certain action, the state of the environment transitions into a new one. After each state transition, the agent receives a "reward," a measure of how good the action was to take in the previous state. Over time, as the agent takes more and more actions in the environment and receives their corresponding rewards, the agent learns the optimal "policy" to take, the set of subsequent actions it should make in order to reach the desired goal. One of the major advantages of reinforcement learning is its ability to learn optimal policies through interaction with the environment, allowing it to handle complex and dynamic systems without requiring explicit programming for every possible scenario.

Another benefit of RL over conventional control methods is that it is largely unbiased in how it approaches a certain problem. As such, RL agents can often come up with novel and out-of-the-box solutions unbiased by human thinking. In addition, RL agents are great at

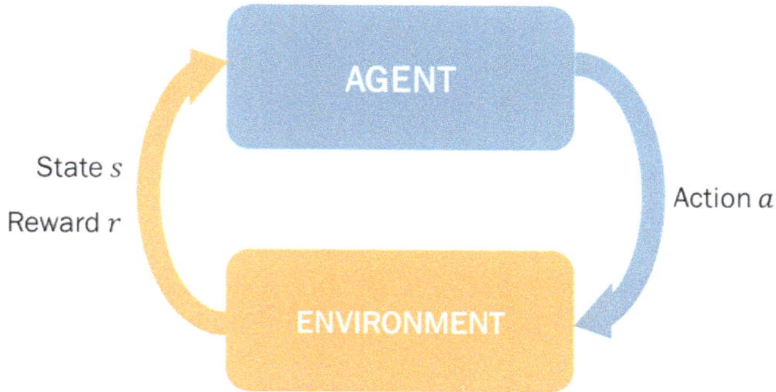

FIG. 3 Schematic of the reinforcement learning agent interacting with the environment through actions, states, and reward.

dealing with delayed rewards, being able to make decisions that do not bring immediate benefits but the cumulative reward is maximized. Finally, while the agent interacts with the system, it also indirectly learns how the system behaves (e.g., which actions can lead to which consequences). Therefore, RL agents are robust at dealing with dynamic systems and can quickly react to changes in the environment. On the other hand, since RL agents learn through trial-and-error, the exploration phase of the agent can be costly and require a lot of iteration and interaction with the environment. RL agents especially suffer from the curse of dimensionality, as they require orders of magnitude more interactions with the environment as the number of actions, control parameters, or states increases. In addition, the right formulation of the problem, in terms of state, action, and rewards can have a major impact on the performance and accuracy of the reinforcement learning.

3.3 Transfer learning

Transfer learning (Pan and Yang, 2009; Weiss et al., 2016) is technically not a training technique in itself, but a way to, as the name implies, transfer learnings (model "hyper" parameters) from one ML model used in a certain application to another one with a similar application. It can thus be applied to supervised, unsupervised, and reinforcement learnings. Transfer learning leverages knowledge from pretrained models to improve performance on related but distinct tasks/applications, reducing the need for extensive training data and computation. Transfer learning is most often used when dealing with problems for which little data are available, but where there are a lot of data available for a different but similar problem or domain. By first training a model on the problem for which a lot of data are available, a general base (source) model can be trained. Then, using the limited data of the original target problem, this source model can be fine-tuned by retraining (part of) the existing model with the new data (Fig. 4). Transfer learning is often done in the field of image recognition, where the initial layers of large convolutional neural networks typically extract underlying structures/features of image data that are shared between all types of images (i.e., lines, shapes, and colors). This feature extraction can be done by training on very large sets of all types of images, even those that might not be of interest to the intended task. Then, once in place, these initial layers can be frozen, and only the last few layers which are used for predicting what the image is can be updated for specific tasks with less labeled data. Another way of employing transfer learning is to move from synthetic model-based data to field data. For instance, one can use a high-fidelity model to generate arbitrarily large datasets for training the initial source model and then fine-tune this model further with a field dataset (which is typically much more limited in size). In many cases, this can help create transfer models that outperform those trained solely on the limited field data (Poort and Bazuin, 2023).

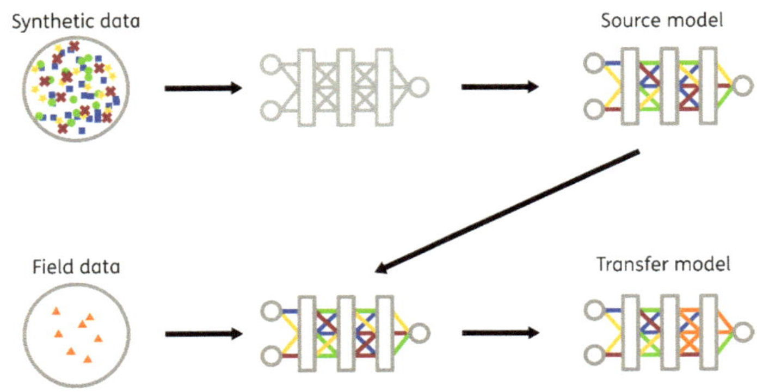

FIG. 4 Schematic depiction of the transfer learning process. An initial source model is trained on one large dataset (for instance, one synthetically generated) and then used as the initial model to be fine-tuned further using the limited target task data (e.g., field data).

4 Examples of machine learning algorithms

In this section, a selection of some of the most used machine learning algorithms will be briefly explained, including the k-means, k-nearest neighbors, random forest, and artificial neural network algorithms.

4.1 K-means

The k-means algorithms are a widely used machine learning methods for unsupervised clustering tasks (Bishop, 2006). The main goal of k-means is to divide an unlabeled dataset into k groups of datapoints with similar features. This is done by dividing a dataset into k partitions such that the total variance within each partition is minimized. These partitions are made by defining k centroids in the data and dividing the data based on their distance to these centroids, as shown in Fig. 5. To find the optimal clusters, the k-means method starts with randomly assigning k centroids within the dataset, before iteratively updating their locations to further minimize the within cluster variance. Once the position of the centroids remains stable over multiple iterations, the process is stopped, and the final clusters are made based on the last centroid locations. Since the method aims to optimize for the variance, the resulting clusters often end up being similar in size, even when this is not the case for the ground truth. The method can be sensitive to the initial placement of the centroids and determining the optimal number of clusters can also be tricky.

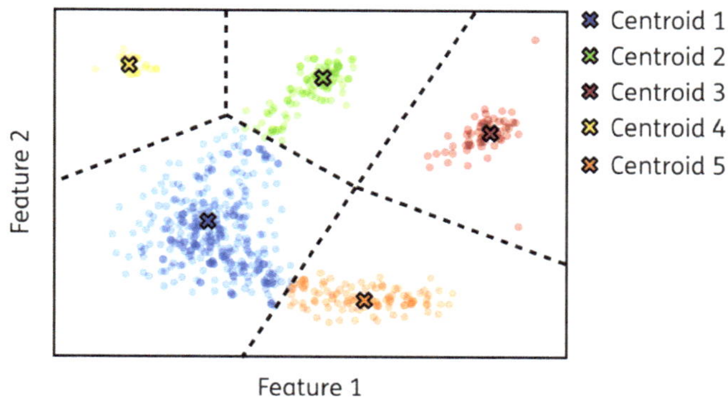

FIG. 5 Visual depiction of the k-means clustering algorithm in which all data points have been divided into partitions based on their proximity to the nearest cluster centroid.

4.2 K-nearest neighbors

The k-nearest neighbor (K–NN) algorithm is used for supervised learning, both in classification and regression. In a k-nearest neighbor algorithm, the label of an unlabeled data sample is predicted by looking at the labels of the k most similar labeled datapoints to that sample. This is shown visually in Fig. 6 for a classification problem. While K-NN is a simple algorithm in principle and at the minimum only requires one parameter to be defined (the number k), it can become cumbersome to use with large datasets, as each time a prediction is made, the entire labeled dataset is processed to find the nearest neighbors to the new sample. This also means that the entire labeled dataset has to be stored in memory, making it a computationally intensive method especially for larger datasets.

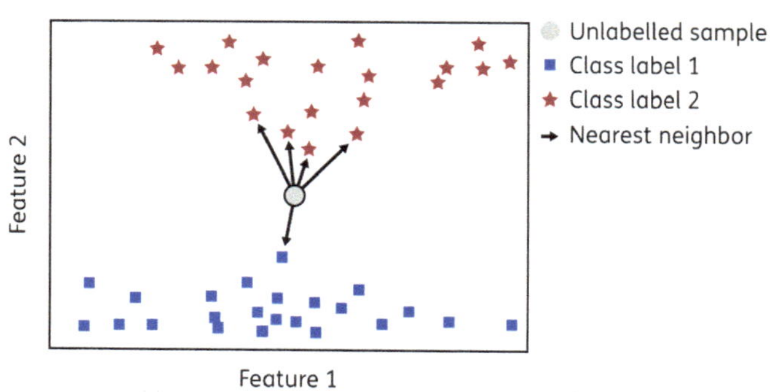

FIG. 6 Visual depiction of the k-nearest neighbor algorithm with $k = 5$. As the majority of the nearest labeled neighbors to the unlabeled sample are *red stars*, it will also be labeled with a *red star*.

4.3 Random forest

The random forest algorithm is a so-called ensemble method, in which a large number of individual models are trained on the same data, and the average (or majority) of their predictions are used as the final overall model prediction. Specifically, the random forest ensemble makes use of a large collection of decision tree models. The aim of a decision tree is to create a set of subsequent divisions in the data that, when applied, gives a prediction of a certain class or value. An illustrative but fictitious example of such a decision tree is given in Fig. 7, in which the goal is to predict the failure risk of an electrical submersible pump (ESP) based on the type of ESP, its operation RPMs, and age (operational hours). By learning from labeled example data, a decision tree model learns how to divide the features such that certain feature combinations lead to the correct label (failure risk of the ESP). One of the main benefits of decision trees is that they provide interpretable predictions as traversing the tree along the decision boundaries can be performed. On the other hand, when the problem becomes very complex with a lot of features, the number of decision boundaries grows rapidly, and the model loses its interpretability. In addition, decision trees tend to be rather unstable and sensitive to data imbalance. Overfitting is another common challenge with random forest models. Even though random forest generally reduces the risk of overfitting by taking an ensemble of multiple trees, the model can still overfit if the trees are too deep. This is also the reason why they are often used in ensembles such as random forests, as it greatly improves their robustness and accuracy (although in an ensemble they again lose their interpretability).

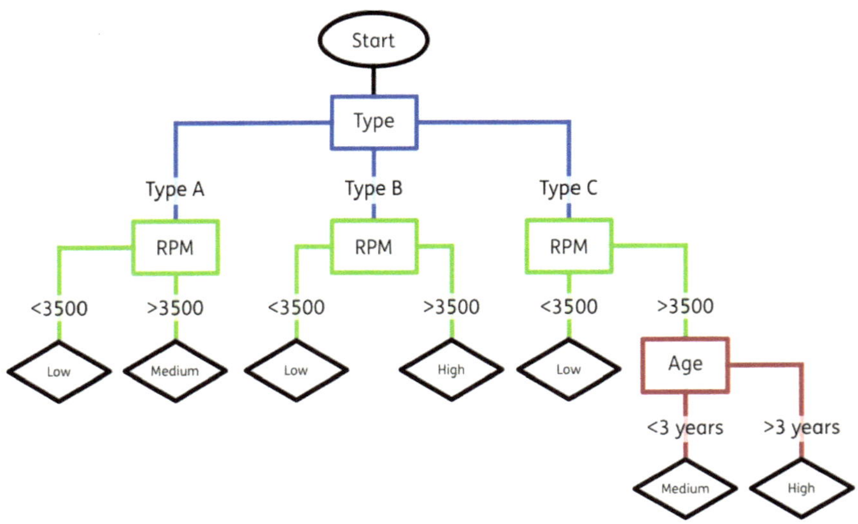

FIG. 7 Fictional example of a decision tree trained to predict ESP failure risks based on the type of ESP, its operating RPMs, and age. A random forest ensemble model would contain a large number of such individual decision trees.

4.4 Artificial neural networks

One of the most used, most versatile, and most powerful machine learning model types is the artificial neural network (ANN). An ANN tries to mimic the structure and functioning of biological neural networks we use to think and reason (Bishop, 2006). ANN comes in many different architectures, each focusing on processing different types of data or problems. For instance, convolutional neural networks (CNNs; Li et al., 2022) have been developed to work with image data; recurrent neural networks (RNNs; Yu et al., 2019) and transformers (Vaswani et al., 2017) specialize in sequential data (text, audio, and time series); graph neural network (GNN; Wu et al., 2021) excel at handling graph-structured data; autoencoders were designed for compressing or denoising data, but have seen many other applications in anomaly detection and generative AI (Li et al., 2023); and generative adversarial networks (GANs; Goodfellow et al., 2020) and, more recently, diffusion models (Ho et al., 2020) are used for generating new data instances. It is even possible to include physical relationships or constraints (equations) when training any ANN type to make it a physics-informed neural network (PINN; Karniadakis et al., 2021). In this subsection, to illustrate the basic principles of the ANN, the focus will be on one of the simplest types of ANN, the multilayer perceptron (MLP). The fundamentals of MLP are applicable to most other ANN variations because of its simplicity.

As the term "network" in artificial neural network indicates, an ANN consists of a graph of nodes, often called neurons, connected by edges. Like biological neural networks, these nodes are divided into layers, where every neuron in one layer is connected to all neurons in the previous and subsequent layer. All layers in between input and output are referred to as the hidden layers. An example MLP network with three inputs, two hidden layers with five neurons each, and two outputs is shown in Fig. 8.

To predict with an MLP, the features (inputs) of the problem are fed into the network at the input layer. These inputs are then propagated to the neurons in the next layer through the connections, where they are combined based on the weight of the individual connections (typically a weighted sum). Once combined, the information is processed at each neuron using a predefined nonlinear activation function. This process is repeated for all the hidden layers in the network until the output layer is reached, where the final processed information is read out as the output (prediction) of the model.

To ensure an ANN predicts the correct outputs, it is "trained" with example data. For a supervised learning problem, the data contains examples of input–output combinations. Then, in the training process, an optimization scheme is used to change the values of the weights (and an additional "bias" term) until the MLP predicts the correct outputs for the example input–output combinations. ANNs are versatile and have been proven to theoretically model any nonlinear function (Hornik, 1991). This assumes that sufficient data are available to learn from, which is often not the case in industrial applications. When little data are available, ANN might not be the most suitable method to choose, and it is sometimes suggested to use other machine learning models, such as random forest regression.

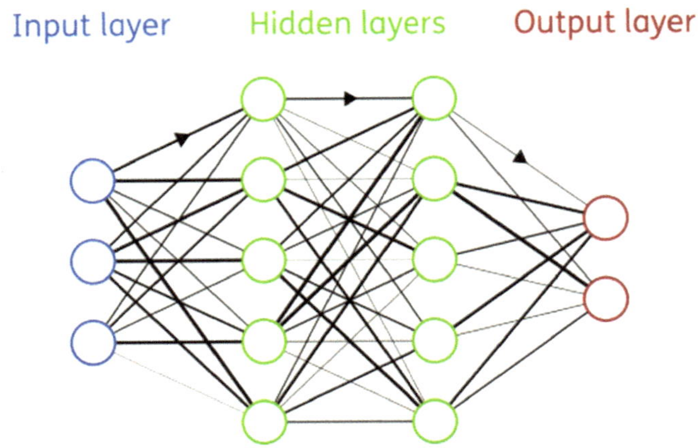

FIG. 8 Diagram of a simple multilayer perceptron-type artificial neural network with three inputs, two hidden layers with five neurons each, and two outputs. Width of the connecting edges has been scaled according to their weight. Base architecture generated using https://alexlenail.me/NN-SVG/; colors, arrows, and labels added by authors themselves.

4.5 Examples from other sectors

AI and ML models have seen an exponential growth in the number and types of applications. This subsection will discuss several concrete uses of AI/ML for real-world problems before going deeper into geo-energy–specific applications in the next section.

One of the earliest widespread uses of ML models is recommender systems, which have existed for many years at almost every large company that sells products or services. In these recommender systems, ML models are used to learn from people's preferences and (online) behavior to recommend new products in web shops, friends in the case of social media, and shows or movies for streaming services. ML models are particularly well suited for these systems, as human preferences are complex to define in conventional rule-based approaches, are person-specific, and might change over time. A comprehensive overview of the topic is given by Zhang et al. (2021).

The recent advances in generative models have also garnered widespread attention and use. Particularly, the automated generation of images through diffusion models (Ho et al., 2020) has become one of the clearest demonstrations of the potential of AI in practice, allowing for the creation of dozens of high-quality images in various styles with simple text-based prompts. That being said, disputes about authorship and copyright issues regarding the original training material for these models have made it a controversial topic (Shaffi, 2023; Greenberger., 2023).

In the medical field, ML, specifically feature detection in images, has seen a rapid growth in supporting the detection of diseases (mostly cancer) from various medical imaging techniques (Erickson et al., 2017). Perhaps less immediately practical to real-world applications, but still very interesting in terms of technological developments, is the application of reinforcement learning to games. By now, all of the highest-ranked chess algorithms are ML-based, and a couple of years ago, the world champion of Go, a game notoriously difficult to solve with computer algorithms due to its sheer number of possible moves, was beaten by a deep RL agent in three out of four matches (Silver et al., 2016). In addition, RL has been extensively shown to far exceed human performance on many classical NES games and has achieved great skill on more complicated games such as StarCraft II (Vinyals et al., 2019). The recent rise of large language models (e.g., ChatGPT, Bard, and Gemini Ultra) should also be mentioned here. The last section of this chapter is specifically dedicated to this topic.

The use of machine learning in the geothermal sector will be discussed in more detail in the following subsections. Before that, it is important to note that the similar workflows applicable to geothermal systems were also developed and deployed in other industries such as petroleum, process, and chemical. Daoutidis has provided an extensive overview of the application of machine learning in chemical and process industry including on process, material, and catalysis design as well as process monitoring, operations, and control (Daoutidis et al., 2024). A comprehensive overview of the application of machine learning in the petroleum sector in several fields, such as predictive maintenance, soft sensors, failure detection, and performance forecasting, is outlined by Bangert (2021). A lot of these studies can inspire geothermal industry to build on existing knowledge and tools from other industries and maximize the utilization of AI and machine learning tools in our sector.

5 Machine learning in geothermal production systems

In this section, we provide an overview of examples from the literature, demonstrating the use of machine learning in various aspects of geothermal energy systems, including exploration and characterization of geothermal sources, drilling of wells, production optimization and reservoir management, and equipment and facilities maintenance. Each section describes the possibilities of ML in each discipline, plus few examples from literature on how ML techniques were used in practice. Where possible, examples will be taken from the geothermal industry, but due to the novelty of ML in the geothermal sector, in many cases some of the best examples come from the oil and gas industry. Nonetheless, as there is significant overlap in the challenges faced by both industries, most examples will be applicable to both.

5.1 Resource exploration and characterization

The first step in exploiting a geothermal system is to find a suitable site for the system. In geoenergy engineering, this phase is typically referred to as "exploration and characterization" and focusses on finding areas that have a high potential for geothermal power and heat with reduced risks of operational challenges. This comprises activities such as locating geothermal reservoirs and estimating their temperature, characterization of the reservoir through seismic interpretation and inversion, and geochemical analyses of the reservoir's brine. In these activities, machine learning is taking on a bigger role due to its strengths in finding relations in large datasets and modeling complex relations in computationally efficient ways.

In exploration, a distinction is often made between visible and hidden or "blind" geothermal systems. As the name implies, blind geothermal systems have no (or minimal) information of the thermal features or characteristics of the reservoir (hot springs, fumaroles, mineral deposits, etc.; Dobson, 2016). As such, discovery, mapping, and characterization of blind systems and reservoirs is typically more difficult. For this reason, many machine learning methods focus primarily on application to these types of systems.

One of the main applications of machine learning in the exploration phase is related to the analysis of many sources of data, including field measurements from water samples, heat flow, gravity, magnetics, and electrical resistivity surveys, (sub)surface geological maps, and satellite images with the goal of mapping the geothermal potential in a given area. This is often also referred to as play fairway analysis (PFA; US Department of Energy, 2023), a term and methodology adapted from the oil and gas industry. When additional data from simulators or other sites are combined and machine learning algorithms are used, the method is often specifically referred to as ePFA (Vessilinov et al., 2022).

In an example of (e)PFA, Ahmmed and Vessilinov used geochemical and temperature data of over 14,000 shallow water sources in the Great Basin area, United States, to predict the potential for geothermal energy in different areas (Ahmmed and Vesselinov, 2022). For this, they made use of a non-negative matrix factorization combined with a nearest-neighbor algorithm (NMFk). NMF (Lee and Seung, 1999) is a matrix factorization method used to decompose a given matrix into two separate matrices with only positive entries. NMF is used a lot in data analytics as it has an inherent clustering property and under certain restrictions is equal to k-means clustering (Ding et al., 2005). Thus, by combining the NMF method with a nearest-neighbor algorithm, Ahmmed and Vessilinov were able to create groups in the sites from which data were obtained, associated with modest, moderate, and high geothermal potential.

Smith performed a study on the application of different machine learning methods to geothermal ePFA (Smith et al., 2023). In this study, they used various data features such as magnetic and gravity gradients, heat flow, fault slip rate, geodetic strain rate, and many others to train two model types, a supervised Bayesian neural network (BNN) and unsupervised principal component analysis coupled to k-means clustering (PCAk). A Bayesian neural network works generally on the same principles as a regular ANN, not only predicting an output value but also learning a probability distribution on the output variable, allowing it to indicate how confident it is in the predictions it makes (Jospin et al., 2022). The BNN was trained to classify sites into those with high and low geothermal potential. The PCAk method was used to cluster different geographical areas together in order to (combined with the BNN predictions) provide additional insights into which areas are geothermally most favorable for exploitation.

A similar goal was achieved with completely different methods, where convolutional neural networks were used on satellite images of land surface temperatures, shapefiles of fault maps, mineral markers from hyperspectral images, and deformation based on synthetic aperture radar measurements in order to automatically label geothermically favorable parts on the map of the Brady and Desert Peak areas in Nevada, United States (Moraga et al., 2022). In the study, they showed that the trained model could accurately label the appropriate areas. In addition, transfer of the Desert Peak model to the Brady area still showed accurate results, demonstrating that a model trained on one area can still be used in a different area. However, the inverse transfer of the Brady model to the Desert Peak area showed less accurate results, indicating that one must be careful when blindly applying ML models to input domains they were not trained on. Other (earlier) examples of ePFA include Vesselinov et al. (2020), Smith (2021), and Vesselinov (2021).

Next to geothermal potential mapping, characterization of the geological structure of the subsurface is another key aspect in the initial phases of a geothermal project, as it helps in assessing potential risks for drilling and operation of wells. Characterization can be done in different ways, with seismic inversion and well log analysis being prime examples. In seismic inversion, seismic data is used to determine the structure of the local subsurface by relating impedances derived from the reflection data to rock properties, fault locations, horizons, porosities, and liquid properties. Due to the scale and complexity of such seismic data, seismic inversion has seen more and more use of machine learning methods for processing the data and predicting associated subsurface properties. For example, Feng, Balling, and Grana used a combination of artificial neural networks and hidden Markov models (HMMs) to classify subsurface lithofacies and porosities using seismic data (Feng et al., 2020). The HMM was used to provide the conditional probabilities of transition between lithofacies. These lithofacies' predictions were than used as inputs for the ANN to predict the subsequent porosities.

As most seismic data has an image-like nature, an often-used type of machine learning model for seismic inversion is the convolutional neural network (CNN). These types of models use sequential convolutional and pooling layers to extract high-level features (lines, blocks, and rudimentary shapes) from images in order to make prediction on the structures or objects in an image. For instance, Das and Mukerji used two CNN to predict petrophysical properties directly from the seismic data (Das and Mukerji, 2020). In this, they first used one CNN to predict the elastic properties of the subsurface from the seismic data, before using these plastic properties in the second CNN to predict petrophysical properties with. Two clear and comprehensive comparisons between the performances of various types of CNN models as applied to seismic inversion are given respectively in Pintea et al. (2022) and Zhang et al. (2022).

Next to seismic inversion, well log analysis is another widely used method for characterization of the subsurface. Well logs are a great source of information on the subsurface due to the fact that they contain direct measurements of the subsurface itself taken during the drilling of exploration wells. In addition, well logs typically have a high vertical resolution, which provides great insights into the evolution of the lithology by depth (Soltani et al., 2016). Due to this high resolution, machine learning has been widely applied to well log data. In another study, well log features such as temperature, pressure, resistivity (in mud, deep, and shallow measurements), various wave speeds, and radioactive isotope presence (thorium, uranium, and potassium) were used to classify the types of rock (basalt, claystone, sandstone, etc.) present in the subsurface (Thongsamea et al., 2018). After comparing three different methods, k-nearest neighbor, support vector machines, and extreme gradient boosting (a decision tree ensemble), they concluded that the gradient boosting model performed best with over 90% accuracy.

Since drilling geothermal wells is more challenging than oil and gas wells due to the high temperatures and corrosive fluids present (more information in the next sections), certain logging sensors might not be present during drilling. To still provide information on such measurements, other data from sensors that are present during drilling can be used to estimate the values for non-present sensors (often referred to as virtual metering). For instance, Vivas and Salehi used drilling data that included rotary speed, weight on bit, torque, flow rate, rate of penetration, and standpipe pressure data to generate associated synthetic compressional sonic log responses (Vivas and Salehi, 2021a). They showed that a k-nearest neighbor algorithm could generate a sonic log with good accuracy (R^2 score of over 0.9). These sonic logs could then be used to further determine the geological structure of the subsurface. Other examples of machine learning applied to various tasks involving drilling data are given in the next subsection.

Due to the high costs associated with large-scale surveys, field data are often available in limited quantities. In this case, transfer learning can help in creating accurate models of the entire geothermal systems based on small-scale lab tests or data from other, similar areas of interest with more data available. For example, Sepehrinezhad used lab-scale ultrasonic seismicity data to train a machine learning model to predict fault shear stress (Sepehrinezhad et al., 2021). They then used transfer learning to fine-tune the model on much smaller datasets of other experiments, demonstrating that models trained using transfer learning can achieve much better performance than standalone models trained from scratch.

5.2 Drilling

A major contributor to the cost of a geothermal project is the cost associated to drilling and completion of wells. Some of the unique feature of geothermal formations, such as hardness of the rock, high temperatures, corrosive fluids, and large fractures can greatly impact the drilling operation and cost. One of the main issues of drilling is the challenge of predicting the rate of penetration (ROP). ROP is the rate that the drilling bit breaks the formation rock, and the drill string moves further downward. ROP prediction is challenging especially in geothermal formations like granite, basalt, and volcanic formations which are characterized by high hardness. In additions, these formations are characterized by high temperatures and contains corrosive fluids and gases which limit the drilling materials and raise safety concerns. One of the issues in the drilling operation is the high bit vibration which will limit the rate of drilling. A typical ROP in geothermal formations (numbers provided mainly from granite) is reported in the range of 3–6 m/h in shallower section and 2–5 m/h in deeper sections (Baujard et al., 2017). The prediction of ROP is even more challenging due to uncertainties in geological formation leading to tripping and nonproductive time (NPT). Another common issue is lost circulation which is caused by formation fracturing while drilling. Lost circulation is costly, impaired by high temperatures and the nature of the fractures. Cementing challenges arise when sealing fractured zones, and "live" production zones necessitate careful well control measures.

As an opportunity, a large volume of drilling data at high frequency is continuously captured during the drilling process. The use of machine learning algorithms can help identify anomalous responses in data and to differentiate between drilling problems and operational difficulties and provide insights and improvements in ROP prediction based on historical data of drilling in other similar or neighboring sites. In addition, this data can be simultaneously interpreted with petrophysical logs to predict formation properties.

There is an extensive review paper by Zhong et al. (2022) demonstrating the application of machine learning in drilling, focusing on the topics such as drilling fluid, drilling hydraulics, and drilling dynamics challenges. Machine learning methods such as artificial neural networks have been employed to predict the molecular and viscoelastic properties of the drilling mud such as mud density (Ahmadi et al., 2018) and apparent and plastic viscosity (Elkatatny et al., 2016). Gul and van Oort (2020) successfully showed the application of ML to predict API and filtrate loss of water-based muds and oil-based muds based on fluid properties such as rheology, density, and temperature for real-time monitoring and automation of the drilling process.

The drilling hydraulics and fluid flow prediction in the drilled borehole is important to control the drilling pressure, predict the cement displacement and cutting transport to the surface, and even select suitable bit nozzle size. One of the frequently performed analysis is the prediction of pump pressure and pressure loss in the drilled borehole (Rooki, 2015; Ozbayoglu et al., 2018). Features such as fluid velocity, density, and rheological parameters together with hole geometrical parameters, such as depth, diameters, dogleg severity, were used to predict pressure losses. For pump pressures prediction incorporating additional features such as pump RPM and torque showed an improvement in the accuracy of the trained model (Wang and Salehi, 2015). Another common application of machine learning is in the prediction of cutting transport during the cleaning of the drilled holes.

In the geothermal application, ROP prediction with ML methods is one of the most investigated areas. Diaz et al. (2018, 2019) have applied ML techniques to predict the rate of penetration (ROP) using controllable drilling parameters like bit weight, bit diameter, bit speed, and fluid rheological properties and further refining it with pore pressure gradient and surface drilling data. By employing linear regression and artificial neural networks, they demonstrated improved accuracy in ROP prediction, emphasizing the significance of data filtering and feature selection. Additionally, Diaz and Kim (2020) assessed the feasibility of utilizing nearby wells to improve the prediction of ROP. Their deep neural network, incorporating a filtration process and resampling, accurately predicted ROP by considering data points from neighboring wells. A similar approach was employed by Arbi Ben Aoun and Madarasz (2022) to predict ROP in Utah Forge site with the random forest regression, SVM, and ANN methods.

Kiran et al. (2022) developed a machine learning and deep learning workflow to identify lost circulation zones applied to FORGE well data. The workflow was able to detect lost circulation zones during drilling by an accuracy of 83%. In addition to the use of drilling data to predict improved drilling operation by predicting ROP or lost circulation, new research trends are rising to use surface drilling data to predict formation properties and lithology classifications. As an example, Vivas and Salehi (2021b) used several machine learning models to predict formation thermal conductivity from drilling operation data. Carbonari et al. (2021) showed the improvements in lithology classification based on the drilling cutting. All the developed work and algorithms can provide additional insights and support the drilling operation by optimizing the control parameters, detection of anomalies and events during drilling to have a proactive response to the problems and avoid unnecessary nonproductive time. As an additional opportunity, the drilling data can provide real-time lithology estimation of the formation.

5.3 Production optimization and reservoir management

Geothermal energy production relies heavily on the performance of the geothermal reservoir. Production and reservoir engineering is a critical field that involves analyzing field data and models to make informed decisions on optimizing the safe performance of geothermal assets, both for short-term decisions (operational time scales) and long-term decisions (reservoir management). Machine learning, which was traditionally developed in the petroleum sector and can be further extended to the geothermal energy sector, has emerged as a powerful tool in this area (Nyokabi et al., 2023). The areas of reservoir management that were investigated by machine learning models are production rate, pressure and temperature forecasting, production–injection management, and induced seismicity.

Machine learning for reservoir engineering and management discipline in the petroleum sector has already been established, and there are a significant number of publications and studies on this topic covering different aspects of reservoir engineering. The topics on applying ML methods in reservoir engineering and simulation can be categorized into reservoir surrogate models, phase stability, and PVT properties (Samnioti and Gaganis, 2023). In addition, history matching, which is one of the most common practices for reservoir engineers to calibrate reservoir parameters, has also been combined with machine learning methods to improve their computational speed and/or accuracy (Esmaili and Mohaghegh, 2016). One of the areas is to estimate and forecast of reservoir conditions such as reservoir temperature, pressure (Aydin et al., 2020) and production rates (Gudmundsdottir and Horne, 2020; Beckers et al., 2021). Most of these studies are using historical production data to estimate reservoir conditions. However, it was shown that including the reservoir parameters and operational constraints as features to the machine learning models can improve the prediction of the models (Shi et al., 2021). In addition, deep learning models were applied on geo-chemistry data to estimate geothermal fluid temperature (Haklidir and Haklidir, 2019). In other studies, production data such as pressure and temperature were used to predict the reservoir properties, such as permeability distributions (Suzuki et al., 2021).

As a final note, optimization of production and long-term reservoir management needs to be done in a safe and responsible manner. Due to an increase in the number of enhanced geothermal systems, the concern regarding the induced seismicity due to an increase in the number of enhanced geothermal systems and reinjection of the water in the conventional geothermal systems needs to be addressed. Production and injection management can have a significant impact on the induced seismicity risk (He et al., 2020). Next to the subsurface and geomechanical models to address the induced seismicity issues, machine learning is also addressed as a potential methodology. The main applications which were addressed by machine learning are fault detection and microseismicity identification. Some of the works were focused on Geysers geothermal field in California which experiences subsidence due to geothermal activities and seismic. Holtzman et al. (2018) studied the structural changes in the cyclic behavior of the seismic source in the Geysers field. Yazbeck and Rundle (2023) developed and demonstrated a method to combine InSAR data with geothermal data and use ML models, in this case convolutional neural networks (CNNs), to predict enhanced deformation at the Geysers. They have found a high correlation between injection rate and number of earthquake events, and integrating all of these sources of data improve the prediction significantly. Wamriew et al. (2021) employed CNN to predict the location of microseismic events using distributed acoustic sensing (DAS) data. An automated microseismic-specific phase picker for monitoring purposes using machine learning algorithms was developed and tested (Okamoto et al., 2024). The developed method accurately detected P- and S-wave arrivals across four distinct geothermal fields, particularly showcasing its effectiveness in revealing subsurface fine structures in the Okuaizu geothermal field.

5.4 Equipment and facilities maintenance

Proper monitoring and maintenance of equipment and facilities are critical to the optimal performance of geothermal systems. Physical phenomena like scaling, corrosion, and erosion can significantly impact the integrity of the main wellbore, pipelines, and individual components such as heat exchangers. Furthermore, filter clogging and ESP failure contribute heavily to the costs of operating a geothermal system (Octaviano et al., 2022). While many methods exist to prevent or mitigate these adverse effects, such as regular inspections and maintenance, determining exactly when or how often to intervene is challenging. Do it too little, and you might not be on time to prevent failures; do it too often, and you needlessly increase the OPEX costs of a plant.

Machine learning can help operators maintain their systems in two ways: through improved condition monitoring and predictive maintenance. In condition monitoring, machine learning methods aid in diagnostics of the system's current state. For instance, virtual metering can help replace expensive or broken sensors by estimating the process values based on the existing sensors. In addition, anomaly detection helps to automate the identification of off-normal behavior of a system based on trends in sensor data. For predictive maintenance, as the name implies, the goal is to predict when a component is expected to fail based on prior incidents. Data from previous failures is used to train machine learning models to predict the remaining useful lifetime (RUL) of components, which can then be used to plan when to schedule maintenance activities.

While virtual metering has seen significant interest in the petroleum sector (Shoeibi Omrani et al., 2018; Bikmukhanetov and Jäschke, 2020), it has as of yet seen less use in the geothermal sector during the production and operation. Nonetheless, virtual metering has potential to predict downhole conditions such as pressure and temperature, similar to the machine learning–based geothermometer used for exploration. A similar concept can be used to detect anomalies in the production and operational data.

Data-driven anomaly or general "event" detection is a powerful tool to aid in the condition monitoring of geothermal systems as it allows for quick identification of off-normal behavior of the system. Here, models can be trained using both supervised and (semi)supervised learnings. However, since anomalies are inherently rare (otherwise, they would not be anomalies), examples of prior incidents are typically scarce, and not enough data are available for training data-driven models. In this case, a semisupervised approach can be used in which a model is trained only on detecting regular operating conditions. Then, when the model encounters data that are not recognized as normal, it can provide the operator with an early warning that something might be changing in the system. As for virtual metering, due to the relative novelty of the widespread application of geothermal systems and machine learning techniques, few examples of these techniques being applied in practice are available in the literature. However, again, a lot of experience and inspiration can be drawn from the O&G industry. For instance, for the detection of leaks in pipelines due to corrosion (Aljameel et al., 2022), classification of anomalous events in oil wells (Ferreira Machado et al., 2022), event detection in gas wells (Poort et al., 2020), and semisupervised novelty detection of well events using autoencoders (Alatrach et al., 2020).

When it comes to applying machine learning techniques to the modeling and monitoring of ESP (electric submersible pump) systems, significantly more research and applications are already being undertaken. This is because they are one of the most used artificial lift methods for O&G and geothermal systems and are one of the main contributors to CAPEX and OPEX of geothermal production (among others, due to frequent failures; Shoeibi Omrani et al., 2021). In addition, ESPs are often equipped with numerous sensors (temperature, pressure, frequency, voltage, current, etc.), which allow for more complex machine learning methods. On the other hand, the large number of potential failure modes associated with ESP systems (Shoeibi Omrani et al., 2021) makes condition monitoring and predictive maintenance challenging. Thus, machine learning can play a significant role in understanding the complex trends in the many sensor data streams coming from ESPs. For instance, Bayesian neural networks were used to predict the degradation of ESPs' in oil wells for different forecasting horizons (Dussi et al., 2022). The authors explicitly encoded the causality between the different input features to improve the model's performance, eventually reaching error scores as low as 1.1% for the next 48h and 2.5% for a 10-day prediction horizon.

Taking inspiration from the way in which operators detect anomalies by looking at the visual trendlines in the data, Lastra and Tulbah applied convolutional neural networks (CNN) to real-time detection of ESP failures (Lastra and Tulbah, 2021). As a CNN treats the time series data as "images" and mainly learns from the shapes of the data, they can be used to detect similar patterns taking place over different timeframes. As such, they can see the same anomaly or event over a few hours or multiple days. This approach was used in a supervised learning method, where a pre-trained CNN was updated using transfer learning to classify (anomalous) ESP events such as pump wear, scale accumulations, mechanical problems, and gassy conditions. In the end, their best (and largest) model was able to classify events in the test set with 96% accuracy. Other examples of machine learning applied to ESP systems include predictive maintenance (Abdalla et al., 2022), failure detection and diagnostics (Lastra and Jinjiang, 2022), and the development of a digital ESP monitoring framework (Lastra, 2019).

6 Large language models in geothermal systems

Next to the other machine learning methods described in the previous section, large language models have a significant impact in the geothermal industry. Large language models (LLMs) are a group of models that belong to natural language processing (NLP). In order to better understand LLMs, first, an introduction to a language model (LM) is given. LLMs are built on the definition of tokens which are the smallest units of text that the model processes. In natural language processing, a token can be as short as one character and as long as one word. Tokens are the building blocks that the model uses to understand and generate text. A classical definition of a language model is a probability distribution over sequences of tokens. The probability assigned by the model reflects how likely a sequence of tokens is to appear in a natural language context. This probability is calculated based on the patterns the model has learned during training. Suppose you have a set of tokens. The language model assigns probabilities (between 0 and 1) to each sequence of these tokens to estimate how likely a particular sequence is to occur. For example:

$$\begin{aligned}&\mathrm{P}(\text{AGS refers to advanced geothermal systems})\\&\quad= \mathrm{P}(\text{AGS}) \times (\text{refers}|\text{AGS}) \times \mathrm{P}(\text{to}|\text{AGS refers}) \times \mathrm{P}(\text{advanced}|\text{AGS refers to})\\&\quad\times \mathrm{P}(\text{geothermal}|\text{AGS refers to advanced})\\&\quad\times \mathrm{P}(\text{systems}|\text{AGS refers to advanced geothermal})\end{aligned}$$

LLMs have the capability to solve many NLP tasks due to their massive architecture, vast number of parameters to be trained and a significantly large training dataset used to train them.

6.1 The evolution toward the transformers revolution

Language models have been around for a while. The MIT professor Joseph Weizenbaum developed the first Chatbot in 1966 for a psychology use case (Natale, 2018). This model—called

Eliza—was based on a set of predetermined rules, and it did not have the ability to truly comprehend language. After ELIZA, the successful early language models relied on statistical methods, i.e., n-grams and hidden Markov models, to analyze natural language. These models predicted the next word in a sequence by considering the words that came before it, within a certain time frame. These models struggled with data scarcity and could not capture long-term relationships between words. With the growth of neural networks, early language models were mainly replaced by recurrent neural networks (RNNs). RNNs made major improvements in natural language modeling; however, the revolution in understanding human language happened only after the transformers were introduced by Ashish Vaswani et al. (2017). While RNNs are sequential by design, transformers use parallelization, i.e., they process all tokens in the input sequence simultaneously. This makes the training process a lot faster on GPUs. Besides, transformers are less affected by vanishing and exploding gradient problems that were persistent issues for RNNs. Moreover, due to the use of self-attention mechanisms, transformers are superior at handling long-term dependencies and therefore better understanding of context. Transformers' high scalability has led to emergence of very large language models, such as generative pretrained transformer (GPT) models, bidirectional encoder representations from transformers (BERT: Devlin et al., 2018), pathways language model (PaLM; Chowdhery et al., 2023), and large language model meta AI (Llama; Touvron et al., 2023). After the successful implementation of ChatGPT by OpenAI (2023), multiple attempts have been made by large tech companies to introduce new large language models.

Apart from emergence of new LLMs from large tech companies, scientists across the globe are fine-tuning large language models and performing prompt engineering to make LLMs more applicable on specific domains. On top of that, retrieval augmented generation (RAG; Lewis et al., 2020) systems have obtained a lot of attention recently. RAG systems allow users to benefit from the language capabilities of LLMs to extract information from their own documents. Considering the huge interest from both businesses and academia in LLMs and the capabilities provided by the transformers' revolution as well as the sharply increasing computation power provided by IT companies, it is expected that LLMs will change the world in many ways and geothermal systems will not be an exception.

6.2 Application of LLMs in geothermal systems

Several articles have been published employing artificial intelligence techniques in the study of geothermal systems (Okoroafor et al., 2022). However, the use of large language models in this domain has remained notably scarce to date. LLMs can reshape the landscape of data analytics in various fields, including geothermal systems. Their capability to analyze vast quantities of textual data, both in isolation and in conjunction with other AI techniques, has opened a new range of opportunities to enhance research, operations, safety measures, and education in geothermal plants. In this section, a selection of examples and potential ideas that may be applicable in the realm of geothermal systems, have been provided.

The geothermal energy domain is characterized by a vast array of textual data, comprising thousands of pages of documents that encompass various aspects of exploration (seismic reports and licenses), operations (maintenance logs), and research and knowledge compilation. This collection of data represents decades of scientific inquiry, empirical observations, and practical experience in the field. By leveraging the capabilities of large language models (LLMs), it is possible to extract meaningful patterns, relationships, and insights from this extensive dataset. This can provide valuable information and guidance for researchers, operators, and educators working in the geothermal energy sector. Since, until now, there is almost no example on the use of LLM in the geothermal sector, in the following section, we will provide an overview of the potential of using LLM in the geothermal sector as an inspiration for future work and research.

6.2.1 Applications for research

- **Literature review**: The rise of web mining capabilities, especially through application programming interfaces (APIs), has made it easier to compile extensive scientific databases. A researcher might use APIs or develop a web scraper to gather new publications and news articles on geothermal energy. Once compiled, a RAG can be used on the database to answer domain-specific research questions.
- **Trend analysis:** LLMs have become invaluable tools for researchers navigating scientific geothermal databases. For example, a researcher might be investigating new trends and topics for improving the operation of geothermal assets or finding new methods to explore and characterize geothermal reservoirs. These steps require dealing with a vast amount of literature and text data in the geothermal sector. In addition, there are other sectors such as petroleum which also provide a rich database of information and learning. Instead of manually sifting through thousands of articles, LLMs can be used for extraction of topics, classification of articles, and trend analysis. As an example, Eckroth recently developed a question-answering prototype tool that answers questions using Petrowiki (n.d) (Eckroth et al., 2023).

6.2.2 Operational excellence through natural language queries

- **Querying technical documents**: Operational, maintenance, and safety documents frequently contain valuable information that can address pressing issues in geothermal energy production. By utilizing large language models (LLMs) to interrogate these documents, operators can gain insights into potential causes of anomalies, such as root cause of pump failures or injectivity decline. This can be achieved by posing specific questions to the LLM, which can then draw upon historical data and past occurrences to provide informed answers. This approach can help identify patterns, trends, and relationships that might not be immediately apparent, enabling operators to make more informed decisions and optimize geothermal energy production.
- **Shift reports as data goldmines**: These reports, routinely generated in geothermal plants, encapsulate vital information related to operations and maintenance. By leveraging LLMs, stakeholders, including the safety, health, and environment (SHE) team, can extract historical trends, tried-and-tested solutions, and potential red flags. By querying LLMs, plant managers can get a quick overview of a certain operational problem and get a consolidated report instead of going through each shift report manually. This significantly accelerates decision-making and issue rectification.
- **Categorization and data segmentation**: LLMs can categorize issues based on various parameters and devices involved, type of problem (e.g., scaling and corrosion), and more. Such categorization facilitates focused analysis and intervention. For instance, if there is a

repeated issue with pump seals across several plants, LLMs can categorize and provide a breakdown of all incidents related to it, helping in pinpointing whether it is a supplier issue, an operational error, or something else.

- **Integrative analysis**: Feature extraction is a key step in the development of machine learning models. LLMs can be integrated into statistical and machine learning developments by being a strong tool to extract features from text data.

6.2.3 *Application of LLMs for risk management*

- **Creating HAZOPs**: By analyzing safety reports, incident logs, and safety audits, LLMs aid in formulating hazard and operability studies (HAZOPs) essential for the safe design and operation of geothermal plants. Feng and Dai used a BERT model to perform HAZOP analysis for chemical plants (Feng et al., 2021). In another study, a fine-tuned BERT model was used to support quantitative risk analysis (QRA) related to the operation of an oil refinery plants (Macêdo et al., 2022). These studies can be inspirational to the geothermal sector to use LLM for HAZOP studies.
- **Incident pattern recognition**: Recognizing recurring patterns in incident reports can be pivotal in preempting major accidents. LLMs can be used for scanning these reports, cross-reference patterns with other sources of data such as shift and maintenance reports, and detect anomalies that may otherwise go unnoticed.

6.3 Recommendations

LLMs function as versatile tools with the ability to interpret complex and massive texts in the geothermal industry for a wide range of applications. Researchers can leverage LLMs to gain valuable insights, operators can optimize performance, safety teams can mitigate risks, and educators and trainees can achieve comprehensive learning. The integration of LLMs with other AI techniques and their capacity for continuous learning further highlights their critical role in the advancement of the geothermal field. LLMs have the potential to improve the way we approach geothermal energy production, paving the way for increased efficiency, safety, and sustainability. The collaboration of experts in large language models and geothermal systems is crucial for the development of practical applications in the domain of geothermal energy. By fine-tuning an LLM with a wealth of geothermal-related text and integrating it into a RAG system, researchers and operators can query a vast database comprising geothermal scientific and technical sources. This approach has the potential to yield valuable insights and benefits for both research and operational aspects of geothermal energy production.

7 Concluding remark

This chapter delves into the world of artificial intelligence and machine learning in geothermal energy. It is clear that with the massive data currently available and more to be available in the near future, AI and ML methods can support the geothermal sector in unlocking several potentials from the data. The potential impact of AI and ML in the geothermal sector, spanning exploration, drilling, and operations, is substantial. Through real-life examples in this chapter, we showcase how these technologies are impacting the geothermal sector. It is important to note that there are several existing AI and ML workflows that the geothermal industry can adapt and deploy from other sectors such as petroleum, process, and chemical industries. Thus, there is no need to reinvent the wheel!

References

Abdalla, R., Samara, H., Perozo, N., Carvajal, C., Jaeger, P., 2022. Machine learning approach for predictive maintenance of the electrical submersible pumps (ESPs). ACS Omega 7 (21), 17641–17651.

Ahmadi, M.A., Shadizadeh, S.R., Shah, K., et al., 2018. An accurate model to predict drilling fluid density at wellbore conditions. Egypt J. Petrol. 27, 1–10. https://doi.org/10.1016/j.ejpe.2016.12.002.

Ahmmed, B., Vesselinov, V.V., 2022. Machine learning and shallow groundwater chemistry to identify geothermal prospects in the Great Basin, USA. Renew. Energy 197, 1034–1048.

Alatrach, Y., Mata, C., Shoeibi Omrani, P., Saputelli, L., Narayanan, R., Mohammad, H., 2020. Prediction of well production event using machine learning algorithms. In: Abu Dhabi International Petroleum Exhibition and Conference, Abu Dhabi.

Aljameel, S.S., Alomari, D.M., Alismail, S., Khawaher, F., Alkhudhair, A.A., Aljubran, F., Alzannan, R.M., 2022. An anomaly detection model for oil and gas pipelines using machine learning. Comput. Des. 10 (8).

Amazon Web Services, What is Natural Language Processing (NLP). Amazon Web Services, [Online]. Available https://aws.amazon.com/what-is/nlp/. Accessed 20 Decemer 2023.

Amazon Web Services, What is Data Analytics? Amazon Web Services, [Online]. Available https://aws.amazon.com/what-is/data-analytics/. Accessed 19 December 2023.

Amini, M.-R., Feofanov, V., Pauletto, L., Hadjadj, L., Devijver, E., Maximov, Y., 2023. Self-Training: A Survey. arXiv.

Arbi Ben Aoun, M., Madarasz, T., 2022. Applying Machine learning to predict the rate of penetration for geothermal drilling located in the Utah FORGE site. Energies 15 (12), 4288.

Aydin, H., Akin, S., Senturk, E., 2020. A proxy model for determining reservoir pressure and temperature for geothermal wells. Geothermics 88.

Bangert, P., 2021. Machine Learning and Data Science in Oil and Gas Industry, Best practices, Tools and Case Studies.

Baujard, C., Hehn, R., Genter, A., Teza, D., Baumgärtner, J., Guinot, F., Martin, A., Steinlechner, S., 2017. Rate of penetration of geothermal wells: a key challenge in hard rocks. In: Proceedings, 42nd Workshop on Geothermal Reservoir Engineering Stanford University, Stanford, California, February 13–15, 2017.

Beckers, K., Duplyakin, D., Martin, M., Johnston, H., Siler, D., 2021. Subsurface characterization and machine learning predictions at Brady Hot Springs. In: Proceedings of the Forty-sixth Workshop on Geothermal Reservoir Engineering. Stanford University.

Bikmukhanetov, T., Jäschke, J., 2020. First principles and machine learning virtual flow metering: a literature review. J. Pet. Sci. Eng. 184.

Bishop, C.M., 2006. Pattern Recognition and Machine Learning. Springer Science+Business Media, Singapore.

Carbonari, R., Ton, D., Bonneville, A., Bour, D., Cladouhos, T., Garrison, G., Horne, R., Petty, S., Rallo, R., Schultz, A., Sørlie, C., 2021. First year report of EDGE project: an international research coordination network for geothermal drilling optimization supported by deep machine learning and cloud based data aggregation. In: Proceedings, 46th Workshop on Geothermal Reservoir Engineering. Vol. 1. Stanford University, Stanford, CA.

Chowdhery, A., Narang, S., Devlin, J., Bosma, M., Mishra, G., Roberts, A., Fiedel, N., 2023. Palm: scaling language modeling with pathways. J. Mach. Learn. Res. 24 (240), 1–113.

Copeland, B.J., Artificial Intelligence. Encyclopedia Britannica, [Online]. Available https://www.britannica.com/technology/artificial-intelligence. (Accessed 19 December 2023).

Daoutidis, P., Lee, J.H., Rangarajan, S., Chiang, L., Gopaluni, B., Schweidtmann, A.M., Harjunkoski, I., Mercangöz, M., Mesbah, A., Boukouvala, F., Lima, F.V., del Rio Chanona, A., Georgakis, C., 2024. Machine learning in process systems engineering: challenges and opportunities. Comput. Chem. Eng. 181.

Das, V., Mukerji, T., 2020. Petrophysical properties prediction from prestack seismic data using convolutional neural networks. Geophysics 85 (5).

Devlin, J., Chang, M.W., Lee, K., Toutanova, K., 2018. Bert: Pre-Training of Deep Bidirectional Transformers for Language Understanding. arXiv preprint arXiv:1810.04805.

Diaz, M.B., Kim, K.Y., 2020. Improving rate of penetration prediction by combining data from an adjacent well in a geothermal project. Renew. Energy 155, 1394–1400.

Diaz, M.B., Kim, K.Y., Kang, T.H., Shin, H.S., 2018. Drilling data from an enhanced geothermal project and its pre-processing for ROP forecasting improvement. Geothermics 72, 348–357.

Diaz, M., Kim, K., Shin, H., Zhuang, L., 2019. Predicting rate of penetration during drilling of deep geothermal wells in Korea using artificial neural networks and realtime data collection. J. Nat. Gas Sci. Eng., 225–232. https://doi.org/10.1016/j. jngse.2019.05.004.

Ding, C., He, X., Simon, H.D., 2005. On the equivalence of nonnegative matrix factorization and spectral clustering. In: Proceedings of the 2005 SIAM International Conference on Data Mining (SDM). Newport Beach, California.

Dobson, P.F., 2016. A review of exploration methods for discovering hidden geothermal systems. Geotherm. Resour. Counc. Trans. 40.

Dussi, S., Octaviano, R., Shoeibi Omrani, P., 2022. Bayesian networks applied to ESP performance monitoring and forecasting. In: SPE Annual Technical Conference and Exhibition. Houston.

Eckroth, J., Gipson, M., Boden, J., Hough, L., Elliott, J., Quintana, J., 2023. Answering natural language questions with OpenAI's GPT in the petroleum industry. In: Paper Presented at the SPE Annual Technical Conference and Exhibition, San Antonio, Texas, USA. https://doi.org/10.2118/214888-MS.

Elkatatny, S., Tariq, Z., Mahmoud, M., 2016. Real time prediction of drilling fluid rheological properties using artificial neural networks visible mathematical model (white box). J. Pet. Sci. Eng. 146, 1202–1210. https://doi.org/10.1016/j.petrol.2016.08.021.

Erickson, J., Korfiatis, P., Akkus, Z., Kline, T.L., 2017. Machine learning for medical imaging. Radiographics 37 (2).

Esmaili, S., Mohaghegh, S.D., 2016. Full field reservoir modeling of shale assets using advanced data-driven analytics. Geosci. Front. 7, 11–20.

Feng, R., Balling, N., Grana, D., 2020. Lithofacies classification of a geothermal reservoir in Denmark and its facies-dependent porosity estimation from seismic inversion. Geothermics 87.

Feng, X., Dai, Y., Ji, X., Zhou, L., Dang, Y., 2021. Application of natural language processing in HAZOP reports. Process Saf. Environ. Prot. 155, 41–48.

Goodfellow, I., Pouget-Abadie, J., Mirza, M., Xu, B., Warde-Farley, D., Ozair, S., et al., 2020. Generative adversarial networks. Commun. ACM 63 (11), 139–144.

Google, What is Artificial Intelligence (AI)? Google, [Online]. Available https://cloud.google.com/learn/what-is-artificial-intelligence. Accessed 19 December 2023.

Google, What is Natural Language Processing? Google, [Online]. Available: https://cloud.google.com/learn/what-is-natural-language-processing. Accessed 20 December 2023.

Greenberger., 2023. AI-generated artwork that spurred controversy Can't be copyrighted, US officials say. ARTnews.

Gudmundsdottir, H., Horne, R., 2020. Prediction modeling for geothermal reservoirs using deep learning. In: Proceedings of the Forty-fifth Workshop on Geothermal Reservoir Engineering. Stanford University.

Gul, S., van Oort, E., 2020. A machine learning approach to filtrate loss determination and test automation for drilling and completion fluids. J. Pet. Sci. Eng. 186.

Haklidir, F., Haklidir, M., 2019. The fluid temperature prediction with hydro-geochemical indicators using a deep learning model: a case study Western Anatolia (Turkey). In: Proceedings of the Forty-fourth Workshop on Geothermal Reservoir Engineering. Stanford University.

He, M., Li, Q., Li, X., 2020. Injection-induced seismic risk management using machine learning methodology – a perspective study. Front. Earth Sci. 8. 22 June 2020 Sec. Solid Earth Geophysics.

Ho, J., Jain, A., Abbeel, P., 2020. Denoising Diffusion Probabilistic Models. arXiv.

Holtzman, B.K., et al., 2018. Machine learning reveals cyclic changes in seismic source spectra in geysers geothermal field. Sci. Adv. 4 (5), 1–8.

Hornik, K., 1991. Approximation capabilities of multilayer feedforward networks. Neural Netw. 4 (2), 251–257.

IBM, What is Machine Learning? IBM, [Online]. Available: https://www.ibm.com/topics/machine-learning. (Accessed 19 December 2023).

IBM, What is Deep Learning? IBM, [Online]. Available: https://www.ibm.com/topics/deep-learning. (Accessed 19 December 2023).

Jospin, L.V., Laga, H., Boussaid, F., Buntine, W., Bennamoun, M., 2022. Hands-on Bayesian neural networks - a tutorial for deep learning users. IEEE Comput. Intell. Mag. 17 (2), 29–48.

Jurafsky, D., Martin, J.H., 2023. Speech and Language Processing, third ed.

Karniadakis, G., Kevrekidis, I., Lu, L., Perdikaris, P., Wang, S., Yang, L., 2021. Physics-informed machine learning. Nat. Rev. Phy. 3, 422–440.

Kiran, R., Dansena, P., Salehi, S., Kumar Rajak, V., 2022. Application of machine learning and well log attributes in geothermal drilling. Geothermics 101, 102355.

Lastra, R., 2019. Electrical submersible pump digital twin, the missisng link for successful condition monitoring and failure prediction. In: Abu Dhabi International Petroleum Exhibition and Conference, Abu Dhabi, 2019.

Lastra, R., Jinjiang, X., 2022. Machine learning engine for real-time ESP failure detection and diagnostics. In: SPE Middle East Artificial Lift Conference and Exhibition.

Lastra, R., Tulbah, F., 2021. Computer vision for real-time ESP failure detection and diagnostics. In: World Petroleum Congress (WPC), Houston.

Lee, D., Seung, H., 1999. Learning the parts of objects by non-negative matrix factorization. Nature 401, 788–791.

Lewis, P., Perez, E., Piktus, A., Petroni, F., Karpukhin, V., Goyal, N., Küttler, H., et al., 2020. Retrieval-augmented generation for knowledge-intensive nlp tasks. Adv. Neural Inf. Proces. Syst. 33, 9459–9474.

Li, Z., Liu, F., Yang, W., Peng, S., Zhou, J., 2022. A survey of convolutional neural networks: analysis, applications, and prospects. IEEE Trans. Neural Netw. Learn. Syst. 33 (12), 6999–7019.

Li, P., Pei, Y., Li, J., 2023. A comprehensive survey on design and application of autoencoder in deep learning. Appl. Soft Comput. 138.

Macêdo, J.B., das Chagas Moura, M., Aichele, D., Lins, I.D., 2022. Identification of risk features using text mining and BERT-based models: application to an oil refinery. Process Saf. Environ. Prot. 158, 382–399.

Machado, P.F., Vaz Vargas, R.E., Marques Ciarelli, P., Jose Munaro, C., 2022. Improving performance of one-class classifiers applied to anomaly detection in oil wells. J. Pet. Sci. Eng. 218.

Mathar, R., Alirezaei, G., Balda, E., Behboodi, A., 2020. Fundamentals of Data Analytics. Springer Nature Switzerland.

Moraga, J., Duzgun, H.S., Cavur, M., Soydan, H., 2022. The geothermal artificial intelligence for geothermal exploration. Renew. Energy 192, 134–149.

Natale, S., 2018. If software is narrative: Joseph Weizenbaum, artificial intelligence and the biographies of ELIZA. New Media Soc. 21 (3), 712–728.

Nyokabi, J., Oyugi, N., Mbugua, G., Hiuhu, A., Nyoni, K.J., 2023. An in-depth review of machine learning applications in geothermal reservoir engineering. In: Proceedings World Geothermal Congress 2023.

Octaviano, R., Dussi, S., de Zwart, H., Shoeibi Omrani, P., van Pul-Verboom, V., Elewaut, K., van Schravendijk, B., 2022. Model-Based Monitoring of Geothermal Assets. (WarmingUP program final report).

Okamoto, K., Mukuhira, Y., Darisma, D., Asanuma, H., Moriya, H., 2024. Machine learning automatic picker for geothermal microseismicity analysis for practical procedure to reveal fine reservoir structures. Geothermics 116, 102832.

Okoroafor, E.R., Smith, C.M., Ochie, K.I., Nwosu, C.J., Gudmundsdottir, H., Aljubran, M.J., 2022. Machine learning in subsurface geothermal energy: two decades in review. Geothermics 102, 102401.

OpenAI, 2023. ChatGPT (Mar 14 version) [Large language model]. https://chat.openai.com/chat.

Ozbayoglu, E.M., Erge, O., Ozbayoglu, M.A., 2018. Predicting the pressure losses while the drillstring is buckled and rotating using artificial intelligence methods. J. Nat. Gas Sci. Eng. 56, 72–80. https://doi.org/10.1016/j.jngse.2018.05.028.

Pan, S.J., Yang, Q., 2009. A survey on transfer learning. IEEE Trans. Knowl. Data Eng. 22 (10), 1345–1359.

Petrowiki. https://petrowiki.spe.org/PetroWiki.

Pintea, S., Sharma, S., Vossepoel, F., van Gemert, J., Loog, M., 2022. Seismic inversion with deep learning. Comput. Geosci. 26, 351–364.

Poort, J., Bazuin, R., 2023. Using transfer learning to construct accurate data-driven compressor models on limited datasets of field measurements. In: Proceedings of the 13th EFRC conference, Zagreb.

Poort, P.J., Shoeibi Omrani, P., Vecchia, A.L., Visser, G., Janzen, M., Koenes, J., 2020. An automated diagnostic analytics workflow for the detection of production events—application to mature gas fields. In: Abu Dhabi International Petroleum Exhibition and Conference, Abu Dhabi.

Rooki, R., 2015. Estimation of pressure loss of Herschel-Bulkley drilling fluids during horizontal annulus using artificial neural network. J. Dispers. Sci. Technol. 36, 161–169. https://doi.org/10.1080/01932691.2014.904793.

Samnioti, A., Gaganis, V., 2023. (2023), applications of machine learning in subsurface reservoir simulation—a review—part I. Energies 16, 6079.

Schmidhuber, J., 2015. Deep learning in neural networks: an overview. Neural Netw. 61, 85–117.

Sepehrinezhad, S., Shreedharan, S., Marone, C., Shokouhi, P., 2021. Learning from laboratory earthquakes: a transfer learning approach to active source seismic monitoring. In: 55th U.S. Rock Mechanics/Geomechanics Symposium.

Settles, B., 2009. Active Learning Literature Survey. University of Wisconsin-Madison Department of Computer Sciences, Madison.

Shaffi, S., 2023. 'It's the Opposite of Art': Why Illustrators Are Furious About AI. The Guardian.

Shi, Y., Song, X., Song, G., 2021. Productivity Prediction of a Multilateral-Well Geothermal System Based on Long Short-Term Memory and Multi-Layer Combinational Neural Network. Energy, Appl, p. 282.

Shoeibi Omrani, P., Dobrovolschi, I., Belfroid, S., Kronberger, P., Munoz, E., 2018. Improving the accuracy of virtual flow metering and back-Allocation through machine learning. In: Abu Dhabi International Petroleum Exhibition and Conference, Abu Dhabi.

Shoeibi Omrani, P., Van der Valk, K., Bos, W., Nizamutdinov, E., Van der Sluijs, L., Eilers, J., et al., 2021. Overview of opportunities and challenges of electrical submersible pumps ESP in the geothermal energy production systems. In: SPE Gulf Coast Section Electric Submersible Pumps Symposium, Virtual and The Woodlands.

Silver, et al., 2016. Mastering the game of go with deep neural networks and tree search. Nature 529, 4840489.

Smith, C.M., 2021. Machine Learning Techniques Applied to the Nevada Geothermal Play Fairway.

Smith, C.M., Faulds, J.E., Brown, S., Coolbaugh, M., DeAngelo, J., Glen, J.M., Ayling, B. F., 2023. Exploratory analysis of machine learning techniques in the Nevada geothermal play fairway analysis. Geothermics 111.

Soltani, S., Kordestani, M., Karim Aghaee, P., 2016. New estimation methodologies for well logging problems via a combination of fuzzy Kalman filter and different smoothers. J. Pet. Sci. Eng. 145, 704–710.

Sutton, R.S., Barto, A.G., 2018. Reinforcement Learning: An Introduction, second ed. The MIT Press, Cambridge, MA.

Suzuki, A., Konno, M., Watanabe, K., Inoue, K., Onodera, S., Ishizaki, J., Hashida, T., 2021. Machine learning for input parameterestimation in geothermal reservoir modeling. In: Proceedings of the World Geothermal Congress. Reykjavik, Iceland.

The European Commission's High-Level Expert Group On Artificial Intelligence, 2018. A Definition of AI: Main Capabilities and Scientific Disciplines. European Commission, Brussels, p. 2018.

Thongsamea, W., Kanitpanyacharoena, W., Chuangsuwanichb, E., 2018. Lithological classification from well logs using machine learning algorithms. BEST 9, 31–43.

Touvron, H., Lavril, T., Izacard, G., Martinet, X., Lachaux, M.A., Lacroix, T., et al., 2023. Llama: Open and Efficient Foundation Language Models. arXiv preprint arXiv:2302.13971.

U.S. Department of Energy, 2023. Play Fairway Analysis. Retrieved from https://www.energy.gov/eere/geothermal/play-fairway-analysis](https://www.energy.gov/eere/geothermal/play-fairway-analysis.

Vaswani, et al., 2017. Attention Is All You Need. arXiv.

Vesselinov, V.V., 2021. Cloud Fusion of Big Data and Multi-Physics Models Using Machine Learning for Discovery, Exploration, and Development of Hidden Geothermal Resources.

Vesselinov, V., Mudunuru, M., Ahmmed, B., 2020. Discovering signatures of hidden geothermal resources based on unsupervised learning. In: Proceedings of the 45th Workshop on Geothermal Reservoir Engineering, Stanford, California.

Vessilinov, V., Mudunuru, M., Ahmmed, B., O'Malley, D., 2022. Machine learning to discover, characterize, and produce geothermal energy. In: Machine Learning Applications in Subsurface Energy Resource Management. CRC Press, Boca Raton.

Vinyals, O., et al., 2019. Grandmaster level in StarCraft II using multi-agent reinforcement learning. Nature 575, 350–354.

Vivas, C., Salehi, S., 2021a. Machine learning approach to generate synthetic sonic logs in geothermal wells. In: GRC Transactions. San Diego.

Vivas, C., Salehi, S., 2021b. Real-time model for thermal conductivity prediction in geothermal wells using surface drilling data: a machine learning approach. In: Proceedings, 46th Workshop on Geothermal Reservoir Engineering. Vol. 1. Stanford University, Stanford, CA.

Wamriew, D., Pevzner, R., Maltsev, E., Pissarenko, D., 2021. Deep neural networks for detection and location of microseismic events and velocity model inversion from microseismic data acquired by distributed acoustic sensing Array. Sensors 21 (19), 6627.

Wang, Y., Salehi, S., 2015. Drilling hydraulics optimization using neural networks. In: Paper Presented at SPE Digital Energy Conference and Exhibition, the Woodlands, Texas, USA, 3–5 March. SPE-173420-MS.

Weiss, K., Khoshgoftaar, T.M., Wang, D., 2016. A survey of transfer learning. J. Big Data 3 (9).

Wu, Z., Pan, S., Chen, F., Long, G., Zhang, C., Yu, P.S., 2021. A comprehensive survey on graph neural networks. IEEE Trans. Neural Netw. Learn. Syst. 32 (1), 4–24.
Yazbeck, J., Rundle, J.B., 2023. A fusion of geothermal and InSAR data with machine learning for enhanced deformation forecasting at the geysers. Land 12 (11).
Yu, Y., Si, X., Hu, C., Zhang, J., 2019. A review of recurrent neural networks: LSTM cells and network architectures. Neural Comput. 31 (7), 1235–1270.
Zhang, Q., Lu, J., Jin, Y., 2021. Artificial intelligence in recommender systems. Complex Intell. Syst. 7, 439–457.
Zhang, S.-B., Si, H.-J., Wu, X.-M., Yan, S.-S., 2022. A comparison of deep learning methods for seismic impedance inversion. Pet. Sci. 19, 1019–1030.
Zhong, R., Salehi, C., Johnson, R., 2022. Machine learning for drilling applications: a review. J. Nat. Gas Sci Eng. 108. https://doi.org/10.1016/j.jngse.2022.104807.
Zhu, X., 2005. Semi-Supervised Learning Literature Survey. University of Wisconsin, Madison.

Key terms and definitions

Artificial intelligence related terms:

Anomaly detection a data analytics problem in which the goal is to predict or identify anomalous data points within a given dataset based on their features.

Artificial intelligence the field of computer science aimed at modeling and/or mimicking human intelligence through computational means.

Classification a machine learning problem in which the goal is to predict what class/group a given data sample belongs based on its features.

Clustering a machine learning problem in which the goal is to divide a set of data samples into groups with similar features.

Deep learning a machine learning that makes use of artificial neural network–type models with complex architectures and/or large numbers of internal parameters.

Data analytics the analysis of (large) datasets using statistics, machine learning, and/or visualization methods.

Feature (measured) variable describing one aspect of a given data sample, used as the input to a machine learning model.

Label variable describing the entirety of a given data sample, such as class or associated regression value. Typically used as the output of a machine learning model in which the goal is to predict a data sample label based on its features.

Large language model a subset of deep learning models trained on processing, understanding, and generating text-based data.

Machine learning the application of advanced mathematical models that can (automatically) learn complex and nonlinear relations from data.

Natural language processing the subfield of artificial intelligence focused on processing, understanding, and modeling natural language (text and speech).

Regression a machine learning problem in which the goal is to predict a continuous output variable associated with a given data sample based on its features.

Token the smallest units of text that a natural language model processes.

Geothermal-related terms:

Blind geothermal system a geothermal system with no surface expression of its reservoir such as springs or fumaroles.

(Well) Completion the process of making a geothermal or oil and gas well ready for production after the initial drilling phase. Also used to refer to the specific design of the completed well itself.

Condition monitoring the use of data analysis software or models for monitoring the (production) status of assets in a system. Particularly aimed at identifying issues in such systems from production data and or reports.

Exploration the characterization of geographical areas with the aim of identifying location of high geothermal potential.

Geothermal potential the potential for a particular location or well to provide effective and efficient geothermal energy.

Play fairway analysis using multiple data sources on geological characteristics of geographical areas to identify the potential for geothermal energy in these areas.

Predictive maintenance the use of physics-based, statistical, or machine learning methods for aiding in planning and scheduling maintenance of production assets or facilities. Primarily focused on predicting the evolution of the performance of an asset over time and/or when it is expected to break down.

Production data time series data measured with physical sensors at high frequency associated with the production of a geothermal well such as wellhead/bottomhole pressures, temperatures at various locations, flowrates, and ESP-related variables.

Production engineering the engineering discipline that involves analyzing field data and models to make informed decisions on how to optimize the performance of geothermal assets.

Seismic inversion using seismic reflection data to predict properties of the subsurface.

Unstructured data data not arranged in a predefined data model or formats such that it cannot be stored in conventional databases. Examples include text documents, logs, and reports.

Virtual metering using physics-based or machine learning models for predicting one or more sensor values based on the values of other sensor data. Typically used to predict values of variables for which sensors are expensive or unreliable (e.g., using topside sensor data to predict expensive downhole measurements).

Well logs data acquired during drilling of a well.

Nomenclature

AI artificial intelligence
AE autoencoder
ANN artificial neural network
API application programming interface
BNN Bayesian neural network
CNN convolutional neural network
ESP electrical submersible pump
DA data analytics
DL deep learning
GAN generative adversarial autoencoder
GNN graph neural network
HMM hidden Markov model
K-NN k-nearest neighbor
ML machine learning
PCA principal component analysis
PINN physics-informed neural network
LLM large language model
NLP natural language processing
PFA play fairway analysis
RAG retrieval augmented generation
RL reinforcement learning
RNN recurrent neural network

CHAPTER

12

Global screening for superhot rock geothermal energy: Geodynamic settings, prospective heat endowment and extraction techniques

Philip J. Ball[a,b], *Graham Banks*[c], *Mikenna Montgomery*[d], *and Juan Carlos Afonso*[e]

[a]School of Geography, Geology and the Environment, Keele University, Newcastle, United Kingdom [b]Geothermal Energy Advisors, Houston, TX, United States [c]Route To Reserves Consulting Inc, Vancouver, BC, Canada [d]M3 Groundwork, San Diego, CA, United States [e]University of Twente, Enschede, the Netherlands

1 Introduction

Humans have interacted with geothermal heat for more than one million years (Lund, 1999); however, it was not until 1904 that geothermal steam was used to demonstrate electrical power production. Following this demonstration, the first geothermal power plant was completed in Larderello, Italy, in 1913 (Lund et al., 2022). Despite the global potential to provide society with heat and power, geothermal energy has remained underutilized (Ball, 2021a, b). To date, geothermal only generates less than 1% of the world's electricity and 0.4% in the USA (IRENA and IGA, 2023; EIA, 2023). In recent years, there have been indications that geothermal could evolve to be an important energy vector and source of baseload energy. This evolution is in-part to an increased awareness regarding the need for low carbon energy sources but also due to innovations within the industry that have changed the potential landscape. Innovations such as engineered or enhanced geothermal (EGS), advanced

Geothermal Energy Engineering
https://doi.org/10.1016/B978-0-443-21662-6.00009-8

Copyright © 2025 Elsevier Inc. All rights are reserved, including those for text and data mining, AI training, and similar technologies.

geothermal systems (AGS), closed-loop geothermal (CLG), ground sourced heat pumps (GSHP), and hybrid geothermal systems (HGS) can be considered Next Generation geothermal technologies (e.g., Ball, 2022; Banks and Ball, 2023). Interest surrounding these new technologies is evidenced through recent reports highlighting the new potential size of the market. Recent market size estimations have indicated that at depths of 1–3 km, the global potential of the shallow geothermal market is on the order of $100 billion USD (BCG, 2022). At depths of 3–7 km, the market could be at the trillion-dollar scale (Wood Mackenzie, 2023). Finally, at the 7–12.5 km, superhot geothermal could offer a multitrillion-dollar market potential (CATF, 2023). These market size estimates are fueling an increased optimism and expectation that drilling advances and a plethora of new engineering concepts could enable geothermal to attain a critical mass and penetrate the heat and power markets at scale. However, as the recent DOE LiftOff report indicates, there are challenges that need to be tackled, including up to $5 billion in funding to develop first-of-a-kind deployments across varied geologies (Blankenship et al. (2024).

Regarding market scalability, the question often arises, how will geothermal scale and reduce costs? Industry advocates think that oil and gas will be the savior of the geothermal industry, because of the alignment between skills and technology (De La Garza and Johnson, 2023; Beard and Jones, 2023). One might argue, therefore, that oil and gas workflows need to be applied systematically across the geothermal industry. If so, portfolios need to be built, and prospects and heat extraction methodologies need to be ranked against each other. Within the oil and gas industry, one of the tools popularized, to the point of mythical status, is the concept of the exploration play (Doust, 2010; de Jager, 2021). If the play concept is employed in day-to-day activities, it can inform risk management and planning, in particular the areas and trends to investigate (Doust, 2010). To date, within the geothermal industry, there are many studies conducted at the play-fairway level which highlight geological trends at a basin or subregional scale (e.g., Ward et al., 1981; Lautze et al., 2017; Faulds et al., 2021; Hannah et al., 2023; Shervais et al., 2024). There are, however, only a few studies that look at exploration concepts from a global scale. Allen and Allen (2013) highlighted an important geological factor behind geothermal drilling, highlighting the underlying geology of the reservoir with increased success in sedimentary sections, not basement. Moeck (2014) introduced the concept of geological controls and especially of geothermal plays, focusing on the heat-transfer mechanism, distinguishing between convection or conduction dominated, with a focus on the porosity and permeability of the reservoir. Jolie et al. (2021) explored which geological factors contribute to the profitable utilization of intermediate-temperature to high-temperature geothermal resources for power generation. They observed that heat flow was focused on regions of active magmatism or where crustal thinning occurred and that permeable structures (e.g., complex fault zones) were the primary influence for fluid flow. This study was heavily biased to conventional geothermal projects and failed to place the resources into a geodynamic setting. Kolker et al. (2022) and Taverna et al. (2024) highlight a relationship between tectonic setting to de-risk exploration in magmatic environments for EGS projects focusing on play-fairway analysis (PFA) approach. Doran et al. (2021) have similarly looked at the benefits of closed-loop projects in these magmatic provinces, such as the Newberry Volcano, USA, but have not applied this to global exploration concepts. Ball (2022) explored the tectonic setting for wells that drilled into temperatures greater than 350°C. One aspect that is lacking in all the earlier analyses is that despite play analyses, there are no yet-to-find estimate analyses for the play concepts which can be easily used to provide estimates of risk by geodynamic setting or tectonic/physiographic province or by type of play. Without a global view, we cannot systematically tackle the technology issues offered by different geodynamic settings or tectonic/physiographic province which if not addressed may limit the first-of-a-kind and next-of-a-kind deployment across varied geological systems.

As consequence of innovation and growth, there has been a rapid widening of the geothermal vocabulary. It is now common to hear terms such as conventional, hydrothermal, traditional, petrothermal, unconventional, blind geothermal, superhot rock (SHR), hot dry rock (HDR), deep hot dry rock geothermal (DHDRG), and next generation (e.g., Faulds et al., 2016; Ball, 2022; Banks and Ball, 2023) and engineering terms such as EGS, AGS, CLG, GSHP, and HGS (e.g., Ball, 2022; Banks and Ball, 2023). This proliferation of terms and acronyms is misleading and confusing nongeothermal audiences, principally because there has not been an equally swift categorization of depth-related terms, engineering approaches, heat extraction methods, and the thermal states of Earth's crust. In contrast, other subsurface natural resource industries, i.e., petroleum, hydrogen, and mineral deposits, center their workflows around geology before economics and engineering. They also organize their vocabulary and entities within a Linnaean hierarchy of categories. For example, the petroleum industry organizes its exploration and production realms into a geological and then economic sequence of analyses: through provinces, basins, facies, plays, prospects, discoveries, and fields (e.g., Doust, 2010; de Jager, 2021; Fig. 1). The main focus of this paper revolves around the creation of maps and tables (Sections 5 and 8, Figs. 4–9 and 21–32, Tables 3–15, and Supplementary Tables S1–S6, S7A, and S7B in the online version at https://doi.org/10.1016/B978-0-443-21662-6.00009-8). These visual aids highlight global-scale relationships between SHRs and geodynamic setting, all grounded in the globally consistent tectonic model of Hasterok et al. (2022) (Fig. 2). In addition to these primary deliverables, this paper also offers (a) a retrospective examination of geothermal exploration, encompassing hydrothermal, EGS, CLG, and wells that have approached superhot conditions (Figs. 13–20); (b) a glossary featuring commonly used geothermal terms and definitions (Section 11), drawn from the more extensive geothermal glossary of Banks and Ball (2023) and SPE (2018); and (c) a framework introducing the Geothermal Entities Classification System (Table 1), which will be expanded upon in Banks and Ball (n.d.).

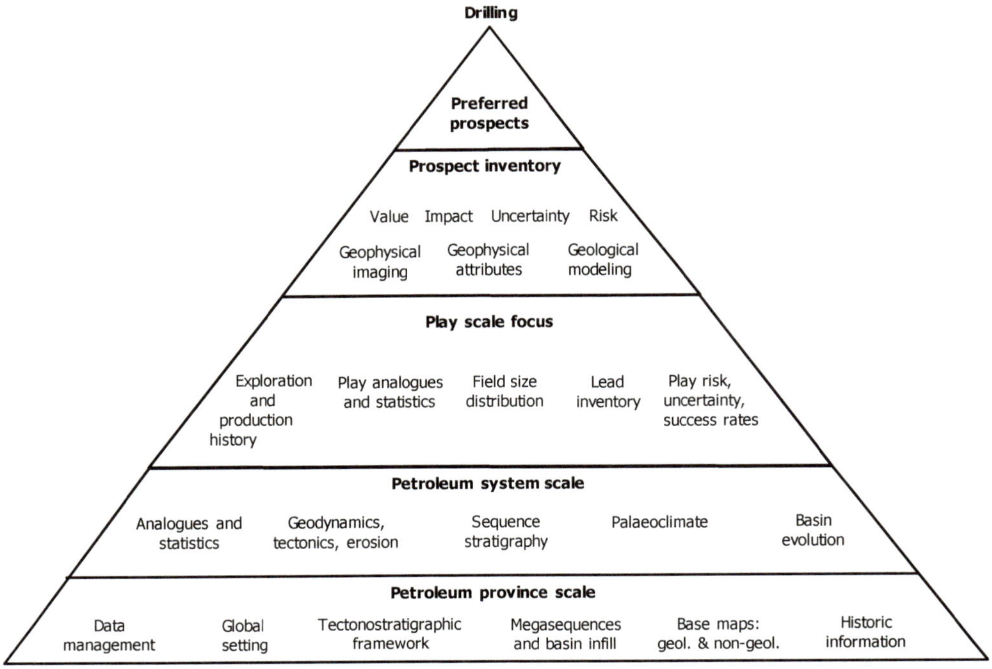

Fig. 1 The typical play-based exploration process for petroleum. The play-based exploration pyramid, used for zooming in from a regional basin focus to a local and more specific prospect focus. *Modified from Royal Dutch Shell, 2013. Play Based Exploration: A Guide for AAPG's Imperial Barrel Award Participants, https://iba.aapg.org/Portals/0/docs/iba/Play_Based_ExplorationGuide.pdf. (Accessed 10 October 2019).*

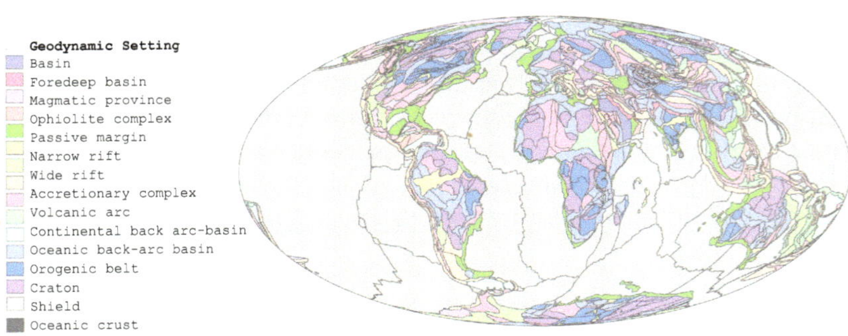

Fig. 2 Tectonic map of the world organized by province type after Hasterok et al. (2022). Note: This classification does not separate paleo- and present-day crustal stress state. Map lines delineate study areas and do not necessarily depict accepted national boundaries.

2 Differentiating geothermal, petrothermal, and hydrothermal types of geological system

The terms geothermal, petrothermal, and hydrothermal are sometimes used interchangeably and sometimes used erroneously throughout the geothermal literature and discussions. However, they are not synonyms because they incorporate different Earth's crust settings, processes, and conditions. They need to be clearly distinguished and specified when designing geoscience, engineering, or economic strategies and programs. Geothermal is merely an adjective relating to heat within Earth (Banks and Ball, 2023). Hydrothermal is an adjective relating to hot water, the action of hot water, or the products of hot water action (Banks and Ball, 2023). Petrothermal is an adjective relating to (very) HDR where an introduced fluid is pumped into the thermal reservoir (Ilyasov et al., 2010; Breede et al., 2013; Moeck, 2014). These adjectives are used to qualify different types of geological heat-transfer system below Earth's surface.

Hydrothermal system and petrothermal system are subdivisions of the broader term geothermal system (e.g., Moeck, 2014). Hydrothermal systems circulate heat toward Earth's surface via a naturally occurring geofluid (Banks and Ball, 2023). They are also termed conventional geothermal system (CGS), presumably because they are the first known and first commercially viable type of geothermal system. All hydrothermal systems are geothermal systems; not all geothermal systems are hydrothermal systems (Banks and Ball, 2023). Meanwhile, the petrothermal system is assumed to lack producible geofluids. These systems are dominantly located in [assumed to be] impermeable rock that would often be below a separate, shallower, hydrothermal system/groundwater aquifer (Banks and Ball, 2023). SHR plays and exploration prospects are assumed to be within the petrothermal system type and will need to be engineered, to extract the heat from the thermal reservoir. However, the presence of naturally circulating fluids in an SHR play remains plausible, as indicated by the commonly held view within the mining industry that some hydrothermal mineral deposits form in the petrothermal realm. This implies that a potentially exploitable rock permeability might exist even in regions traditionally perceived as impermeable, such as petrothermal zones (Marchesini et al., 2019; Bischoff et al., 2024).

Humans can convert petrothermal systems and other low permeability, geothermal systems into hydrothermal systems with engineering interventions and an introduced working fluid, i.e., an EGS (Banks and Ball, 2023). Examples of engineered processes which have extracted heat from petrothermal settings are demonstrated by EAVOR (Zatonski and Brown, 2023), GreenFire (Scherer et al., 2020), and FERVO (Norbeck and Latimer, 2023). Emerging hybrid examples have not been demonstrated beyond the laboratory, but a demonstration is planned in 2024 (Jacobs, 2024). For those new to the topic, the definitions of system, CGS, EGS, geofluid, hydrothermal system, impermeable, introduced working fluid, permeable, petrothermal system, play, prospect and reservoir are also provided in Section 11.

3 The superhot rock type of play

A SHR geothermal play is the segment of a chosen petrothermal system that is superhigh enthalpy, i.e., at very high temperature and high-pressure conditions (Banks and Ball, 2023), at depths that the organization could be capable of exploiting. There will be many of them globally. As a rule they are assumed to be able to deliver supercritical fluids or superheated fluids—with a very high energy density—to Earth's surface. The benefits of targeting superhigh-enthalpy plays is that these systems could provide much higher enthalpy in the produced fluids than that which is produced in conventional hydrothermal systems (e.g., Suárez-Arriaga, 2019; Scott et al., 2023; Rivera et al., 2023; Rivera, 2023). See definitions of play, geothermal play, supercritical state, SHR geothermal play, and superhigh enthalpy in Section 11 and Banks and Ball (2023). The rest of this section summarizes how the SHR geothermal type of play is categorized in relation to the myriad of geothermal-related entities (Table 1 and Fig. 3).

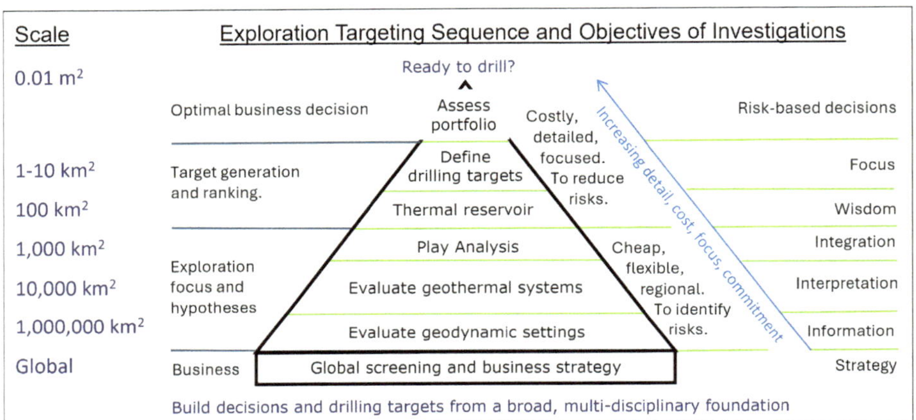

Fig. 3 The exploration triangle sketch is a simple depiction of the hierarchical sequence of tasks to discover underground SHR resources and their scales of investigation. *Modified from Banks, G.J., 2018. Towards a holistic exploration workflow: adapting play-based exploration for minerals. In: EIT Raw Materials Exploration Workshop, DTU, Denmark.*

There is a wide span of geothermal-related entities and vocabulary that encompasses many natural and human constructs, e.g., geological processes, rock types, categories of enthalpy, engineering techniques, stages of project development, economic categories, etc. This paper's authors have not seen a framework or analysis sequence that organizes them all. Therefore, we introduce a hierarchical Geothermal Entities Classification System that arranges geothermal-related entities and provides a sequence of tasks when assessing the viability of any geothermal region/project (Table 1). This system should enable all geothermal industry stakeholder groups globally to understand, organize, and compare geothermal-related entities—in a systematic way across jurisdictions and organizations—based on their genetic relationships and $ value. This Geothermal Entities Classification System parallels the framework and investigation sequence that is internationally standard practice for petroleum-related entities (e.g., Doust, 2010, 2017; Royal Dutch Shell, 2013) (Table 1) and mimics the Linnaean classification system (that groups species into a hierarchy or "ranking" based on their anatomical similarities and "family" relationships) (Table 1).

The Geothermal Entities Classification System contains seven levels of ranking, where each level (rank) is a group of entities and a member of the rank below. As an example we give two examples: (1) SHR is a type of play (level 3) and a member of the petrothermal type of

TABLE 1 The seven levels of categories in the Geothermal Entities Classification System, with the seven levels (ranks) of petroleum and Linnaean classification systems for comparison. Note superhot wells are omitted from level 7 deliberately, since they are often accidental penetrations.

Level of classification and analysis	Sequence of analysis	Footprint (km^2)	Types of geothermal entities	Examples	Petroleum entity analogs	Linnaean classification analogs
Commodity			Thermal energy in water, rock, or sediment		Thermal energy in petroleum in rock or sediment	
Engineering and economic entities and human desires						
Seventh level	Last	Smallest	Energy extraction technique (how humans get the heat to the surface)	Closed-loop circuit, open-loop circuit, next-generation circuit, hydrothermal (i.e., conventional geothermal) system, hydraulic fracturing, hybrid system, storage (mechanical)	Petroleum extraction technique	Seventh rank; species (*Homo sapiens*)
Sixth level			Category of thermal resource (economic and risk categories)	Production, reserves, contingent resource, prospective resources, unrecoverable resources, exploration result	Category of petroleum resource/reserve	Sixth rank: genus (*Homo*)
Fifth level			Stage of project development (i.e., effort, knowledge, and yield)	Naturally producing geothermal field, engineered geothermal field, artificially fractured reservoir, discovery, postdrill, dry well, flowing well, prospect, exploration lead, exploration concept	Stage of project development	Fifth rank; family (hominids)
Natural entities that occur regardless of humans						
Fourth level			Thermal energy reservoir (a portion of a play that stores the thermal energy; the types are categorized by lithology and porosity-permeability)	Sedimentary rock, crystalline rock, naturally fractured rock, weathered rock, sediment (e.g., soil, sand, gravel), flooded mine shaft	Reservoir	Fourth rank; order (primates)

Continued

TABLE 1 The seven levels of categories in the Geothermal Entities Classification System, with the seven levels (ranks) of petroleum and Linnaean classification systems for comparison. Note superhot wells are omitted from level 7 deliberately, since they are often accidental penetrations—cont'd

Level of classification and analysis	Sequence of analysis	Footprint (km^2)	Types of geothermal entities	Examples	Petroleum entity analogs	Linnaean classification analogs
Third level			Play (a portion of a system that hosts a group of genetically similar fields, discoveries, prospects, and exploration concepts)	Superhot rock, hot dry rock, dry rock, volcanic, magmatic, hydrothermal, meteoric hydrothermal, etc.	Play	Third rank; class (mammals)
Second level			Geological system (and category of enthalpy, pressure, and depth)	Petrothermal (dry rock) or hydrothermal (with water) Enthalpy: very low, low, medium, high, superhigh enthalpy Pressure: high, low, overpressured Depth: shallow, <400 m; moderate depth, 400 m to 5 km; deep >5 km	Petroleum system	Second rank: phylum (chordates)
First level	First	Largest	Geodynamic setting (geological source of heat; segments of Earth to focus our efforts to discover anomalously high heat flow)	Continental crust, oceanic crust, rift system, volcanic arc, back-arc, mantle hot spot, active igneous province, cooled igneous province, sedimentary basin	Petroleum province	First rank: kingdom (animals)

geological system (level 2) that could occur in several types of geodynamic setting (level 1) (Table 1); (2) A closed loop circuit that is extracting SHR with economic viability could be classified as follows from level 1 up to 7: geological rift system: petrothermal heat system: SHR type of play: crystalline rock: geothermal field: Reserves: closed-loop circuit. Further explanations and use of this Geothermal Entities Classification System, using actual examples, will be provided in Banks and Ball (n.d.). A sample set of project hierarchies are supplied in the supplementary files, Supplementary Table S1 in the online version at https://doi.org/10.1016/B978-0-443-21662-6.00009-8.

4 Global screening of SHR geothermal energy and the exploration triangle

SHR geothermal exploration contains several scales and objectives of investigation (Fig. 3). Organizations need to apply the right level of detail to each task. The work should be proportionate to the scale of geological understanding and the scale of the business decision that the organization is trying to make. Done properly, it is providing a logically structured sequence of steps that allows professionals to meticulously organize and refine geological data at each stage of investigation. This ensures a systematic analysis and the effective transfer of information across scales of investigation. Moreover, the precise identification of potential sources and rigorous assessment of their economic viability leads to the formulation of informed decisions regarding exploration and subsequent development strategies. This "system type" of analysis benefits an organization through the organization of information and analysis (Table 1 and Fig. 3).

Considerable benefits can be realized by committing time and effort to do basic concept-/prospect-scale evaluations at the regional scale. Such analysis could reveal positive trends, differences, or fundamental flaws in a given geothermal system. To date, however, the geothermal industry is not taking this big-picture view. Instead, focusing on local intricacies is caused by an inability to view critical issues at an appropriately large scale: to "see the forest rather than the trees." Such mistakes are observed across other industries. For example, 90% of mineral exploration publications are about district- to meter-scale details (less than 5% have global application), even though mineral deposits are formed by tectonic-scale processes (McCuaig et al., 2010, 2018; Groves and Santosh, 2015). The geothermal industry is in this same, paradoxical, property-scale mindset. This does not give full assistance to government policymakers and deep geothermal organizations who need to make intelligent decisions on the jurisdiction scale and the geothermal province scale.

Global screening and the hierarchical thinking of the exploration triangle (Fig. 3) provide the technique for gaining the "big-picture" context about the location of the optimal SHR plays. Global screening should automatically be the first task of an organization's exploration triangle workflow (Fig. 3). This phase of work creates the robust foundation for the subsequent sequence of work phases, namely, geodynamic setting evaluation, play analysis, prospect generation and maturation, prospect risk and volume, and engineering integration (Fig. 3). This philosophy for deep geothermal exploration parallels the standard geology-engineering-economic assessment sequence of petroleum, carbon storage, and rare earth elements exploration sectors. Hierarchical thinking, using the exploration triangle (Fig. 3), clarifies what the organization should be working on, promotes focus, facilitates planning, and enables the best use of its valuable time and resources.

Global screening is an integral part of a business or research decision-making process for SHR evaluations because it.

(a) Creates the foundation for all prospectivity investigations and mapping.
(b) Sets individual business opportunities into context.
(c) Includes organization strategy, country screening, and database building stages of the holistic SHR exploration process.
(d) Provides an early filter to the breadth of ideas.
(e) Assesses the quality of any global region "belowground" (i.e., subsurface geology and engineering, including the exploration history) and "aboveground" (i.e., country/commercial/fiscal background).
(f) Identifies which global regions to progress a project up the exploration triangle into maturer ideas or opportunities.
(g) Ensures an organization invests time and budget on the best SHR global locations and plays that are aligned with the organization's strategic intent (e.g., Route To Reserves, 2018; Banks et al., 2020).

Global screening is therefore inextricably linked to SHR strategic, economic, and portfolio analyses (e.g., Banks et al., 2020; Royal Dutch Shell, 2013; Route To Reserves, 2018).

The global screening person/team in an organization provides consistent and auditable global basin screening. This is to compare the fundamentals (quality of geodynamic settings, play size, depth to heat, heat endowment, exploration history, success rates, energy already yielded, viable extraction techniques, etc.) of SHR geodynamic settings and plays worldwide. The most cornerstone of SHR geothermal energy global screening lies in comprehensive global maps and statistical data showcasing geodynamic settings. This paper delivers these global-scale maps and statistics (Section 8.5, Figs. 21–30, and Supplementary Tables S2–S6, S7A, and S7B in the online version at https://doi.org/10.1016/B978-0-443-21662-6.00009-8) to the geothermal community.

Global screening is a simple and low-cost way for junior exploration companies and research and development groups to generate knowledge that will generate a global-scale impact on the geothermal sector. Global screening is also valuable during other business activities, e.g., to identify upside around geothermal fields, for asset acquisitions, or for joint venture evaluations (Banks et al., 2020).

5 Superhot rock mapping

As introduced in Section 3, SHR geothermal plays are assumed to be superhigh enthalpy. Pure water would be a supercritical fluid above 374°C and 22.1 MPa. However, groundwater and "deep groundwater" represent a vast body of saline water that exists down to at least 10 km (Ferguson et al., 2021). "Average seawater" becomes a supercritical fluid above 407°C and 30 MPa (e.g., Hovland et al., 2014). Groundwater systems can be more saline than "average" seawater. Therefore, maps showing depth to SHR geothermal plays and reservoirs, or locations where an SHR heat reservoir's fluid would be in the supercritical state, should be based upon the 450°C isotherm (isotherm: depth to the same contour of temperature across a region), as shown in Fig. 4. This section explains how the depth to the 450°C isotherm was modeled globally for the purpose of geothermal screening/prospecting, and the resultant "steady state conductive" maps (see definition of steady state in Section 5.1) were constructed (Figs. 21–28). By steady state we mean that the temperature within the lithosphere is assumed to be in equilibrium and time-independent (i.e., not changing with time).

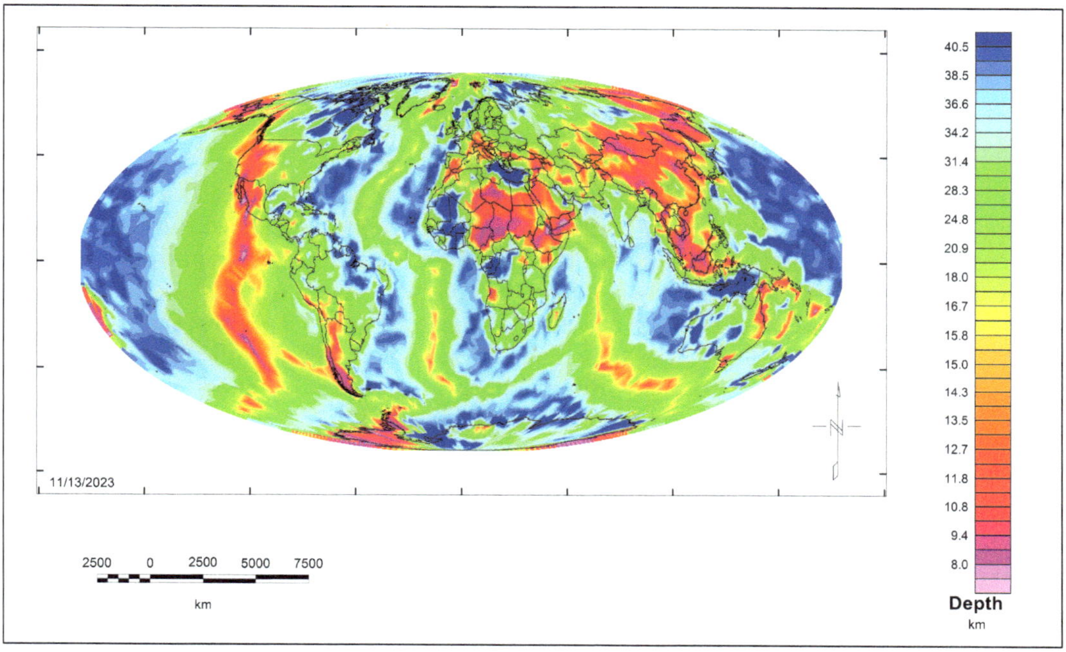

Fig. 4 LithoRef18 computed 450°C isotherm calculated, Mollweide Projection. This is the first global representation of the 450°C steady-state conductive isotherm. Map lines delineate study areas and do not necessarily depict accepted national boundaries.

The maps essentially reveal where shallow heat may be predicted assuming that (a) conduction is the principal process within the lithosphere and that (b) lithospheric thickness is a first-order factor controlling the local geotherm. The first-order influence that the lithosphere has on the geotherm is the reason why we embarked on lithospheric-scale modeling.

In contrast to the regional, steady-state conductive conditions within the Earth's lithosphere, there are areas that exhibit transient or advection-dominated heat transport related to shallow heat-transfer phenomena occurring in the upper 5–8 km of the Earth's crust. It's worth noting that the depth of these phenomena is currently poorly constrained and may extend deeper than currently assumed. The regions with evidence of either unsteady or advection-dominated anomalies identified in this study may indicate local manifestations of vigorous flow of hot groundwater or magma through fractures or highly permeable zones. These convective/advective processes disrupt the normal steady-state heat conduction within the lithosphere (Figs. 8–9 and 27–28). This rapid transport of hot fluid to shallow aquifers along fault zones and fracture zone does not necessarily result in the surrounding rock volume being heated up, and care is needed because shallow heat resulting from these processes may not be linear below in areas where the steady-state conductive gradient resumes.

In this analysis, we attempt to separate the two processes. Exploration and production of SHR geothermal and other categories of deep geothermal should target regional-scale locations of steady-state, heat conditions rather than nonsteady-state heat flow anomalies, to ensure long-term maintenance of geothermal heat source and operations. Conductive steady-state geotherms within the Earth's lithosphere were obtained from an updated version of the global lithospheric model LithoREF18 (Afonso et al., 2019; Ball and Afonso, 2023) (Section 5.1).

5.1 The steady-state 450°C isotherm

LithoREF18 is a 2° × 2° global lithospheric reference model that also includes the sub lithospheric mantle down to 410 km. Using an updated version of LithoREF18, we use this low-resolution lithospheric model to calculate the first insight into the depth of the 450°C, steady-state conductive isotherm (Fig. 4). With the absence of deep well penetrations or other data, the model was obtained through a formal joint inversion of 3D gravity anomalies, geoid height, satellite-derived gravity gradients, and absolute elevation complemented with seismic, thermal, and petrological prior information. The model includes full lithospheric geotherms, under the steady-state assumption, via the solution of the heat-transfer equation with internal heat production (in the crust). Both the oceanic and continental lithospheres are modeled to provide a seamless continent-ocean model (see Afonso et al., 2019 and Ball and Afonso, 2023 for details).

Since surface heat flow measurements were not included as input data in the joint inversion, the computed SHF is a true prediction of LithoREF18, and it is influenced by the choice of thermal parameters in the shallow crust (Fig. 5). Measurements, on the other hand, are

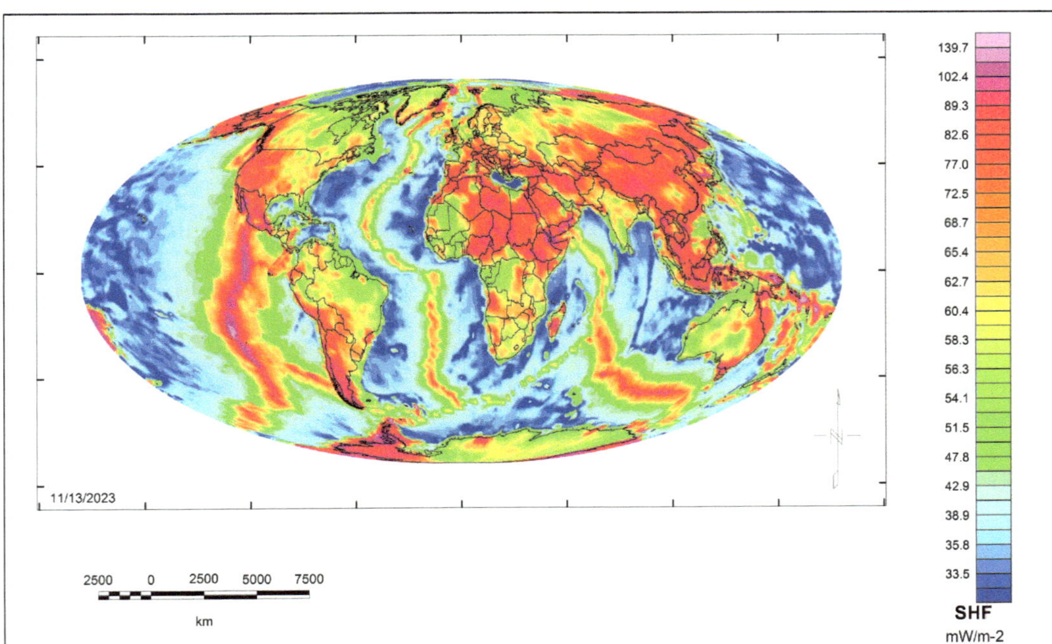

Fig. 5 Calculated surface heat flow derived from the LithoRef18 Earth reference model (Afonso et al., 2019). Mollweide Projection. Map lines delineate study areas and do not necessarily depict accepted national boundaries.

affected by local, shallow perturbations (e.g., underground water circulation) that are commonly difficult to remove/correct in order to isolate the large-scale or "whole lithosphere" thermal structure.

With these caveats in mind, we note that a comparison with real measurements (e.g., the recent compilation of Fuchs et al., 2023) (Fig. 6) and/or models based on real SHF measurements (e.g., Lucazeau, 2019) (Fig. 7) reveals a close agreement in the general spatial pattern of SHF variation. Furthermore, considering typical uncertainties in SHF measurements (10%–15%), our predicted SHF values in continental areas are comparable to those from global compilations in many parts of the world, validating the first-order lithospheric structure of LithoREF18. In the offshore domains, Lucazeau (2019) proposes and introduces a new thermal cooling model for young oceanic crust, resulting in significantly higher surface heat flow values (Fig. 7).

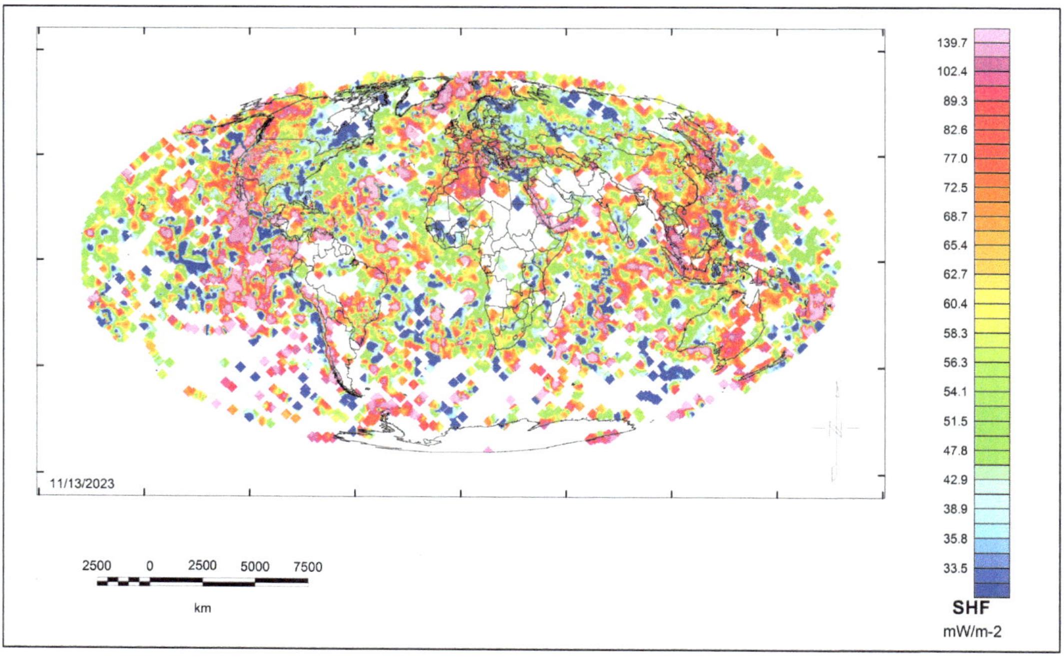

Fig. 6 Measured surface heat flow, after, Fuchs et al. (2023). Mollweide Projection. This figure shows the actual distribution of surface heat flow measurements and the range of heat flows mapped at the surface. Map lines delineate study areas and do not necessarily depict accepted national boundaries.

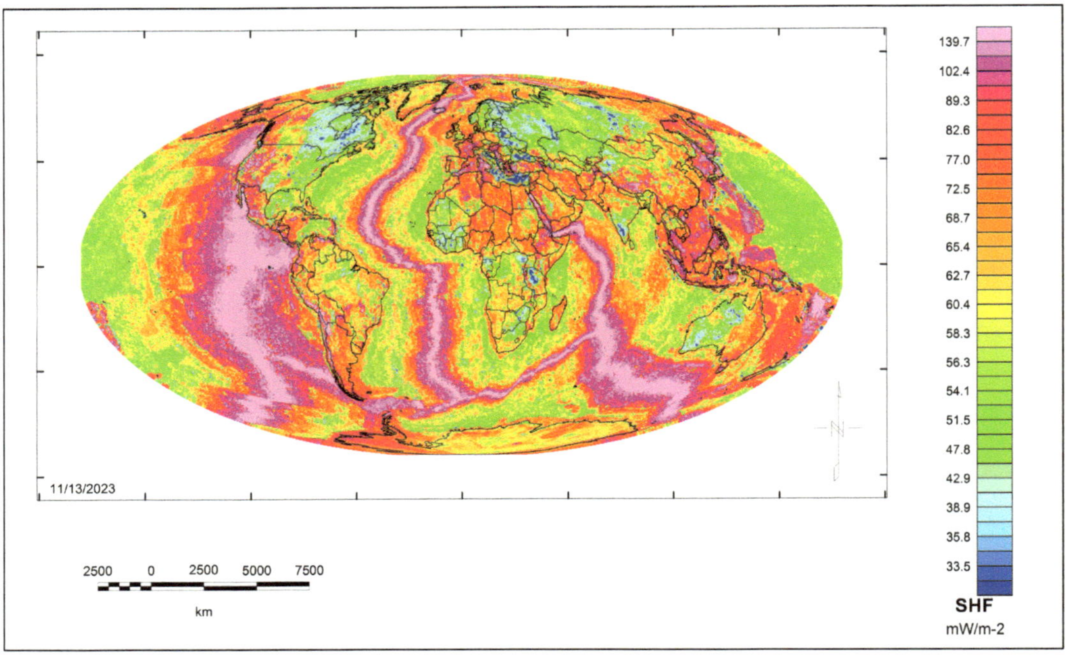

Fig. 7 Global surface heat flow model; 14 observables model (Lucazeau, 2019). Lucazeau (2019) presents a global surface heat flow model which combines global heat flow measurements with the generalized similarity method for extrapolation into areas of similar geology. The surface heat flow model is constructed using a global 0.5° × 0.5° grid, which translates to scale of approximately 55 km × 55 km. Map lines delineate study areas and do not necessarily depict accepted national boundaries.

There are some limitations to the approach used to model the depth to the 450°C isotherm: First, with a surface discretization of 2° × 2° (ca. 230 km × 230 km) grid, the model is low resolution and only captures the large-scale, first-order temperature variations within the crust/lithosphere. Small wavelength structures can be below resolution, for example, the back-arc rifting in Tuscany or the Taupo volcanic line in New Zealand. Second, the model does not compute the convective or advective heat flow which locally perturbs the shallow crustal geotherms—this includes recent igneous activity linked to subduction processes, partial melting, or adiabatic decompression. Third, the current version of LithoREF18 does not include an optimization of the thermophysical parameters (e.g., conductivity, volumetric heat production) of the crust, which have a significant effect on the shallow temperature structure. Fourth, the model does not include an explicit treatment of igneous activity linked to deep mantle plumes. Therefore, volcanic centers such as Hawaii and Iceland are not explicitly modeled. Finally, the estimated depth uncertainties associated with the 450°C isotherm are ~20%. In summary, the conductive model calculation will likely underestimate local temperatures where vigorous convective processes and/or shallow magmatic systems exist. These are of particular relevance to geothermal potential. Therefore, considering the abovementioned uncertainties and the modeling bias, our analysis and maps include a small but realistic correction of +5 km.

5.2 Nonsteady state domains

To account for convective and advective process due to hydrothermal or magmatic activity, it is assumed that this information can be indirectly extracted by calculating the difference between the measured surface heat flow and the modeled surface heat flow, resulting in a residual anomaly, or "hydrothermal proxy" map. While the assumptions represent a significant simplification of geological processes, it enables us to assume, until further investigation is completed, that residual anomalies are regions where nonsteady-state process (i.e., hydrothermal activity) occurs within the Earth's crust. It is acknowledged that this requires simplifying assumptions, which is not without caveats. It is important to remember that the difference between calculated heat flow and the measured heat flow is a convolution of a set of possible solutions: (1) error in reported heat flow measurement; (2) hydrothermal activity; (3) shallow magmatic activity; (4) natural variations in assumed U, Th, and K of rocks; (5) variations in assumed thermal conductivities of rocks; (6) errors in assumed lithospheric structure; (7) thermal perturbation or intrusions that are below resolution of the model; (8) the model does not correct for transient thermal effects due to sedimentary loading in basins; (9) gridding artefacts; (10) errors in the IHFC global heat flow database. Which is still limited in its data acquisition and density, and itself may also be subject to errors in temperature reading and quality; (11) gaps in the difference model represent areas with no-data.

Acknowledging the limitations behind the assumptions, there are two ways in which this calculation can be completed. The first is to compare the LithoREF18 computed surface heat flow with the global database of heat flow measurements (Fig. 8). The International Heat Flow

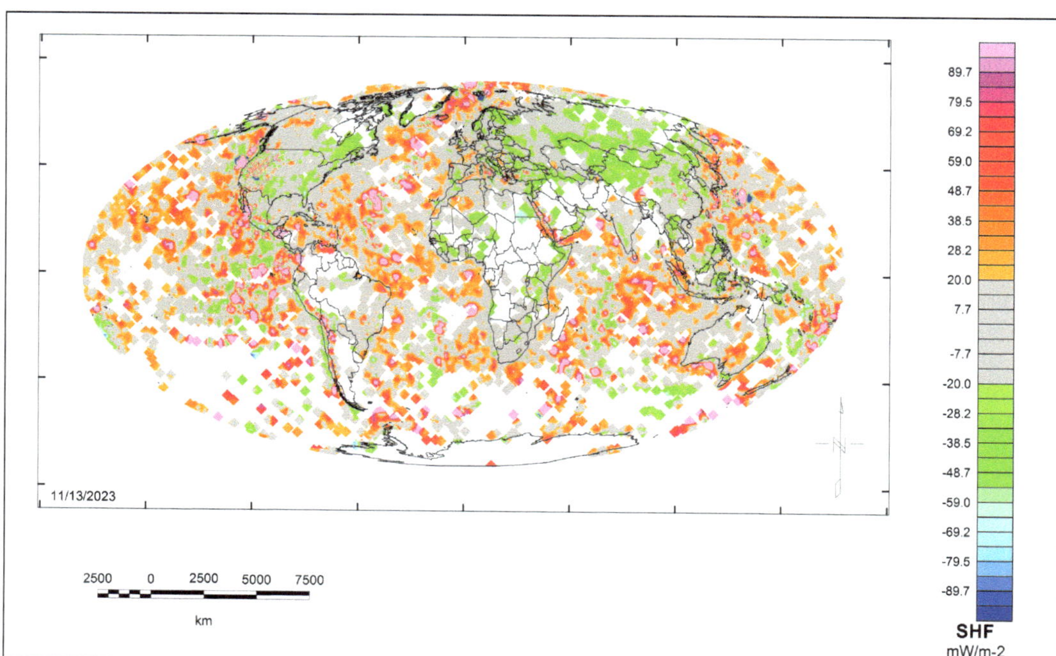

Fig. 8 The global residual or difference in surface heat flow (SHF) from Fuchs et al. (2023), minus the SHR computed from the LithoRef18 model (Afonso et al., 2019). The result shows pinks/reds indicating the measured surface heat flow model is hotter than the predicted LithoRef18 model and blues/greens indicating LithoRef18 model is hotter than the measured surface data. White spaces are data gaps. Mollweide Projection. Map lines delineate study areas and do not necessarily depict accepted national boundaries.

Commission (IHFC) model in the 2023 data release contains 73,033 data points (40,082 from the continental domain and 32,951 from the oceanic domain). Because the IHFC database is not consistent globally and there are data gaps, this forced us to resort to the modeled surface heat flow method of Lucazeau (2019), where measured data coverage is inadequate. Constraining a model with another model is an imperfect solution, but in the absence of data, it represents the next best option (Fig. 9). Note that the difference observed offshore at the mid-ocean ridges and in young oceanic crust is due to a new method of estimating hydrothermal heat flow loss for oceanic crust (Lucazeau, 2019).

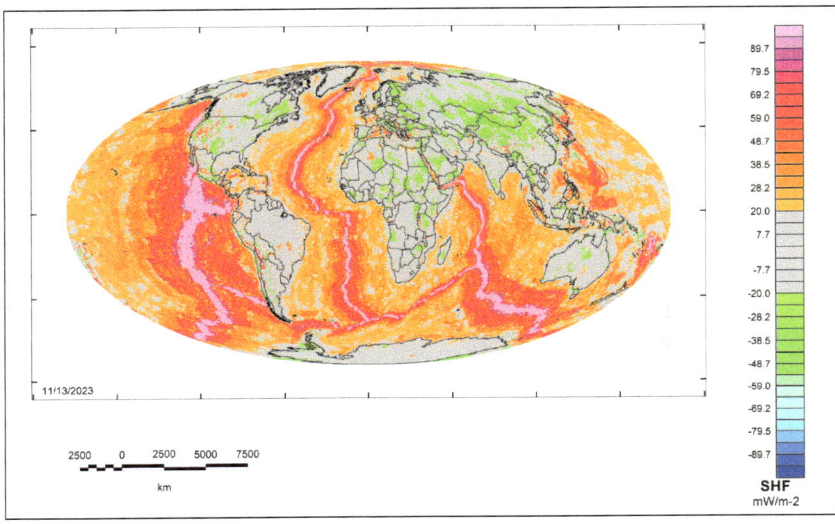

Fig. 9 The global residual or difference in surface heat flow (SHF) from Lucazeau (2019), minus the SHR computed from the LithoRef18 model (Afonso et al., 2019). This model infills the gaps from measured heat flow using a generalized similarity method to predict the heat flow. The result: pinks/reds indicate the Lucazeau surface heat flow model is hotter than the predicted LithoRef18 model, and blues/greens are where LithoRef18 model is hotter. Mollweide Projection. Map lines delineate study areas and do not necessarily depict accepted national boundaries.

In order not to overestimate the potential areas of SHR, only the areas that had a residual anomaly greater than +40 MW/m^2 were selected, allowing us to select only the most anomalous or hottest regions. The workflow presented in Fig. 10 outlines how the hydrothermal proxy map was created. The steps and assumptions are as follows (i) using Oasis Montaj, grid math tool there were two options; Option 1: the LithoREF18 gridded data was subtracted from the IHFC measured surface heat flow data Fig. 5, (Fuchs et al., 2023). In Option 2, the LithoREF18 model was subtracted from the Lucazeau (2019) heat flow model (Fig. 7). (ii) To consider the areas where superhot could be shallow, a "hydrothermal layer" is created by selecting regions where the residual anomaly is greater than +40 mW/m^2. The areas of interest are contoured and extracted from Oasis Montaj and turned into closed polygons in ArcGIS. (iii) To investigate the validity of the proposed "hydrothermal" areas and a data bias in ArcGIS we plotted the measured data from the IHFC database. If the data density was too low, and there was a risk of overestimation or no data returned due to data gaps, in this case it is recommended that the result from Option 2 is selected.

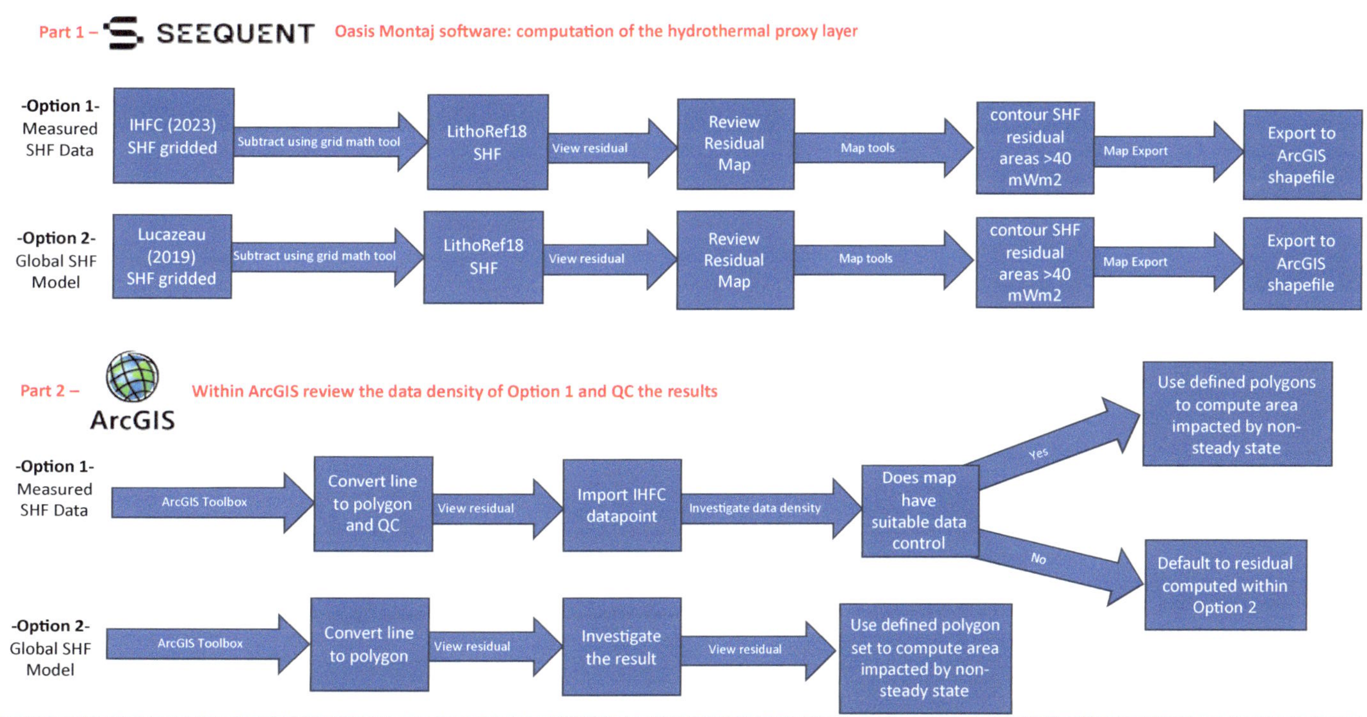

Fig. 10 Nonsteady-state (hydrothermal) workflow—this flowchart details the steps for creating the regions of nonsteady state (hydrothermal proxy maps).

Comparison of the LithoREF18 SHF model with the IHFC database is the preferred approach; however, in regions where the measured heat flow data is minimal (Fig. 8), it was necessary to resort to the modeled surface heat flow method of Lucazeau (2019) (Fig. 9). The resulting maps are not directly relatable to a given depth. For simplicity, it is assumed they represent the 4–7.5 km range. In the future, this may be refined using a 1D analytical solution to the convective–diffusive problem, but further analysis is needed. One assumption that is forced here is that if the anomaly pattern is not observed in deeper depth slices with the conductive model, the hydrothermal proxy areas are added to the 10 and 12.5 km slices, since it could be considered anomalous to have heat at shallow locations and not at depth.

6 Prospective heat and power endowment methodology

Heat endowment is the quantity and quality of geothermal heat in a particular geographic region or depth. It can be reported in MW-thermal (MWth) and/or MW-electric (MWe) units. Heat endowment can be used to compare and rank the natural heat "wealth" of different regions above project-specific cutoffs (Banks et al., 2020; Banks and Ball, 2023). The methodology and inputs to estimate the possible endowment of an SHR region include heat density, depth to heat reservoir, geothermal play type, and available technology. The process of estimating a region's possible SHR endowment is similar to the yet-to-find process used in petroleum exploration, except that for SHR it is not yet conducted at the play scale, due to the absence of practice in the geothermal industry. A petroleum yet-to-find number is an estimate of the amount of petroleum that could be undiscovered within a petroleum play, basin, province, or country. Yet-to-find estimation has played a key role in exploration success worldwide (Schenk et al., 2012; Lattoroli et al., 2018). Despite the data constrained approach, lookbacks of historical yet-to-find studies show that this methodology was useful, and some yet-to-find maps made >20 years ago were accurate in identifying a region's petroleum "sweet spots" (Lattoroli et al., 2018). Due to the difficulties of achieving the YTF resolution, we present a prospective endowment methodology by geodynamic setting using the tectonic model of Hasterok et al. (2022).

The prospective SHR endowment estimates are initially uncertain because they are made before a period of prolonged exploration, i.e., in a sparse data environment. Despite the uncertainty, there are numerous reasons why organizations and governments should conduct an SHR endowment evaluation to a scoping level of detail, as many have for petroleum yet-to-find estimations. An SHR endowment estimate is a crucial input to

- Identify which regions a limited exploration or technology budget would be best focused at for a potential return on investment.
- Plan for long-term energy policy, strategy, land use, and location of infrastructure.
- Determine the questions that decision-makers need to ask and the information they need.
- Avoid areas that would present environmental/social challenges.
- Strategic planning for technology development and access by geodynamic type.

There are several screening methods that can be applied to estimate a region's deep heat endowment. Each method has its own strengths and weaknesses (Foum, 2018). The key, however, is to complete the analysis consistently globally and with similar data quality. If done, the workflows can give you a relative indicator of the potential resource in any given area.

The rest of this section summarizes the first attempt to estimate Earth's prospective heat and power endowment within the emerging SHR play type (Section 3). The depth to the 450°C conductive isotherm was derived from the LithoREF18 reference earth model (Section 5). The "Area That Is Developable" method was selected. This approach proposes simply that 1% of SHR heat is developable because competing surface and subsurface land rights and distance to infrastructure could mean that 99% of SHR resources may not initially be available for development. This method is easy to apply and geological input to the estimate is minimal (which aligns with the currently sparse global dataset). This analysis is "evergreen," i.e., able to be updated easily as new data becomes available. One limitation that is glaring at this time is that the current analysis is based on an EGS development scenario with a flash power plant installation. Energy density data for CLG or hybrid developments does not exist within the literatures base.

The workflow, illustrated in Fig. 11, is as follows:

- **(i)** The 450°C isotherm from LithoREF18 was projected and gridded in Oasis Montaj, and a 5-km depth adjustment was applied globally, to account for an optimistic scenario.
- **(ii)** Using Oasis Montaj, contours were extracted at depths to create the conductive depth slices at 5, 7.5, 10, and 12.5 km for the conductive, dry rock resources. Using the hydrothermal proxy layer, the area in an assumed 4–7.5-km depth range is extracted, as discussed earlier.
- **(iii)** The extracted contours per depth maps were imported into ArcGIS and converted into closed polygons.
- **(iv)** Using the ArcGIS "Intersect" tool, the area of the hydrothermal proxy and the depth slices were extracted, based on the tectonic map of Hasterok et al. (2022). This step is repeated for each of the maps, including the hydrothermal proxies at 5.0, 7.5, 10.0, and 12.5 km.
- **(v)** Finally, a table is exported—using the simplified developable area method, the area that can be developed is reduced to an assumed 1% developable.
- **(vi)** The power potential for each geodynamic setting, globally, using the calculated 1% areas was calculated. The required inputs are (1) mapped area and (2) an estimation of the power density per km^2. For the different scenarios, this power density is calculated using results from the CATF technoeconomic model (Herter, 2023). While the power density retrieved is dependent on the flowrate assumed, a simple physical constraint of 1 km^2 for four production wells was assumed—this requires a 0.5-km well spacing including the required injection wells at a 2:1 ratio of injecting well to producing well.
- **(vii)** Calculate SHR potential for 500, 250, 100, and 50 MWe geothermal power plant scenarios, enabling the reader to consider the cost associated with the size of planned geothermal plant design. Due to the limited data on geological success, commercial success of SHR geothermal projects we do not apply a risk number at this time.

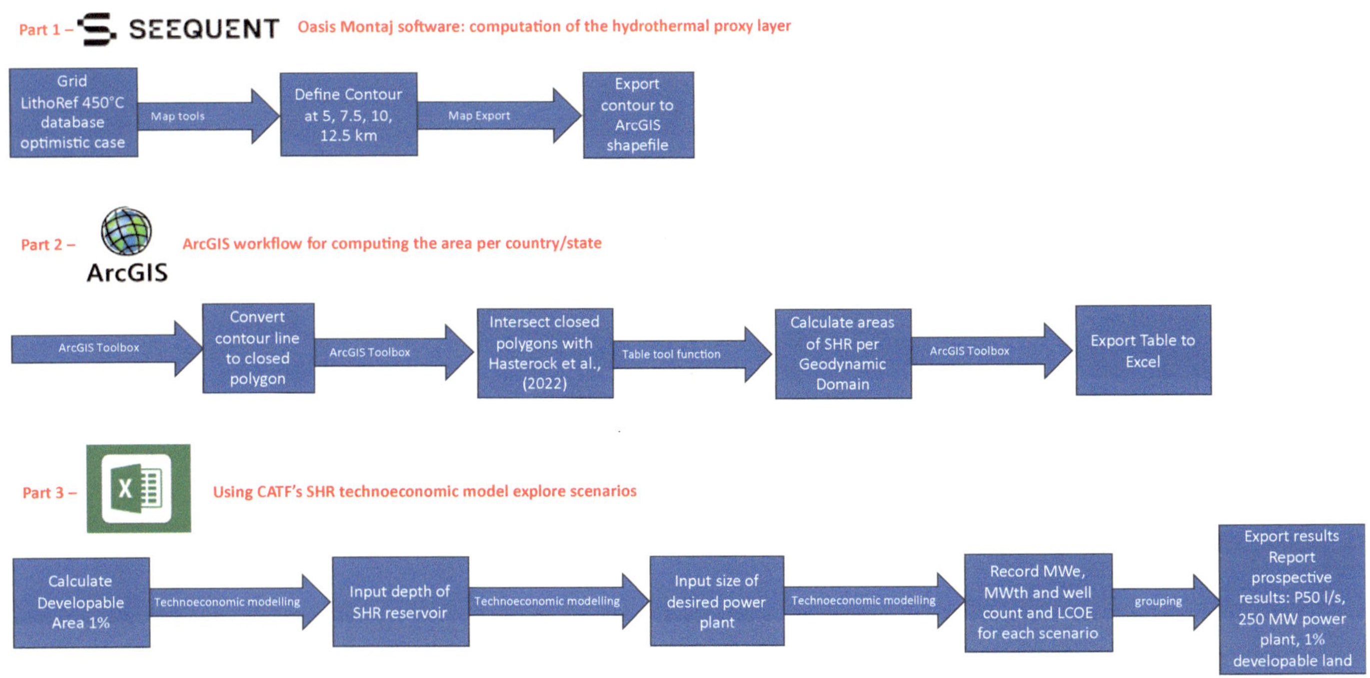

Fig. 11 The sequence of steps used to compute and model SHR endowment globally using the 1% area is developable method.

For simplicity, the results presented for each geodynamic setting are the P50 flowrate, 55 L/s, and an assumed 250 MWe power plant. We do not speculate here over what is drillable with today's technology. We do however compile a list of deep wells (Section 8.1) and superhot wells (Section 8.2, Supplementary Table S2 in the online version at https://doi.org/10.1016/B978-0-443-21662-6.00009-8) to inform the reader what has been achieved to date in the drilling world. The resource size presented highlights the potential prospective resource that could be unlocked if governments around the world invested in research toward drilling, well completion, reservoir/permeability creation, and power plant design.

7 Energy extraction and production techniques and technologies

This section summarizes the techniques that have brought heat from SHR reservoirs up to Earth's surface and the emerging geothermal engineering techniques that could produce commercially viable SHR reservoirs, namely, a superhot well, a CGS, an EGS, a CLG circuit, and a HGS. Low-temperature geothermal applications and technologies are not included. Compilations of the respective databases for well or geothermal project are available as supplementary files (Supplementary Tables S3–S6, S7A, and S7B in the online version at https://doi.org/10.1016/B978-0-443-21662-6.00009-8). Each datapoint was georeferenced and cross-correlated with the tectonic map of Hasterok et al. (2022) to extract the geodynamic setting within which the project occurred.

The deliberate targeting of SHR has been a relatively recent activity for the geothermal industry with many of the early well penetrations encountered superhot conditions while exploring for conventional resources. We observe, globally, that 53 wells have penetrated or approached those temperature and pressure conditions (Supplementary Table S3 in the online version at https://doi.org/10.1016/B978-0-443-21662-6.00009-8). To date, few geothermal projects have successfully extracted heat from reservoirs that host superhigh-enthalpy heat or supercritical brine. Some geothermal power plants, however, do produce high-enthalpy (superheated) steam, for example, at the Geysers (USA), Larderello (Italy), and Nesjavellir (Iceland) geothermal sites.

The global maps of past and planned superhot drilling penetrations, CGS, EGS, CLG, and HGS projects, are presented in Figs. 13–20. These maps are vital to (a) chart the geothermal industry's past and present focus locations and jurisdictions and (b) visualize the optimal locations to target SHR energy. In Section 8, the locations of all these projects are analyzed in relation to the distribution of theoretically accessible (4–15 km) steady-state and nonsteady-state geothermal heat identified within Earth's crust.

7.1 Deep drilling capability and historic success

An obvious set of questions about extracting SHR geothermal energy are about the drilling industry's capability and previous experience to drill down to SHR depths. Table 2 presents the depths (in true vertical depth: "TVD") of a subset of the deepest drilled wells globally. This dataset is limited to publicly accessible data of commercial and nationally strategic boreholes. Table 2 confirms the capability to drill vertically deeper than 7500 m (TVD). The deepest

TABLE 2 Compilation of deep wells drilled globally.

Well name	Country	Year and organization name	Rock type/ lithology at total depth (TD)	True vertical depth (TVD) meter (ft)	Temperature at bottom °C (°F) If offshore well, depth to seabed meter (ft)	References
Maustrenk ÜT1a	Austria	1982–1984	Sedimentary	6563 [21,532]	n/a	Strasser (2015)
Aderklaa UT1A	Austria	1982–1985	Garnet-mica schist	6630 [21,752]	>170	Strasser (2015)
Zistersdorf UT1A	Austria	1977–1980	Sedimentary	7544 [24,751]	n/a	Strasser (2015)
Zistersdorf UT2A	Austria	1981–1983	Sedimentary	8553 [28,061]	211 [421]	Strasser (2015)
SG-1, Saatlyn	Azerbaijan, Russian Geological Institute	1977	n/a	8324 [27,310]	180 [356]	Hirshmiller and Riva (2022)
West Aladag-1 (AZ)	Azerbaijan, Russian Geological Institute	n/a	n/a	9620 [31,562]	n/a	Hirshmiller and Riva (2022)
KTB	Germany	1990–1994, Kontinentale Tiefbohrung	Gneiss, feldspar, granite	9101 [29,859]	265 [509]	Strasser (2015) and Hirshmiller and Riva (2022)
SG-7, En-Yakhin	Novy-Urengoy, Russian Geological Institute	2000	n/a	8250 [27,953]	n/a	Hirshmiller and Riva (2022)
SG-3, Kola Peninsula	Russia, Russian Geological Institute	1970–1989	Granite	12,262 [40,230]	180 [356]	Khakhaev et al. (1996), Juhlin et al. (1988), and Strasser (2015)

SG-6, Tymen	Siberia, Russian Geological Institute	n/a	Basalts and tuffs	7502 [24,613]	200 [392]	Juhlin et al. (1988) and Survok et al. (1993)
Stenberg-1, Siljan Ring	Sweden	1993	Granite	6529 [21,421]	n/a	Balling (2023)
Gravberg-1, Siljan Ring	Sweden	1989	Granite	6957 [22,825]	116 [240.8]	Aldahan et al. (1991)
31-10X, Tenneco	USA, California	(n/a) Great Basins Petroleum	Sedimentary	6595 [21,640]	ca. 230 [ca. 450]	Patterson et al. (1973)
Tiber	USA, GoM, Keathley Canyon 101	2009, BP	Sedimentary	10,685 [35,055]	n/a, water depth, 1259.4 [4132]	Hirshmiller and Riva (2022)
Kaskadia	USA, GoM, Keathley Canyon 292	2010, BP	Sedimentary	9906 [32,500]	n/a, water depth 1786.1 [5860]	Strasser (2015)
Chevron #1	USA, GoM, Walker Ridge 758	2015, Chevron	Sedimentary	8686 [28,497]	n/a, water depth, 2121.1 [6959]	Hirshmiller and Riva (2022)
Howell 1–2	USA, Oklahoma	1991 (?) Exxon	Sedimentary	7792 [25,565]	n/a	Koenig (1991)
Walne 1–34	USA, Oklahoma	1991 (?) Exxon	Sedimentary	7872 [25,825]	n/a	Koenig (1991)
Easley No. 1	USA, Oklahoma	1971–1973, El Paso Natural Gas Company	Sedimentary	8245 [27,050]	n/a	Patterson et al. (1973)
#1–1 Robinson	USA, Oklahoma	1981–1984, GHK Corporation	Sedimentary	8913 [29,241]	n/a	Dyman et al. (1990) and Hirshmiller and Riva (2022)
#1–34 Duncan	USA, Oklahoma	1981–1983, GHK Corporation	Sedimentary	8934 [29,312]	n/a	Dyman et al. (1990) and Hirshmiller and Riva (2022)
Earnest R. Baden No. 1	USA, Oklahoma	1970–1972, Loan Star Gas Producing Company	Sedimentary	9159 [30,050]	>260 [>500]	Patterson et al. (1973) and Dyman et al. (1990)
Bertha Rogers No. 1	USA, Oklahoma	1972–1974, Loan Star Gas Producing Company	Sedimentary	9583 [31,441]	246 [474.8]	Strasser (2015); Dyman et al. (1990)

Continued

TABLE 2 Compilation of deep wells drilled globally—cont'd

Well name	Country	Year and organization name	Rock type/ lithology at total depth (TD)	True vertical depth (TVD) meter (ft)	Temperature at bottom °C (°F) If offshore well, depth to seabed meter (ft)	References
Centurion #1	USA, Texas	1979–1981, Napeco Inc.	Sedimentary	8235 [27,019]	n/a	Dyman et al. (1990); Hirshmiller and Riva (2022)
University #1–17	USA, Texas	1972, Ralph Lowe Estate	Sedimentary	8687 [28,500]	n/a	Dyman et al. (1990); Hirshmiller and Riva (2022)
University #17	USA, Texas	1982, Kimball Production	Sedimentary	8687 [28,500]	n/a	Hirshmiller and Riva (2022)
University #1	USA, Texas	1978–1981, Chevron	Sedimentary	8762 [28,747]	n/a	Dyman et al. (1990)
2 Emma Lou Unit 1	USA, Texas	1978–1980, Gulf Oil Corporation	Sedimentary	9029 [29,622]	n/a	Dyman et al. (1990); Hirshmiller and Riva (2022)
1–9 Cerf Ranch	USA, Texas	1979–1982, Hunt Energy Corporation	Sedimentary	9037 [29,649]	n/a	Dyman et al. (1990); Hirshmiller and Riva (2022)
Cerf Ranch	USA, Texas	1994, Raita Energy	Sedimentary	9043 [29,669]	n/a	Hirshmiller and Riva (2022)
Yuejin 3-3XC	China, Tarim Basin	2023, Sinopec	Sedimentary	8797 (TVD?) to 9432 (MD) [28,862 to 30,945]	ca. 200 [392] estimate	Sinopec (2023)

vertical borehole—the Kola deep well—reached 12,262 m deep (assumed TVD) (Table 2). Meanwhile, 4655 wells have reached deeper than 5000 m TVD, and 59 wells have reached deeper than 7500 m TVD in North America (Enverus, 2023; Fig. 12). Table 2 highlights if the wells reached crystalline or sedimentary rock types, because this may have implications on rock mass porosity and permeability and drilling rate. Table 2 and Fig. 12 do not include well lengths with horizontal sections. For comparison with vertical wells, Axis Rig 55, in the USA, completed a lateral well that measured depth of 10,374 m length (Axis, 2022), and the longest well globally—Sakhalin O-14 well drilled by ExxonMobil—was 14,935 m [49,000 ft] long (BW, 2011).

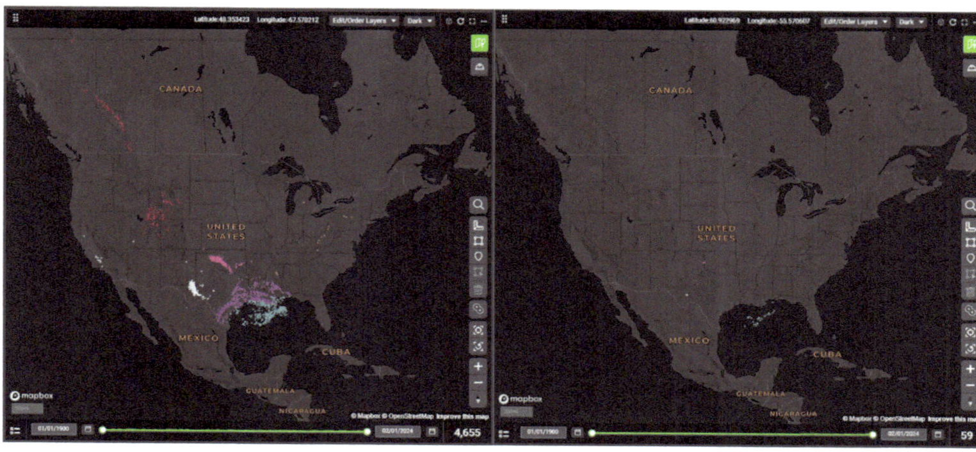

Fig. 12 The locations and number of vertical wells drilled in North America deeper than 5000 m TVD and 7500 m TVD, respectively (Enverus, 2023).

7.2 Superhot drilling penetrations (>350°C)

"Superhot" conditions are defined, in Section 11, as 374°C or above, i.e., the supercritical temperature of deionized water (Banks and Ball, 2023). Supercritical conditions of deionized water are defined as temperature >374°C and pressure >21 MPa. However, the superhot geothermal project database (Fig. 13, Supplementary Table S3 in the online version at https://doi.org/10.1016/B978-0-443-21662-6.00009-8) contains the 53 wells globally that reported a minimum temperature of 350°C rather than 374°C. There are two justifications for this: (a) The public domain literature of wells that approached supercritical temperatures and pressures does not contain accurate bottom-hole temperature readings due to temperature gauge issues, well blowouts caused by fluid overpressure or natural hydrogen, or other mechanical failures (Ball, 2022; Kruszewski and Wittig, 2018) and (b) Supplementary Table S2 in the online version at https://doi.org/10.1016/B978-0-443-21662-6.00009-8 is an update of a database of 48 wells that penetrated rock >350°C compiled by Ball (2022).

Several superhot geothermal wells have penetrated superheated or supercritical reservoirs in several countries (Fig. 13). This demonstrates the repeatable capability to reach and produce superhot conditions. The well IDDP-1 flowed hot superheated steam at 452°C to surface (Fridleifsson et al., 2015; Fridleifsson et al., 2019) with a fluid enthalpy estimated at 3200 kJ/kg. Pressure is critical for production because dissolved silica, in the steam phase, can precipitate at pressures lower than 8 MPa [1160.3] (Fridleifsson et al., 2015). Other tests of IDDP-1 well explored production at pressures up to 14 MPa [2030.5 PSI] (Fridleifsson et al., 2015). Higher fluid flow rates were achieved with lower pressures to a maximum flow of 50 L/s under a pressure of 4 MPa [580 PSI]. Silica was not the only issue faced: gaseous sulfur converted to sulfuric acid upon condensation (Markússon and Hauksson, 2015).

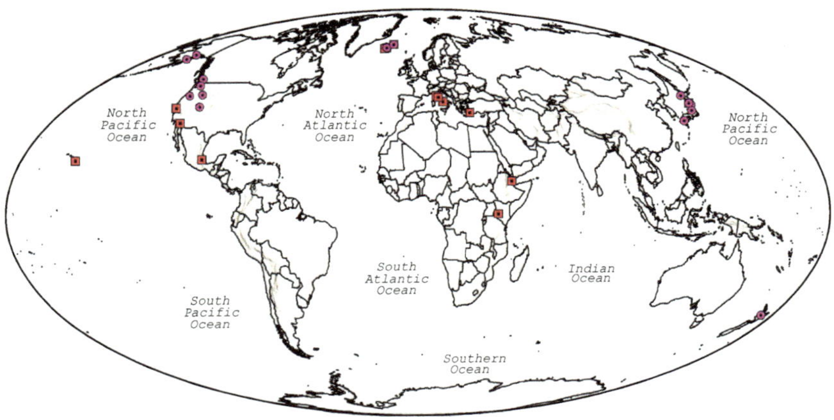

Fig. 13 Global map of the wells drilled and planned into rock at temperatures >350°C. Pink circles: planned superhot rock projects. Red squares: projects where temperatures >350°C compilation is presented in Supplementary Tables S3 and S4 in the online version at https://doi.org/10.1016/B978-0-443-21662-6.00009-8. Mollweide Projection. Map lines delineate study areas and do not necessarily depict accepted national boundaries.

Supercritical conditions have been penetrated in nine wells, namely, Nisyros-1, Greece; NJ-11, KJ-39, IDDP-2, Iceland; San Vito-1, San Pompeo-2, Venelle-02, Italy; and Prati-32 (redrill), Wilson A1, USA (see Supplementary Table S3 in the online version at https://doi.org/10.1016/B978-0-443-21662-6.00009-8, updated and corrected from Ball and Afonso, 2023). Many of Supplementary Table S3 in the online version at https://doi.org/10.1016/B978-0-443-21662-6.00009-8 wells were drilled while prospecting for high-enthalpy conventional hydrothermal systems and accidentally penetrated reservoirs at superhot or supercritical conditions. However, those investigations did not achieve a controlled flow of superheated or supercritical fluids to surface. In addition to the drilled wells database of

Supplementary Table S3 in the online version at https://doi.org/10.1016/B978-0-443-21662-6.00009-8, Supplementary Table S4 in the online version at https://doi.org/10.1016/B978-0-443-21662-6.00009-8 contains drilled superhot wells that are planned. While many of the targets listed aim to reach superhot or supercritical conditions, most of these wells are chasing existing hydrothermal magmatic systems as defined by Kolker et al. (2022) rather than HDR.

7.3 Conventional geothermal systems (hydrothermal)

CGSs produce geothermal energy to Earth's surface from shallow or deep reservoirs within their naturally occurring geofluid, i.e., they do not require a heat-carrying fluid to be introduced by humans (Banks and Ball, 2023; Section 11). There are 489 conventional geothermal plants active in numerous countries globally (Fig. 14; ThinkGeoEnergy, 2024). These projects use advective or convective heat-transfer methods (e.g., Moeck, 2014). A key exploration screening idiom is that these projects only work in reservoirs where naturally occurring water, rock mass permeability, and heat coincide (Fig. 15). It is possible that some future SHR projects will be developed as CGSs, as demonstrated by existing superhot well penetrations, in hydrothermal or magmatic systems (Section 7.1). Supplementary Table S5 in the online version at https://doi.org/10.1016/B978-0-443-21662-6.00009-8 provides a list of existing conventional hydrothermal projects.

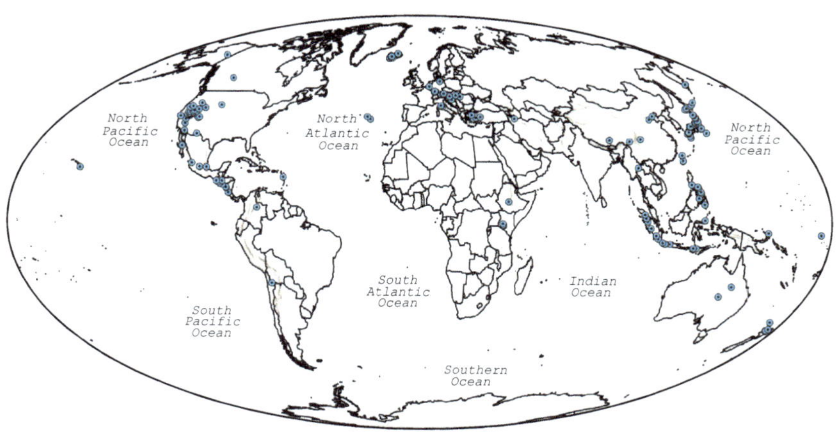

Fig. 14 Global map of the active, conventional geothermal (i.e., hydrothermal) system projects (blue circles). Project locations are modified from ThinkGeoEnergy (2024). See Supplementary Table S5 in the online version at https://doi.org/10.1016/B978-0-443-21662-6.00009-8. Mollweide Projection. Map lines delineate study areas and do not necessarily depict accepted national boundaries.

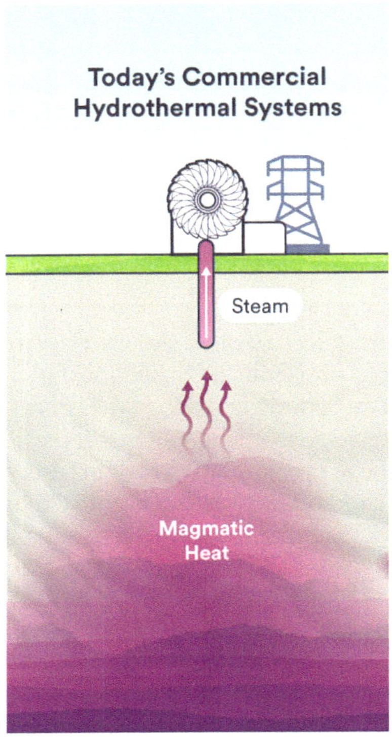

Fig. 15 Conventional hydrothermal systems (reused with permission of CATF).

7.4 Engineered geothermal systems (EGS)

An EGS (and its historic synonym enhanced geothermal system [e.g., Tester et al., 2006; Breede et al., 2013; Buijze et al., 2019; Pollack et al., 2021; DOE, 2016]) has been artificially created or improved by engineering techniques (Banks and Ball, 2023). Engineered and enhanced are deliberately differentiated in this document, with enhanced considered acceptable to categorize projects that have undergone intervention to improve reservoir permeability, using techniques such as acid stimulation or thermal stimulation using cold water, as is common in oil and gas fields. In this review, engineered geothermal circuits (see definition in Section 11) are classified as a network of fluid permeable fractures created through fracking causing the deliberate linkage of one or more wells. Fracking is completed to create a reservoir volume, wherein a circuit of heat-carrying fluid in direct contact with hot rock (i.e., in an open-loop to the reservoir) to the power plant via reinjection and production wells is constructed (Fig. 16). The wells can be vertical, inclined, or horizontal in dip angle and be configured in pairs or triplets, stacked vertically, offset (en-echelon), or arranged horizontally to each other.

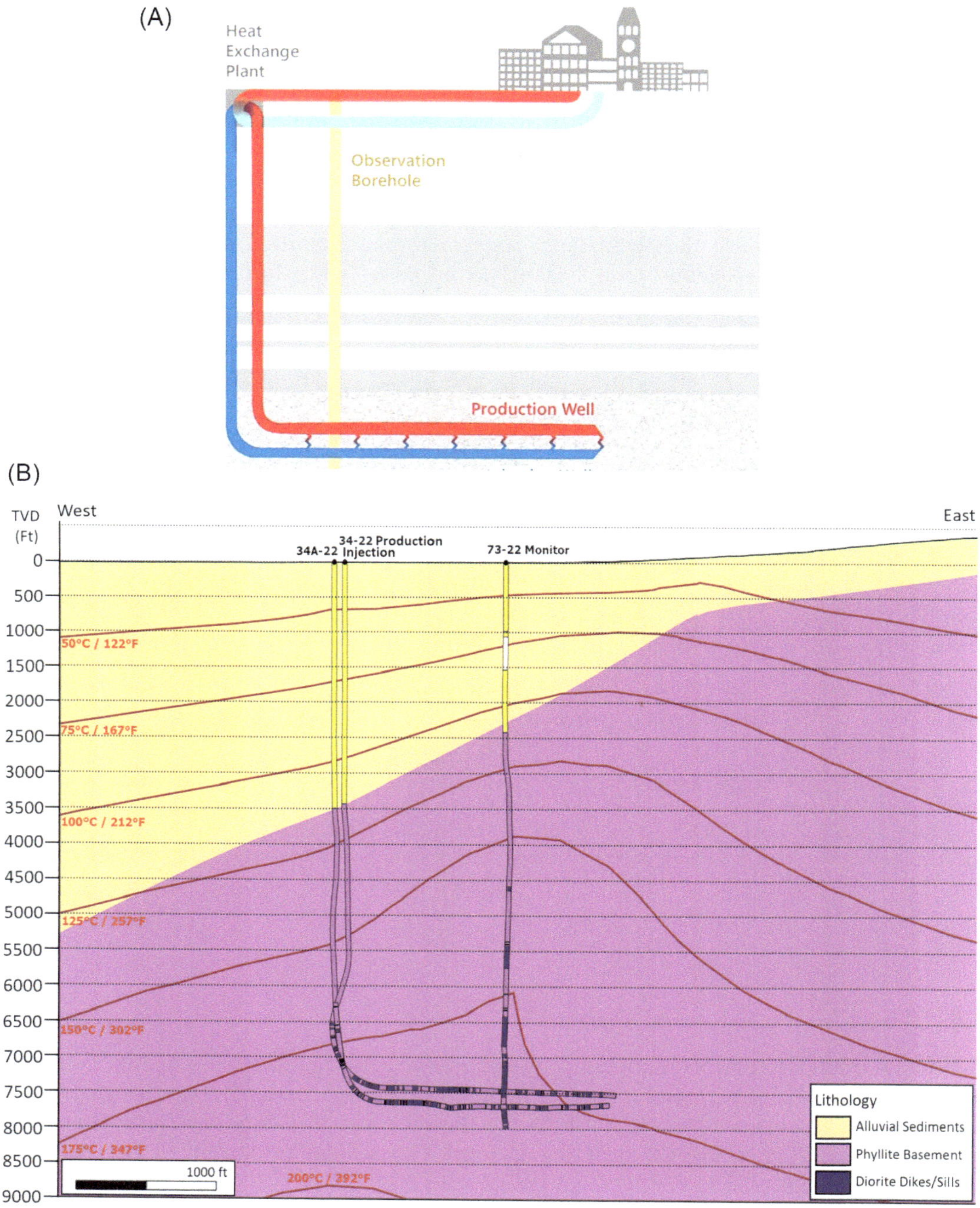

Fig. 16 (A) A schematic representation of an engineered geothermal reservoir that connects a well pair at Cornell University, USA (Bland, LinkedIn online), the original figure was drafted by Hadassah Jahani, Cornell University. It is a 2-mile-deep EGS for geothermal heat to use directly to heat the campus buildings https://www.linkedin.com/feed/update/urn:li:activity:7159922054559846400/. (B) A schematic representation of an engineered geothermal reservoir that connects a well pair in Nevada (Norbeck and Latimer, 2023). It has an estimated maximum flow rate of 61 L/s to produce 3.5 MWe of power.

The database in Supplementary Table S6 in the online version at https://doi.org/10.1016/B978-0-443-21662-6.00009-8 and Fig. 17 only presents petrothermal or petrothermal-hydrothermal projects with wells that are connected by a reservoir fracture network engineered to flow subsurface working fluid between them. Supplementary Table S6 in the online version at https://doi.org/10.1016/B978-0-443-21662-6.00009-8 and Fig. 17 show 33 global projects that fit this criterion. Twenty-four are "pure-play" petrothermal systems with commercial objectives. Six are hydrothermal-petrothermal projects that were included to increase the analysis breadth. Three research project wells are also included. The variety of countries with EGS wells and projects emphasizes the capability in numerous countries to synthesize geothermal systems to improve heat production from deep heat reservoirs.

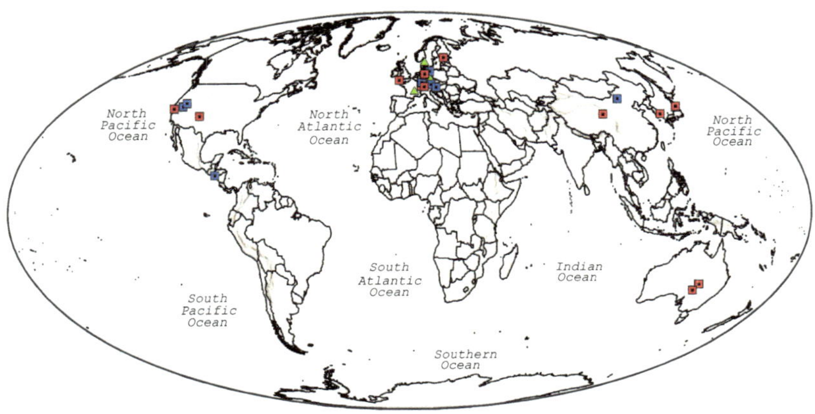

Fig. 17 Global map of the active EGS projects. Green triangle: research wells. Red squares: presently noncommercial projects. Blue squares: commercial engineered projects. The data is presented in Supplementary Table S6 in the online version at https://doi.org/10.1016/B978-0-443-21662-6.00009-8. Mollweide Projection. Map lines delineate study areas and do not necessarily depict accepted national boundaries.

7.5 Closed-loop geothermal systems

A CLG circuit is a type of engineering system with a working fluid that is heated in the rock reservoir without direct contact with rock pores and fractures. Instead, the subsurface working fluid stays inside a closed loop of deeply buried pipes that conduct Earth's heat (e.g., Banks and Ball, 2023). Fig. 18 provides examples of single- or multiwell geometries of closed-loop circuits.

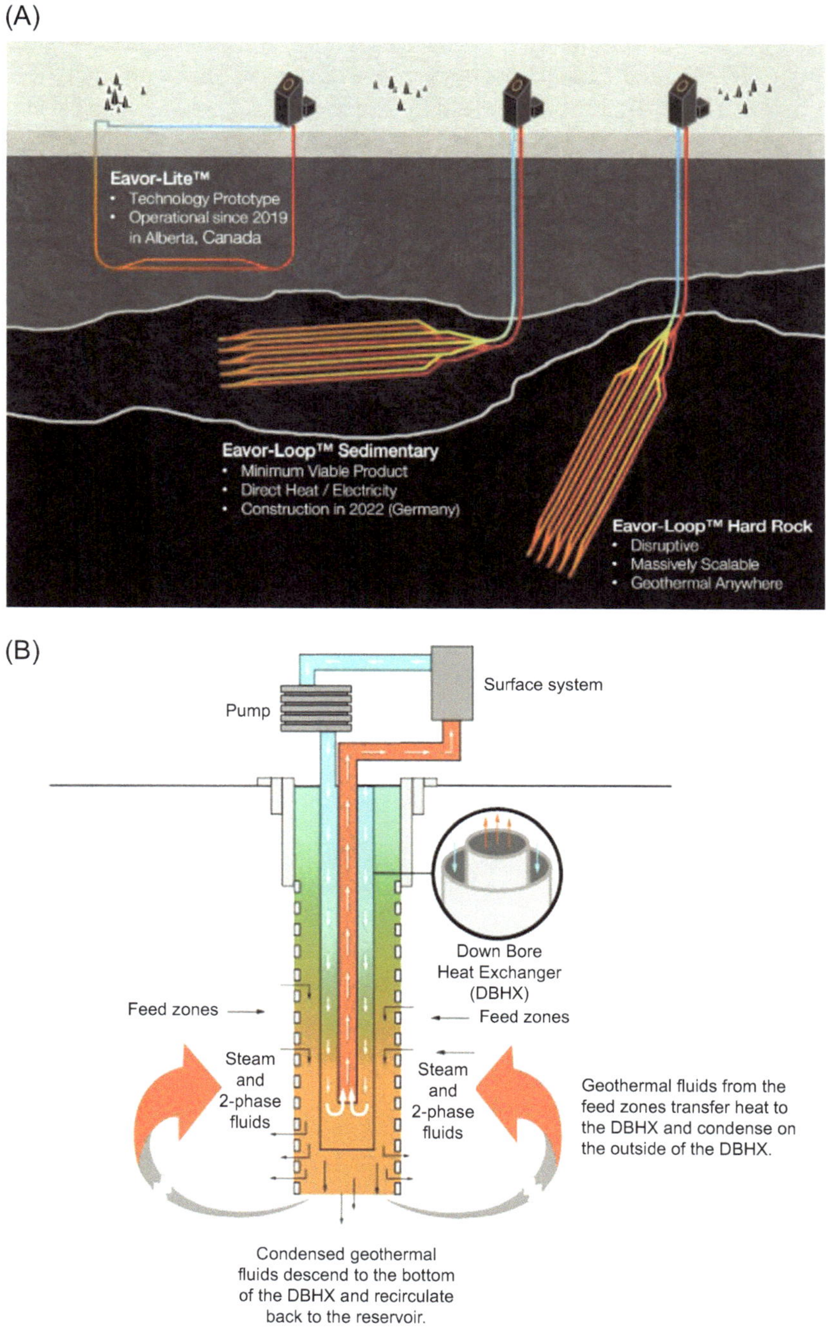

Fig. 18 (A) An example of how closed-loop geothermal circuits work by Eavor (Longfield et al., 2022). (B) An example of how a deep borehole heat exchanger in a closed-loop system could work, by GreenFire (Scherer et al., 2023).

This technology and engineering approach can be applied at different depths, temperatures, and rock types. Supplementary Tables S7A and S7B in the online version at https://doi.org/10.1016/B978-0-443-21662-6.00009-8 and Fig. 19 display and map the nine deep, closed-loop projects globally that have occurred or will occur in the near future. These projects and demonstrations have reached, and are targeting, heat reservoirs between 74°C and 250°C, for both heat and power production. The initial geothermal gradient, inlet (reservoir temperature), and thermal conductivity will have a significant impact on heat transfer from a wellbore (Brown and Gioia, 2024), hence the attraction for their future application to SHR geothermal. Deep, closed-loop circuits are thought to currently be more expensive than EGS systems (e.g., Alimonti, 2023; Beckers and Johnston, 2022). However, one possible advantage is that they may be less likely to trigger induced seismicity that in turn increases the potential geographic area of interest where they can be installed and accepted by local communities (Matthews, 2023).

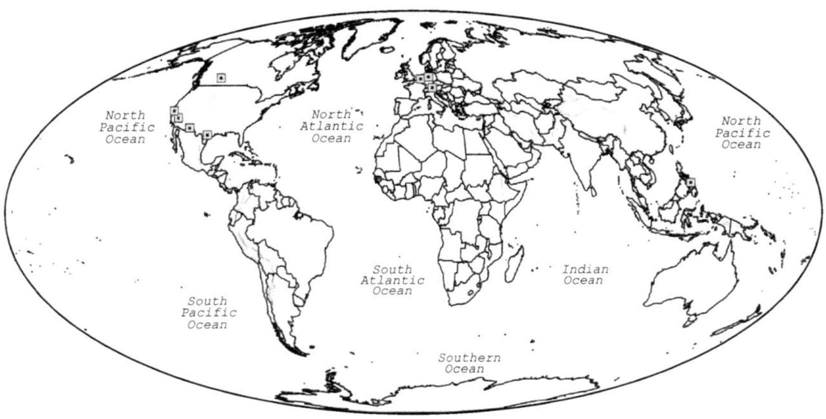

Fig. 19 Global map of the closed-loop projects (green squares). Data table and compilation are presented in Supplementary Tables S7A and S7B in the online version at https://doi.org/10.1016/B978-0-443-21662-6.00009-8. Mollweide Projection. Map lines delineate study areas and do not necessarily depict accepted national boundaries.

7.6 Hybrid geothermal systems

A HGS combines two geothermal technology or system types (Banks and Ball, 2023) to increase heat-transfer efficiency, drilling depth, drilling speed, etc. No commercial or pilot demonstration has been completed in SHR yet. One example of a HGS in development is the thermal reach enhancement (TRE) method for vertical or deviated boreholes (Fig. 20) by a company named XGS (Jacobs, 2024). XGS utilizes a simple, modular pipe-in-pipe closed loop system and leverages a proprietary thermally conductive materials surrounding the well bore to enhance heat extraction. The XGS surry would conduct heat toward a closed loop well pipe and enable the circuit to absorb 30% to 50% more heat compared to standard geothermal wells. This hybrid engineering technique has been tested in the laboratory using granitic rock samples, and demonstrated its lab-scale ability to operate in conditions above 400°C

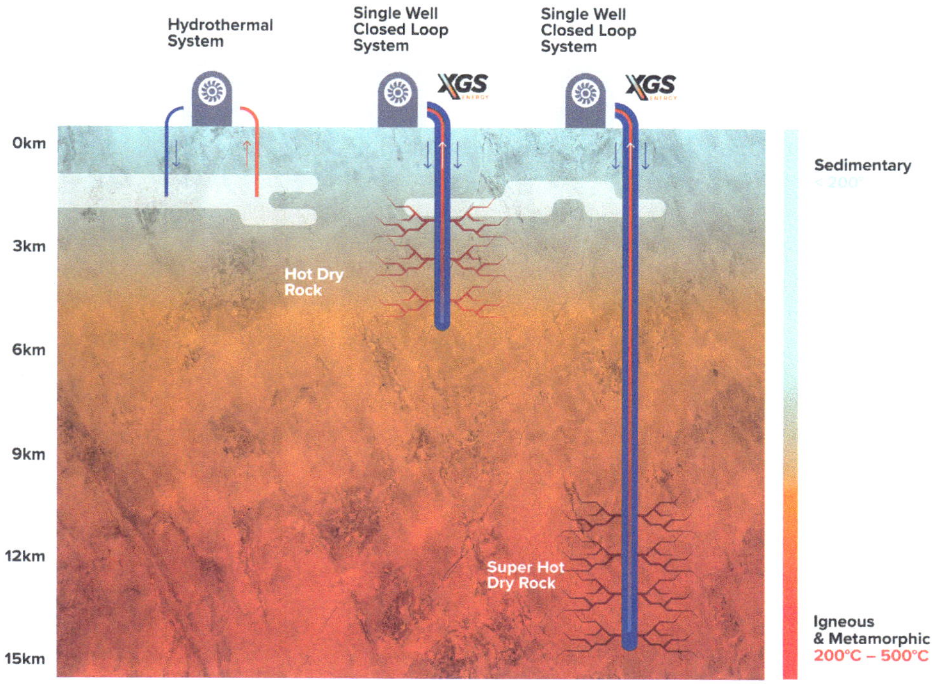

Fig. 20 An example of how a hybrid geothermal system could work using the XGS subsurface concept (Jacobs, 2024).

8 Results: Superhot rock geothermal resources and the influence of geodynamic setting

In this section, using the globally consistent tectonic model of Hasterok et al. (2022), we explore the variation of geodynamic settings at the tectonic/physiographic province and orogenic/tectonic scale. Hasterok divides the world into 15 principle geodynamic settings (see Hasterok et al., 2022 for details). As we analyze the data, we consider a simple classification of these geodynamic settings and group them into tensional [continental back-arc basin (BAB), narrow rift, oceanic BAB, oceanic crust, passive margin, wide rift], compressional (accretionary complex, foredeep basin, ophiolite complex, orogenic belt, volcanic arc), and stable (basin, craton, magmatic province, shield) settings, assuming the term applied to present-day stress. It is noted there are two limitations to this approach—first, the geodynamic classification of Hasterok et al. (2022) does not differentiate present-day stress from paleo-stress. Second, strike-slip is not featured; third, the interpretation in domain is limited if there are errors within the classification scheme. Fixing these issues is not within the scope of the current analysis.

The results are presented as follows: First, we review the distribution of geothermal exploration where (a) superhot (>350°C) has been penetrated and where future projects are currently planned (Section 7.2, Supplementary Tables S3 and S4 in the online version at https://doi.org/10.1016/B978-0-443-21662-6.00009-8). We also review the current distribution of conventional (hydrothermal) geothermal (low to high enthalpy) power plants (Section 7.3, Supplementary Table S5 in the online version at https://doi.org/10.1016/B978-0-443-21662-6.00009-8), engineered, open-loop, geothermal systems engineered, closed-loop, geothermal systems (Section 7.4, Supplementary Table S6 in the online version at https://doi.org/10.1016/B978-0-443-21662-6.00009-8, and Section 7.5, Supplementary Tables S7A and S7B in the online version at https://doi.org/10.1016/B978-0-443-21662-6.00009-8). The intention is to explore the existing distribution and bias of geothermal exploration trends. Second, we explore the influence that geodynamic setting has upon the location of the modeled 450°C steady-state isotherm and nonsteady-state locations (Section 5). Third, we present an inferred heat and power endowment assuming by tectonic/physiographic province using a simplified prospective endowment model (Section 6).

8.1 Superhot geothermal drilling and geodynamic setting

We look at geothermal projects and drilling to get an insight into where geothermal exploration is currently focused. When we look at the database of projects that deliberately or accidentally drilled into superhot and supercritical conditions (>350°C, Fig. 13, Supplementary Table S3 in the online version at https://doi.org/10.1016/B978-0-443-21662-6.00009-8), we observe that 16 projects (30%) were drilled into oceanic crust settings, 6 were drilled in transitional continental crust (11%), and 31 (59%) were located within continental crust settings. Transitional and continental crust together means 70% of projects were within continental crust. One striking observation is that out of the 53 wells drilled, the areas where mantle plumes appear to play a key role (in both oceanic and continental settings) is in 25 wells. This means in the drilled SHR wells category 47% drilled into areas where mantle plumes are documented.

At the geodynamic setting, the drilling is observed to have occurred within five geodynamic settings (Table 3). The dominant geodynamic setting for drilling is into volcanic arcs (23, 43.4%) followed by narrow rifts (16, 30.2%) and oceanic Crust (note this includes Iceland and Hawaii) with 16 wells (30.2%). These distributions are somewhat distorted by the number of wells drilled within individual geothermal fields. If we consider the tectonic/physiographic province, 11 provinces have revealed superhot resources with the Trans-Mexican Volcanic Belt (Mexico) the most drilled, followed by the Pacific Basin (Hawaii) and the Campanian arc (Italy). SHR drilling has occurred in the Corderillian and Alpine–Himalayan orogenic settings. SHR projects are dominantly within the Cordilleran with respect to the last orogenic/tectonic event. Finally, we observe that 25 (47%) of the wells are drilled into what should be a compressional domain and 28 (53%) are drilled into tensional domains.

TABLE 3 Summary table of superhot and projects that drilled into temperatures >350°C.

Superhot rock wells drilled				
Geodynamic setting (present day and paleo)	**Dominant "tectonic" [physiographic province] domain**	**Last orogen/ tectonic event**	**Crustal type**	**Wells [53]**
Accretionary complex	California Coast Ranges (2)	Cordilleran	Transitional	2
Narrow rift (mantle plume)	East African Rift (5)	Afar	Continental	9
	East Ethiopian Rift (4)	Afar	Continental	
Narrow rift (mantle plume)	West Iceland Rift (5)	Atlantic NE	Oceanic	7
	East Iceland Rift (2)	Atlantic NE	Oceanic	
Oceanic crust (mantle plume)	Pacific Basin (9)	Pacific	Oceanic	9
Volcanic arc	Trans-Mexican Volcanic Belt (12)	Cordilleran	Continental	23
	Campanian Arc (9)	Alpine-Himalayan	Continental	
	North Japan Arc (1)	Cordilleran	Continental	
	Aegean Volcanic Arc (1)	Alpine-Himalayan	Transitional	
Wide rift	Gulf of California Coast (3)	Cordilleran	Transitional	3

Organized by geodynamic setting, tectonic/physiographic province, last orogen/tectonic event, and crustal type as defined by Hasterok et al. (2022).

Planned SHR projects/wells, based on those compiled by Ball (2022), indicate currently an inventory of 16 defined projects (Supplementary Table S4 in the online version at https://doi.org/10.1016/B978-0-443-21662-6.00009-8). Only two (12.5%) of the planned wells occur in oceanic crust modified by a mantle plume—both these projects are located in Iceland. Three (19%) of the wells are located within transitional continental crust and 11 (69%) of the wells planned are located in continental crust. The total well count in continental settings is 14 (87.5%). Of the planned wells, only two (12.4%) are located in oceanic crust modified by mantle plumes—both of these are located in Iceland (Table 4).

Based on this limited database, planned SHR exploration is observed to be occurring within seven geodynamic settings (Table 4). Drilling is dominantly located within volcanic arc settings (9, 56%), followed by narrow rifts (2, 12.5%) and wide rifts (2, 12.5%). As with drilled wells, the dominant exploration trend for SHR is within volcanic arcs with Japan dominating the prospect list linked to the existence of the Japan's Supercritical Geothermal Project (Asanuma, 2022). Compared to historical drilling, it is worth noting that new geodynamic settings may be penetrated within continental a BAB setting, for example, in the Taupo Volcanic Zone, New Zealand. A well is also possibly located within the continental shield setting, Utah, USA, although the current location is not disclosed, and this location is likely incorrect. Geographically, we see a dominance of planned SHR wells to North America, Iceland, Japan, and New Zealand which is a smaller footprint than where early exploration has occurred.

TABLE 4 Projects that plan to drill into temperatures >350°C.

Superhot rock wells planned					
Geodynamic setting (present day and paleo)	**Dominant "tectonic" [physiographic province] domain**	**Last orogen/ tectonic event**	**Crustal type**	**Organization planning to drill**	**Projects [16]**
Accretionary complex	Northern Japan Forearc (1)	Cordilleran	Transitional	Japan JBBP	1
Continental back-arc basin	Taupo Volcanic Zone (1)	Terra Australis	Transitional	GNS Research, Taupo Volcanic Zone	1
Narrow rift (mantle plume)	West Iceland Rift (1)	NE Atlantic	Oceanic	IPPD-3	2
	East Iceland Rift (1)	NE Atlantic	Oceanic	Krafla KMT-1	
Shield	Colorado Plateau (1)	Cordilleran	Continental	GeoX-Utah	1
Volcanic arc	North Japan Arc (4)	Cordilleran	Continental	Japan Supercritical Project	8
	Peninsular Terrane (2)	Cordilleran	Continental	GeoAlaska	
	Coast plutonic complex (1)	Cordilleran	Continental	Meager Creek	
	Cascade Arc (1)	Cordilleran	Continental	GeoX-Washington	
Volcanic arc	Ryukyu-Kyushu Volcanic Arc (1)	Cordilleran	Transitional	Japan Supercritical Project	1
Wide rift	Northern Rocky Mountains (1)	Cordilleran	Continental	GeoX-Idaho	2
	Basin and Range (1)	Cordilleran	Continental	Newberry	

Organized by geodynamic setting, tectonic/physiographic province, last orogen/tectonic event, and crustal type as defined by *Hasterok et al. (2022)*.
Note: GeoX locations are approximate since for confidentiality, the exact location was not provided by the company. Similarly, GNS in New Zealand has not supplied a location—only the basin was listed as a potential development site. Note, local interpretations may be different. For example, Meager Creek may be alternatively considered as part of the Cascade Volcanic Arc and more specifically part of the Garabaldi Volcanic Belt *(pers. Comm Craig Dunn, 2024)*. However, for consistency within this review, we use the definitions of Hasterok et al. (2022).

The planned SHR wells exploration occurs across 12 tectonic/physiographic provinces, with repeated SHR drilling occurring only in the West and East Iceland Rift and the North Japan Arc (Tables 3 and 4). When we explore the tectonic event setting up the locations, planned drilling is dominantly with the Cordilleran orogenic terrane with 13 (81.3%). Finally, when looking at tectonic stress, we observe that one project (6.2%) is planned within stable crust, 10 (62.5%) of the projects/wells are planned in what should be a dominantly compressional domain, and five (31.3%) are planned within tensional domains.

8.2 Conventional (hydrothermal) geothermal projects and geodynamic setting

Using the ThinkGeoEnergy database, we explored where the 489 conventional (hydrothermal) geothermal projects are located (Supplementary Table S5 in the online version at https://doi.org/10.1016/B978-0-443-21662-6.00009-8). At the first order, we observe that conventional geothermal projects occur dominantly within continental crust. Four hundred and sixty-two projects of 94.5% are within continental crust. In detail, we observe that 125 (25.5%) are in fact located on transitional continental crust, meaning 337 (68.9%) of projects are located on "normal" continental crust. Only 27 projects (5.5%) are located in oceanic crust settings. Of the conventional projects, we observe that only 33 (6.7%) conventional geothermal projects are linked with present-day mantle plumes.

At the geodynamic setting, we observe that current drilling spans 12 geodynamic settings (Table 5), with only oceanic BABs, ophiolite complexes, and basins omitted by conventional geothermal drilling. The overwhelmingly dominant geodynamic setting for conventional geothermal projects is volcanic arcs representing 250 projects (51.1%) followed by wide rifts with 77 projects (15.7%) and magmatic provinces which account for 64 (13.1%) of projects. When we look at the volcanic arcs and wide rifts in detail, it is worth considering that volcanic arcs in continental crust dominate with 177 projects (70.8%), versus 64 (25.6%) occurring on transitional continental crust, and only 9 (3.6%) of these are oceanic volcanic arcs. For wide rifts, 47 (61.0%) projects are within continental domains, and 27 (35.1%) are transitional continental crust and 3 (3.9%) are within oceanic crust.

TABLE 5 Global conventional power plants, modified from ThinkGeoEnergy (2024).

Geothermal power plants—modified from ThinkGeoEnergy (2024)				
Geodynamic setting (present day and paleo)	**Dominant "tectonic" [physiographic province] domain**	**Last orogen/ tectonic event**	**Crustal type**	**Projects [489]**
Accretionary complex (continental)	Alborz Mountains (1)	Alpine-Himalayan	Continental	3
	Anatolide-Tauride Mountains (1)	Alpine-Himalayan	Continental	
	Chortis Block (1)	Cordilleran	Continental	
Accretionary complex (transitional)	**California Coast Ranges (16)**	Cordilleran	Transitional	16
Back-arc basin (continental-transitional)	Cycladian back-arc (1)	Alpine-Himalayan	Transitional	18
	Sea of Japan (1)	Cordilleran	Transitional	
	Taupo Volcanic Zone (16)	Alpine-Himalayan	Transitional	
Craton	Sichuan Basin (1)	Alpine-Himalayan	Continental	2
	Wyoming Craton (1)	Cordilleran	Continental	

Continued

TABLE 5 Global conventional power plants, modified from ThinkGeoEnergy (2024)—cont'd

Geothermal power plants—modified from ThinkGeoEnergy (2024)				
Geodynamic setting (present day and paleo)	**Dominant "tectonic" [physiographic province] domain**	**Last orogen/ tectonic event**	**Crustal type**	**Projects [489]**
Foredeep basin (continental)	Molasse Basin (12)	Alpine-Himalayan	Continental	12
Magmatic province (continental)	**Menderes Massif (63)**	Alpine-Himalayan	Continental	63
Narrow rift continental (mantle plume)	**East African Rift-east (18)**	Afar	Continental	18
Narrow rift oceanic (mantle plume—Iceland)	West Iceland Rift (2)	NE Atlantic	Oceanic crust	9
	East Iceland Rift (7)	NE Atlantic	Oceanic crust	
Oceanic crust (mantle plume—Hawaii)	Pacific Basin (2)	Pacific	Oceanic crust	2
Oceanic crust	Pacific Basin (1)	Terra Australis	Oceanic crust	4
	Azores Plateau (1)[a]	Central Atlantic	Oceanic crust	
	West European Basin (2)		Oceanic crust	
Orogenic belt (continental)	Arunta Block (1)	Terra Australis	Continental	6
	Saxo-Thuringian Belt (1)	Antilles	Continental	
	Saxo-Thuringian Belt (3)	Variscan-Hercynian	Continental	
	West Thompson Orogen (1)	Terra Australis	Continental	
Passive margin (continental)	North America Platform (1)	Cordilleran	Continental	2
	Sibumasu Terrane (1)	Indosinian	Continental	
Shield (continental)	Buffalo Head Terrane (1)	Wopmay	Continental	7
	Colorado Plateau (4)	Cordilleran	Continental	
	Longgang Block (2)	Alpine-Himalayan	Continental	
Volcanic arc (continental)	Avalonia (2)	Caledonian	Continental	177
	Banda Volcanic Arc (2)	Alpine-Himalayan	Continental	
	Bantam Arc (10)	Cordilleran	Continental	
	Bicol Volcanic Arc (17)	Cordilleran	Continental	

TABLE 5 Global conventional power plants, modified from ThinkGeoEnergy (2024)—cont'd

Geothermal power plants—modified from ThinkGeoEnergy (2024)				
Geodynamic setting (present day and paleo)	**Dominant "tectonic" [physiographic province] domain**	**Last orogen/ tectonic event**	**Crustal type**	**Projects [489]**
	Campanian Arc (35)	Alpine-Himalayan	Continental	
	Central American Arc (9)	Cordilleran	Continental	
	Central Andean Arc (1)	Terra Australis	Continental	
	Java Volcanic Arc (21)	Alpine-Himalayan	Continental	
	Kamchatka Volcanic Arc (3)	Cordilleran	Continental	
	Lhasa Terrane (2)	Cordilleran	Continental	
	Luzon Volcanic Arc (4)	Cordilleran	Continental	
	Nakfa Terrane (1)	East African	Continental	
	Negros Volcanic Arc (4)	Cordilleran	Continental	
	North Japan Arc (19)	Cordilleran	Continental	
	Peninsular Ranges Batholith (2)	Cordilleran	Continental	
	Pontide Mountains (4)	Alpine-Himalayan	Continental	
	Sangihe Volcanic Arc (4)	Cordilleran	Continental	
	South Japan Arc (7)	Cordilleran	Continental	
	Sumatran Volcanic Arc (10)	Alpine-Himalayan	Continental	
	Taranaki Province (5)	Terra Australis	Continental	
	Trans-Mexican Volcanic Belt (13)	Cordilleran	Continental	
	Ventuari-Tapajos Belt (1)	Inner-Nunanian	Continental	
	Yidun Arc (1)	Cratonic	Continental	
Volcanic arc (transitional)	**Ryukyu-Kyushu Volcanic Arc (64)**	Cordilleran	Transitional	64
Volcanic arc (oceanic)	Izu Bonin Arc (3)	Cordilleran	Oceanic	9
	Kuril Arc (3)	Cordilleran	Oceanic	
	Lesser Antilles (1)	Antilles	Oceanic	
	Lesser Antilles (1)	Variscan-Hercynian	Oceanic	
	San Christobal Arc (1)	Terra Australis	Oceanic	

Continued

TABLE 5 Global conventional power plants, modified from ThinkGeoEnergy (2024)—cont'd

Geothermal power plants—modified from ThinkGeoEnergy (2024)				
Geodynamic setting (present day and paleo)	**Dominant "tectonic" [physiographic province] domain**	**Last orogen/ tectonic event**	**Crustal type**	**Projects [489]**
Wide rift (continental)	Columbia Basin and Range (4)	Cordilleran	Continental	47
	Northern Basin and Range (35)	Cordilleran	Continental	
	Pannonian Basin (4)	Alpine-Himalayan	Continental	
	Sierra Madre Oriental (1)	Cordilleran	Continental	
	Southern Basin and Range (3)	Cordilleran	Continental	
Wide rift (transitional)	**Gulf of California Coast (26)**	Cordilleran	Transitional	27
	Gulf of California Coast (1)	Cratonic	Transitional	
Wide rift (oceanic) mantle plume—Iceland	Greenland-Iceland Ridge (2)	NE Atlantic	Oceanic	3
	Iceland-Faroe Ridge (1)	NE Atlantic	Oceanic	

Organized by geodynamic setting, tectonic/physiographic province, last orogen/tectonic event, and crustal type as defined by Hasterok et al. (2022).
[a] *Mantle plume related.*

At the tectonic/physiographic province level, conventional geothermal projects are developed in distinct clusters of activity with 278 (58.9%) projects located in only 9 tectonic provinces. Activity is dominated within the following tectonic provinces: Accretionary (transitional), California Coast Ranges (16); Back Arc Basin (continental-transitional), Taupo Volcanic Zone (16), Magmatic province (continental), Menderes Massif (63), Narrow rift (continental, with mantle plume), East African Rift-east (18), Volcanic arc (continental), Campanian Arc (35), Java Volcanic Arc (21), North Japan Arc (19), Volcanic arc (transitional) Ryukyu-Kyushu Volcanic Arc (64) and Wide rift (transitional), Gulf of California Coast (26). Of note, out of these nine provinces, only the Taupo Volcanic Zone, Menderes Massif, Java Volcanic Arc, and the Ryukyu-Kyushu Volcanic Arc have yet to experience any SHR drilling.

If we consider whether the drilling is tensional or compressional or within stable crust, we observe that 287 (58.7%) of the geothermal plants are developed in a compressional domain and 130 (26.6%) are developed in tensional domains and 72 (14.7%) of projects are developed in stable tectonic domains. Finally, conventional geothermal power plants are dominantly developed in the Cordilleran orogenic terrains with 259 (53%). The next most developed orogenic setting is the Alpine-Himalayan orogen with 157 (32%) followed by the Terra Australis with 26 (5.3%) power plants.

8.3 Engineered geothermal systems and geodynamic setting

There are 33 engineered, open-loop, geothermal systems globally (Fig. 17, Supplementary Table S6 in the online version at https://doi.org/10.1016/B978-0-443-21662-6.00009-8). At the crustal scale, all of the 33 EGS projects have been completed in continental crust settings. Of the 33, only 1 occurred in transitional continental crust. Zero projects have occurred in oceanic crust settings.

At the geodynamic setting, EGS projects have occurred in seven different settings. The focus of EGS exploration has been dominantly within orogenic belt settings, with 15 (45.5%) projects identified. The next most prominent geodynamic settings are volcanic arcs with six (18.2%), shields with four (12.1%), and wide rifts with four (12.1%) projects (Table 6). Exploration adjacent to mantle plumes does not appear to be a priority for EGS projects, and while arc settings have dominated, conventional and SHR geothermal success in this domain has been low for EGS projects.

TABLE 6 Summary of EGS geothermal projects.

Geothermal power plants—EGS projects						
Geodynamic setting (present day and paleo)	**Dominant "tectonic" [physiographic province] domain**	**Last orogen/ tectonic event**	**Crustal type**	**Projects total**	**Projects commercial**	**Success rate %**
Accretionary complex	California Coast Ranges	Cordilleran	Transitional	1	0	0%
Basin	Gyeongsang Basin	Indosinian	Continental	1	0	0%
Foredeep	Molasse Basin	Alpine-Himalayan	Continental	2	2	100%
Orogenic belt	Rheno-Hercynian Belt	Variscan-Hercynian	Continental	2	1	50%
	Moldanubian Belt	Variscan-Hercynian	Continental	4 (includes research well)	1	25%
	Saxo-Thuringian Belt	Variscan-Hercynian	Continental	5 (includes research well)	4	100% (ignoring the research well)
	West Thompson Orogen	Terra Australis	Continental	1	0	0%
	North Delamerian Orogen	Variscan-Hercynian	Continental	1	0	0%
	Bainaimiao-Ondor Sum Belt	Solonker	Continental	1	1	100%
	Teisseyre-Tornquist Zone	Circum-Nunanian	Continental	1	1	100%

Continued

TABLE 6 Summary of EGS geothermal projects—cont'd

Geothermal power plants—EGS projects						
Geodynamic setting (present day and paleo)	**Dominant "tectonic" [physiographic province] domain**	**Last orogen/ tectonic event**	**Crustal type**	**Projects total**	**Projects commercial**	**Success rate %**
Shield	Bergslagen-Livonia	Circum-Nunanian	Continental	1	0	0%
	Colorado Plateau	Cordilleran	Continental	1	0	0%
	Curnamona Province	Circum-Nunanian	Continental	1	0	0%
	Sveconorwegian Belt	Rodinian	Continental	1 (research well)	0	0%
Volcanic arc	Avalonia	Caledonian	Continental	2	0	0%
	Central American Arc	Cordilleran	Continental	1	1	100%
	Kunlun/South Qilian Fold Belt	Alpine-Himalayan	Continental	1	0	0%
	North Japan Arc	Cordilleran	Continental	2	0	0%
Wide rift	Northern Basin and Range	Cordilleran	Continental	3	3	100%
	Pannonian Basin	Alpine-Himalayan	Continental	1	1	100%
	Total EGS projects			**34**	15	44

Organized by geodynamic setting, tectonic/physiographic province, last orogen/tectonic event, and crustal type as defined by Hasterok et al. (2022). Project count and commercial operations included with a current success rate by tectonic/physiographic province.

At the tectonic province/physiographic domain level, the Saxo-Thuringian Belt (5, 15.2%) and Moldanubian Belt (4, 12.1%) and the Northern Basin and Range (3, 9.1%) have seen the most EGS projects. The Molasse Basin, Rheno-Hercynian Belt, and Avalonia volcanic arc are other settings with two EGS projects within each domain.

If we consider the orogenic setting, EGS project development has occurred dominantly in the Corderillian (6 projects, 18.2%), Variscan-Hercynian (4 projects, 12.1%), Alpine-Himalayan (3 projects, 9.1%), and Circum-Nunanian (3 projects, 9.1%) orogenic settings. If we consider whether the drilling is tensional or compressional or within stable crust, we observe that 24 (72.7%) of the projects are drilled into what should be a compressional domain and 4 (12.1%) are drilled into tensional domains, and 6 (18.2%) are developed in stable crust settings.

8.4 Closed-loop geothermal systems (CGS) and geodynamic setting

CLG systems are a new emerging engineered concept within the geothermal industry—as such only nine projects are completed and reached the final investment decision (FID) or is in advanced design phase (Supplementary Tables S7A and S7B in the online version at https://doi.org/10.1016/B978-0-443-21662-6.00009-8). At the crustal scale, all of the nine CGS projects have been completed in continental crust settings, with only one located in transitional continental crust. Zero projects have occurred in oceanic crust settings. At the geodynamic setting, activity is predominantly within volcanic arcs, with four projects (44.4%) followed by wide rifts, with two projects (22.2%) (Table 7). At the tectonic/physiographic province level, the most drilled province is the Avalonia, volcanic arc. CGS projects, to date, are dominantly in Corderillian (four projects; 44.4%) and Caledonian orogenic terrains (two projects; 22.2%). If we consider whether the drilling is tensional or compressional or within stable crust, we observe that two (22.2%) of the wells are drilled into what should be a compressional domain and six (66.7%) are drilled into tensional domains, and one (11.1%) project is located within stable crust.

TABLE 7 Closed-loop projects globally.

Closed-loop geothermal system projects					
Geodynamic setting (present day and paleo)	**Dominant "tectonic" [physiographic province] domain**	**Last orogen/ tectonic event**	**Crustal type**	**Company**	**Projects [9]**
Volcanic arc	Rimbey Arc (1)	Inner-Nunanian	Continental	EAVOR	4
	Avalonia (2)	Caledonian	Continental	EAVOR	
	Bicol Volcanic Arc (1)	Cordilleran	Continental	GreenFire	
Wide rift	Southern Basin and Range (1)	Cordilleran	Continental	EAVOR	2
	Sierra Madre Oriental (1)	Cordilleran	Continental	GreenFire	
Foredeep basin	Molasse Basin (1)	Alpine-Himalayan	Continental	EAVOR	1
Shield	Llano Province (1)	Rodinian	Continental	EAVOR	1
Accretionary complex	California Coast Ranges (1)	Cordilleran	Transitional	GreenFire	1

Organized by geodynamic setting, tectonic/physiographic province, last orogen/tectonic event, and crustal type as defined by Hasterok et al. (2022).

8.5 LithoREF18: Superhot rock and geodynamic settings

In this study, we have extracted the steady-state conductive isotherm at SHR conditions (450°C) globally at 7.5, 10, 12.5, and 15 km (Figs. 21–26). In addition, we have calculated the regions where nonsteady-state heat flow is thought to occur (Figs. 27 and 28). It should be noted that only onshore areas are included in the final prospective resource calculations. For brevity, only the summaries incorporating the P50 flow 55 L/s and a power plant scenario

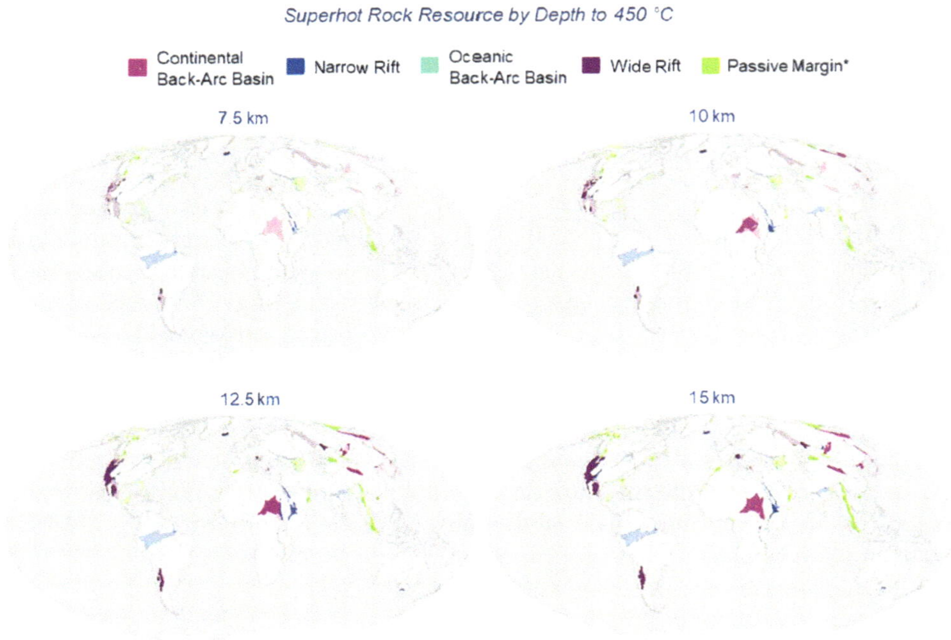

Fig. 21 (Model 1) Distribution of prospective SHR resource at 7.5, 10, 12.5, and 15 km, for tensional geodynamic settings: (i) continental back-arc basin, (ii) narrow rift, (iii) oceanic back-arc basin, (iv) oceanic crust, (v) passive margin, and (vi) wide rift. Bold colors indicate SHR resource as defined by Hasterok et al. (2022). Transparent colors indicate geodynamic setting with no SHR identified at the relevant depth interval. Map lines delineate study areas and do not necessarily depict accepted national boundaries.

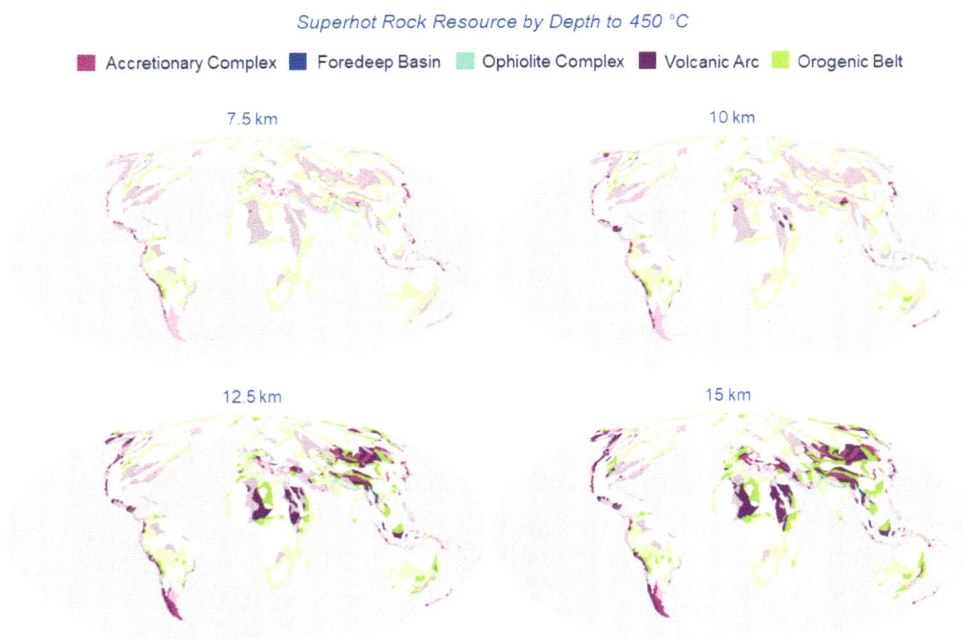

Fig. 22 (Model 1) Distribution of prospective SHR resource at 7.5, 10, 12.5, and 15 km, for compressional geodynamic settings: (i) accretionary complex, (ii) foredeep basin, (iii) ophiolite complex, (iv) volcanic arc, (v) continental back-arc, and (vi) orogenic belt. Bold colors indicate SHR resource as defined by Hasterok et al. (2022). Transparent colors indicate geodynamic setting with no SHR identified at the relevant depth interval. Map lines delineate study areas and do not necessarily depict accepted national boundaries.

Fig. 23 (Model 1) Distribution of prospective SHR resource at 7.5, 10, 12.5, and 15 km, for continental "stable" geodynamic settings: (i) basin, (ii) craton, (iii) magmatic province, and (iv) shield. Bold colors indicate SHR resource as defined by Hasterok et al. (2022). Transparent colors indicate geodynamic setting with no SHR identified at the relevant depth interval. Map lines delineate study areas and do not necessarily depict accepted national boundaries.

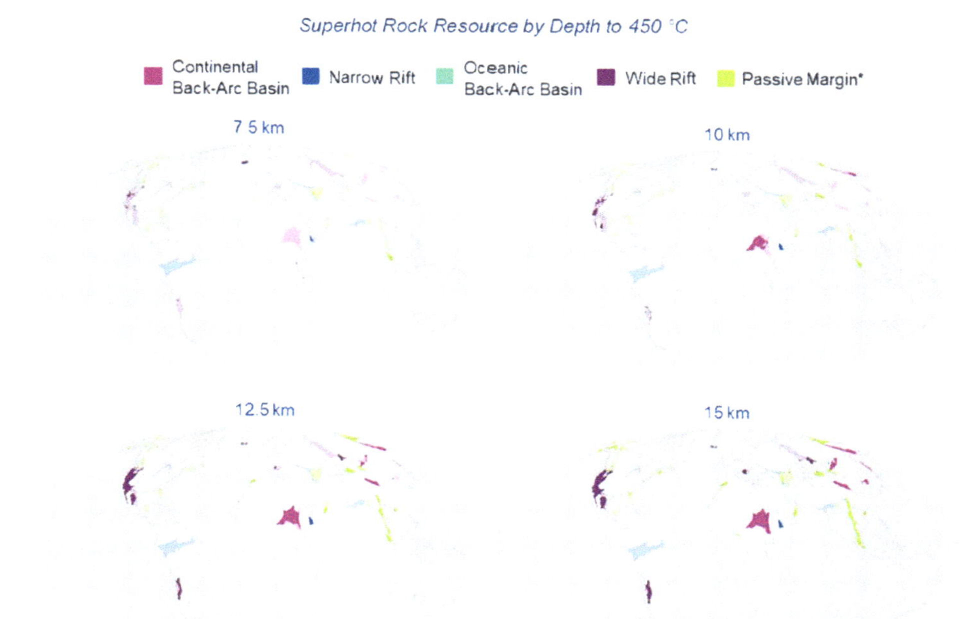

Fig. 24 (Model 2) Distribution of prospective SHR resource at 7.5, 10, 12.5, and 15 km, for tensional geodynamic settings: (i) continental back-arc basin, (ii) narrow rift, (iii) oceanic back-arc basin, (iv) oceanic crust, (v) passive margin, and (vi) wide rift. Bold colors indicate SHR resource as defined by Hasterok et al. (2022). Transparent colors indicate geodynamic setting with no SHR identified at the relevant depth interval. Map lines delineate study areas and do not necessarily depict accepted national boundaries.

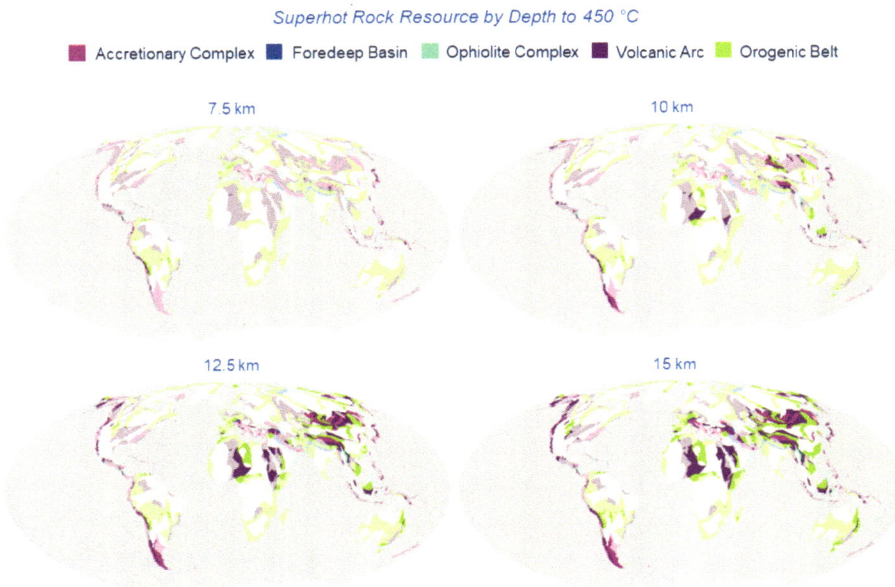

Fig. 25 (Model 2) Distribution of prospective SHR resource at 7.5, 10, 12.5, and 15 km, for compressional geodynamic settings: (i) accretionary complex, (ii) foredeep basin, (iii) ophiolite complex, (iv) volcanic arc, (v) continental back-arc, and (vi) orogenic belt. Bold colors indicate SHR resource as defined by Hasterok et al. (2022). Transparent colors indicate geodynamic setting with no SHR identified at the relevant depth interval. Map lines delineate study areas and do not necessarily depict accepted national boundaries.

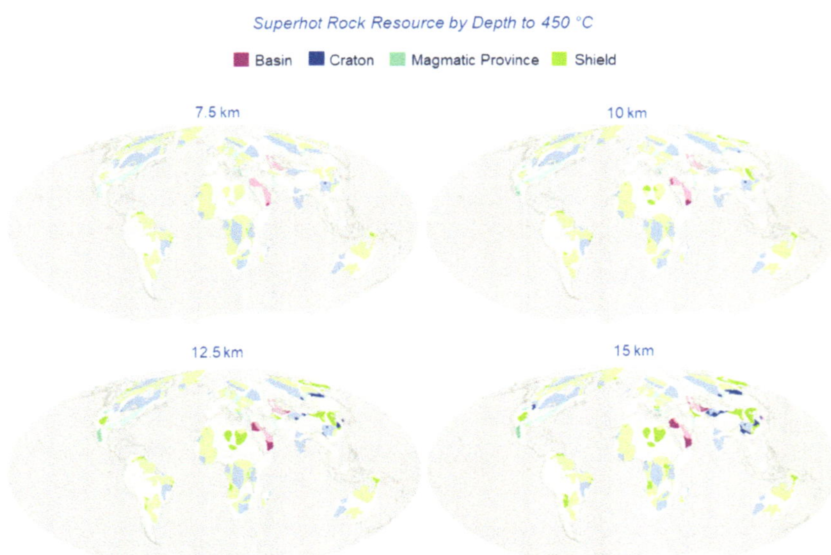

Fig. 26 (Model 2) Distribution of prospective SHR resource at 7.5, 10, 12.5, and 15 km, for continental "stable" geodynamic settings: (i) basin, (ii) craton, (iii) magmatic province, and (iv) shield. Bold colors indicate SHR resource as defined by Hasterok et al. (2022). Transparent colors indicate geodynamic setting with no SHR identified at the relevant depth interval. Map lines delineate study areas and do not necessarily depict accepted national boundaries.

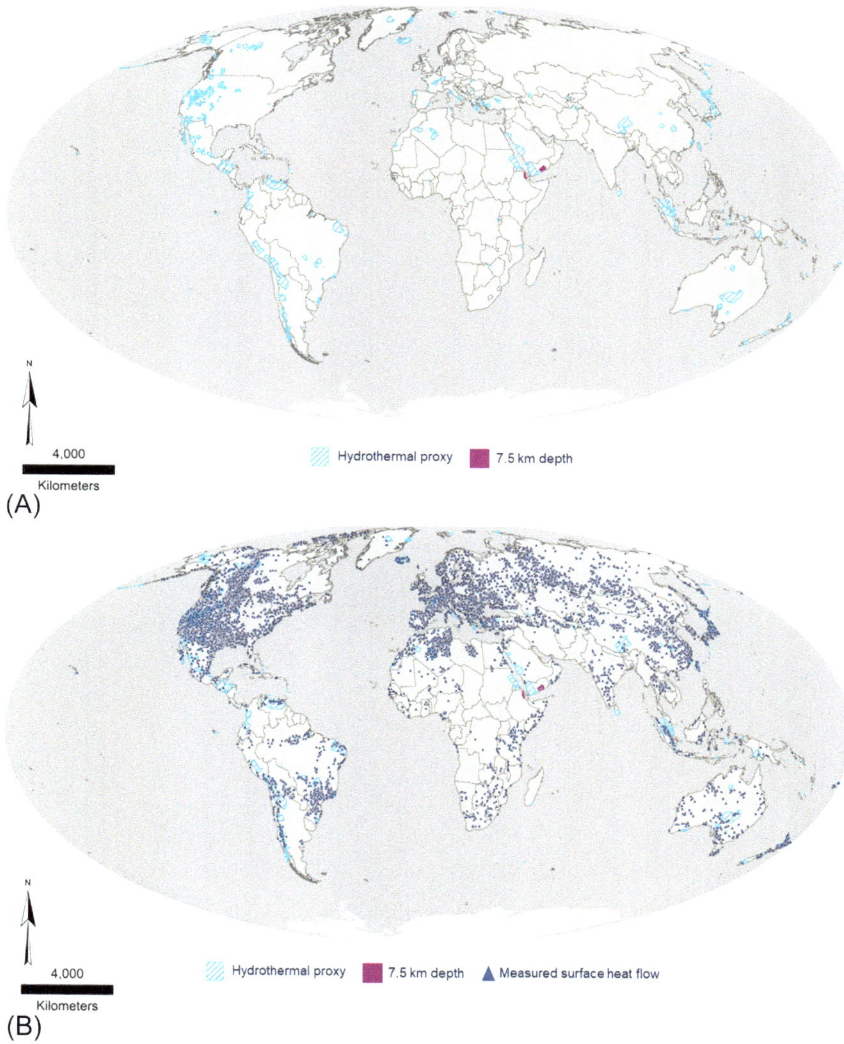

Fig. 27 (A) (Model 1) Distribution of prospective SHR resource at 7.5. Isolating the steady-state from the nonsteady-state prospective resource area. (B) (Model 1) Distribution of prospective SHR resource at 7.5 km. Isolating the steady-state from the nonsteady-state prospective resource area. *Measured data (triangles) from Fuchs, S., Norden, B., Neumann, F., Kaul, N., Tanaka, A., Kukkonen, I.T., Pascal, C., Christiansen, R., Gola, G., Šafanda, J., Espinoza-Ojeda, O.M., Marzan, I., Rybach, L., Balkan-Pazvantoğlu, E., Ramalho, E.C., Dědeček, P., Negrete-Aranda, R., Balling, N., Poort, J., Wang, Y., Jõeleht, A., Rajver, D., Gao, X., Liu, S., Harris, R., Richards, M., McLaren, S., Chiozzi, P., Nunn, J., Madon, M., Beardsmore, G., Funnell, R., Duerrast, H,. Jennings, S., Elger, K., Pauselli, K., Verdoya, M., 2023. Quality-assurance of heat-flow data: the new structure and evaluation scheme of the IHFC global heat flow database. Tectonophysics 863, 229976, https://doi.org/10.1016/j.tecto.2023.229976.* Map lines delineate study areas and do not necessarily depict accepted national boundaries.

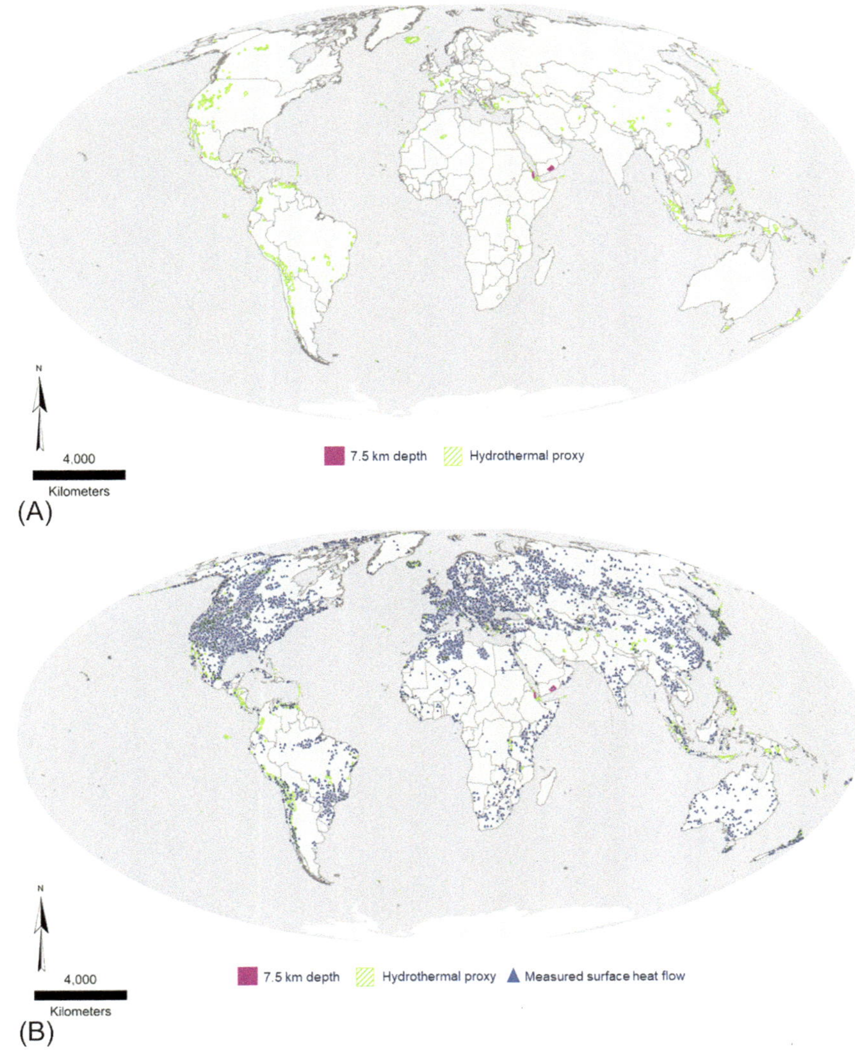

Fig. 28 (A) (Model 1) Distribution of prospective SHR resource at 7.5. Isolating the steady-state from the nonsteady-state prospective resource area. (B) (Model 2) Distribution of prospective SHR resource at 7.5 km. Isolating the steady-state from the nonsteady-state prospective resource area. *Measured data (triangles) from Fuchs, S., Norden, B., Neumann, F., Kaul, N., Tanaka, A., Kukkonen, I.T., Pascal, C., Christiansen, R., Gola, G., Šafanda, J., Espinoza-Ojeda, O.M., Marzan, I., Rybach, L., Balkan-Pazvantoğlu, E., Ramalho, E.C., Dědeček, P., Negrete-Aranda, R., Balling, N., Poort, J., Wang, Y., Jõeleht, A., Rajver, D., Gao, X., Liu, S., Harris, R., Richards, M., McLaren, S., Chiozzi, P., Nunn, J., Madon, M., Beardsmore, G., Funnell, R., Duerrast, H,. Jennings, S., Elger, K., Pauselli, K., Verdoya, M., 2023. Quality-assurance of heat-flow data: the new structure and evaluation scheme of the IHFC global heat flow database. Tectonophysics 863, 229976, https://doi.org/10.1016/j.tecto.2023.229976.* Map lines delineate study areas and do not necessarily depict accepted national boundaries.

of 250 MW are presented here (Tables 8 and 9, Figs. 29 and 30) showing the results from the two model approaches described (Section 5). These areas are converted to GWe (Figs. 31–32) using the methodology outlined within Section 6.

TABLE 8 Model 2, area of superhot rock at selected depth, summarized using the geodynamic mapping defined by Hasterok et al. (2022).

Model 1: LithoREF18 + Fuchs et al. (2023)					
Geodynamic setting	**7.5 km (nonsteady state)**	**7.5 km (steady state)**	**10 km**	**12.5 km**	**15 km**
Accretionary complex	463,463.51	0.00	1,486,660.42	3,532,579.11	4,838,450.06
Basin	4693.80	**40,967.00**	189,655.05	614,841.41	888,224.96
Continental back-arc basin	16,293.43	0.00	1093,434.72	2,117,743.13	2,546,801.96
Craton	75,218.08	0.00	88,506.60	569,058.17	1,265,104.20
Foredeep basin	0.00	0.00	10,843.60	70,100.13	141,754.23
Magmatic province	104,234.69	0.00	216,655.49	421,817.48	493,205.02
Narrow rift	102,143.05	**18,303.00**	261,178.08	351,272.04	429,759.47
Oceanic back-arc basin	4723.05	0.00	4723.05	5113.49	5936.99
Oceanic crust	5643.78	0.00	5643.78	10,302.36	36,594.29
Ophiolite complex	35,827.64	0.00	100,657.77	283,414.92	374,383.53
Orogenic belt	1,037,518.01	0.00	2,862,946.29	6,819,128.50	10,437,852.78
Passive margin	249,484.43	0.00	428,849.60	873,448.52	1,231,877.00
Shield	673,632.20	0.00	1,490,056.74	3,780,223.19	4,854,043.14
Volcanic arc	1,446,416.66	**1922.00**	3,583,356.43	8,579,445.99	11,855,578.74
Wide rift	446,902.87	0.00	707,946.50	1,943,801.75	2,452,496.11

TABLE 9 Model 2, area of superhot rock at selected depth, summarized using the geodynamic mapping defined by Hasterok et al. (2022).

Model 2: LithoREF18 + Lucazeau (2019)					
Geodynamic setting	**7.5 km (nonsteady state)**	**7.5 km (steady state)**	**10 km**	**12.5 km**	**15 km**
Accretionary complex	157,031.20	0.00	1,284,721.69	3,352,284.38	4,679,537.76
Basin	123.85	**40,967.00**	188,019.54	613,205.89	887,107.77
Continental back-arc basin	9170.56	0.00	1,086,311.84	2,110,620.26	2,539,759.81
Craton	26,152.10	0.00	39,440.62	519,992.19	1,216,575.17

Continued

TABLE 9 Model 2, area of superhot rock at selected depth, summarized using the geodynamic mapping defined by Hasterok et al. (2022)—cont'd

Model 2: LithoREF18 + Lucazeau (2019)					
Geodynamic setting	**7.5 km (nonsteady state)**	**7.5 km (steady state)**	**10 km**	**12.5 km**	**15 km**
Foredeep basin	0.00	0.00	10,843.60	70,100.13	141,754.23
Magmatic province	40,181.19	0.00	173,882.83	408,841.66	482,133.56
Narrow rift	61,617.36	**18,303.00**	219,786.71	326,560.77	416,894.98
Oceanic back-arc basin	1246.29	0.00	1246.29	1636.73	2460.23
Oceanic crust	11,645.87	0.00	11,645.87	16,453.23	37,419.10
Ophiolite complex	24,560.75	0.00	89,390.88	272,148.03	363,116.64
Orogenic belt	393,913.40	0.00	2,257,859.90	6,270,287.81	9,981,387.02
Passive margin	106,668.90	0.00	286,034.02	732,971.39	1,106,218.41
Shield	161,576.50	0.00	978,001.05	3,273,924.68	4,346,436.11
Volcanic arc	**794,477.00**	**1922.00**	3040,017.29	8,168,467.13	11,563,010.35
Wide rift	268,093.90	0.00	558,823.75	1,886,261.66	2,413,889.58

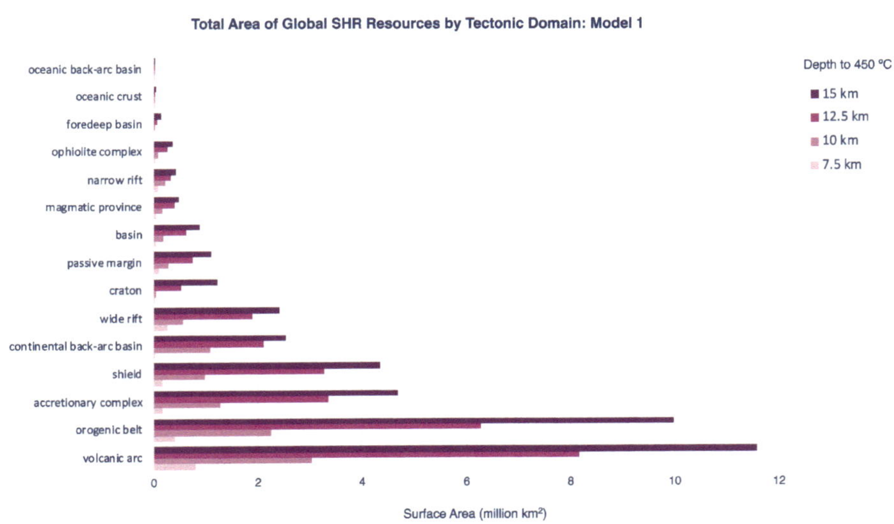

Fig. 29 Area distribution by depth of SHR steady-state and nonsteady state resources. (Model 1) Fuchs et al. (2023), as input for nonsteady-state calculation.

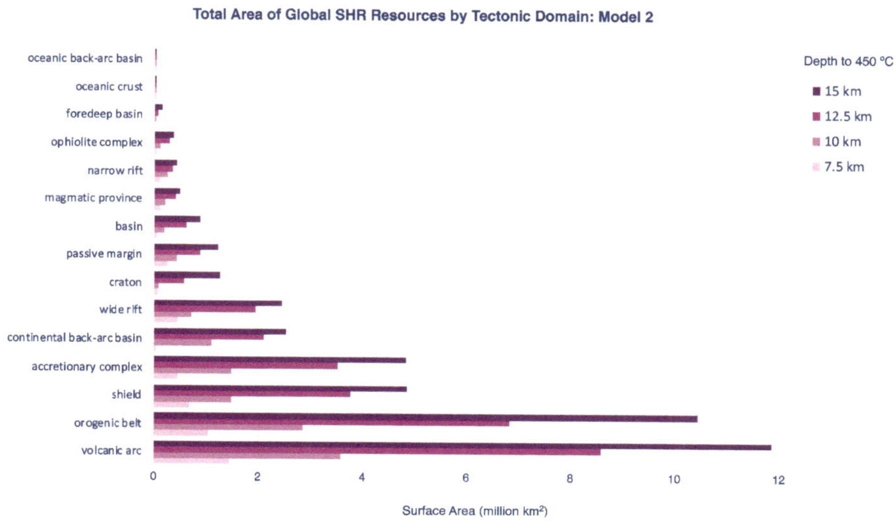

Fig. 30 Area distribution by depth of SHR steady-state and nonsteady state resources. (Model 2) Lucazeau (2019) as input for nonsteady-state calculation.

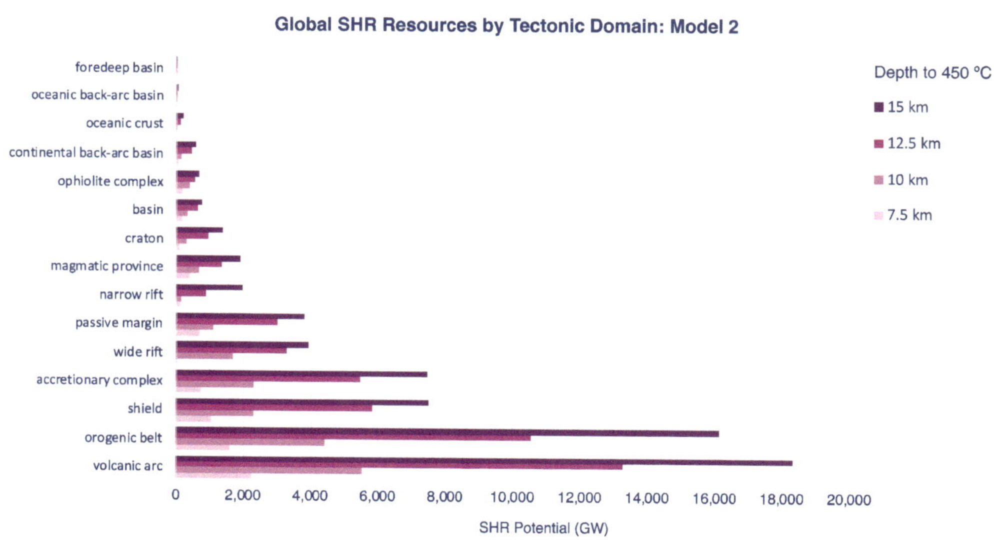

Fig. 31 GW potential based on 1% development model. SHR steady-state and nonsteady-state resources. (Model 1) Fuchs et al. (2023), as input for nonsteady-state calculation.

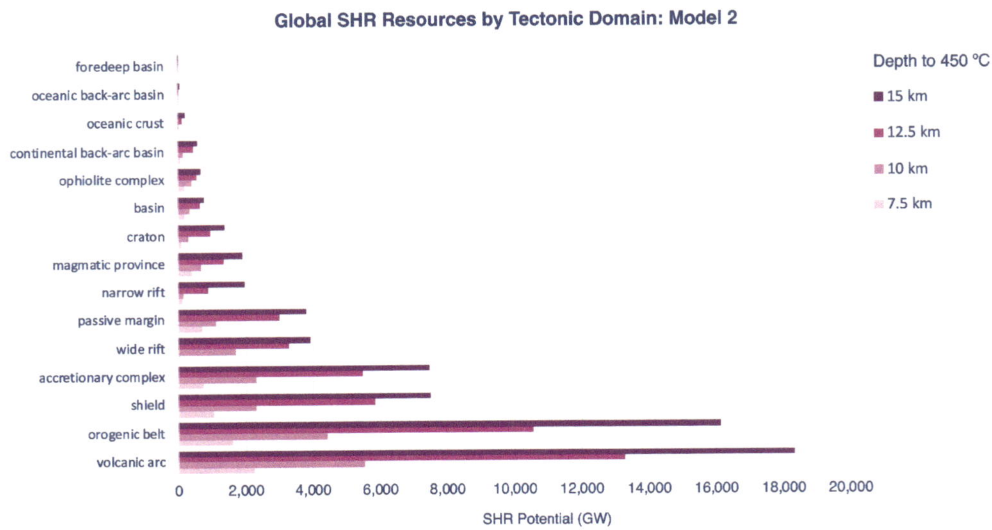

Fig. 32 GW potential based on 1% development model. SHR steady-state and nonsteady-state resources. (Model 2) Lucazeau (2019), as input for nonsteady-state calculation.

If we examine the top 7 geodynamic settings for both models, these are essentially the same although the order is not necessarily the same when the values are ranked (Figs. 29–32). The top 7 are volcanic arc, orogenic belt, shield, accretionary complex, continental BAB, wide rift, and craton. The global distributions can also be explored (Figs. 21–28). It is correct to assert that the opportunity in all geodynamic settings increases with depth, offering larger prospective resource areas for SHR development. Although we note that the foredeep basin does not appear to have a significant nonsteady state neither conductive steady state resource which is observed by the 0 km area at the 7.5 km interval.

At the highest level, a review of the heat found in geodynamic setting reveals that the conductive steady-state isotherm and the nonsteady-state areas are present in 15 of the 15 mapped geodynamic settings, and this is observed for both Models 1 and 2 (Tables 8 and 9, Figs. 29–32). The two models are, at first glance, similar; however, the predictions of shallow SHR within the shield settings are greater in Model 1 than in Model 2. Model 2 places more accessible area into the accretionary complex. If we focus on the steady-state versus nonsteady-state heat flow in shallow region (Figs. 27–28; Tables 8 and 9), we observe that shallow conductive heat is predicted within only three geodynamic settings: basin (Yemen), narrow rifts (Ethiopia), and volcanic arc (Mexico). For the shallow layer, we observe a strong contribution from the nonsteady-state heat flow prediction (Figs. 27–28). For Model 1, the top geodynamic settings are volcanic arc, orogenic belt, and shield. For Model 2, the third place is wide rift. Differences at 7.5 km may be due to the data density and gridding effects resulting from the Fuchs et al. (2023) database versus the surface heat flow model of Lucazeau (2019). A quick comparison of the total area of resource per geodynamic setting reveals that the selection of model impacts the predicted resource with the areas predicted being greater with the Model 1 (Tables 8 and 9). The preference for wide rift versus shield in Model 2 could be due to an improved lithospheric model, but more work is needed to validate these differences. Comparison of the nonsteady-state and steady-state layers can be made which provide additional understanding on the confidence of the results or the assumed depth. The insight gained is looking at the deeper layers: If the 7.5-km layers remain the same to the 10 or 12.5 km depth intervals, this highlights that no conductive heat flow is predicted. Second, it tells us there is an increased uncertainty in the shallow 7.5 km nonsteady state heat flow.

Examination of the maps shows the spatial distribution of the SHR resource. For simplicity, the maps have been divided by the broad topic of tensional, compressional, and stable crustal domains (Figs. 21–26; Tables 10–13). One major observation is that within any particular geodynamic setting, there is considerable variation in the predicted location of heat. For example, it is not possible to conclude that all wide rift settings are good, although by the very nature of this geodynamic classification, heat flow should be high to allow the formation of characteristic deformation styles to form. This internal variability requires more in-depth investigation and study but remains an important observation. This observation is reinforced if we look at Tables 10–13. Table 10 summarizes for each geodynamic setting the percentage of area that is identified to have a SHR resource by depth as a percentage of the mapped area for that setting. In Model 1 at 7.5 km, we observe that the top performing geodynamic settings are volcanic arc (3.95%), wide rift (3.85%), and magmatic province (2.21%). At 10 km, the top performing geodynamic settings are volcanic arc (9.76%), continental BAB (11.95%), and basin (7.6%). At 12.5 km, the top performing geodynamic settings are volcanic arc (23.38%), continental BAB (23.14%), and basin (24.63%). At 15 km, the top performing geodynamic settings are volcanic arc (32.31%), continental BAB (27.82%), and basin (35.58%). The results confirm significant variability within different geodynamic settings. In Model 2 at 7.5 km, we observe that the top performing geodynamic settings are volcanic arc (2.17%), wide rift (2.31%), and basin (1.65%). At 10 km, the top performing geodynamic settings are volcanic arc (8.28%), continental BAB (11.87%), and basin (7.53%). At 12.5 km, the top performing geodynamic settings are volcanic arc (22.26%), continental BAB (23.06%), and basin (24.56%). At 15 km, the top performing geodynamic settings are volcanic arc (31.51%), continental BAB (27.75%), and basin (35.54%). The results confirm significant variability within different geodynamic settings. What likely transpires is that when a tectonic/physiographic province has heat, it is highly prospective over other provinces within the geodynamic setting resource type. For example, wide rifts in the North American Basin and Range is a prolific tectonic province versus other provinces mapped as a wide rift (Figs. 21–26)—why this occurs needs to be understood and is out of the scope of the current paper. The implication of this work however means that we can quickly eliminate >70% of the world for each geodynamic setting with respect to exploration and focus on those with shallow SHR heat predicted.

TABLE 10 Summary of SHR resource, area, and GW potential (P50 scenario; see Section 6 for details)—Model 1: LithoREF18 and Fuchs et al. (2023).

Model 1: LithoREF18 and Fuchs et al. (2023)													
Compressional tectonic domain		**SHR potential at 7.5-km depth**			**SHR potential at 10-km depth**			**SHR potential at 12.5-km depth**			**SHR potential at 15-km depth**		
Type	**Area (km²)**	**Area (km²)**	**% of total area**	**GWe [GWth] potential**	**Area (km²)**	**% of total area**	**GW potential**	**Area (km²)**	**% of total area**	**GWe [GWth] potential**	**Area (km²)**	**% of total area**	**GWe [GWth] potential**
Accretionary complex	21,263,281.07	463,463.51	2.18%	716.14 [3040.95]	1,486,660.42	6.99%	2297.19 [9754.51]	3,532,579.11	16.61%	5458.54 [23,178.52]	4,838,450.06	22.75%	7476.37 [31,746.81]
Foredeep basin	1,704,789.12	0.00	0.00%	0.00 [0.00]	10,843.60	0.64%	16.76 [71.15]	70,100.13	4.11%	108.32 [459.95]	141,754.23	8.32%	219.04 [930.10]
Ophiolite complex	2,392,331.28	35,827.64	1.50%	55.36 [235.08]	100,657.77	4.21%	155.54 [660.45]	283,414.92	11.85%	437.93 [1859.59]	374,383.53	15.65%	578.50 [2456.46]
Orogenic belt	47,340,645.45	1,037,518.01	2.19%	1603.17 [6807.53]	2,862,946.29	6.05%	4423.82 [18,784.82]	6,819,128.50	14.40%	10,536.92 [44,742.75]	10,437,852.78	22.05%	16,128.57 [68,486.50]
Volcanic arc	36,696,024.92	1,448,338.66	3.95%	2237.97 [9503.07]	3,583,356.43	9.76%	5537.00 [23,511.69]	8,579,445.99	23.38%	13,256.96 [56,292.82]	11,855,578.74	32.31%	18,319.24 [77,788.71]
Totals—compressional	109,397,071.85	2,985,147.82	2.73%	4612.65 [19,586.63]	8,044,464.51	7.35%	12,430.31 [52,782.62]	19,284,668.65	17.63%	29,798.67 [126,533.63]	27,648,019.33	25.27%	42,721.72 [181,408.58]
Tensional tectonic domain		**SHR potential at 7.5-km depth**			**SHR potential at 10-km depth**			**SHR potential at 12.5-km depth**			**SHR potential at 15-km depth**		
Type	**Area (km²)**	**Area (km²)**	**% of total area**	**GWe [GWth] potential**	**Area (km²)**	**% of total area**	**GW potential**	**Area (km²)**	**% of total area**	**GWe [GWth] potential**	**Area (km²)**	**% of total area**	**GWe [GWth] potential**
Continental back-arc basin	9,153,714.21	16,293.43	0.18%	25.18 [106.91]	1093,434.72	11.95%	1689.58 [7174.42]	2,117,743.13	23.14%	3272.34 [13,895.27]	2,546,801.96	27.82%	3935.32 [16,710.48]
Narrow rift	5,087,880.58	120,446.05	2.37%	186.11 [790.29]	261,178.08	5.13%	403.57 [1713.68]	351,272.04	6.90%	542.79 [2304.82]	429,759.47	8.45%	664.06 [2819.81]
Oceanic back-arc basin	9,226,125.54	4723.05	0.05%	7.30 [30.99]	4723.05	0.05%	7.30 [30.99]	5113.49	0.06%	7.90 [33.55]	5936.99	0.06%	9.17 [38.95]
Oceanic crust	283,644,125.81	5643.78	0.00%	8.72 [37.03]	5643.78	0.00%	8.72 [37.03]	10,302.36	0.00%	15.92 [67.60]	36,594.29	0.01%	56.55 [240.11]
Passive margin	20,083,354.33	249,484.43	1.24%	385.50 [1636.96]	428,849.60	2.14%	662.66 [2813.84]	873,448.52	4.35%	1349.65 [5731.01]	1,231,877.00	6.13%	1903.50 [8082.79]

Wide rift	11,593,875.53	446,902.87	3.85%	690.55 [2932.29]	707,946.50	6.11%	1093.92 [4645.09]	1,943,801.75	16.77%	3003.56 [12,753.98]	2,452,496.11	21.15%	3789.60 [16,091.71]
Totals: tensional (on- and offshore)	338,789,076.00	843,493.61	0.25%	n/a	2,501,775.72	0.74%	n/a	5,301,681.30	1.56%	8192.16 [34,786.23]	6,703,465.82	1.98%	n/a
Onshore only oceanic and passive margin elements	35,061,595.86	588,365.40	1.68%	1303.37 [5534.46]	2,067,282.34	5.90%	3865.74 [16,415.05]	4,417,930.41	12.60%	16,384.32	5,434,994.53	15.50%	10,358.20 [43,983.84]

Stable tectonic domain		**SHR potential at 7.5-km depth**			**SHR potential at 10-km depth**			**SHR potential at 12.5-km depth**			**SHR potential at 15-km depth**		
Type	**Area (km²)**	**Area (km²)**	**% of total area**	**GWe [GWth] potential**	**Area (km²)**	**% of total area**	**GW potential**	**Area (km²)**	**% of total area**	**GWe [GWth] potential**	**Area (km²)**	**% of total area**	**GWe [GWth] potential**
Basin	2,496,337.84	45,660.80	1.83%	70.56 [299.60]	189,655.05	7.60%	293.05 [1244.39]	614,841.41	24.63%	950.05 [4034.20]	888,224.96	35.58%	1372.49 [5827.96]
Craton	19,610,219.41	75,218.08	0.38%	116.23 [493.53]	88,506.60	0.45%	136.76 [580.72]	569,058.17	2.90%	879.31 [3733.79]	1,265,104.20	6.45%	1954.84 [8300.80]
Magmatic province	4,714,026.68	104,234.69	2.21%	161.06 [683.92]	216,655.49	4.60%	334.78 [1421.55]	421,817.48	8.95%	651.79 [2767.70]	493,205.02	10.46%	762.10 [3236.10]
Shield	34,141,479.60	673,632.20	1.97%	1040.90 [4419.94]	1,490,056.74	4.36%	2302.44 [9776.80]	3,780,223.19	11.07%	5841.20 [24,803.40]	4,854,043.14	14.22%	7500.47 [31,849.12]
Totals—stable	**60,962,063.52**	898,745.78	**1.47%**	1388.74	1,984,873.88	**3.26%**	3067.03 [13,023.47]	5,385,940.25	**8.83%**	8322.35 [35,339.09]	7500,577.32	**12.30%**	11,589.89 [49,213.98]
Totals—all	509,148,211.37	4,727,387.20	2.3%	7304.76 [31,018.08]	12,531,114.10	6.10%	19,363.08 [82,221.14]	29,972,290.20	14.59%	46,313.18 [196,658.95]	41,852,062.46	20.37%	64,669.81 [274,606.40]

Organized by geodynamic setting and dominant stress regime (paleo and present day). Area calculations are derived from the total mapped geodynamic domain.

TABLE 11 Summary of SHR resource, area, and GW potential (P50 scenario—see Section 6 for details)—Model 2.

Model 2: LithoREF18 and Lucazeau (2019)														
Compressional tectonic domain		SHR potential at 7.5-km depth			SHR potential at 10-km depth			SHR potential at 12.5-km depth			SHR potential at 15-km depth			
Type	Area (km^2)	Area (km^2)	% of total area	GWe [GWth] potential	Area (km^2)	% of total area	GW potential	Area (km^2)	% of total area	GWe [GWth] potential	Area (km^2)	% of total area	GWe [GWth] potential	
Accretionary complex	21,263,281.07	157,031.17	0.74%	242.64 [1030.34]	1,284,721.69	6.04%	1985.15 [8429.52]	3,352,284.38	15.77%	5179.95 [21,995.54]	4,679,537.76	22.01%	7230.82 [30,704.13]	
Foredeep basin	1,704,789.12	0.00	0.00%	0.00 [0.00]	10,843.60	0.64%	16.76 [71.15]	70,100.13	4.11%	108.32 [459.95]	141,754.23	8.32%	219.04 [930.10]	
Ophiolite complex	2,392,331.28	24,560.75	1.03%	37.95 [161.15]	89,390.88	3.74%	138.13 [586.53]	272,148.03	11.38%	420.52 [1785.66]	363,116.64	15.18%	561.09 [2382.54]	
Orogenic belt	47,340,645.45	393,913.43	0.83%	608.68 [2584.61]	2,257,859.90	4.77%	3488.85 [14,814.63]	6,270,287.81	13.25%	9688.85 [41,141.61]	9,981,387.02	21.08%	15,423.24 [65,491.46]	
Volcanic arc	36,696,024.92	796,399.04	2.17%	1230.60 [5225.46]	3040,017.29	8.28%	4697.43 [19,946.64]	8,168,467.13	22.26%	12,621.92 [53,596.24]	11,563,010.35	31.51%	17,867.16 [75,869.06]	
Total—compressional	109,397,071.85	1,371,904.39	1.25%	2119.87 [9001.56]	6,682,833.36	6.11%	10,326.31 [43,848.47]	18,133,287.48	16.58%	28,019.56 [118,979.01]	26,728,805.99	24.43%	41,301.35 [175,377.29]	
Tensional tectonic domain		SHR potential at 7.5-km depth			SHR potential at 10-km depth			SHR potential at 12.5-km depth			SHR potential at 15-km depth			
Type	Area (km^2)	Area (km^2)	% of total area	GWe [GWth] potential	Area (km^2)	% of total area	GW potential	Area (km^2)	% of total area	GWe [GWth] potential	Area (km^2)	% of total area	GWe [GWth] potential	
Continental back-arc basin	9,153,714.21	9170.56	0.10%	14.17 [60.17]	1,086,311.84	11.87%	1678.57 [7127.68]	2,110,620.26	23.06%	3261.33 [13,848.54]	2,539,759.81	27.75%	3924.44 [16,664.28]	
Narrow rift	5,087,880.58	79,920.36	1.57%	123.49 [524.39]	219,786.71	4.32%	339.61 [1442.10]	326,560.77	6.42%	504.60 [2142.68]	416,894.98	8.19%	644.19 [2735.40]	
Oceanic back-arc basin	9,226,125.54	1246.29	0.01%	1.93 [8.18]	1246.29	0.01%	1.93 [8.18]	1636.73	0.02%	2.53 [10.74]	2460.23	0.03%	3.80 [16.14]	
Oceanic crust	283,644,125.81	11,645.87	0.00%	18.00 [76.41]	11,645.87	0.00%	18.00 [76.41]	16,453.23	0.01%	25.42 [107.96]	37,419.10	0.01%	57.82 [245.52]	

Passive margin	20,083,354.33	106,668.85	0.53%	164.82 [699.89]	286,034.01	1.42%	441.98 [1876.77]	732,971.39	3.65%	1132.59 [4809.29]	1,106,218.41	5.51%	1709.33 [7258.30]
Wide rift	11,593,875.53	268,093.89	2.31%	414.26 [1759.06]	558,823.75	4.82%	863.49 [3666.64]	1,886,261.66	16.27%	2914.65 [12,376.44]	2,413,889.57	20.82%	3729.94 [15,838.40]
Totals: tensional (on- and offshore)	338,789,076.00	476,745.81	0.14%	n/a	2,163,848.47	0.64%	n/a	5,074,504.04	1.50%	n/a	6,516,642.10	1.92%	n/a
Onshore only oceanic and passive margin elements	35,061,595.86	358,431.09	1.02%	736.67 [3128.10]	1,866,168.59	5.32%	3343.58 [14,197.79]	4,325,079.42	12.34%	7841.12 [33,295.64]	5,373,004.59	15.32%	10,069.52 [42,758.03]
Stable tectonic domain		**SHR potential at 7.5-km depth**			**SHR potential at 10-km depth**			**SHR potential at 12.5-km depth**			**SHR potential at 15-km depth**		
Type	**Area (km²)**	**Area (km²)**	**% of total area**	**GWe [GWth] potential**	**Area (km²)**	**% of total area**	**GW potential**	**Area (km²)**	**% of total area**	**GWe [GWth] potential**	**Area (km²)**	**% of total area**	**GWe [GWth] potential**
Basin	2,496,337.84	41,090.85	1.65%	63.49 [269.61]	188,019.54	7.53%	290.53 [1233.66]	613,205.89	24.56%	947.53 [4023.46]	887,107.77	35.54%	1370.76 [5820.63]
Craton	19,610,219.41	26,152.10	0.13%	40.41 [171.59]	39,440.62	0.20%	60.94 [258.78]	519,992.19	2.65%	803.49 [3411.86]	1,216,575.16	6.20%	1879.85 [7982.39]
Magmatic province	4,714,026.68	40,181.19	0.85%	62.09 [263.64]	173,882.83	3.69%	268.68 [1140.91]	408,841.66	8.67%	631.74 [2682.56]	482,133.56	10.23%	744.99 [3163.45]
Shield	34,141,479.60	161,576.51	0.47%	249.67 [1060.16]	978,001.05	2.86%	1511.21 [6417.02]	3,273,924.68	9.59%	5058.87 [21,481.39]	4,346,436.11	12.73%	6716.11 [28,518.53]
Totals: stable	60,962,063.52	269,000.65	0.44%	415.66 [1765.01]	1,379,344.03	2.26%	2131.36 [9050.37]	4,815,964.42	7.90%	7441.63 [31,599.27]	6,932,252.60	11.37%	10,711.72 [45,485.00]
Totals: all	509,148,211.37	2,117,650.85	1.03%	3272.19 [13,894.67]	10,226,025.87	4.98%	15,801.26 [67,096.62]	28,023,755.93	13.64%	43,302.31 [183,873.92]	40,177,700.70	19.56%	62,082.58 [263,620.31]

Organized by geodynamic setting and dominant stress regime (paleo and present day). Area calculations are derived from the total mapped geodynamic domain.

TABLE 12 Summary of SHR resource, by depth, highlighting distribution of resource in percentage and totals—Model 1.

Model 1: LithoREF18 and Fuchs et al. (2023)							
Tectonic domain of global SHR potential		**% of total SHR surface area by depth**					
Dominant stress regime (paleo and present day)	**Geodynamic setting**	**7.5 km**	**10 km**	**12.5 km**	**15 km**	**% total SHR resource by dominant stress regime (paleo and present day)**	**Total prospective TWe [TWth] endowment by dominant stress regime (paleo and present day)**
Compressional	Accretionary complex	9.8%	11.9%	11.8%	11.6%	65.1%	89.6 [380.3]
	Foredeep basin	0.0%	0.1%	0.2%	0.3%		
	Ophiolite complex	0.8%	0.8%	0.9%	0.9%		
	Orogenic belt	21.9%	22.8%	22.8%	24.9%		
	Volcanic arc	30.6%	28.6%	28.6%	28.3%		
Totals		**63.1%**	**64.2%**	**64.3%**	**66.1%**		
Tensional	Continental back-arc basin	0.3%	8.7%	7.1%	6.1%	17.2%	23.7 [100.7]
	Narrow rift	2.5%	2.1%	1.2%	1.0%		
	Oceanic back-arc basin	0.1%	0.0%	0.0%	0.0%		
	Oceanic crust	0.1%	0.0%	0.0%	0.1%		
	Passive margin	5.3%	3.4%	2.9%	2.9%		
	Wide rift	9.5%	5.6%	6.5%	5.9%		
Totals		**17.8%**	**20.0%**	**17.7%**	**16.0%**		
Stable	Basin	1.0%	1.5%	2.1%	2.1%	17.7%	24.4 [103.5]
	Craton	1.6%	0.7%	1.9%	3.0%		
	Magmatic province	2.2%	1.7%	1.4%	1.2%		
	Shield	14.2%	11.9%	12.6%	11.6%		
Totals		**19.0%**	**15.8%**	**18.0%**	**17.9%**		

Organized by geodynamic setting and dominant stress regime (paleo and present day). Calculations are the percentage of all SHR heat by geodynamic setting and assumed stress regime (paleo and present day).

TABLE 13 Summary of SHR resource, by depth, highlighting distribution of resource in percentage and totals—Model 2.

Model 2: LithoREF18 and Lucazeau (2019)							
Geodynamic setting of global SHR potential		% of total SHR surface area by depth					
Dominant stress regime (paleo and present day)	Geodynamic setting	7.5 km	10 km	12.5 km	15 km	% Total SHR resource by dominant stress regime (paleo and present day)	Total prospective TWe [TWth] endowment by dominant stress regime (paleo and present day)
Compressional	Accretionary complex	7.4%	12.6%	12.0%	11.6%	65.7%	81.8 [347.2]
	Foredeep basin	0.0%	0.1%	0.3%	0.4%		
	Ophiolite complex	1.2%	0.9%	1.0%	0.9%		
	Orogenic belt	18.6%	22.1%	22.4%	24.8%		
	Volcanic arc	37.6%	29.7%	29.1%	28.8%		
Totals		**64.8%**	**65.4%**	**64.7%**	**66.5%**		
Tensional	Continental back-arc basin	0.4%	10.6%	7.5%	6.3%	17.7%	22.0 [93.4]
	Narrow rift	3.8%	2.1%	1.2%	1.0%		
	Oceanic back-arc basin	0.1%	0.0%	0.0%	0.0%		
	Oceanic crust	0.5%	0.1%	0.1%	0.1%		
	Passive margin	5.0%	2.8%	2.6%	2.8%		
	Wide rift	12.7%	5.5%	6.7%	6.0%		
Totals		**22.5%**	**21.2%**	**18.1%**	**16.2%**		
Stable	Basin	1.9%	1.8%	2.2%	2.2%	16.6%	20.7 [87.9]
	Craton	1.2%	0.4%	1.9%	3.0%		
	Magmatic province	1.9%	1.7%	1.5%	1.2%		
	Shield	7.6%	9.6%	11.7%	10.8%		
Totals		**12.7%**	**13.5%**	**17.2%**	**17.3%**		

Organized by geodynamic setting and dominant stress regime (paleo and present day). Calculations are the percentage of all SHR heat by geodynamic setting and assumed stress regime (paleo and present day).

If we look at the distribution of resources by dominant stress regime (paleo and present day) (Tables 12 and 13), the analyses are only for onshore settings. For the heat mapping, we did not subdivide the analysis to tectonic domain/physiographic province. What we observe is the following: Within Model 1 (65.1%) and Model 2 (65.7%) of the modeled resource is located within geodynamic settings that are currently classified as compressional domain. If we focus only on the 7.5-km interval, we see some variability in the shallow heat predicted. The shallow conductive heat is dominantly in stable settings (65%)—in Model 1, the nonsteady-state heat falls within compressional (63.9%), but in Model 2, the nonsteady-state heat is identified within a tensional environment (45.6%). This clearly translates into massive prospective resources based on the method used herein (Section 6 and Tables 10–13).

In detail, starting with Model 1, the area where superhot resources are mapped is prevailing within compressional domains, with 63.1%–66.1% of the mapped resource area from 7.5 to 15 km depth. Within tensional settings, there is a fluctuation of the resource from 16% to 20%. Interestingly the greatest share of the resource for tensional domains is at the 10-km depth interval. The range for stable crust is 15.8%–19% of the superhot resource. A counterintuitive trend is also observed in stable domains with the greatest area within this class occurring at 7.5 km. Within the compressional geodynamic settings, the volcanic arc and orogenic belt contain more than half of the mapped resource. In the tensional domain, the shallow heat is most prevalent in the passive margin and wide rift geodynamic settings, indicating these areas have a strong nonsteady-state component to the mapped resource. From 10 to 15 km, the dominant geodynamic setting is however the continental BAB and the wide rift. It is interesting to note that within narrow rifts, the resource is greatest at 7.5 km and 10 km but declines with depth, indicating a nonsteady-state component to the shallow heat. In stable settings, we observe that the shield is the dominant area for SHR resource.

In Model 2, the area where superhot resources are mapped is prevailing within compressional domains, with 64.8%–66.5% of the mapped resource area from 7.5 to 15 km depth. Within tensional settings, there is a fluctuation of the resource from 22.5% to 16.2%. Interestingly, the greatest share of the resource for tensional domains is at the 7.5- and 10-km depth intervals. The range for stable crust is 12.7%–17.3% of the superhot resource. As with Model 1, the compressional geodynamic domains, the volcanic arc, and orogenic belt contain more than half of the mapped resource. In the tensional domain, the shallow heat is most prevalent in wide rift, passive margin, and narrow rift settings, indicating these areas have a strong nonsteady-state component to the mapped resource. From 10 to 15 km, the dominant geodynamic setting is however the continental BAB and the wide rift. It is interesting to note as with Model 1 that within narrow rifts, the resource is greatest at 7.5 km and 10 km but declines with depth, indicating a nonsteady-state component to the shallow heat. In stable settings, we observe that the Shield is the dominant area for SHR resource.

9 Discussion

The world has finally realized the potential that recent innovations have brought to the geothermal industry. Next-generation technologies (e.g., EGS, CLG, and HGS) could evolve into a 100 billion in the 1- to 3-km depth range and a trillion-dollar market down to 7 km and possibly a multitrillion market if depths of 12.5 km can be unlocked (BCG, 2022; Wood Mackenzie, 2023; CATF, 2023; Blankenship et al., 2024). Next-generation innovations have redefined how we can explore and extract heat from the subsurface. Adopting these innovations, this study exposes the massive potential these technologies could bring as resource plays and characterizes the opportunity by geodynamic setting. As such, it is hoped that this work will contribute to that understanding that heat in the western USA is not the same as the heat in western Saudi Arabia or western Australia. Heat is still heat, but the geodynamic processes deriving the heat are different—the stress regimes are different, and at the finer scales of observation, the variances in the geology, structure, rocks, minerology, gases, and fluids available could be different as we develop an improved understanding of the geological systems and plays' scale of investigation, down to the prospect level. The key is in simplifying the world down to learn the successful ingredients of an economic, and repeatable, geothermal project. Within this paper, we aimed to provide a framework for which the subsequent scales of investigation and the characterization of risk can be systematically developed, and understood, thus reducing exploration and development risk. This work is only the beginning; more needs to be done to understand the intricacies of exploration within each geodynamic setting and the success and failure modes between the different geodynamic settings. For example, it is unlikely that approaches can be "cut and pasted" from wide rifts to orogenic belts, volcanic arc settings, or shield settings, without some engineering modifications due to the change in stress regime acting upon these settings and the changing geological conditions. Understanding the geological variance is key as Blankenship et al. (2024) stated, true success of next-generation technologies will be measured by success of first-of-a-kind (FOAK) developments in varied geologies.

In Section 1, we highlighted how another energy sector—the petroleum sector—starts its global screening by using a play-based exploration pyramid and global regions organized by geodynamic settings (which influence the thermal parameters of tectonic/physiographic provinces). In Section 2, we introduced the difference between Geothermal System, Petrothermal System and Hydrothermal System and we highlighted the correct vocabulary and the geological conditions. The geothermal community needs this clarity because most SHR plays and exploration prospects are likely to be within the petrothermal system type and will need to be engineered to some degree. Section 3 and Table 1 describe SHR geothermal plays and where they are ranked within the Linnaean-style Geothermal Entities Classification System that is original to this paper (yet parallels the globally standardized petroleum entities classification scheme where many SHR scientists and engineering techniques will be transferred from). Targeting this superhigh-enthalpy play type is crucial, as it is expected to produce significantly more heat and/or power for society compared to conventional hydrothermal systems; this is because the supercritical or superheated fluids extracted to the Earth's surface will possess a high energy density.

Scaling geothermal efficiently from a niche industry to a global industry is important for our collective decarbonization efforts. Decision-makers, governments, and geothermal organizations are interested in regional (nationwide) results so they can make decisions on investment according to their local or national potential. Knowledge of where the heat is predicted to be and within which geodynamic setting, and its likely stress state can influence how much effort we need to commit to at each level of our geological understanding. Such information can therefore impact the associated business decisions at each layer of the Geothermal Entities Classification System. To meet that need, Section 4 and Fig. 3 introduced a global screening technique, and the hierarchical logic of the SHR exploration triangle (Fig. 3) providing the technique for gaining the "big-picture" context about the location of the optimal SHR plays.

Section 5 and Figs. 4–9 continue to build the optimal locations to explore for SHR theme by presenting the first global map of the depth to the 450°C steady-state conductive isotherm and a new hydrothermal-proxy map which defines nonsteady-state opportunities. These maps represent the first attempt to define depths to supercritical temperatures for geothermal exploration. These maps therefore need to be considered the first pass, remembering they are derived from low-resolution lithospheric models. In their current form, these maps, however, give tremendous insight to the spatial location and depth of the potential target at the regional scale. The results presented herein require further research and development to refine the predictions and to develop prospect-scale maps. The nonsteady-state maps are also first-of-a-kind estimation, and these maps allow us to isolate areas where heat flow is high due to hydrothermal convection, in its most simple explanation. The areas defined as nonsteady state may have both conventional and EGS, CLG, and HGS geothermal opportunities.

As project developers advance projects from leads to prospects to economically producing geothermal fields, and as we develop an understanding of the resource and the heat in place, and heat extractable over the lifetime of a project, we argue that organizations could compare their portfolio or new opportunities to the basic level of the geodynamic setting and tectonic/physiographic province to understand whether their projects are comparable in their tectonic or geological setting. This may help build an understanding of how to characterize risk variance and geological uncertainty between projects. For example, an organization may wish to select its preferred geodynamic settings to explore for SHR heat. At the highest level, we ask the question are the risks behind exploration and development the same in wide rifts versus orogenic belts, volcanic arcs, and shields?

Section 6 proves the industry's deep drilling capability and historic successes and then summarizes the SHR current and upcoming energy extraction and production techniques and technologies, namely, a superhot well, a CGS, an EGS, a CLG circuit, and a HGS. A fundamental question for an SHR heat investigation in any geodynamic setting should be its possible heat endowment that is the natural heat "wealth" and how to compare and contrast the heat endowment of different regions.

Section 7 describes a sequence of steps used by these authors to estimate, compute, and model the SHR prospective endowment globally, using a similar process to the yet-to-find process for petroleum exploration. The "Area That Is Developable" method was selected. This approach proposes simply that 1% of SHR heat is developable because of competing surface and subsurface land rights and distance to infrastructure. In reality, maps need infrastructure added, but the knowledge of which areas have heat and the value of this heat which could be developed will help geoscientists and economists to

- Identify which regions have limited exploration or which technology would be best focused at for a potential return on investment.
- Plan for long-term energy policy, strategy, land use, and location of infrastructure.
- Determine the questions that decision-makers need to ask and the information they need.
- Avoid areas that would present environmental/social challenges.

Section 8 summarizes the various geothermal exploration trends and extraction techniques with respect to the geodynamic setting that we analyze the distribution of mapped superhot resources (Section 8.5). We placed the existing and upcoming exploration into a geodynamic setting, tectonic/physiographic province, linking the heat to the orogenic or tectonic/

physiographic province at a global scale and reviewing the dominant stress regime (paleo and present day). This analysis combined with the distribution of superhot geothermal resources by geodynamic setting represents the world's first deep dive into understanding where SHR heat may be located by geodynamic setting, providing resource area and numbers of MWe and MWth by geodynamic setting. The observations in Section 8 provide the first fundamental step to understanding the geodynamic drivers of heat in the shallow lithosphere. The results presented here for SHR are biased toward an open-loop, engineered (EGS) approach in the subsurface, utilizing water as the geofluid and flash power plant at surface. Further work is needed to define the CLG or HGS contributions, if they are indeed different from an EGS approach. As we grow and develop geothermal projects globally, we may see those different technologies complement (i.e., be deployed in different settings and locations, increasing the footprint of accessibility) or compete with each other for land, resulting in land congestion. With respect to the working fluid, there are other geofluids that may be considered, for example, molten salt or supercritical CO_2. In the case of supercritical CO_2, the working temperature is considerably lower than that, outlined here, for water. The maps carry an inherent bias, with a specific filter that only looks for areas where our model predicts the heat at a shallow location (<15 km). Finally, it is important to understand the white spaces are not redundant for geothermal applications at lower temperatures, although to reach temperatures for heat or power deployment, more than a several kilometers will likely need to be drilled. However, as this review has shown, the technology exists to unlock 200°C, at depths of 7–10 km. Finally, low-temperature geothermal was not the focus of this study; therefore, these data have been deliberately omitted from the maps.

9.1 Existing geothermal projects and geodynamic settings

Within Section 8, it is observed that geothermal projects, globally, to date, are dominantly located in continental crust. Using the simple stress-grouping outlined within Figs. 21–26 (compressional, tensional and stable) we see that SHR-drilled and CLG projects are dominantly developed in tensional settings, whereas compressional settings dominate planned SHR projects, EGS and conventional geothermal projects. Finally, we also observed that exploration is heavily biased to the Corderillian orogenic terrain across all subtypes analyzed.

A look back at existing geothermal exploration indicates that arc settings are dominating exploration. SHR projects totaled (23; 43.4%) and SHR planned (8; 50%), CLG (4; 44%), and conventional geothermal projects (250; 51.1%) (note, numbers reflect the total of wells/projects for each classification). This observation aligns with the fact that arc setting are the largest prospective resource endowment based on our geodynamic categorization. To date, however, EGS projects, are dominantly located in orogenic belt settings (16 projects; 48.5%), which was the second most prolific geodynamic setting. Further differences in exploration trends are observed when we look at the second ranked geodynamic settings for exploration. Here, we see more deviation from the dominant geodynamic settings where SHR resource areas are identified. For example, of the historical SHR projects, 16 (30%) were drilled within the Narrow Rift setting. Within this categorization we observe 7 projects, 43.8% are drilled within narrow rifts that are actually located within oceanic crust settings. Furthermore, both the oceanic domains and the narrow rifts are anomalous because they are both modified by a mantle plume, representing 25 projects or 47% of all historical SHR wells. With respect to planned SHR projects, we observe that wide rifts (2; 12.5%) and narrow rifts (2; 12.5%) are the next most popular geodynamic settings. It should be noted the narrow rifts are in fact both within oceanic crust and are influenced by a mantle plume. Finally, we can conclude that of the planned SHR wells, we see they are not focused to the same extent on regions with a mantle plume overprint, as early wells did, indicating a shift in exploration setting.

The second and third most dominant geodynamic setting for exploration of EGS projects is observed to be in volcanic arcs (six projects; 18.2%), shields (four projects; 12.1%), and wide rifts (four projects; 12.1%). The completion of the FORGE and Cape Projects in Utah (e.g., FERVO, 2023; Moore et al., 2020) will see wide rifts jump to six projects, but success or failure of projects was not available at the time of writing. The second tier for CLG is within wide rift settings with two projects, 22.2%. The geodynamic categorization conventional geothermal wells revealed an unexpected trend. While project development has dominantly occurred in volcanic arc settings, the next most prolific settings are wide rift (77 projects; 15.7%) and magmatic provinces (64 projects; 13.1%). Further surprises are yielded if we drill down further to the tectonic/physiographic province level. For conventional geothermal, we observe significant clustering. At the physiographic province, we observe a remarkable bias or success with 58.9% of the existing projects located within only nine tectonic/physiographic provinces. Most of the conventional geothermal projects are highly focused, and the top 4 of the nine provinces are located in the Kyushu volcanic arc (64; 13.1%); the Menderes Massif, Turkey (63; 12.9%); the Campanian arc (35; 7.2%); and the Northern Basin and Range, USA (35; 7.2%).

We believe the lookback summarized here provides a deep understanding of how geothermal exploration has been conducted over time. Further work ought to be completed to ensure that analysis in future compares project location by geodynamic setting or tectonic/physiographic province to learn from experiences in drilling, completion, water chemistry, or other geological variances or complicating factors so that learnings and successes are effectively carried to new projects. We argue that communication of results whether conventional, CLG, EGS, or HGS is learned and applied to the geothermal resource regardless of the play that is being developed.

9.2 Superhot geothermal and geodynamic setting

The results presented in Section 8 indicate that with depth, the resource area increases (Figs. 29–30). This is in line with other global studies (e.g., Aghahosseini and Breyer, 2020). Aghahosseini and Breyer (2020) explored temperatures on a global basis from 150°C to >400°C, presenting the thermal energy (heat content) as a function of depth for a given temperature, down to 10 km. Cladouhos et al. (2018), Garrison et al. (2020), and Hill et al. (2021, 2023) presented similar maps, outlining the potential global playing field for superhot resources, also mapping a 450°C isotherm. These studies however assumed a constant conductivity with depth and a linear geothermal gradient. None of the aforementioned studies linked the occurrence of shallow heat to the geodynamic setting. This study differs specifically in how we constrain the depth to heat. We model the steady-state heat flow and approximate the areas where nonsteady-state processes my occur. Furthermore, we limit out maps to show only shallow SHR (<15 km). The cutoff was selected presently based on the understanding of existing technology and the fact that the maximum true vertical depth achieved presently is 12.5 km with Kola deep well (see Section 5). Finally, we attempt to relate the distribution of heat by geodynamic setting to develop the foundational elements of a geothermal play-based exploration concept.

Based on the results presented in Section 8, we observed that the SHR resources are dominantly located within the volcanic arc, orogenic belt, and shield settings (Model 1) and volcanic arc, orogenic belt, and accretionary complex settings (Model 2). It is worth mentioning that wide rifts and passive margins are geodynamic settings that are closely following. Examining the global SHR maps, we do see some interesting trends that are worthy of discussion:

(i) The maps show that the volcanic arc and orogenic setting are the dominant resource areas. This may "loosely" equate to the "ring of fire" which conventional geothermal projects have historically exploited. But in fact, the maps presented in Figs. 21–26 show that the distribution of heat is also linked to "paleo," not just "present-day" stress. This observation will require further work to understand the role of stress and its relation to the mapped geodynamic setting and represents a weakness of the current analysis.

(ii) While the volcanic arc and orogenic belt geodynamic settings dominate the resource, an interesting anomaly is observed in the shallow <7.5 km steady-state maps. At these depths, the largest areas of shallow heat are observed in the basin and narrow rift and volcanic arc geodynamic setting (Figs. 27 and 28). The regions in Yemen and Ethiopia may however have modifying plume activities that explain the anomalously shallow heat and may be contributing to excess hydrothermal activity linked to fluids associated with magmatic activity. Mexico and the volcanic arc are located in the upper plate of a subduction system which is well known geologically for volatiles due to the subducting plate (e.g., Gibson and McKenzie, 2013). Staying with shallow heat, it is also worth commenting that shield and wide rift geodynamic settings also ranked as areas of interest where greater areas of nonsteady-state heat are observed—the driving mechanisms of nonsteady-state heat need to be investigated further and are outside of the scope of this study.

(iii) An important result of this work that warrants further investigation is that the Hasterok et al. (2022) classification of geodynamic settings and tectonic/physiographic provinces does not properly factor present-day stress. The mapped domains as used did not separate the interpretations based on paleo- or more importantly present-day stress. This means that, for example, domains are characterized by their forming mechanism rather than their present-day stress setting. Furthermore, the geodynamic classification orogenic belt does not properly address the angle of slab subduction (flat vs. dipping vs. retreating). In addition, the geodynamic settings do not account for strike-slip activity. The result is that the geodynamic settings as defined represent a simplified classification of geothermal resources. As a result, in the present analysis, it is not possible to fully characterize the importance of plume, fertile mantle, depleted mantle, subducting slab dip, or strike-slip from the resource maps. We believe it is important to characterize the paleo versus present-day stress regime. Such detailed analysis may highlight what could be considered the most critical aspect: Globally, none of the geodynamic settings—whether orogenic belt, volcanic arc, basin, continental BAB, wide rift, or any other defined geodynamic setting—exhibited a maximum penetration of SHR resources exceeding 35.6% (e.g., basin at 15-km depth). Most of the dominant geodynamic settings had a maximum SHR penetration, at 15 km, equating to 22%–35.6% of the mapped geodynamic setting area on a global basis. The implications of this are significant, meaning we can quickly eliminate >70% of the world for each geodynamic setting in terms of SHR exploration, since the resource does not exist, based in the maps presented here. Why two-thirds of a geodynamic resource do not contain heat requires further investigation. The results lead us to question what are the subcontrolling factors at the geodynamic setting scale? Further work is required to understand the distribution of heat in geodynamic settings and their successful physiographic provinces. This might be linked to a need to better characterize the geological variance within the geodynamic setting. The a priori hypothesis prior to getting the results was that wide rift geodynamic settings due to their unique geodynamic properties should all be hot in the present day. The fact that wide rifts and other geodynamic settings reveal similar trends or thermal heterogeneities means further work is needed to understand the cause of this heterogeneity.

(iv) Finally, it may be worth noting that this work did not include water availability—leading to a key question, does mapping water systems impact nonsteady-state polygons or provide a risk to petrothermal plays?

9.3 Order-of-magnitude prospective heat endowments for superhot rock geodynamic and tectonic/physiographic provinces

Our results have revealed there is considerable variability in the spatial distribution of SHR heat within each geodynamic setting. The observed internal variability may originate from classification error or the need to differentiate present-day stress from past stress in geodynamic settings that have a paleo-interpretation applied (as discussed earlier). At a simplified stress categorization, Tables 12 and 13 indicate that 65%–66% of the SHR resource (across all mapped intervals) is in compressional regimes. Figs. 29–32 illustrate that resource distribution is heavily skewed to the volcanic arc, orogenic belts, shields, accretionary complexes, wide rifts, and passive margins, as discussed earlier. Figs. 29 and 30 indicate that the area within which SHR resources can be found is enormous. The GWe potential that SHR could offer, assuming an EGS approach (based on our preliminary estimations) is more power than which needed today, assuming development of only 1% of the potential resource. If we explore this further in Tables 10–13, we see that for each depth, compressional regimes dominate. If we break down the totals by stress type, we see the following: At 7.5 km of the 7304 GWe, 4613 GWe is within compressional domains. 1303 GWe is within tensional domains and 1389 in stable domains. At 10 km, the total is 19,363–12,430 GWe is within compressional domains. 3865 GWe is within tensional domains and 3067 is within stable domains. At 12.5 km, the resource pool balloons to 46,313 GWe. 29,799 GWe is within compressional domains. 8192 GWe is within tensional domains and 8322 GWe in stable domains. At 15 km, the total resource is 64,670 GWe. 42,721 GWe is within compressional domains. 10,3358 GWe is within tensional domains and 11,589 in stable domains. One question arises which is, are current exploration RD&D budgets within the geothermal industry appropriately reflecting this resource bias? It ought to be noted that the GWe and GWth results presented here are strongly biased toward EGS technologies and flash power plants for the topsides, using a P50 (55 L/s) and 250 MW power plant scenario. It is important to note the results do not consider how CLG or hybrid engineering solutions could contribute to developing the superhot resource play. The characteristics of these projects versus EGS are still to be determined with respect to the flowrate, cost, and the estimated MWth and MWe that may be returned.

Finally, going forward and currently out of the scope of this work, further research is needed to define more robustly the EGS, CLG, and hybrid prospect and field concepts. Because these technologies are new and potentially developable as resource plays, how do we define the edges of a prospect or field? What is the heat in-place numbers versus heat recoverable, over the lifetime of the projects, and what are the uncertainties of these numbers? Can these new geothermal resource concepts be modeled within the framework of the SEC to define what are the heat reserves? These analyses are clearly of interest to the energy, private equity debt, insurance, and banking industries, and this represents a clear gap in the analysis. Before we can get to that level of detail, it is clear that more data is needed regarding the chance of success of exploration and development projects. What are the uncertainty drivers and probabilistic confidences of success for technical and economic development? In addition, further work is needed to understand if there are advantages in project development using EGS, CLG, or HGS concepts. Such analysis may reveal that some areas may only be developed using CLG or HGS concepts.

9.4 Chance of success EGS for projects

In the data collection for this paper, we observed that there are 33 EGS geothermal systems globally (Fig. 17, Supplementary Table S6 in the online version at https://doi.org/10.1016/B978-0-443-21662-6.00009-8). Based on the analysis conducted here, we observe that these projects have occurred in seven geodynamic settings. The largest geodynamic setting, the focus of EGS exploration has been dominantly within orogenic settings, with 15 (45.5%) projects identified. The next most prominent geodynamic settings are volcanic arcs with six (17.6%), shields with four (12.1%), and wide rifts with four (12.1%) projects (Table 6). Unlike conventional, SHR, or CLG, for EGS projects globally, we have been able to compile a database that contains research and commercial and noncommercial projects. While it is hard to know if this is a complete dataset, it is worth further discussion.

With the mapped/listed projects, if we explore which of these are research, commercial or noncommercial projects we see a discrete picture emerge. First, by commercial, we simplify this to only commercial if the project is in use today. Note, there is no further analysis, and we accept this is highly simplified and it does not account for subsidies or other tax breaks. Given these assumptions, what we observe is that in the orogenic belt setting, two projects were research wells, and eight are commercial and six are noncommercial. In volcanic arcs, only one of the six projects has declared commercial success, to date. Within shield settings, one project was a research project, and the other three were declared as noncommercial. Within wide rift settings of the four projects, all were declared commercial. A second observation is that most of the exploration has been within crystalline rock. Here, we observe that of the nonresearch projects, there are 24 projects that tried to link wells within crystalline rock, at a success of 46%. Within sedimentary systems of the six projects, only three were successful, or 50%, based on the data available (Table 14). The third trend we observe is revealed when we step toward the tectonic/physiographic province (Table 6)—here we observe current development success versus failure is interesting. The numbers are still not statistically meaningful, and project bias is hard to deconstruct, but what we see if we ignore research wells is that certain provinces are showing success. For example, within the foredeep setting, the Molasse Basin

TABLE 14 Summary of engineered geothermal projects studied highlighting the commercial and noncommercial operations.

Geodynamic setting	**Projects all**	**Research wells only**	**Sedimentary**	**Crystalline rock**	**Crystalline commercial**	**Crystalline noncommercial**	**Sed commercial**	**Sed noncommercial**
Accretionary complex (transitional)	**1**	0	0	**1**	0	1	0	0
Basin (continental)	**1**	0	0	**1**	0	**1**	0	0
Foredeep (continental)	**2**	0	**2**	0	0	0	1	1
Orogenic belt (continental)	**15**	**2**	**1**	**12**	**7**	**5**	1	**0**
Shield (continental)	**4**	**1**	0	**3**	0	**3**	0	0
Volcanic arc (continental)	**6**	0	**2**	**4**	**1**	**3**	0	2
Wide rift (continental)	**4**	**0**	**1**	3	3	**0**	**1**	0
Totals	**33**	**3**	**6**	**24**	**11**	**13**	**3**	**3**
Success rates all %	**47%**	(not including research wells)						
Success rate hard rock %	**46%**	(not including suspended wells or research wells)						
Success rate sedimentary %	**50%**	(not including suspended wells or research wells)						

With limited data, an inferred chance of success is assumed.

has seen 100% success (two out of two); in the orogenic belt setting, the Saxo-Thuringian Belt has also seen 100% success (four out of four); and the Teisseyre-Tornquist Zone has seen 100% success (one out of one). In the volcanic arc, the Central American Arc recorded 100% success (one out of one), and finally within the wide rift setting, the Northern Basin and Range has 100% success (three out of three) and the Pannonian Basin also has 100% success (one out of one). This suggests that there may be not only geodynamic controls but also significant variation at the tectonic/physiographic province level. Looking at the domain exploration trends at a higher level (number of projects, % of all EGS projects), we see that the exploration is favored within the Saxo-Thuringian Belt (5; 14.3%), Moldanubian Belt (4; 11.8%), and the Northern Basin and Range (3; 8.8%). Finally, one interesting observation is that the early research EGS projects were located in shield (x1) and orogenic belt settings (x2), whereas success in EGS projects has been found in other geodynamic settings with the exception of the orogenic belt settings which have received the most exploration and development. It is worth contemplating that the FORGE and FERVO's Cape project in Utah were omitted from the analysis because at the time of writing they were not completed (FERVO, 2023; Moore et al., 2020). But they would add one research well and potentially one commercial project to the wide rift, Northern Basin and Range, geodynamic setting. The results here for EGS projects emphasize that historical and ongoing research is not in alignment with the greatest geodynamic resource opportunity.

9.5 Technological next steps and future work

The databases and SHR mapping presented here represent the first steps in understanding how play-based exploration can influence and benefit the geothermal industry. These models are a first of a kind and much refinement is needed. Here we highlight some key areas for future work.

9.5.1 Model resolution and accuracy

As described in Section 5, the LithoREF18 model is a low-resolution, first attempt at calculating the depth to superhot resources. At approximately a $2° \times 2°$ or 230 km × 230 km scale, it is only giving us trends. Future work and investment are needed to better refine these models and others like them, for example, WinterC (Fullea et al., 2021). As Ball and Afonso (2023) indicated, future models will benefit from research that explores joint convection and conduction modeling in the global inversions. The models presented here are low resolution, $2° \times 2°$ (~200 km × 200 km). Improvements in modeling resolution toward $0.5° \times 0.5°$ (~50 × 50 km) could dramatically improve our understanding of the distribution of heat and the internal variances of shallow superhot resources. New models should incorporate crustal stress conditions to further explore the role stress has upon the variance of geothermal resources. Another aspect that limited the analysis presented herein was the accuracy of the tectonic model of Hasterok et al. (2022).

9.5.2 World stress mapping and geomechanics

In addition to higher resolution models, further work is needed to increase data relating to our understanding of crustal stress conditions. In Fig. 33, it is observed that significant parts of the world are lacking any stress data. As discussed earlier, key data relating to the stress

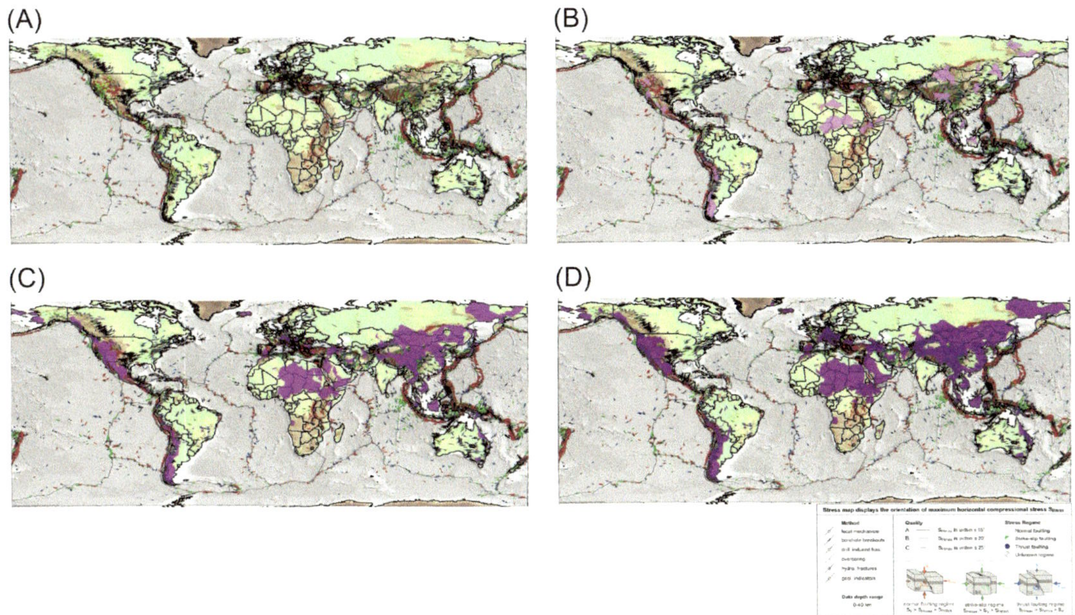

Fig. 33 SHR geothermal resource overlaying the present-day WSM2016 (Heidbach et al., 2016). Map lines delineate study areas and do not necessarily depict accepted national boundaries. (A) 7.5 km, (B) 10 km, (C) 12.5 km, (D) 15 km.

regimes within geodynamic settings and the related tectonic/physiographic province is missing. This data is essential for planning successful exploration of geothermal resource opportunities. How crustal stress changes with depth, particularly for deep SHR geothermal projects, will be essential for well planning of injector and producer wells or successfully linking fractures to the boreholes. In addition to understanding the role of preexisting joints or faults, understanding the stress regime will benefit the creation of new fracture sets. The propagation of fractures will change according to the local stress regime. For example, within strike-slip and tensional settings, fractures may propagate vertically, whereas within compressional stress regimes, hydraulic fractures tend to form horizontally (e.g., Maxwell et al., 2009). Understanding the stress regime and frackability is therefore critical to the design of the well-pair injector and producer. Fig. 33 presents the different depth slices here with the WSM2016 draped on top of our results. Data does not exist to define changes in stress with depth, and key questions remain for SHR projects at the predicted depths, temperatures, and pressures. For example, how can we drill the target and maintain open fractures or boreholes in the brittle or ductile domain? In addition to stress, it is also necessary to think about the pressure while drilling to target. Experience from the Habanero wells, in Australia, revealed that due to complex pressure settings, the wells needed to be drilled using managed pressure drilling, "exactly balancing the pressure from the well with a combination of heavy weight drilling fluid and back pressure" (Toralde and Nas, 2010).

10 Conclusions

The maps presented here have delineated the 450°C steady-state, conductive isotherm and potential nonsteady-state geothermal regions. The distribution of thermal anomalies and their depths are examined by geodynamic setting. At this scale of observation, we observe that SHR resources are predicted in all 15 defined geodynamic settings, but that at best, only 20%–35% of the area of any one geodynamic setting is prospective based on the analysis of the 0–15-km depth resource maps. These results highlight considerable heterogeneity within mapped geodynamic settings. The most prospective geodynamic settings for exploration are volcanic arc and orogenic belts, followed by shield and accretionary, BABs, and wide rift settings. As such, these results should be used to drive research into drilling, completion, reservoir development/management, and project citing criteria to take advantage of the lion's share of the resource opportunity.

If we consider stress state of the resource distribution, it is clear further work is needed. In the steady-state mapping, we observe that the resource is dominant in stable settings. But conversely in the nonsteady-state mapping, we observe that in the shallow <7.5-km resource within Model 1, the resource is located within in compressional settings, 63.9% of the resource. However, in Model 2, we observe 45.6% of shallow resource is located in tensional settings. The understanding of the nonsteady-state resource requires further work which is out of the scope of the existing study.

Table 15 is used to summarize the results from this study incorporating the LithoREF18 steady-state and nonsteady-state maps. Regarding the current project development, we observe that exploration is dominantly in continental crust, in compressional domains linked to Corderillian tectonics. For conventional geothermal, we observe that hydrothermal project development as spanned 12 geodynamic settings, that exploration is dominantly within

TABLE 15 Summary table of key observations from geothermal well/projects (SHR, conventional, EGS, CLG) and the new SHR heat maps, Model 1 and Model 2.

Dataset	Dominant three geodynamic settings (present day and paleo)	Dominant stress regime (paleo and present day)	Dominant "tectonic" [physiographic province] terrain explored [orogen explored]	Other comments/observations
SHR wells drilled (53)	(Current exploration observed within five geodynamic settings) **Volcanic arc** (23; 43.4%) Narrow rift (16; 30.2%) Pacific oceanic crust (9; 17.0%)*	Tensional (28; 53%)	(Current exploration observed within 11 tectonic provinces) *trans*-Mexican Volcanic Belt (12; 22.6%) Campanian Arc (9; 17%) Pacific Basin (9; 17%) Corderillian (18, 34%)	Continental crust, continental (31) and transitional (6), total (37; 69.8%). *16 wells in total are on oceanic crust (30.2%). In fact, we observe that the oceanic crust in all cases is modified by a mantle plume. This includes a subset of wells drilled in narrow rift (Iceland) 25 wells total have a mantle plume influence; 47% of projects if we include continental and oceanic crust projects
SHR wells planned (16)	(Current exploration observed within six geodynamic settings) **Volcanic arc** (8; 50%) Wide rift (2; 12.5%) Narrow rift (mantle plume) (2; 12.5%)	Compressional (10; 62.5%)	(Current exploration observed within 12 tectonic provinces) North Japan Arc (4; 25%) Corderillian (13; 81.3%)	Continental crust, continental (11) and transitional (3), total (14, 87.5%). Only two projects (12.5%) are planned in oceanic crust, modified by a mantle plume. These both occur in Iceland
EGS (33)	(Current exploration observed within seven geodynamic settings) **Orogenic settings** (16; 48.5%) Volcanic arc (6; 18.2%) Shields (4; 12.1%) Wide rifts (4; 12.1%)	Compressional (24; 72.7%)	(Current exploration observed within 20 tectonic provinces) Saxo-Thuringian Belt (5; 15.2%) Moldanubian Belt (4; 12.1%) Northern Basin and Range (3; 9.1%) Corderillian (6; 18.2%)	Continental crust, continental (32) and transitional (1), total (33; 100%) Exploration success is mixed. Most successful geodynamic setting, wide rift. Mixed results in orogenic settings. But tectonic province trends are important

TABLE 15 Summary table of key observations from geothermal well/projects (SHR, conventional, EGS, CLG) and the new SHR heat maps, Model 1 and Model 2—cont'd

Dataset	Dominant three geodynamic settings (present day and paleo)	Dominant stress regime (paleo and present day)	Dominant "tectonic" [physiographic province] terrain explored [orogen explored]	Other comments/observations
Closed-loop LG (9)	(Current exploration observed within five geodynamic settings) **Volcanic arc** (4; 44.4%) Wide rift (2; 22.2%)	Tension (6; 66.7%)	(Current exploration observed within eight tectonic provinces) Avalonia (2; 22.2%) Corderillian (4; 44.4%)	Continental crust, continental (8) and transitional (1), total (9, 100%)
Conventional (489)	(Current exploration observed within 12 geodynamic settings) **Volcanic arc** (250; 51.1%) Wide rift (77; 15.7%) Magmatic provinces (63; 12.9%)	Compressional (287; 58.7%)	(Current exploration observed within 62 tectonic provinces) Kyushu Volcanic Arc (64; 13.1%) Menderes Massif (63; 12.9%) Campanian Arc (35; 7.2%) Northern Basin and Range (35; 7.2%) Corderillian (259; 53%)	Continental crust, continental (337) and transitional (125), total (462; 94.5%). Oceanic crust accounts for 27 or 5.5% of projects
450°C maps Model 1 (combined analysis [Fuchs et al. (2023)] constrained model)	(SHR resource observed within 15 geodynamic settings) **Volcanic arc (28.3%–30.6%)** Orogenic belt (21.9%–24.9%) Shield (11.6%–14.2%)	Compressional (65.1%) of mapped resource at depth of <15 km	Not analyzed at present	The analysis was only completed for onshore domains. Offshore data was not considered in this analysis Note Percentages are for area covered by shallow SHR, the range represents the spread from 7.5 to 15 km Note: Maximum coverage of SHR by geodynamic setting occurs at 15 km within the basin setting at 35.6% followed by volcanic arc setting at 32.31%. The major geodynamic setting's maximum coverage of SHR at <15 km equates to 22.1%–35.6% of the mapped geodynamic setting area on a global basis

Continued

TABLE 15 Summary table of key observations from geothermal well/projects (SHR, conventional, EGS, CLG) and the new SHR heat maps, Model 1 and Model 2—cont'd

Dataset	Dominant three geodynamic settings (present day and paleo)	Dominant stress regime (paleo and present day)	Dominant "tectonic" [physiographic province] terrain explored [orogen explored]	Other comments/observations
450°C maps Model 2 (combined analysis—Lucazeau constrained model)	(SHR resource observed within 15 geodynamic settings) **Volcanic arc (28.8%–37.6%)** Orogenic belt (18.6%–24.8%) Accretionary complex (7.4%–12.6%)	Compressional (65.7%) of mapped resource at depth of <15 km	Not analyzed at present	The analysis was only completed for onshore domains. Offshore data was not considered in this analysis Note: Percentages are for area covered by shallow SHR, the range represents the spread from 7.5 to 15 km. Note: Maximum coverage of SHR by geodynamic setting occurs at 15 km within the basin setting at 35.5% followed by volcanic arc settings at 31.5%. The major geodynamic setting's maximum coverage of SHR at <15 km equates to 21%–35.5% of the mapped geodynamic setting area on a global basis
450°C steady-state model LithoREF18 at 7.5 km	(SHR resource observed within three geodynamic settings) **Basin** Narrow rift Volcanic arc	Stable (65%)	Basin anomaly is identified to be in Yemen (Arabian Platform) Narrow rift anomaly (Ethiopian Rift) Volcanic arc anomaly is found within the Trans-Mexican Volcanic Belt (Corderillian)	LithoREF18 heat (conductive) shallower than 7.5 km is not predicted in many places
450°C maps Model 1 (Fuchs et al., 2023 nonsteady state)	(SHR resource observed within 15 geodynamic settings) **Volcanic arc** Orogenic belt Shield	Compressional (63.9%)	Not analyzed at present	This approach contains errors due to gridding of data but also suffers from absence of data (large gaps)
450°C maps Model 2 (Lucazeau nonsteady state)	(SHR resource observed within 15 geodynamic settings) **Volcanic arc** Orogenic belt Wide rift	Tensional (45.6%)	Not analyzed at present	

Note: Maximum penetration of SHR resource by geodynamic setting is basin at 35.6% followed by volcanic arc at 32.31%. Of the major geodynamic settings, this range of SHR at <15 km equates to 22%–35.6% of the mapped geodynamic setting area on a global basis—meaning at the current definition, more work is required to understand the distribution of heat in geodynamic settings and their successful tectonic/physiographic provinces.

compressional regimes. We also observe that projects in Volcanic arc settings dominate (51%). At the physiographic province, we observe a remarkable bias or success with 58.9% of the existing projects located within only nine tectonic/physiographic provinces. For EGS projects, we see exploration is distributed across seven geodynamic settings. We observe that to date, exploration has been dominantly located within compressional regimes with a focus on Orogenic belt settings. The 33 projects are distributed across 20 tectonic/physiographic provinces, which is in stark contrast to the clustering of conventional projects. CLG projects while limited in number are being developed across five geodynamic settings—that exploration is dominantly within tensional regimes, and projects are preferentially located in volcanic arcs. At the tectonic/physiographic province, eight provinces are being explored. Of the 53 wells that approached SHR conditions, the drilling occurred in five geodynamic provinces (six is continental vs oceanic are separated). The greatest number of wells was in Volcanic arcs, biased by the Mexican drilling. Overall, existing SHR wells occurred dominantly in tensile domains. In comparison, planned SHR wells are located in six geodynamic settings. Planned SHR wells are dominantly located within compressive settings, with a strong project bias to Volcanic arcs. For SHR wells, historical trends do not match present-day planned SHR exploration, with exploration biases shifting from mantle plume dominated rift and oceanic (tensile) geodynamic settings to dominantly compressional domains with today's projects. It is also observed that planned SHR wells, while few in number are planned across a greater range of geodynamic settings. In conclusion, the existing exploration is dominantly located within volcanic arcs and orogenic belts—the two most prolific geodynamic settings based on the models presented here. Finally, we observe when comparing steady-state and nonsteady-state heat we see a clear spatial difference in the location of heat. This requires further investigation in order to improve exploration and development of SHR geothermal and other geothermal play types.

The work presented here represents a global prospective endowment workflow. It is enough to inform stakeholders what the high-level spatial and depth distribution of heat is globally. It informs us where in the world SHR geothermal might be located. Further work is needed to take this to a prospective, contingent, or proven resource at a prospect and portfolio scale as outlined in Fig. 3. The key aim of this work is that we hope that it will lead to an improved understanding how geodynamic settings influence geothermal systems and plays. Our analysis provides value, highlighting where the potentially transformational resource represented by superhot geothermal exists. It is hoped that geothermal exploration can benefit from this introductory analysis where we have shown that geothermal and SHR geothermal can be developed within a Linnaean-style Geothermal Entities Classification System. The very first building block, outlined here in this paper, is the geodynamic setting, followed by the tectonic/physiographic province. As geothermal projects are drilled and databases are developed, we should use this scheme become better at exploring and developing commercial projects. The scheme here links all projects enabling the next projects to benefit from the knowledge gained, rather than projects existing as boutique independent projects. As project developers within geothermal organizations move projects from leads to prospects to geothermal fields, we propose that organizations will see which geodynamic setting they are successful within and with which technology. At the highest level, this method allows the developer/investor or other to ask the right question: What are the exploration and development risks for our operations/projects in the wide rift or orogenic belt versus shield geodynamic setting, and how do our current projects compare to success for failure in similar settings?

Acknowledgments

This work was supported by CATF, Boston, USA. Special thanks are saved for Dr. Farshad Salajeghe (University of Sydney, Australia) for assistance in creating the 450°C isotherm. Thanks are extended to Josh Sellars and Kathleen Gould from Seequent who provided software and workflow assistance with Geosoft, Oasis Montaj.

References

Afonso, J.C., Salajegheh, F., Szwillus, W., Ebbing, J., Gaina, C., 2019. A global reference model of the lithosphere and upper mantle from joint inversion and analysis of multiple data sets. Geophys. J. Int. 217 (3), 1602–1628. https://doi.org/10.1093/gji/ggz094.

Aghahosseini, A., Breyer, C., 2020. From hot rock to useful energy: a global estimate of enhanced geothermal systems potential. Appl. Energy 297, 115769. https://doi.org/10.1016/j.apenergy.2020.115769.

Aldahan, A.A., Castañ, J., Collini, B., Gorody, T., Juhlin, C., Sandstedt, H., 1991. Scientific Summary Report of the Deep Gas Drilling Project in the Siljan Ring Impact Structure. Vattenfall RD&D Rep, p. 257.

Alimonti, C., 2023. Technical performance comparison between U-shaped and deep borehole heat exchangers. Energies 2023 (16), 1351. https://doi.org/10.3390/en16031351.

Allen, P.A., Allen, J.R., 2013. Basin Analysis-Principles and Application to Petroleum Play Assessment, third ed. Wiley and Blackwell, Hoboken.

Asanuma, H., 2022. Status of Japanese Supercritical Geothermal Project in 2022. GRC Transactions. https://www.geothermal-library.org/index.php?mode=pubs&action=view&record=1034583.

Axis, 2022. Axis Completes Longest Lateral in North America. Online Resource https://www.axisofs.com/axis-completes-longest-lateral-in-north-america/#:~:text=Axis%20Rig%2055%20recently%20completed,well%20with%20equally%20challenging%20conditions. (Accessed January 2024).

Ball, P.J., 2021a. Macro energy trends and the future of geothermal within the low-carbon energy portfolio ASME. J. Energy Resour. Technol. 143. https://doi.org/10.1115/1.4048520.

Ball, P.J., 2021b. A review of geothermal technologies and their role in reducing greenhouse gas emission. ASME J. Energy Res. Technol. 143. https://doi.org/10.1115/1.4048187.

Ball, P.J., 2022. Factsheets Superhot Rock Projects and Drilled Provinces: Summary of Wells Drilling Into >350°C. CATF. Online Resource https://www.catf.us/shrmap/. (Accessed November 2023).

Ball, P.J., Afonso, J.C., 2023. Superhot rock for sustainable power generation. Geosc. Mag. Online, Geological Society of London, September https://geoscientist.online/sections/features/superhot-rock-for-sustainable-power-generation/.

Balling, N., 2023. Thermal structure of two deep boreholes in the Siljan Ring, Central Sweden, and paleoclimatic implications. Tectonophysics 872 (1), 230149. https://doi.org/10.1016/j.tecto.2023.230149.

Banks, G.J., Ball, P.J., 2023. Superhot Rock Energy Glossary. CATF. Online Resource https://www.catf.us/superhot-rock/glossary/. (Accessed November 2023).

Banks, G.J., and Ball, P.J., (n.d.). Application of Play-Based Exploration for Deep Geothermal: Mapping, Comparing Resource and Decision Quality.

Banks, G.J., Peacock, D.C.P., 2020. Basement highs: definitions, characterisation and origins. Basin Res. 32, 1685–1710. https://onlinelibrary.wiley.com/doi/full/10.1111/bre.12448.

Banks, G.J., Olsen, S.D., Gusak, A., 2020. A method to evaluate REE-HFSE mineralised provinces by value creation potential, and an example of application: Gardar REE-HFSE province, Greenland. Geosci. Front. 2020 (6), 2141–2156. https://doi.org/10.1016/j.gsf.2020.05.019.

BCG, 2022. Accelerating Decarbonization with Next-Generation Geothermal. Boston Consulting Group. Energy Article, Online Resource https://www.bcg.com/publications/2022/the-advantages-of-geothermal-energy. (Accessed November 2023).

Beard, J.C., Jones, B., 2023. The Future of Geothermal in Texas, Contemporary Projects and Perspectives. The Mitchell Foundation. https://energy.utexas.edu/research/geothermal-texas.

Beckers, K.F., Johnston, H.E., 2022. Techno-economic performance of Eavor-Loop 2.0. In: Proceedings, 47th Workshop on Geothermal Reservoir Engineering Stanford University, Stanford, California, February 7–9, 2022, SGP-TR-223.

Bischoff, A., Heap, M.J., Mikkolaa, P., Kuvaa, J., Reuschle, T., Jolisa, E.M., Jon Engstroma, J., Reijonena, H., Leskela, T., 2024. Hydrothermally altered shear zones: a new reservoir play for the expansion of deep geothermal exploration in crystalline settings. Geothermics 118, 102895. https://doi.org/10.1016/j.geothermics.2023.102895.

Blankenship, D., Gertler, C., Kamaludeen, M., O'Connor, M., Porse, S., 2024. Pathways to Commercial Liftoff: Next-Generation Geothermal Power. U.S. Department of Energy. https://liftoff.energy.gov/wp-content/uploads/2024/03/LIFTOFF_DOE_NextGen_Geothermal_v14.pdf.

Breede, K., Dzebisashvili, K., Liu, X., Falcone, G., 2013. A systematic review of enhanced (or engineered) geothermal systems: past, present and future. Geotherm. Energy 1, 4. https://doi.org/10.1186/2195-9706-1-4.

Brown, C.S., Gioia, F., 2024. Investigating heat transmission in a wellbore for low-temperature, open-loop geothermal systems. Therm. Sci. Eng. Prog. 48, 102352. https://doi.org/10.1016/j.tsep.2023.102352.

Buijze, L., van Bijsterveldt, L., Cremer, H., Paap, B., Veldkamp, H., Wassing, B.B.T., van Wees, J.-D., van Yperen, G.C.N., ter Heege, J.H., Jaarsma, B., 2019. Review of induced seismicity in geothermal systems worldwide and implications for geothermal systems in the Netherlands. Neth. J. Geosci. 98, e13. https://doi.org/10.1017/njg.2019.6.

BW, 2011. Sakhalin-1 Project Drills World's Longest Extended-Reach Well. Online Resource https://www.businesswire.com/news/home/20110128005586/en/Sakhalin-1-Project-Drills-World%E2%80%99s-Longest-Extended-Reach-Well. (Accessed January 2024).

CATF, 2023. The Potential for Superhot Rock Energy in the Middle East, Case Study. Online Resource https://www.catf.us/superhot-rock/middle-east/. (Accessed November 2023).

Cladouhos, T.T., Petty, S., Bonneville, A., Schultz, A., Sorlie, C.F., 2018. Super Hot EGS and the Newberry Deep Drilling Project. In: Proceedings, 43rd Workshop on Geothermal Reservoir Engineering Stanford University, Stanford, California, February 12–14, SGP-TR-213. https://pangea.stanford.edu/ERE/pdf/IGAstandard/SGW/2018/Cladouhos.pdf.

de Jager, J., 2021. Play-Based Exploratie Handboek. GEODE Atlas to Explore Hydrocarbon Opportunities in the Dutch Offshore. Online Resource https://www.geodeatlas.nl/pages/play-based-exploration. (Accessed December 2023).

De La Garza, A., Johnson, A., 2023. Geothermal Energy Could be Huge. But We Need Oil and Gas Companies to Build It. Time, Online Resource https://time.com/collection/time-co2-futures/6338438/oil-gas-geothermal-energy/. (Accessed November 2023).

DOE, 2016. What is an Enhanced Geothermal System (EGS)? U.S. Department of Energy, Geothermal Technologies Office. Online Resource https://www.energy.gov/sites/default/files/2016/05/f31/EGS%20Fact%20Sheet%20May%202016.pdf. (Accessed June 2023).

Doran, H.R., Renaud, T., Falcone, G., Pan, L., Verdin, P.G., 2021. Modelling an unconventional closed-loop deep borehole heat exchanger (DBHE): sensitivity analysis on the Newberry volcanic setting. Geotherm. Energy 9 (1), 1–24.

Doust, H., 2010. The Exploration Play—What Do We Mean By It? AAPG, Search and Discovery Article #40486. https://www.searchanddiscovery.com/documents/2010/40486doust/ndx_doust.pdf.

Doust, H., 2017. Petroleum systems in southeast Asian tertiary basins. Bull. Geol. Soc. Malaysia 64, 1–16. https://archives.datapages.com/data/geological-society-of-malaysia/bulletins/064/064001/pdfs/1.pdf.

Dyman, T.S., Nielson, D.T., Obuch, R.C., Baird, J.K., Wise, R.A., 1990. Summary of Deep Oil and Gas Wells and Reservoirs in the US. USGS. Open-File Report 90-305.

EIA, 2023. What is U.S. Electricity Generation by Energy Source? Online Resource https://www.eia.gov/tools/faqs/faq.php?id=427&t=3. (Accessed April 2024).

Enverus, 2023. Wells Drilled >5 km and >7.5 km, TVD. Data provided by Graham Bain, Enverus.com.

Faulds, J.E., Hinz, N.H., Coolbaugh, M.F., Shevenell, L.A., Siler, D.L., 2016. The Nevada Play Fairway Project—Phase II: Initial Search for New Viable Geothermal Systems in the Great Basin Region, Western USA. GRC Transactions. vol. 40, p. 2016. https://www.osti.gov/servlets/purl/1724099.

Faulds, J.E., Hinz, N.H., Coolbaugh, M., Ayling, B., Glen, J., Craig, J.W., McConville, E., Siler, D., Queen, J., Witter, J., Hardwick, C., 2021. Discovering Blind Geothermal Systems in the Great Basin Region: An Integrated Geologic and Geophysical Approach for Establishing Geothermal Play Fairways: All Phases. DE-EE0006731, Phases 1–3—Final Report BOR on behalf of UNR, DOE-UNR-06731-01, https://doi.org/10.2172/1724080.

Ferguson, G., McIntosh, J.C., Warr, O., Sherwood Lollar, B., Ballentine, C.J., Famiglietti, J.S., Kim, J.-H., Michalski, J.R., Mustard, J.F., Tarnas, J., McDonnell, J.J., 2021. Crustal groundwater volumes greater than previously thought. Geophys. Res. Lett. 48, e2021GL093549. https://doi.org/10.1029/2021GL093549.

FERVO, 2023. Collaboration and Innovation Power Geothermal Energy Project on Public Lands. Online Resource https://www.blm.gov/blog/2023-10-13/collaboration-and-innovation-power-geothermal-energy-project-public-lands. (Accessed November 2023).

Foum, A., 2018. Estimating Yet to Find in Petroleum Exploration. LinkedIn. Online Resource https://www.linkedin.com/pulse/estimating-yet-find-petroleum-exploration-alan-foum/. (Accessed October 2023).

Fridleifsson, G.Ó., Pálsson, B., Albertsson, A., Stefánsson, B., Gunnlaugsson, E., Ketilsson, J., Gíslason, Þ., 2015. IDDP-1 drilled into magma—world's first magma-EGS system created. In: Proceeding, World Geothermal Congress, Melbourne, Australia, 2015.
Fridleifsson, G.Ó., Albertsson, A., Stefánsson, A., Þórólfsson, G., Mesfin, K.G., Sigurdsson, K., Sigurdsson, Ó., Gislason, Þ., 2019. The Reykjanes DEEPEGS demonstration well – IDDP-2. In: European Geothermal Congress (EGC-2019), Den Haag, The Netherlands, 11–14 June, 2019. https://europeangeothermalcongress.eu/wp-content/uploads/2019/07/51.pdf.
Fuchs, S., Norden, B., Neumann, F., Kaul, N., Tanaka, A., Kukkonen, I.T., Pascal, C., Christiansen, R., Gola, G., Šafanda, J., Espinoza-Ojeda, O.M., Marzan, I., Rybach, L., Balkan-Pazvantoğlu, E., Ramalho, E.C., Dědeček, P., Negrete-Aranda, R., Balling, N., Poort, J., Wang, Y., Jõeleht, A., Rajver, D., Gao, X., Liu, S., Harris, R., Richards, M., McLaren, S., Chiozzi, P., Nunn, J., Madon, M., Beardsmore, G., Funnell, R., Duerrast, H., Jennings, S., Elger, K., Pauselli, K., Verdoya, M., 2023. Quality-assurance of heat-flow data: the new structure and evaluation scheme of the IHFC global heat flow database. Tectonophysics 863, 229976. https://doi.org/10.1016/j.tecto.2023.229976.
Fullea, J., Lebedev, S., Martinec, Z., Celli, N.L., 2021. WINTERC-G: mapping the upper mantle thermochemical heterogeneity from coupled geophysical-petrological inversion of seismic waveforms, heat flow, surface elevation and gravity satellite data. Geophys. J. Int. 226, 146–191. https://doi.org/10.1093/gji/ggab094.
Garrison, G.H., Uddenberg, M., Petty, S., Watz, J., Hill, B., 2020. Resource potential of super hot rock. GRC Trans. 44. https://www.geothermal-library.org/index.php?mode=pubs&action=view&record=1034285.
Gibson, S.A., McKenzie, D., 2013. On the role of Earth's lithosphere mantle in global volatile cycles. Earth Planet. Sci. Lett. 602, 117946. https://doi.org/10.1016/j.epsl.2022.117946.
Groves, D.I., Santosh, M., 2015. Province-scale commonalities of some world-class gold deposits: implications for mineral exploration. Geosci. Front. 6 (3), 389–399. online https://doi.org/10.1016/j.gsf.2014.12.007.
Hagen, K.D., 1999. Heat Transfer With Applications. Prentice Hall, New Jersey. www.eng.fsu.edu/~shih/succeed-2000/roadmap/map.htm.
Hager, B.H., Clayton, R.W., 1989. Constraints on the structure of mantle convection using seismic observations, flow models, and the geoid. In: Peltier, R.W. (Ed.), Mantle Convection; Plate Tectonics and Global Dynamics. Gordon and Breach Science Publishers, New York, pp. 657–763. https://web.gps.caltech.edu/~clay/PDF/Hager1989.pdf. (Chapter 9).
Hannah, P., Taverna, N., Trainor-Guitton, W., Witter, E., Kolker, A., Warren, I., Robins, J., Rhodes, G., 2023. Geothermal Play Fairway Analysis Best Practices. National Renewable Energy Laboratory, Golden, CO. NREL/TP-570086139 https://www.nrel.gov/docs/fy23osti/86139.pdf.
Hasterok, D., Halpin, J.A., Collins, A.S., Hand, M., Kreemer, C., Gard, M.G., Glorie, S., 2022. New maps of global geological provinces and tectonic plates. Earth Sci. Rev., 104069. https://doi.org/10.1016/j.earscirev.2022.104069.
Heidbach, O., Rajabi, M., Cui, X., Fuchs, K., Müller, B., Reinecker, J., Reiter, K., Tingay, M., Wenzel, F., Xie, F., Ziegler, M.O., Zoback, M.-L., Zoback, M., 2016. The world stress map database release 2016: crustal stress pattern across scales. Tectonophysics. https://doi.org/10.1016/j.tecto.2018.07.007.
Herter, J., 2023. A Preliminary Techno-Economic Model of Superhot Rock Energy. Prepared for Clean Air Task Force by Lucid Catalyst and Hot Rock Research Organization. Clean Air Task Force. November 2022. 24 pp. https://www.catf.us/resource/preliminary-techno-economic-model-superhot-rock-energy/.
Hill, L.B., Chaisson, J., Herter, J., Ingersoll, E., Uddenberg, M., Hill, J., Grossman, D., 2021. Superhot Rock Geothermal A Vision for Zero-Carbon Energy "Everywhere". CATF. Online report https://www.catf.us/wp-content/uploads/2021/09/CATF_SuperhotRockGeothermal_Report.pdf. (Accessed March 2024).
Hill, L.B., Rogers, T., Herter, J., Ingersoll, E., Uddenberg, M., Grossman, D., Chaisson, J., Hill, J., Ball, P.J., Garth, A., Montgomery, M., 2023. Superhot Rock Energy: A Vision for Firm, Global Zero-Carbon Energy. CATF, Boston, p. 27. https://www.catf.us/resource/superhot-rock-energy-a-visionfor-firm-global-zero-carbon-energy/.
Hirshmiller, J., Riva, D., 2022. Super Deep Geothermal Potential in the Oil Sands Region of Alberta. COSIA. 74 pp. https://cosia.ca/sites/default/files/attachments/Super%20Deep%20Geothermal%20Potential%20in%20the%20Oil%20Sands%20Regions%20of%20Alberta.pdf.
Hovland, M., Rueslåtten, H., Johnsen, H.K., 2014. Buried hydrothermal systems: the potential role of supercritical water, "ScriW", in various geological processes and occurrences in the sub-surface. Am. J. Anal. Chem. 5, 128–139. https://doi.org/10.4236/ajac.2014.52016.

Ilyasov, M., Ostermann, I., Punzi, A., 2010. Modeling deep geothermal reservoirs: recent advances and future problems. In: Freeden, W., Nashed, M.Z., Sonar, T. (Eds.), Handbook of Geomathematics. Springer, Berlin, pp. 679–711. https://doi.org/10.1007/978-3-642-01546-5_22.

IRENA and IGA, 2023. Global Geothermal Market and Technology Assessment, International Renewable Energy Agency; Abu Dhabi. International Geothermal Association, The Hague. https://www.irena.org/Publications/2023/Feb/Global-geothermal-market-and-technology-assessment.

Jacobs, T., 2024. Hot rock slurry: developer of emerging geothermal tech readies for field tests. J. Pet. Technol. SPE https://jpt.spe.org/hot-rock-slurry-developer-of-emerging-geothermal-tech-readies-for-field-tests.

Jolie, E., Scott, S., Faulds, J., Chambefort, I., Axelsson, G., Gutiérrez-Negrín, L.C., Regenspurg, S., Ziegler, M., Ayling, B., Richter, A., Zemedkun, M.T., 2021. Geological controls on geothermal resources for power generation. Nat. Rev. Earth Environ. 2, 324–339. https://doi.org/10.1038/s43017-021-00154-y.

Juhlin, C., Wallroth, T., Smellie, J., Eliasson, E., Ljunggren, C., Leijon, B., Beswick, J., 1988. The Very Deep Hole Concept - Geoscientific Appraisal of Conditions at Great Depth. SKB Technical Report 98-05.

Khakhaev, B.N., Lipilin, A.V., Mazur, V.B., Militenko, N.V., Pevzner, L.A., 1996. Superdeep drilling in Russia: progress and outlooks. In: 8th International Symposium on the Observation of the Continental Crust through Drilling: Drilling into Active Island Arcs. Tsukuba, Japan, pp. 4–9.

Koenig, R.L., 1991. An extraordinary drilling challenge in the Anadarko Basin. In: Proceedings of SPE Annual Technical Conference and Exhibition. https://onepetro.org/SPEATCE/proceedings-abstract/91SPE/All-91SPE/SPE-22575-MS/52636.

Kolker, A., Taverna, N., Dobson, P., Benediksdóttir, A., Warren, I., Pauling, H., Sonnenthal, E., Hjörleifsdóttir, V., Hokstad, K., Caliandro, N., 2022. Exploring for superhot geothermal targets in magmatic settings: developing a methodology. Geothermal rising conference: using the earth to save the earth. Geotherm. Resour. Counc. Trans. 46, 285–307. https://www.geothermal-library.org/index.php?mode=pubs&action=view&record=1034597.

Kruszewski, M., Wittig, V., 2018. Review of failure modes in supercritical geothermal drilling projects. Geotherm. Energy 6, 28. https://doi.org/10.1186/s40517-018-0113-4.

Laske, G., Masters, G., 1997. A global digital map of sediment thickness. EOS Trans. AGU vol. 78, F483. http://igppweb.ucsd.edu/~gabi/sediment.html#sed.

Lattoroli, F., Craig, J., Cozzi, A., 2018. Evaluating a vintage play fairway exercise using subsequent exploration results: did it work? Pet. Geosci. https://doi.org/10.1144/petgeo2016-150.

Lautze, N., Thomas, D., Hinz, N., Apuzen-Ito, G., Frazer, N., Waller, D., 2017. Play fairway analysis of geothermal resources across the state of Hawaii: 1. Geological, geophysical, and geochemical datasets. Geothermics 70, 376–392. https://doi.org/10.1016/j.geothermics.2017.02.001.

Longfield, S., Schwarz, B., Hodder, M., Stuebing, T., Holmes, M., Vany, J., Mölk, D., 2022. Eavor loop™ commercial project at Geretsried, Molasse Basin, Germany. In: European Geothermal Congress 2022, Berlin, Germany, 17–21 October 2022. www.europeangeothermalcongress.eu.

Lucazeau, F., 2019. Analysis and mapping of an updated terrestrial heat flow data set. Geochem. Geophys. Geosyst. 20, 4001–4024. https://doi.org/10.1029/2019GC008389.

Lund, J.W., 1999. From the beginning, life in the great East African Rift. In: Cataldi, R., Hodgson, S.F., Lund, J.W. (Eds.), Stories From a Heated Earth: Our Geothermal Heritage. Geothermal Resources Council, The University of California (Chapter 3).

Lund, J.W., Huttrer, G.W., Toth, A.N., 2022. Characteristics and trends in geothermal development and use, 1995 to 2020. Geothermics 105, 102522. https://doi.org/10.1016/j.geothermics.2022.102522.

Marchesini, B., Garofalo, P.S., Menegon, L., Mattila, J., Viola, G., 2019. Fluid-mediated, brittle-ductile deformation at seismogenic depth—Part 1. fluid record and deformation history of fault veins in a nuclear waste repository (Olkiluoto Island, Finland). Solid Earth 10, 809–838. https://se.copernicus.org/articles/10/809/2019/se-10-809-2019.pdf.

Markússon, S.H., Hauksson, T., 2015. Utilization of hottest well in the world, IDDP-1 in Krafla. In: Proceedings World Geothermal Congress 2015, Melbourne, Australia, 19–25 April 2015.

Matthews, D., 2023. Is the Future of Energy Pouring Water on Hot Rocks in the Ground? Vox.com. Online Resource https://www.vox.com/future-perfect/23825844/geothermal-enhanced-fervo-demonstration-superhot. (Accessed 29 February 2024).

Maxwell, S.C., Zimmer, U., Gusek, R., Quirtk, D., 2009. Evidence of a horizontal hydraulic fracture from stress rotations across a thrust fault. SPE, 312–319. SPR-22-696-PA https://doi.org/10.2188/110696-PA.

McCuaig, T.C., Beresford, S., Hronsky, J., 2010. Translating the mineral systems approach into an effective exploration targeting system. Ore Geol. Rev. 38 (3), 128–138. https://doi.org/10.1016/j.oregeorev.2010.05.008.

McCuaig, T.C., Scarselli, S., O'Connor, T., Busuttil, S., McCormack, N., 2018. The power of a systems approach to mineral and petroleum exploration in sedimentary basins. In: Arribas, A.M., Mauk, J.L. (Eds.), Metals, Minerals, and Society, Special publications of the Society of Economic Geologists, p. 21, https://doi.org/10.5382/SP.21.03 (Chapter 3).

Moeck, I.S., 2014. Catalog of geothermal play types based on geologic controls. Renew. Sust. Energ. Rev. 37, 867–882. https://doi.org/10.1016/j.rser.2014.05.032.

Moore, J., McLennan, J., Pankow, K., Simmons, S., Podgorney, R., Wannamaker, P., Jones, C., Rickard, W., Xing, P., 2020. The Utah Frontier Observatory for Research in Geothermal Energy (FORGE): a laboratory for characterizing, creating and sustaining enhanced geothermal systems. In: Proceedings, 45th Workshop on Geothermal Reservoir Engineering Stanford University, Stanford, California, February 10–12, 2020 SGP-TR-216. https://gdr.openei.org/files/1146/Moore_UtahFORGE_overview_Stanford_2020.pdf.

Norbeck, J., Latimer, T., 2023. Commercial-Scale Demonstration of a First-of-a-Kind Enhanced Geothermal System. EarthArXiv, https://doi.org/10.31223/X52X0B.

Patterson, S.O., Sabels, B.E., Kooharian, A., 1973. Ultra-Deep Drilling of Geothermals. Defence Advanced Research Projects Agency. TetraTech Report, TT-A-74-339-005.

Pollack, A., Horne, R., Mukerji, T., 2021. What are the challenges in developing enhanced geothermal systems (EGS)? In: Observations from 64 EGS Sites. Proceedings World Geothermal Congress 2020+1, Reykjavik, Iceland, April–October 2021.

Rivera, J., 2023. Comparison of Icelandic supercritical well forecasts with subcritical geothermal production well outputs. In: Proceedings 45th New Zealand Geothermal Workshop 15–17 November 2023, Auckland, New Zealand. ISSN 2703-4275.

Rivera, J.M., Carey, B., Chambefort, I., 2023. Characterisation of supercritical fluid production in Taupo volcanic zone (TVZ) through wellbore modelling and simulation. GRC Trans. 47, 2023.

Route To Reserves, 2018. Regional Risking, Ranking and Recommendations: Prototype Methodology for Cobalt Exploration Targeting in Norway. Unpublished Report.

Royal Dutch Shell, 2013. Play Based Exploration: A Guide for AAPG's Imperial Barrel Award Participants. https://iba.aapg.org/Portals/0/docs/iba/Play_Based_ExplorationGuide.pdf. (Accessed 10 October 2019).

Schenk, C., Brownfield, M.E., Charpentier, R.R., Cook, T.A., Klett, T.R., Kirschbaum, M.A., Pitman, J.K., Pollastro, R.M., Tennyson, M.E., 2012. Assessment of Undiscovered Conventional Oil and Gas Resources of South America and the Caribbean, 2012. USGS. https://pubs.usgs.gov/fs/2012/3046/fs2012-3046.pdf.

Scherer, J., Higgins, B., Muir, J., Amaya, A., 2020. Closed-Loop Geothermal Demonstration Project. California Energy Commission. Publication Number: CEC-300-2020-007 https://www.energy.ca.gov/sites/default/files/2021-05/CEC-300-2020-007.pdf.

Scherer, J., Golla, G., Chandrasekar, H., Amaya, A., Chin, H., Brown, S., Sonnenthal, E., Rutqvist, J., Oldenburg, C., 2023. The GreenLoop closed-loop geothermal system: optimizing power generation and evaluating its impact on the reservoir. In: The 9th Indonesian International Geothermal Convention and Exhibition. 20–22 September 2023, Jakarta Convention Centre.

Schlumberger, 2019. Oilfield Glossary. Online Resource http://www.glossary.oilfield.slb.com/. (Accessed June 2023).

Scott, S., Yapparova, A., Weis, P., Houde, M., 2023. Numerical modelling of heat mining at 10–25 km depth. GRC Trans. 47, 2023.

Shervais, J.W., DeAngelo, J., Glen, J.M., Nielson, D.L., Garg, S., Dobson, P., Gasperikova, E., Sonnenthal, E., Liberty, L.M., Newell, D.L., Siler, D., Evans, J.P., 2024. Geothermal play fairway analysis. Part 1. Example from the Snake River Plain, Idaho. Geothermics 117, 102865. https://doi.org/10.1016/j.geothermics.2023.102865.

Sinopec, 2023. Sinopec Completes Asia's Deepest Onshore Oil, Gas Well. Online Resource http://www.sinopecgroup.com/group/en/Sinopecnews/20231116/news_20231116_351055127835.shtml. (Accessed 30 January 2024).

SPE, 2018. Petroleum Resources Management System. PRMS. Version 1.03, ISBN 978-1-61399-994-3.

Strasser, E., 2015. Applicability of Ultra-Deep Vienna Basin Drilling Experience for Future Exploration Requirements. MSc Thesis, Montanuniversität Leoben, Austria. https://pure.unileoben.ac.at/files/3921017/AC13051420n01.pdf.

Suárez-Arriaga, M.C., 2019. Thermodynamics of deep supercritical geothermal systems. IOP Conf. Ser. Earth Environ. Sci. 249. https://iopscience.iop.org/article/10.1088/1755-1315/249/1/012019.

Survok, V.S., Devyatov, V.P., Zhero, O.G., Kazakov, A.M., Kramnik, V.N., Smirnov, L.V., 1993. Structure of the Earth's crust in the Tyumen superdeep borehole zone. Russian J. Geol. Geophys. 34, 120–126.

Taverna, N., Pauling, H., Trainor-Guitton, W., Kolker, A., Mibei, G., Dobson, P., Sonnenthal, E., Tu, X., Schultz, A., 2024. De-risking superhot EGS development through 3D play fairway analysis: methodology development and application at Newberry volcano, Oregon, USA. Geothermics 118, 102909. https://doi.org/10.1016/j.geothermics.2023.102909.

Tester, J.W., Anderson, B., Batchelor, A., Blackwell, D., DiPippo, R., Drake, E., Garnish, J., Livesay, B., Moore, M.C., Nichols, K., Petty, S., Toksz, M.N., Veatch Jr., R.W., 2006. The Future of Geothermal Energy: Impact of Enhanced Geothermal Systems (EGS) on the United States in the 21st Century. Massachusetts Institute of Technology, Cambridge, MA, United States. https://www1.eere.energy.gov/geothermal/pdfs/future_geo_energy.pdf.

ThinkGeoEnergy, 2024. Global Geothermal Power Plant Map. ThinkGeoEnergy. Online Resource https://www.thinkgeoenergy.com/map/. (Accessed January 2024).

Toralde, J.S.S., Nas, S., 2010. Controlled pressure drilling applications for enhanced geothermal systems. In: Proceedings World Geothermal Congress, Bali, Indonesia, 25–29 April 2010. https://www.geothermal-energy.org/pdf/IGAstandard/WGC/2010/3162.pdf.

Turcotte, D., Schubert, G., 2002. Geodynamics, second ed. Cambridge University Press, New York, ISBN: 978-0-521-66186-7.

van Zelst, I., Crameri, F., Pusok, A.E., Glerum, A., Dannberg, J., Thieulot, C., 2021. 101 geodynamic modelling: how to design, carry out, and interpret numerical studies. Solid Earth Discuss. https://doi.org/10.5194/se-2021-14.

Ward, S.H., Ross, H.P., Nielson, D.L., 1981. Exploration strategy for high temperature hydrothermal systems in the Basin and Range Province. Am. Assoc. Pet. Geol. Bull. 65 (1), 86102. https://doi.org/10.1306/2F919779-16CE-11D7-8645000102C1865D.

Wood Mackenzie, 2023. Geothermal Energy: The Next Trillion Dollar Low-Carbon Solution? The Market Outlook for Enhanced (EGS), Advanced (AGS) and Co-Located Geothermal Systems. Wood Mackenzie https://www.woodmac.com/reports/energy-markets-geothermal-energy-the-next-trillion-dollar-low-carbon-solution-150186009.

Zatonski, V., Brown, C., 2023. Eavor-lite update after four years of operation. GRC Trans. 47, 2023.

Key terms and definitions

Accretionary complex, active/subduction Margin consisting of sedimentary wedges built on oceanic or continental crust (Hasterok et al., 2022).

Advection Rapid heat transfer of heat due to magmatic or hydrous fluids (Allen and Allen, 2013).

Basement Crystalline rock is sometimes called "basement," "basement rock," or "basement reservoir." However, "basement" is not a recommended term for the geothermal industry unless it is used with qualifiers, e.g., economic basement, seismic basement, naturally fractured basement reservoir, etc. (Banks and Peacock, 2020; Banks and Ball, 2023).

Basin, intracontinental sedimentary Cover built on preexisting continental crust with uncertain or unknown basement provenance (Hasterok et al., 2022).

Closed-loop [geothermal circuit] A circuit containing subsurface working fluid that is heated in the reservoir without direct contact with rock pores and fractures. Instead, the subsurface working fluid stays inside a closed loop of deeply buried pipes that conduct Earth's heat (e.g., Banks and Ball, 2023).

Conduction Heat transfer through rocks or fluids that are in contact, but not moving. Conduction is the most common type of heat transfer in Earth, from a higher temperature region to a lower temperature region (e.g., Earth's surface) (Banks and Ball, 2023).

Continental back-arc basin A hyperextended basin either transitional continental or oceanic crust created as a result of **upper plate extension** in response to subduction rollback (Hasterok et al., 2022).

Contingent resource The quantities of heat estimated, as of a given date, to be potentially extractable from known thermal reservoirs by application of development projects, but which are not currently considered to be commercially recoverable owing to one or more contingencies (modified from SPE (2018)).

Convection Heat transfer by the physical movement of molecules (i.e., fluid). The word is used in two contexts: (a) for fluids flowing in a circular path and (b) heat transfer out of rock and into a moving fluid that comes into contact with the rock (Banks and Ball, 2023).

Conventional geothermal system [energy generation] Delivery of thermal energy from shallow or deep hydrothermal reservoirs to Earth's surface without an introduced working fluid. The technique is well-established with technologically mature and commercially available methods (e.g., Banks and Ball, 2023).

Craton Predominantly **Archean core**, contains granite-greenstone belts and other undifferentiated terranes with relatively small area (Hasterok et al., 2022).

Critical isotherm The critical isotherm for pure water is ~374°C (705°F) and above 22 MPa (3205 PSI) for water.

Dry rock Dry rock geothermal is one category of petrothermal systems. Dry rock means the rock is at low to medium enthalpy, but water is not naturally present at the site.

Engineered geothermal system (EGS) A geothermal system that engineers have artificially created or improved because there are hot rocks but insufficient reservoir quality. Human intervention enhances the heat exchange from the hot rock to the subsurface working fluid (e.g., Banks and Ball, 2023).

Enthalpy Is used to relate the energy of a system, heat transfer, and work done. Geothermal systems can be low, medium, high, or superhigh enthalpy (after Banks and Ball, 2023).

Foredeep basin (Foreland basin) thick intracontinental sedimentary basin created during continent-continent **collision**, basement uncertain (Hasterok et al., 2022).

Geodynamic Combines the Greek word *geo* meaning "Earth" and the term "dynamics" which is a discipline of physics concerned with forces acting on a body and the subsequent motions and energy flow they produce (modified after van Zelst et al., 2021). Geodynamics is generally concerned with processes that move materials throughout the Earth (Turcotte and Schubert (2002). The main motive force behind stress in the Earth is provided by thermal energy from radioisotope decay, friction, and residual heat from the formation of Earth (Hager and Clayton, 1989).

Geofluid Naturally occurring brine, water, steam, or supercritical fluid within Earth or at Earth's surface. Geofluid is an essential ingredient of a hydrothermal system because it transfers geothermal energy to Earth's surface. It has little monetary value. The value is in the geothermal energy that it transfers. An exception is geothermal brine that may contain minerals such as lithium in high enough concentrations to be extracted and sold. Geofluid is sometimes referred to as groundwater or formation water (e.g., Moeck, 2014), but these terms are discouraged because they do not communicate that geofluids may be supercritical vapor, steam, or brine (after Banks and Ball, 2023).

Geothermal discovery An underground geological structure or region in which a hydrothermal reservoir or geothermal reservoir has been proven, usually with an exploration well. A geothermal discovery provides insufficient information to state the size and type of geothermal resource and whether it has economic potential (after Banks and Ball, 2023).

Geothermal field A localized volume of Earth's crust where a geothermal resource can be extracted from a reservoir by subsurface working fluids and then delivered to a place of use (modified after Moeck, 2014). See Banks and Ball (2023) for further description.

Geothermal play The portion of a geothermal system that contains a heat reservoir, its subsurface working fluid (either naturally present geofluid or introduced fluid), and/or a cluster of geothermal fields, geothermal discoveries, geothermal prospects, and/or geothermal leads—which resemble each other geologically and share common risks (modified after Doust, 2010)—and is surrounded by lower permeability rocks. The play is a useful construct to describe the spatial locations and monetary value of a portfolio of similar geothermal fields. The play concept enables geothermal exploration strategy planning, risk management, land evaluation, prediction of future geothermal exploration success, and ultimately the optimal strategic decisions about groups of geothermal fields and their geothermal resources (Banks and Ball, 2023).

Geothermal prospect An underground, geological structure or region with indications that it could contain a hydrothermal reservoir or geothermal reservoir, which could be economically exploited (after Banks and Ball, 2023).

Geothermal resource The portion of a reservoir's total geothermal energy that is technically recoverable from the Earth (Banks and Ball, 2023).

Geothermal An adjective relating to heat within Earth (e.g., Banks and Ball, 2023).

Heat endowment The quantity and quality of geothermal heat in a particular geographic region or depth. Heat endowment can be used to compare and rank the natural heat "wealth" of different regions above project-specific cutoffs. Assessment inputs include heat density, depth to heat reservoir, geothermal play type, and available technology (Banks et al., 2020; Banks and Ball, 2023).

Heat transfer Is the transfer of thermal energy from a body, at a high temperature, to another at a lower temperature. This transfer of thermal energy may occur under steady or unsteady state conditions. Under "steady-state" conditions, the temperature within the system does not change with time. Conversely, under nonsteady-state conditions, the temperature within the system does vary with time. No system can remain under nonsteady-state conditions perpetually. The temperature of the system will eventually reach the temperature of the heat source, and once this happens, the system will be at steady state (Hagen, 1999).

Hot dry rock Hot dry rock geothermal is one category of petrothermal systems. Hot dry rock means the rock is at high enthalpy, but water is not naturally present at the site.

Hybrid geothermal XGS is a company developing this concept. Simply put, a hybrid system is a closed-loop borehole surrounded by fractures filled with high conductivity cement/slurry (Jacobs, 2024).

Hydrothermal system The category of geothermal system with convective heat transfer in a naturally occurring fluid. All hydrothermal systems are geothermal systems. Not all geothermal systems are hydrothermal systems (Banks and Ball, 2023).

Hydrothermal An adjective relating to hot water, the action of hot water, or the products of this action (e.g., Banks and Ball, 2023 and references therein).

Impermeable A rock that does not permit the passage of fluids through it (after Banks and Ball, 2023).

Lithosphere The lithosphere is the solid, outer part of Earth, including the brittle upper portion of the mantle and the crust. Its thickness changes from about 40 km to perhaps 280 km depending on whether the lithosphere is cratonic or extensional. Continental lithosphere is inherited whereas oceanic lithosphere is pristine. Oceanic lithosphere thickness can range from <20 to 120 km depending on its age. The base of the lithosphere is conventionally defined as the 1300°C isotherm.

Magmatic province A large **intraplate magmatic** terrane not clearly associated with subduction or extension processes; and oceanic crust, typical oceanic crust not created in a back-arc setting (Hasterok et al., 2022).

Mantle plume Mantle plumes are areas where buoyant asthenosphere is rising. Where a mantle plume intersects with the crust, it can result in uplift and thinning, and it is often associated with excess magmatism and heat.

Narrow rift Focused **tensional terrane** with continental basement (Hasterok et al., 2022).

Nonsteady state (assumption: heat linked to dominantly convective processes) Nonsteady-state thermal contributions are transient—due to hydrothermal convection or magmatic processes.

Oceanic back-arc basin, a back-arc basin Where seafloor spreading has been sustained, creating enriched basaltic compositions relative to mid-ocean ridge basalt (Hasterok et al., 2022).

Ophiolite complex, abducted oceanic Crust of some variety, excluding volcanic arc-type, but including suprasubduction zone oceanic crust (Hasterok et al., 2022).

Orogenic belt Fold and thrust belts created during **accretionary, collisional,** and intracontinental settings that may incorporate a variety of preexisting terrane types, often commingled, making them difficult to differentiate at the regional scale (Hasterok et al., 2022).

Passive margin Sediment accumulation built on transitional crust between continental and oceanic crust marking half of a tectonically inactive fossil rift (Hasterok et al., 2022).

Permeable The ability of a rock to permit the passage of fluids without rupture or displacement of pores. The antonym is impermeable. An impermeable rock is sometimes informally called "dry rock" (from Banks and Ball, 2023).

Petrothermal system A type of geothermal system that is assumed to lack producible geofluids because the subsurface heat exchanger is in impermeable rock below a hydrothermal system/groundwater aquifer. To extract Earth's heat, the geothermal reservoir needs to be engineered into a hydrothermal reservoir, i.e., an engineered geothermal system (Banks and Ball, 2023).

Prospective resources Are those quantities of heat estimated, as of a given date, to be potentially recoverable from undiscovered accumulations by application of future development projects. Prospective resources have both an associated chance of geologic discovery and a chance of development. Prospective resources are further categorized in accordance with the range of uncertainty associated with recoverable estimates, assuming discovery and development, and may be subclassified based on project maturity (modified from SPE (2018)).

Province A regional-scale geographic area, i.e., an arc, an orogenic belt, or a sedimentary basin, that is notable for a commodity (geothermal heat, petroleum, etc.) also incorporates present-day topography and regional-scale geological terranes. A province needs to be large enough to encompass a geothermal system's footprint (modified after Banks et al., 2020).

Reserves Those quantities of heat anticipated to be commercially extractable by application of development projects to known thermal reservoirs from a given date forward under defined conditions. Reserves must satisfy four criteria: They must be discovered, recoverable, commercial, and remaining (as of a given date) based on the development project(s) applied (modified from SPE (2018)).

Screening A preliminary assessment of a province by comparing its characteristics to criteria that have influenced success or failure in similar projects (adapted from Schlumberger Oilfield Glossary, 2019).

Sedimentary basin Sedimentary basins can vary in size and shape and can be formed by various geodynamic processes such as tectonic activity, erosion, and sedimentation. Sedimentary basins can be 0 to >20 km thick (Laske and Masters, 1997).

Shield Similar to a **craton**, predominantly Meso- to Paleoproterozoic lithosphere, undifferentiated (Hasterok et al., 2022).

Steady state (conductive geotherm) By steady state, we mean that the temperature within the lithosphere is assumed to be in equilibrium and time-independent (i.e., not changing with time).

Supercritical [state] A substance at a temperature and pressure where distinct liquid and gas phases do not exist. This would be above 374°C (705°F) and above 22 MPa (3205 PSI) for water (e.g., Banks and Ball, 2023).

Superhot rock A subsurface geologic rock resource existing in situ at or above the supercritical temperature of water, 374°C in deionized water (or higher in brine). From Banks and Ball (2023).

Thermal conductivity A measurement of the ability of a material (fluids, rocks, and soils) to conduct or transfer heat. This is extremely important for the design of geothermal wells and underground energy storage systems (Banks and Ball, 2023).

Unrecoverable resources That portion of either discovered or undiscovered heat initially in place evaluated, as of a given date, to be unrecoverable by the currently defined project(s). A portion of these quantities may become recoverable in the future as commercial circumstances change, technology is developed, or additional data are acquired. The remaining portion may never be recovered because of physical/chemical constraints represented by subsurface interaction of fluids and reservoir rocks (modified from SPE (2018)).

Volcanic arc Predominantly magmatic arc crust related to **subduction** but may contain crust predating the arc and/or interspersed accretionary material in island arcs and in seaward migrating arcs due to retreating trenches (Hasterok et al., 2022).

Wide rift Distributed **tensional** terrane with continental basement (Hasterok et al., 2022).

CHAPTER

13

Geothermal energy and fluid properties with a special focus on geothermal noncondensable gases

Niyazi Aksoy[a], *Silviu Livescu*[b,c], *Ram Ratnakar*[d], *and Birol Dindoruk*[e]

[a]NTU Geothermal Consulting Ltd., Ankara Cad, Izmir, Türkiye [b]Bedrock Energy, Austin, TX, United States [c]Department of Petroleum Engineering, University of Texas at Austin, Austin, TX, United States [d]Shell, Houston, TX, United States [e]Department of Petroleum Engineering, University of Houston, Houston, TX, United States

1 Introduction

There is enough geothermal energy beneath the surface of the earth to supply more power globally than the entire oil and gas industry. Geothermal energy is renewable, abundant, and has a minimal carbon footprint. However, the current geothermal energy use is only 1% of the global energy production. To scale it up, several technical problems need to be solved, addressing how, where, and when to harness it safely, efficiently, and cost-effectively. In addition, using some of the opportunities brought up from the intense drilling campaigns from the unconventional hydrocarbon resources can also be a plus in harnessing the geothermal energy and changing the exit strategies for mature wells and assets. Geothermal systems can be divided into various categories, including synergistic categories:

(a) Hydrothermal systems: hydrothermal systems have been the main area of focus for worldwide traditional geothermal energy generation. Most of the large geothermal power plants in the world use steam and hot water from volcanically active regions (New Zealand, Turkey, Indonesia, Italy, the US, Iceland, etc.) to generate electric power.

(b) Hot-igneous, conduction-dominated systems: hot-igneous systems and conduction-dominated systems may contain significant amounts of energy in the form of heat, yet extraction of heat from such systems has been challenging. They normally need enhancements in reservoirs as well as implementation/selection of various power fluids.

Geothermal Energy Engineering
https://doi.org/10.1016/B978-0-443-21662-6.00013-X

Copyright © 2025 Elsevier Inc. All rights are reserved, including those for text and data mining, AI training, and similar technologies.

(c) Geothermal heat/energy extraction from existing oil/gas wells: this category basically will take advantage of the existing wells and leverage via the sunken cost due to primary reasons for the existence of these wells, o/g production.

Hydrothermal systems consist of high-temperature water, steam, and water-steam (either as liquid-dominant systems or vapor-dominant fluids), as well as porous and permeable rocks. As a result, a hydrothermal system leads to convective heat transfer from the network of permeable rocks. Producing geothermal energy from existing hydrocarbon wells, as electricity and/or low-temperature waste heat, can yield significant advantages over traditional geothermal wells, especially in terms of reduced capital expenditure.

The use of geothermal energy extracted from underground as thermal energy for heating and/or cooling is called direct utilization. Direct use has many perspectives, from swimming pools to greenhouses, buildings, and other applications (Kapusta et al., 2023; Livescu, 2024). Rejkavik (Iceland) and Balçova-Narlıdere (Turkey) district heating systems are the world's largest known geothermal heating systems. Geothermal resources are also widely used in greenhouse heating. In Turkey, geothermal greenhouse heating has exceeded 1000 ha. Natural geothermal systems (hydrothermal systems) include waters that can contain various non-condensable gases (NCGs) as well as salts. Existence of components beyond water, which is being the main fluid contained within, can generate various complexities for the extraction of energy from such systems. This can also happen in enhanced geothermal systems as well as in open systems (Ratnakar et al., 2024a, b, c); the working fluid(s) can dissolve various compounds from the formation contacted. For each use case, direct or indirect use of geothermal energy needs to consider the existence of NCGs. NCGs and dissolved species such as NaCl, SiO_2, KCl, and carbonates will have an impact on the temperature-enthalpy variations for heat exchange as well as P-T-enthalpy-entropy relations for turbine expansions. In general, the presence of large amounts of dissolved NCGs in geothermal brines may prevent the use of condensing steam turbine cycles because of the excessive extractive pump work required to remove the gas from the condenser. Dissolved salts not only present potential scaling issues but also reduce the specific enthalpy of the brine and, subsequently, the amount of the steam produced in a flash step (Edwards et al., 1982).

Furthermore, the working fluids to extract the heat from geothermal systems (or heat sources in general) may require different fluids based on the power cycle used coupled with the system pressure and temperature. Accurate data on the thermodynamic properties of proposed fluids is required to calculate electric power cycle performance (Edwards et al., 1982). Depending on the fluid type parallel to the system parameters, estimation methodologies and tools and techniques will vary. For example, variants of Martin and Hou's (1955) equations of state have been used for the calculation of PVT properties of various organic compounds, such as isobutane, fluorocarbons, and ammonia.

District heating is a significant example of the direct use of geothermal energy (Fig. 1). Hot water produced from a geothermal source is sent to a separator to separate it from gases, and the hot water is pumped through a heat exchanger into a reinjection well. In the heat exchanger, heat is transferred to water that has been conditioned against corrosion and scaling in a closed loop, and sent to buildings to be heated. The cooler water returning from the buildings reenters the heat exchangers (Fig. 1). If the geothermal fluid produced from the wells is pumped and its pressure is kept above the gas separation pressure, no NCG is released, and there is no need to use separators. The geothermal fluid being produced directly from subsurface rocks contains a variable amount of NCG, consisting primarily of CO_2 and H_2S and trace amount of NH_3, H_2, N_2, and CH_4. NCGs originally stem from degassing magma and degradation of organic sediments and carbonated rocks (Ármannsson et al., 2005).

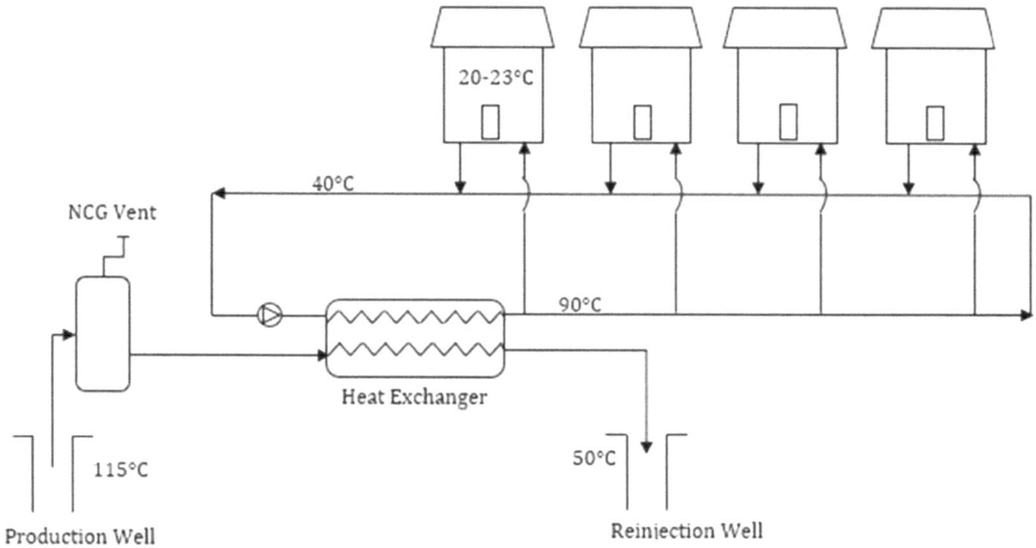

FIG. 1 Geothermal district heating system, schematic representation.

After geothermal waters are brought to the earth's surface through wells, they are pumped back to the reservoir, where they are produced after taking their thermal energy. This is called reinjection. Reinjection helps to maintain the reservoir pressure, allows the heat in the rock that stores a significant portion (>80%) of the thermal energy to be transferred to the water, and helps to protect the environment.

NCGs are dissolved in the geothermal fluid in water-dominated fields. In steam-dominated fields, it moves with steam in the reservoir. In water-dominated fields, as it rises from production wells to the surface, NCG is released at the point where the pressure drops below the saturation pressure of the gases. This point is known in the industry as the flash point. The depth of the flash-point depends on production flow rate, productivity index of the well, temperature, NCG concentration, and water salinity. At the flash-point, the chemical structure of the water changes very rapidly due to decomposing gases. This causes scaling in wells. Knowing the flash point depth is important since inhibitors to prevent scaling should be pumped below the flash point depth. If a pump is needed in a well, the pump should be set below the flash point depth (Aksoy, 2007).

If the production wells are under-pressurized or due to low NCG or temperature, the desired amount of production may not be achieved. In these cases, artificial lifting methods are required. The most well-known of these are pumps, gas lifts, and jet pumps. There are two types of pumps commonly used for geothermal wells. These are LSP (line shaft pump) and ESP (electrical submersible pump). For LSPs, power is transmitted to the pump by an electric motor on the ground through shafts. Expansion due to temperature and rotation of very long shafts at high speed and lubrication cause problems. Although they can be used up to depths of 700 m, common depths of use are 400 m and shallower. ESPs, also routinely used in oil and gas wells, have no depth limit. Since the pump motor runs in the well, cables transmit the required power to the motor. Cooling and insulation of the electric motor is the biggest problem in geothermal wells. However, ESPs operating at higher temperatures are being developed day by day. The average operating life of ESP pumps at temperatures around 200°C is close to 2 years. As the temperature decreases, the service life is longer, and the service life is shorter as the temperature increases. The total efficiency of these pumps (motor, pump, transformer, drivers, and cable losses) is between 50% and 60%. Initial investment costs are high. Pumps can operate more economically in wells with a high productivity index. In wells with low productivity index, their energy may not meet their consumption. In the calculation of pump set depth, reservoir pressure, temperature, NCG ratio (flash point) production index, flow rate, well diameter, and desired wellhead pressure are taken into consideration (Aksoy, 2007).

Geothermal waters are generally prone to precipitation. Calcium and silica minerals, especially iron and sulfur compounds, cause scaling. Precipitations cause clogging in pipes and energy loss. It is possible to predict scaling and corrosion and take precautions by examining the temperature, pressure, evaporation, and gas decomposition phases of the produced water throughout the plant (Fig. 2).

FIG. 2 Scaling and corrosion of a pipe.

Some of the precipitation-related issues are also related to the pH, pressure, and temperature of the system. Generally, pressure and temperature-related aspects of the scaling are well known, while pH in relation to the acid gas dissolution (i.e., CO_2) in mixed salt systems is more challenging as the solubility of such gases are both functions of P-T and as well as the compositions of the formation waters (Ratnakar et al., 2023).

1.1 Use of salts and minerals as geothermometer

Chemical geothermometers: ationic relationship: Based on cation proportions (ppm), semiempiric equations are developed for temperatures for which reactions are in equilibrium. For example, the expression by Fournier (1979).

$$T(^{\circ}C) = \frac{1{,}217}{log\,(Na/K) + 1.483} - 273 \tag{1}$$

An extensive study on geothermometric applications for Dutlu-Beypazari geothermal field was conducted by Sumnu (1989), where surface waters are utilized to estimate the temperatures of the source. A summary of various geothermometers can also be found in the manuscript written by Lajwe (2013).

The most common indirect use of geothermal energy is electricity generation. In conventional steam turbines, the pressure of the steam is up to 300 bar, the temperature is up to 500–600°C, and the efficiency of these turbines is less than 50%. Geothermal sources have lower temperatures and pressures, so their efficiency is incomparably lower than that of coal or gas-fired plants. The main advantage of geothermal resources is the lower cost of fuel. However, it should not be forgotten that exploration and drilling costs to reach geothermal resources are high and risky.

Two main types of power plants are used to generate electricity from geothermal resources. For those that produce electricity with classical steam turbines by using steam directly, the geothermal resource temperature is 200°C and above, and the geothermal fluid at the surface is separated as hot water and steam at low pressures. The steam is sent directly to the turbine to generate electricity (Fig. 3). If the geothermal resource temperature is high,

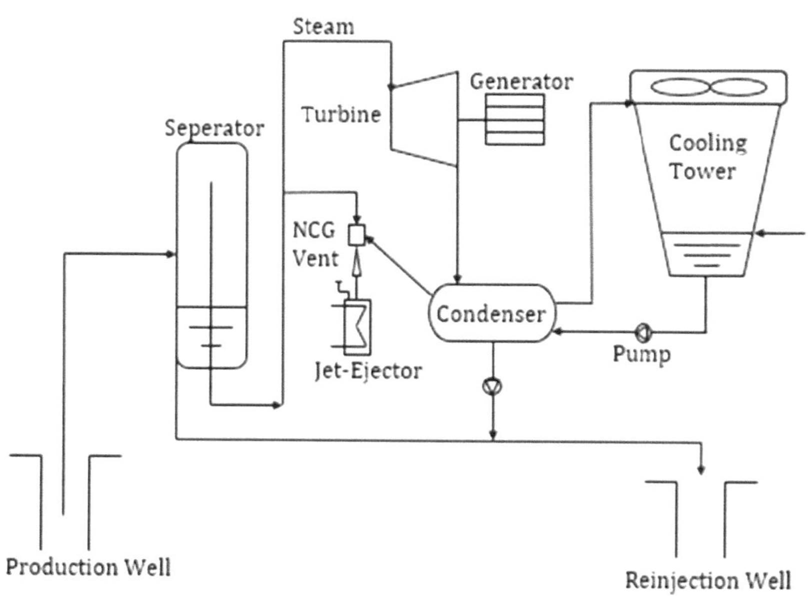

FIG. 3 Single flash steam turbine plant.

the geothermal fluid is flashed twice or three times to increase the efficiency of the geothermal plant (Fig. 4). This type of power plant's efficiency is up to 25%, depending on temperature. In such plants, NCG moves with the steam, and the partial pressure created by NCG at the exit of the steam turbine reduces the turbine's efficiency. NCG is withdrawn from the turbine outlet by an ejector or compressor to increase production. This causes a significant internal consumption. For example, according to Gokcen and Yildirim (2008), the compressors in the Kizildere-I unit in the Kizildere (Turkey) geothermal field have a power of 2.8 MWe and consume 18.3% of the gross capacity.

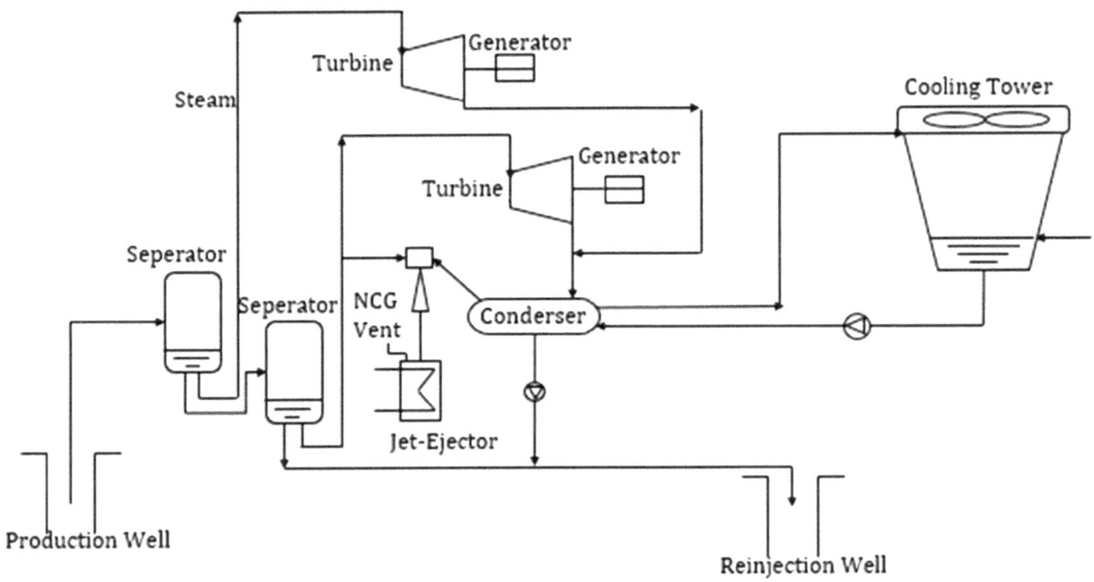

FIG. 4 Double flash steam turbine plant.

Under exceptional conditions, electricity generation from a geothermal resource with a temperature of 80°C may not be economical, even if technically possible. Resource temperatures of 150°C and above are commonly used in electricity generation. The binary cycle is preferred for electricity generation from geothermal resources with temperatures between 150°C and 200°C. In these binary cycle systems, fluids that can vaporize more easily are used in the secondary circuit instead of sending water vapor to the turbine. For binary power plants, some correlations predict efficiency based on source temperature or enthalpy of the source. The most famous of these are arguably the studies by Tester (2006) (Eq. 2) and Moon and Zarrouk (2012) (Eq. 3). These studies established simple correlations using the resource temperature and efficiency of operating geothermal power plants.

$$n = 0.0935*T - 2.3266 \quad (2)$$

$$n = 13.40*\ln(h_i) - 74.60 \quad (3)$$

In these empirical equations, n is the net efficiency, T is the temperature in °C, and h_i is the inlet enthalpy of the fluid in kJ/kg.

A simple binary cycle is shown in Fig. 5. Geothermal water pumped from the production well transfers its energy to the heat exchanger's secondary fluid (Penthane, Buthane, Isopenthane, etc.). The vaporized secondary fluid turns the turbine, and the generator generates electricity. The fluid leaving the turbine is cooled to approach the vaporization temperature at operating conditions. The geothermal fluid from the heat exchanger is pumped to the reinjection well.

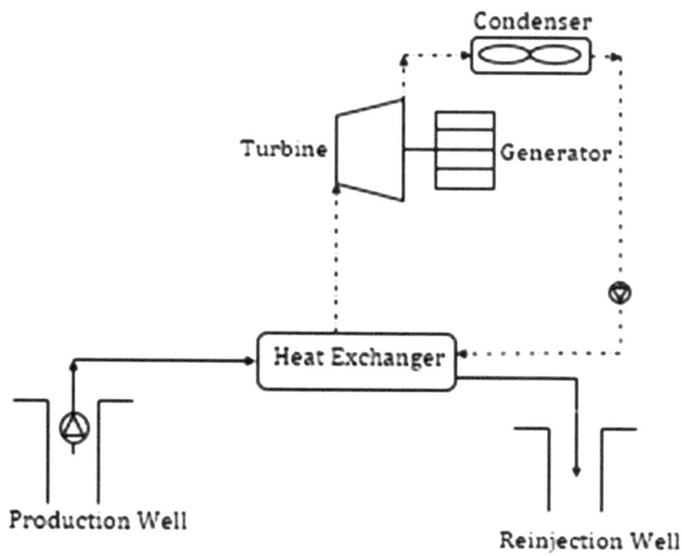

FIG. 5 Simplified representation of a binary cycle.

If the geothermal fluid is produced from an artesian well, or if NCG, vapor, and liquid are separated under surface operating conditions, the geothermal fluid is passed through a separator. While the steam and NCG move together, the hot water is passed through separate heat exchangers. The NCG is released from the condensing steam into the atmosphere. The condensed steam and hot water are sent together to the reinjection well. The secondary cycle continues in the same way as before (Fig. 6).

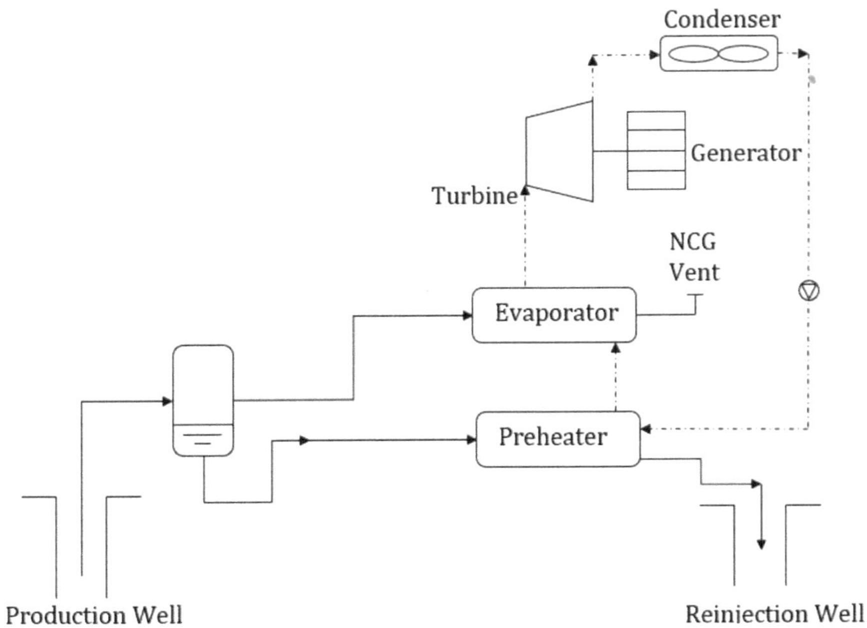

FIG. 6 Binary cycle representation using steam and hot water together.

In order to obtain more energy from high-enthalpy geothermal fluids, there are also types of power plants that combine steam and binary turbines. In combined GPPs, steam from the steam turbine and hot water from the separator enter the heat exchangers of the binary GPP before being sent for reinjection (Fig. 7). In another type (Fig. 8), a steam and a binary are simply combined. The hot water separated from the steam is sent to the steam turbine and sent to a binary GPP. This type of GPP is called "bottoming binary."

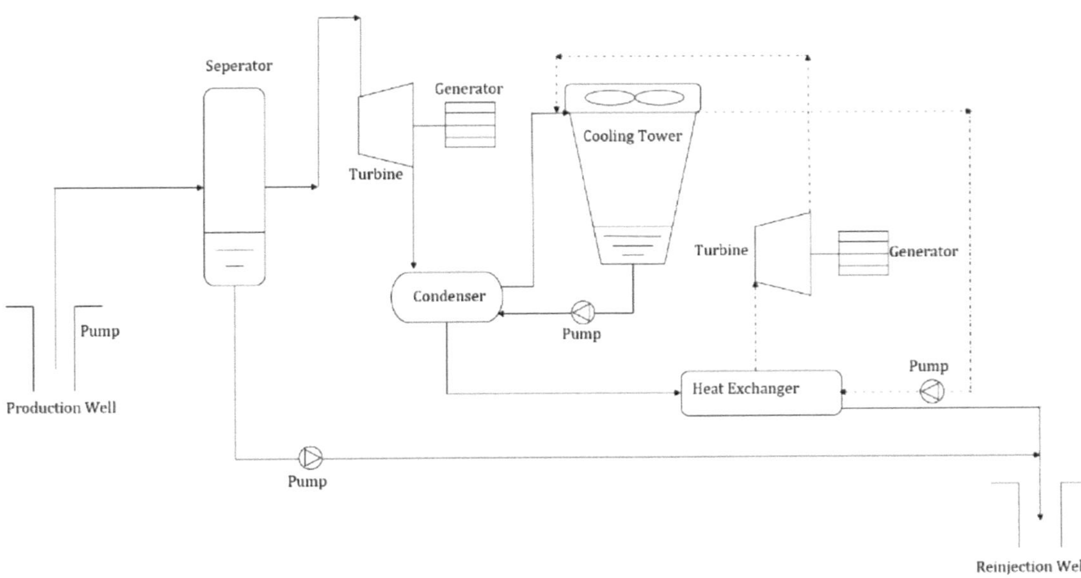

FIG. 7 Combining vapor and binary GPP.

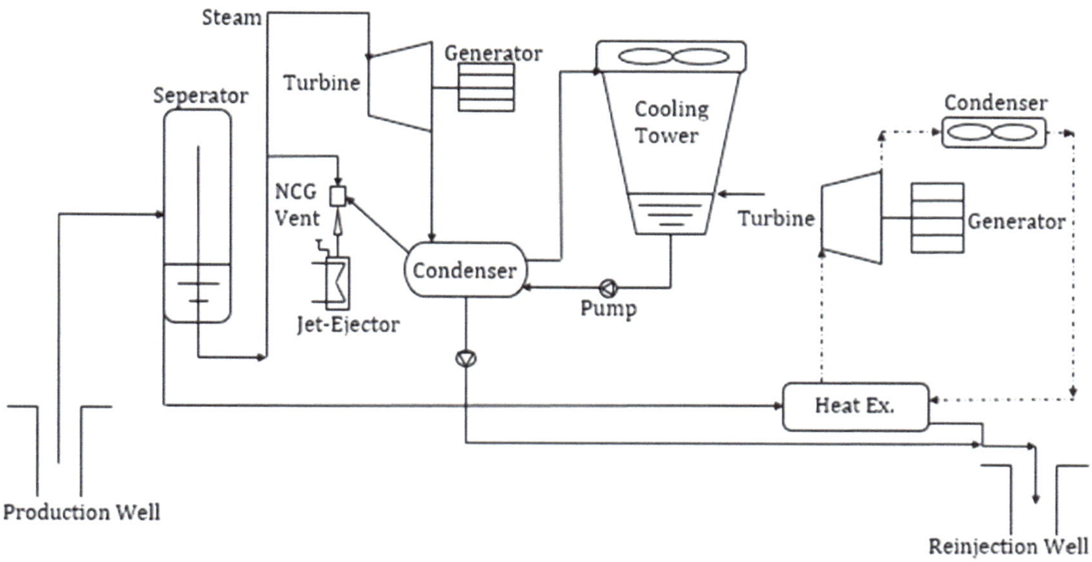

FIG. 8 Steam and bottoming binary GPP.

Steam turbines are generally preferred when the source temperature exceeds 200°C. However, the limits at all temperature transitions given above are not precise. In addition to the temperature of the geothermal resource, many other factors, such as tariff price and environmental temperature, affect the preferences in transition zones and the economic feasibility of the work. Even if the geothermal resource temperature is suitable for steam turbines, if the geothermal source contains a high percentage of NCG, choosing binary over a steam turbine may be more advantageous. This is because the net production of steam turbines may be lower than the net production of binary turbines due to NCG extraction (Kaplan and Serpen, 2010).

2 NCG's contribution to production

The pressure of most geothermal reservoirs is below normal hydrostatic pressure. For wells to flow spontaneously, the average fluid density in the well must decrease. NCG and temperature change the fluid from a single liquid phase to a two-phase flow by reducing the hydrostatic pressure of the fluid rising towards the surface along the wellbore. NCG and temperature create a natural "gas-lift" impact in the well.

NCGs are mainly in the range of 1%–3% by mass. Even if the mass loss in the geothermal reservoir due to NCG emission is minimal, the production performance decreases very quickly. When waters with negligibly low NCG content are reinjected into the geothermal reservoir, the average NCG rate decreases (Fig. 9). If some of the injected water reaches the production wells, the average NCG content of the production wells drops even below the average reservoir concentration. For example, in a well with a temperature of 175°C, when the NCG ratio is 3%, the flash point pressure is 107 bar, whereas when the NCG ratio is 1%, the flash pressure is only 41 bar. In many fields with similar temperatures in Turkey, the NCG content has decreased to 0.2%, and the flash pressure has dropped to around 14 bar (calculated using Ellis and Golding (1963), assuming that NCG is composed of 100% CO_2). Due to NCG depletion, some wells have stopped producing completely or are losing significant artesian production capacity, even if the reservoir pressure is maintained. Fig. 10 shows the loss of production performance of the GM1 well in the Germencik Geothermal field in Turkey. In this well, where the reservoir pressure and temperature did not change, the NCG decreased from 1.7% at the beginning to 0.25% over time. A decrease in NCG decreases production. This well produces 350 t/h with an LSP from a depth of 350 m. Wellhead pressure is 10 bar. Pump power consumption is around 500 kW. There is a binary-type power plant in the field. The contribution of GM1 well is 5.9 MW under power plant operating conditions. Approximately 0.5 MW is consumed by the pump. The initial investment cost of the pump is around 500,000 USD. It requires an average renewal and maintenance of 150,000–200,000 USD every 3 years.

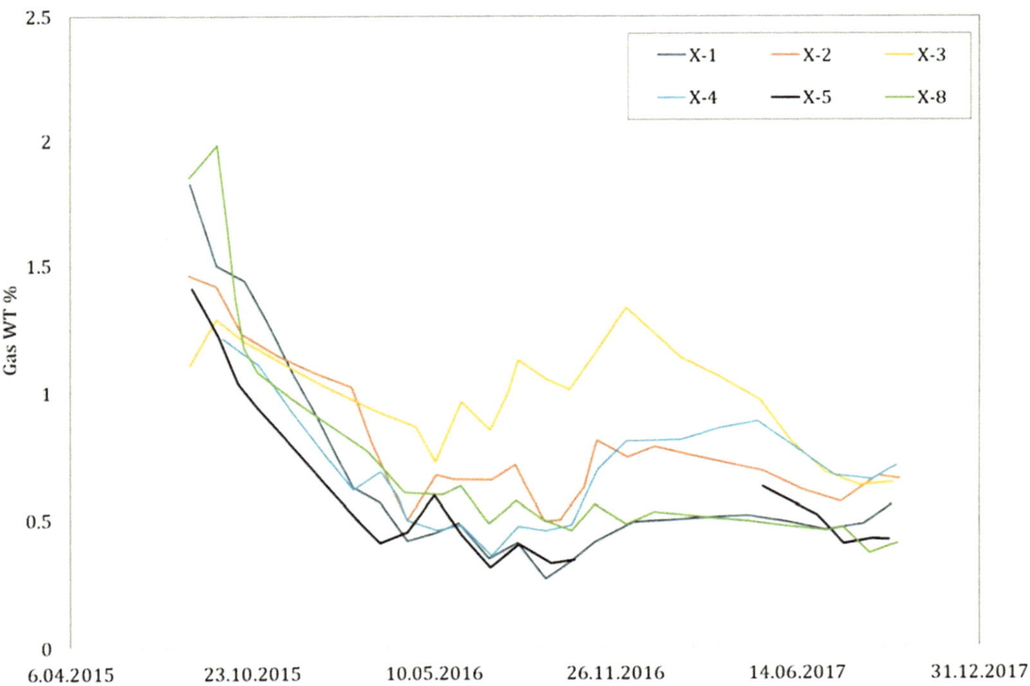

FIG. 9 NCG decline of geothermal wells in Alaşehir field (Aydın et al., 2018).

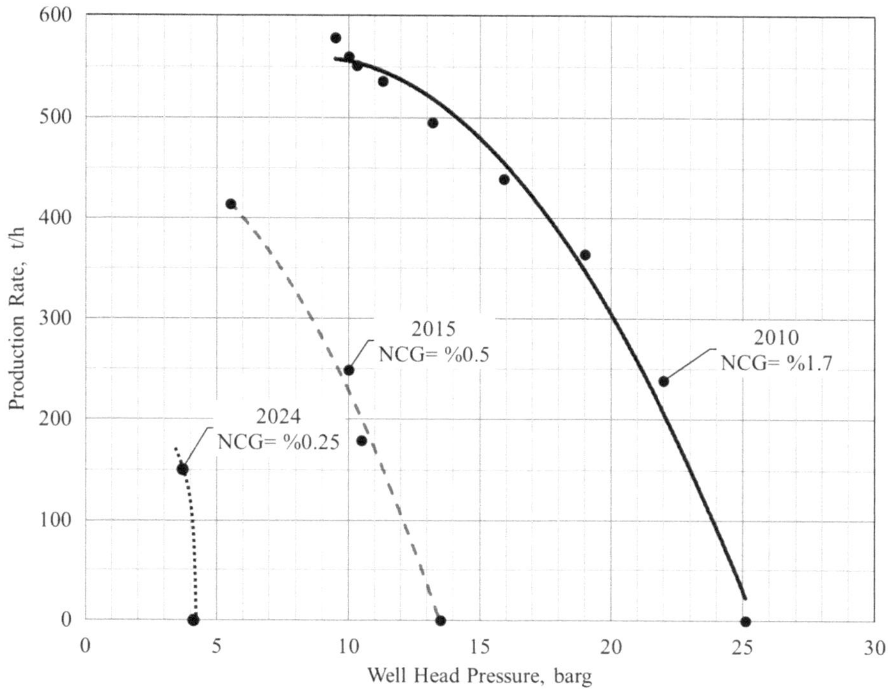

FIG. 10 Variation of flow performance of GM1 well with NCG.

Production can be increased by lowering the pump to the wells with decreasing production. Depending on the well's productivity index, the temperature and the total pressure that the pump should produce, if we assume that the geothermal resource is used in electricity generation, the pump consumes 10%–20% of the energy produced. Pump investment, maintenance costs and, assuming that the well is inoperable in case of pump failure, pumping costs together with the increasing domestic consumption deteriorate the financial situation of geothermal facilities considerably.

3 The problems created by the NCGs

Most of the NCG content tends to be the CO_2 in geothermal reservoirs. Even small percentage of CO_2 can alter the properties of the boiling water. In addition, NCGs can lead to issues in compressors and power generation cycles. Therefore, the nature and the amount of the NCGs need to be determined. For example, small amounts of H_2S can generate issues from health and safety risks to corrosion-related outcomes. Aydin et al. (2020) studied the effects of

CO_2 on the production performance of geothermal wells using a wellbore flow simulator in a proxy model approach. They showed that geothermal wells might not flow naturally after a sharp decline of noncondensable gas (NCG) content in the produced fluid resulting in less pronounced partial pressure support. Satman et al. (2017) studied the effect of CO_2 on the performance of the Kızıldere geothermal field wells. They also reported that CO_2 is the main factor controlling wellhead flowing pressure, flow rate, and flashing depth of the producing wells.

NCG emitted to the atmosphere in geothermal plants is one of several factors that negatively affect the public acceptability of these projects. Others are water and soil pollution, land use, seismicity, light, noise, and visual pollution. These concerns need to be addressed for the public acceptability of geothermal projects. Geothermal power plants sometimes emit 2–3 times more carbon emissions than coal-fired power plants (Aksoy, 2014). Carbon taxes pose a severe threat to many geothermal projects. NCG emission per unit kWh of energy increases in reservoirs produced from carbonated reservoirs with low source temperatures. This negatively impacts the reputation of geothermal energy, which is considered an environmentally friendly resource.

In the last 20 years, the installed capacity of geothermal-based electricity generation in Turkey has increased from 15 to 1800 MWe. Geothermal energy investments have come to a standstill for the last few years. This is because geothermal energy has largely lost public support. Some bad practices had an impact on this. A few blow-outs during drilling operations and limited water-soil contamination were greatly exaggerated. All of these could have been prevented with simple measures. The most severe complaint about operating installations is the NCG discharged into the air. The NCG emission of geothermal power plants in Turkey varies between 0.2 and 1.4 kg/kW. More than 99% of NCG is CO_2. H_2S rate is 0.5% and below. H2S emissions vary between 1 and 10 g/kW, and although its concentration in the air falls below the permitted limits, the odor felt from time to time disturbs the public. Another issue is the use of fertile agricultural lands for geothermal projects. Unfortunately, most of the geothermal resources in Turkey and the most productive agricultural areas of the country are located in grabens. Drilling multiple directional wells from one drilling pad can reduce land use. However, the negative impact of NCG emission still persists.

4 NCG injection

Gases dissolve in water depending on pressure and temperature. Dissolution pressures are determined by the Henry's Law (Eq. 3).

$$P = K_h{*}X \tag{4}$$

Here, P is the dissolution pressure of the gas in water, K_h is the Henry constant, and X is the ratio of gas to water by mass. Henry's constant for each gas and solvent depends on the temperature and salinity of the liquid. Except in some special cases, solubility decreases with increasing temperature. For a newer and more comprehensive way of computing the dissolution amounts of the gases in multisalt solutions can be found in Ratnakar et al., 2024a, b, c.

For the common geothermal gases, the order of solubility in water is (starting at the least soluble) $N_2 < O_2 < H_2 < CH_4 < CO_2 < H_2S < NH_3$ (most soluble) (Nicholson, 1993). Of these gases, CO_2 and CH_4 are the greenhouse gases that contribute most to global warming.

Numerous studies have been conducted to capture gases released from geothermal sources. Prominent examples include the solidification of gases in basalt (Carbfix Project, Aradóttir et al., 2014), reinjecting NCG-water mix with water into the reservoir (GECO, 2021). ORMAT, a leading binary plant producer and operator, has been injecting water and NCG together since 1990 at the 38 MW Puna Hawaii, 2006 at the 84 MW Galena-Nevada, and 2020 at the 13 MW North Brawley. The amount of NCG pumped at these sites is 0.3–1.9 t/h. Similar studies are being carried out in New Zealand and Turkey.

Kaya and Zarrouk (2017) extensively covered past NCG injection experiments in geothermal fields. In their research on numerical models with and without NCG injection, they showed that NCG injection could contribute to reservoir feeding, reservoir pressure maintenance, production increase, and environmental benefits. They also concluded that NCG injection should be carefully planned, and the location, reservoir boundary conditions, and natural nutrition should be taken into account in NCG injection planning.

Below we present the methods and results of our NCG reinjection trials in geothermal fields in Turkey.

Low-temperature fields with low NCG content: in these fields, the geothermal fluid is in liquid phase. The pump is lowered to a depth where the NCG will not separate from the liquid phase, and the pressure is always kept above the NCG break-out pressure. The brine will cool down after passing through the heat exchangers, so it will have more NCG dissolving capacity. In the liquid phase, it can be pumped to reinjection wells. This method has been applied at the 8 MW Babadere GPP (MTN Energy Çanakkale-Turkey) since 2018 and operates with zero emissions. Similar precedents are quite common in many countries. Due to the energy required to keep the fluid in the liquid phase, its use is common for low temperature and low NCG fields (Fig. 5). They are good examples for monitoring the potential impacts of NCG injection. No record of them causing problems in injection wells or the reservoir exists.

The NCG emission pressure is higher than reinjection pressure. In this case, the NCG and reinjection line are connected and NCG is dissolved in water and pumped to the reinjection wells. It can be applied mainly in areas with NCG ratio between 0.2% and 0.3% and low reinjection pressure. Two GPPs with this type of operation can be exemplified. Dora-II (Aydın—Turkey) has a capacity of 9.5 MW, the NCG emission pressure is around 6.5 bar, and the reinjection pressure is around 4 bar. 2 t/h NCG and 850 t/h brine are pumped to the reinjection line at 70–90°C. NCG combines with its own pressure in the injection line and goes to the suction of the reinjection pumps (Fig. 11). Since NCG is dissolved in water, no gas is released even at the reinjection pump suction, which goes to the reinjection wells. A second similar example is the 12.3 MW RSC GPP (Seferihisar—Turkey). In this application, 2 t/h NCG and 800 t/h brine are pumped to the reinjection wells. In this project, the NCG emission pressure is 6.5 bar and the reinjection pressure is 4 bar.

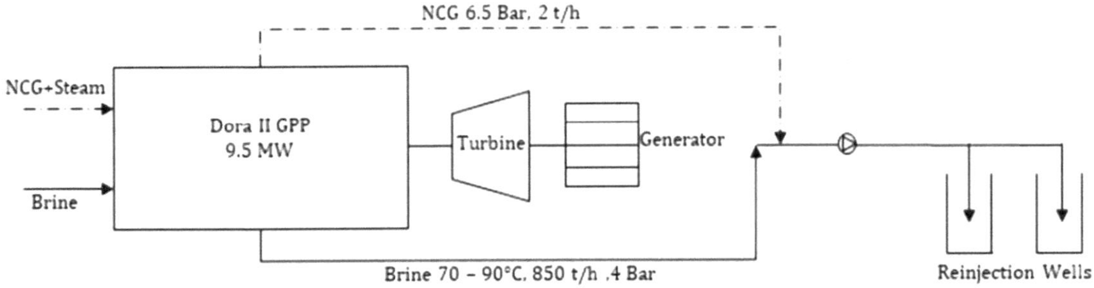

FIG. 11 NCG emission pressure is higher than reinjection pressure.

Let us suppose the pressure at the reinjection pump suction is not sufficient to dissolve all NCG at the reinjection pump suction. In that case, the reinjection pumps will operate under cavitation conditions, and it will not be possible to pump NCG. The pressure at the reinjection pump outlet should be checked. If the pressure at this point is below the NCG emission pressure, NCG injection is possible. An example of this practice at Dora-3a GPP (Aydin—Turkey) was realized in early 2023. At the 20 MW GPP, the reinjection pump suction pressure is 4.6 bar. The discharge pressure is 6.5 bar. The NCG emission pressure is 7.5 bar and the flow rate is 4.5 t/h. In practice, the NCG was first supplied from the suction side of the pump, but there was a problem. Then, the NCG was supplied to the pump's discharge line. The reinjection wells are 1.5 km away. NCG and brine flow in phase in the pipeline and dissolve immediately after entering the well.

Another example of NCG injection is the 24 MW Nezihe Beren GPP (Germencik—Turkey). Initially, the temperature of the production wells was measured at 175°C and the NCG content was 0.5% by mass. Therefore, a GPP was designed as described in Fig. 5, keeping all NCG in the liquid phase, without releasing NCG.

However, shortly after the plant was commissioned, the temperature of the wells increased to 180°C and the NCG content increased to 0.6%. The problem is most likely due to insufficient production after the drilling of the wells. It is generally observed that the temperature rises for a while in newly drilled wells. Due to the increased temperature and NCG, contrary to the plan, NCG entered the heat exchangers as gas. In addition to carbonate scaling, the passage of gas through the heat exchangers instead of water prevented the transfer of enough energy to the motivated fluids in binary conversion. Due to the pressure limits of the heat exchangers, the pressure of the ESPs could not be increased. To solve this problem, the fluid was sent to a separator operating at 20 bar pressure. While the hot water was sent to the GPP, 7 t/h of NCG and 2 t/h of steam were released to the atmosphere at a pressure of 20 bar. The steam and NCG were then connected to a 9-bar reinjection line at the outlet of the GPP (Fig. 12). In the last example, if planning had been done from the beginning, 7 t/h of NCG and 2 t/h of steam could have been sent to an independent heat exchanger, and the energy could have been used.

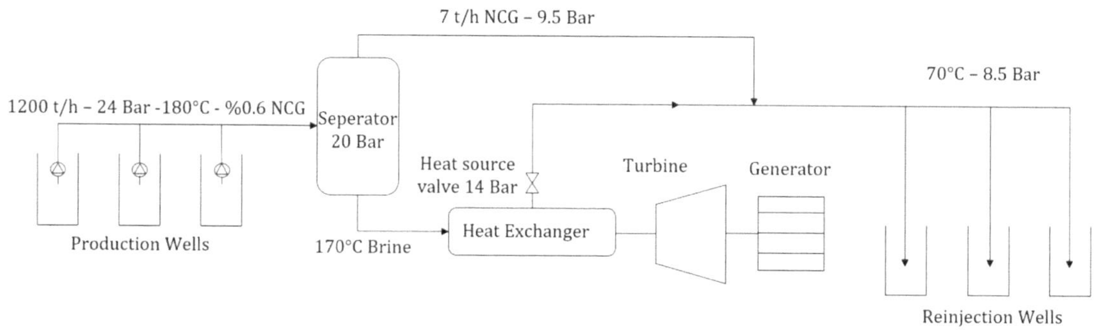

FIG. 12 A model for NCG injection when using a downhole pump.

Since 2022, at 1200 t/h, 80°C brine, 7 t/h NCG, and 2 t/h steam have been sent to reinjection wells 2 km away and 30 m below from the GPP. The reinjection line pressure at the junction point of NCG and water increased from 8 to 9.5 bar due to NCG injection.

Three reinjection wells are located in the same pad, and the wellhead pressures are 8.2 bar. Wellhead pressures increased by 2 bar due to NCG injection. The pressure at the point where the NCG injection started and at the wellhead is lower than 12.2 bar (for 0.6% NCG at 80°C), which is the dissolution pressure of NCG (NCG dissolution pressure is calculated from Ellis and Golding (1963), where all NCG is assumed to be CO_2). The dynamic pressure profile and pressure gradient graphs taken from the well during NCG injection (Fig. 13) show that while 400 t/h brine and 2.34 t/h of NCG were pumped into the HB18 reinjection well, the pressure gradient between 0 and 56 m was less than 9 kPa/m and remained constant around 9 kPa/m after 56 m. From this, we can conclude that the NCG and water flow in two phases between 0 and 56 m and after 56 m NCG is dissolved in water. The pressure at 56 m depth is 12.2 bar from the dynamic pressure profile. The pressure gradient decreases very little in the area of the filtered pipe. This is estimated to be related to heating water and reducing its density.

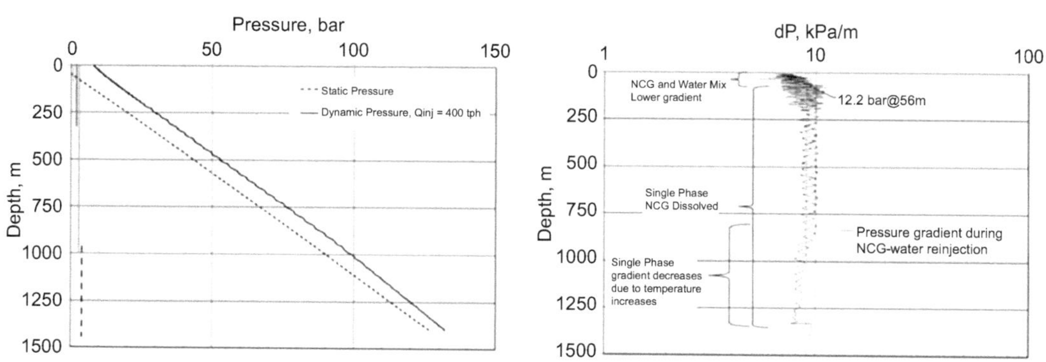

FIG. 13 Pressure profile and pressure gradient during 400 t/h water and NCG injection.

The common feature of the above examples is that NCG emission pressures are higher than reinjection pressures. The well pumps can increase the pressure if the NCG emission pressure is low. If the pumps are not preferred, the NCG pressure can be increased with a compressor. To prevent water from entering the inlet of the compressor, the gas must first be cooled and excess water removed. Compressor injection of acid gases (such as CO_2, H_2S) is very common in the oil industry.

NCG and reinjected water can be combined at the GPP outlet and transported together to reinjection wells. It is possible for water and NCG to be combined at the reinjection wellhead in separate pipes, or for NCG to be pumped through a pipe lowered into the well while the water to be reinjected is pumped through two pipe annuli and dissolved in the well.

According to Yucetas et al. (2018), 700 m of pipe was lowered to the ASK-1 well in the Umurlu (Aydın—Turkey) geothermal field. They pumped 2.76 t/h of NCG and some cold water through the pipe and reinjection water from the annulus. NCG was pumped with the help of a compressor. The NCG pressure was 52 bar, the cold water (16°C) flow rate was 9.3 m^3/h, and the brine flow rate was 72 t/h. It is reported that the system worked smoothly but was stopped after a few weeks due to compressor failure. The total internal consumption is 250 kWh.

A larger-scale trial was conducted at a geothermal field in Salihli, Manisa, Turkey. The GPP has an installed capacity of 72 MW and NCG emissions of 15 t/h. With a compressor, the pressure of the 5 t/h NCG was increased to 40 bar and NCG was pumped into a reinjection line carrying a flow rate of 700 t/h at 38 bar pressure and 80°C fluid (Fig. 14). NCG was pumped to all reinjection wells in the field for 45 days without any reported incidents, and then a serious compressor failure was encountered. The contribution of the installed NCG injection system to domestic consumption is 260 kWh. It was realized that approximately 800 kWh of internal consumption would be required for a full-capacity reinjection in the field. No increase in wellhead pressures was observed during the trial period of NCG injection. Therefore, the NCG reinjection cannot be said to have caused a permeability decrease in the reservoir during this period. Corrosion was also not a problem.

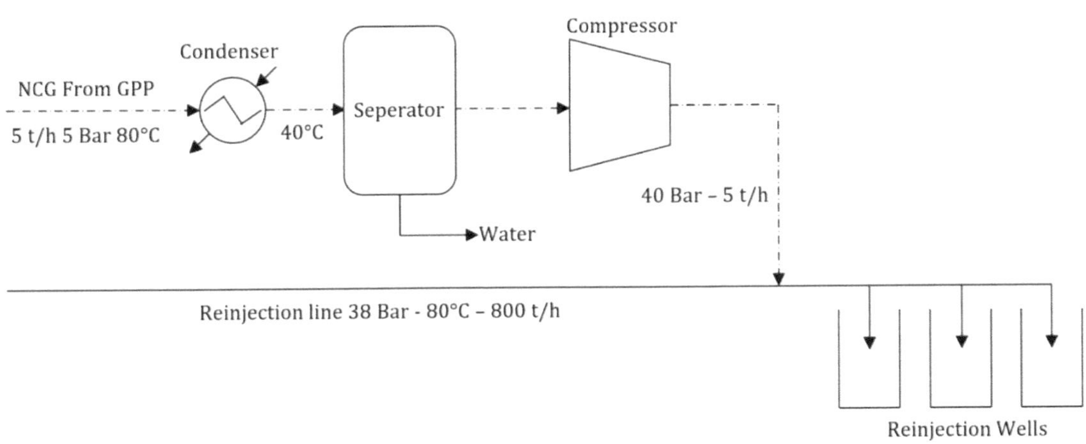

FIG. 14 NCG injection using a compressor.

When two phases, NCG and water, flow together in a pipe together, flow regimes develop depending on the volumetric ratio of gas to liquid. The compressor and reinjection pump pressures have to be recalculated as friction losses increase during flow in the pipeline. The calculation can be done backward from the reinjection well. The measured downhole injection pressure (Pwf) does not change for a constant flow rate in the liquid phase. However, as can be seen in Fig. 15, with the injection of NCG, there is a two-phase flow up to the point where NCG dissolves in the well. The difference between WHP2 (NCG-water) and WHP1 (water only) will increase the wellhead pressure. These values will determine the compressor and reinjection pump pressures. In addition to the compressor power, the injection pumps will need to be increased by the wellhead pressure difference for the NCG injection.

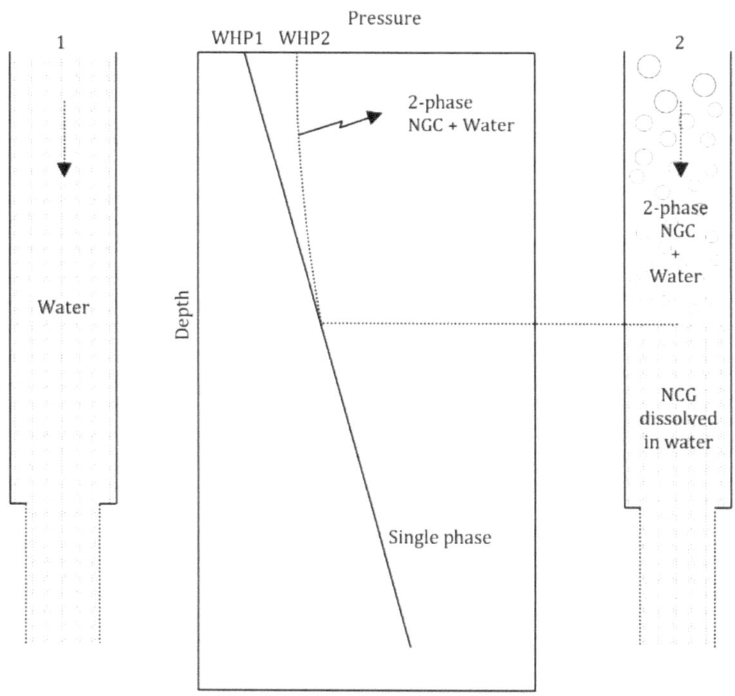

FIG. 15 Downhole pressure profile during NCG injection.

Recycling/reinjection of NCGs can have a significant impact on the production performance of the reservoir. Fig. 16 indicates steam and NCG production flow rates for a closed model for no reinjection and for reinjection of mixed NCG-water at various rates. It can be seen from this figure that, for the overall production period, the higher reinjection rate results in higher long-term steam production. It also shows that, after the first 5 years of production, NCG breakthrough occurred at the production wells and NCG production continued with an increasing flow rate.

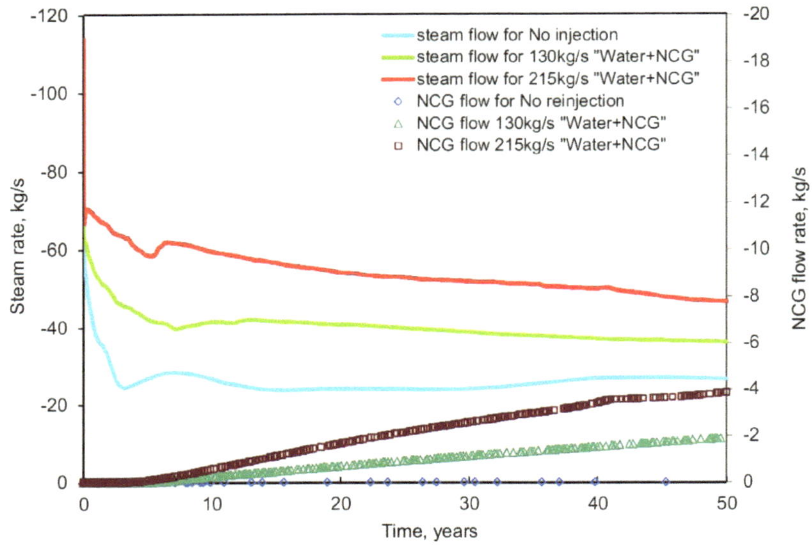

FIG. 16 Steam and NCG production rates for no-reinjection, and the 130 and 215 kg/s water-NCG mixture injection cases (Kaya and Zarrouk, 2017).

5 Summary

- Noncondensable gases (NCGs), such as CO_2, play an important role in both performance and production in geothermal systems. The presence of CO_2 in the reservoir increases the efficiency of production wells as CO_2-saturated brine moves towards them. This enhances fluid flow and boosts well performance, contributing to better overall system productivity.
- Power station requirements (steam quality, etc.) and scale considerations: composition of the geothermal waters and geothermometry: some of the geochemical techniques enable one to understand the nature of the geothermal system easily using certain chemical indicators. Quantitative temperature indicators are also called geothermometers: silica, Na, Ca, K, Li, etc.
- Selection of power fluids is a function of the P–T and the composition of the system. Among those the temperature of the system is the key parameter. The use of CO_2 as the heat transmission fluid increases the efficiency of geothermal energy extraction from low-enthalpy resources such as depletion oil and gas reservoirs. In the resulting so-called CO_2 plume geothermal energy generation.
- CO_2 sequestration and geothermal power generation: CO_2 sequestration into geological formations has been considered a reliable and sustainable solution to mitigate CO_2 emissions and global warming. Typical candidates for structural, mineral, and solubility trapping of CO_2 are depleted oil and gas reservoirs, deep saline aquifers, and coal seams. Geothermal reservoirs have tremendous potential for CO_2 sequestration. Solubility estimates (of CO_2) in geothermal waters.

References

Aksoy, N., 2007. Optimization of downhole pump setting depths in liquid-dominated geothermal systems: a case study on the Balcova-Narlidere field, Turkey. Geothermics 36, 436–458.

Aksoy, N., 2014. Power generation from geothermal resources in Turkey. Renew. Energy 68, 595–601.

Aradóttir, E.S.P., Gunnarsson, I., Didérikssen, K., Gislason, S.R., Sigfusson, B., Grandia, F., Oelkers, E., Snaebjornsdottir, S.O., 2014. CarbFix Final Report. Project No. 281348.

Ármannsson, H., Fridriksson, T., Kristjánsson, B.R., 2005. CO2 emissions from geothermal power plants and natural geothermal activity in Iceland. Geothermics 34 (3), 286–296.

Aydın, H., Akın, S., Tezel, S., 2018. Practical experiences about reservoir monitoring in Alaşehir geothermal field. In: Proceedings of the 43rd Workshop on Geothermal Reservoir Engineering (SGP-TR-213). Stanford University, Stanford, California. February 12–14.

Aydin, H., Akin, S., Senturk, E., 2020. A proxy model for determining reservoir pressure and temperature for geothermal wells. Geothermics 88, 101916.

Edwards, L.M., Chilingar, G.V., Rieke III, H.H., Fertl, W.H. (Eds.), 1982. Handbook of Geothermal Energy. Gulf Publishing Company, Houston.

Ellis, A., Golding, R.M., 1963. The solubility of carbon dioxide above 100°C in water and in sodium chloride solutions. Am. I. Sci. 261, 47–60.

Fournier, R.O., 1979. A revised equation for the Na/K geothermometer. Geotherm. Resour. Counc. 3, 221–224.

GECO (Geothermal Emission Control), 2021. Final Report: Innovations in Geothermal Emission Control. GECO Project No. 818169. Retrieved from https://www.geco-h2020.eu.

Gokcen, G., Yildirim, N., 2008. Effect of non-condensable gases on geothermal power plant performance. Case study: Kizildere geothermal power plant-Turkey. Int. J. Exergy 5 (5/6), 684–695.

Kaplan, U., Serpen, U., 2010. Developing geothermal power plants for geothermal fields in Western Turkey. In: Proceedings World Geothermal Congress 2010, Bali, Indonesia, 25–29 April 2010.

Kapusta, S., Livescu, S., Dindoruk, B., Schulz, R., Webber, M., 2023. Direct use applications: decarbonization of industrial processes, and heating and cooling scenarios. In: Beard, J., Jones, B. (Eds.), The Future of Geothermal in Texas. The Energy Institute, The University of Texas at Austin. https://doi.org/10.26153/tsw/44082.

Kaya, E., Zarrouk, S.J., 2017. Reinjection of greenhouse gases into geothermal reservoirs. Int. J. Greenh. Gas Control 67 (2017), 111–129.

Lajwe, G., 2013. Comparison, characterization, and interpretation of geothermal fluid geochemistryin the sedimentary environments of Kibiro, Panyimur, and Oxarfjordur. In: Omarsdottir, M. (Ed.), Geothermal Training in Iceland.

Livescu, S., 2024. Engineering the geothermal heating and cooling of buildings. CHOA J. Retrieved on August 11, 2024 from: https://www.linkedin.com/pulse/engineering-geothermal-heating-cooling-z2jqc/?trackingId=uCvUaAWDS62uYQTeslde7Q%3D%3D.

Martin, J.J., Hou, Y.C., 1955. Development of an equation of state for gases. AICHE J. 1, 142–151.

Moon, H., Zarrouk, S.J., 2012. Efficiency of geothermal power plants: a worldwide review. In: New Zealand Geothermal Workshop 2012 Proceedings 19–21 November 2012 Auckland, New Zealand.

Nicholson, K., 1993. Geothermal Fluids: Chemistry and Exploration Techniques. Springer-Verlag, Berlin Heidelberg. 263 pp.

Ratnakar, R.R., Chaubey, V., Dindoruk, B., 2023. A novel computational strategy to estimate CO2 solubility in brine solutions for CCUS applications. Appl. Energy 342, 121134. https://doi.org/10.1016/j.apenergy.2023.121134.

Ratnakar, R.R., Livescu, S., Dindoruk, B., Gautam, S., 2024a. A comparative study of the impact of the CO2 properties on the thermal output of a geothermal well. Geoenergy Sci. Eng. 243, 213321. https://www.sciencedirect.com/science/article/pii/S2949891024006912.

Ratnakar, R.R., Gupta, S.S., Hackbarth, J., Livescu, S., Dindoruk, B., 2024b. Assessment of geothermal and hydrocarbon wells using supercritical fluids for heat production. In: Paper SPE-220844-MS to be presented at the 2024 SPE Annual Technical Conference and Exhibition to Be Held 23–25 September, New Orleans, LA, USA.

Ratnakar, R.R., Chaubey, V., Gupta, S.S., Hackbarth, J., Rui, Z., Dindoruk, B., 2024c. Physics augmented machine learning models for determining gas solubility in formation brines for CCS and gas processing applications.

In: Paper SPE-220752-MS to be presented at the 2024 SPE Annual Technical Conference and Exhibition to Be Held 23–25 Septembe, New Orleans, LA, USA.

Satman, A., Tureyen, O.I., Basel, E.D.K., Guney, A., Senturk, E., Kindap, A., 2017. Effect of carbon dioxide content on the well and reservoir performances in the Kizildere geothermal field. In: 42nd Workshop on Geothermal Reservoir Engineering.

Sumnu, M.D., 1989. Geochemical Survey of Dutlu-Beypazari Geothermal Field and Geothermometry Applications. MSc. Thesis, Middle East Technical University, September.

Tester, J.W. (Ed.), 2006. The Future of Geothermal Energy. Massachusetts Institute of Technology.

Yucetas, I., Ergicay, N., Akın, S., 2018. Carbon Dioxide Injection Field Pilot in Umurlu Geothermal Field, Turkey. vol. 42 GRC Transactions, p. 2018.

Index

Note: Page numbers followed by *f* indicate figures and *t* indicate tables.

D

H

I

J

K

L

M

N

T

U

V

W

X